BIOPHYSICAL DISCUSSIONS

Fast Biochemical Reactions
in Solutions, Membranes, and Cells

ORGANIZING COMMITTEE

ROBERT L. BERGER, *Chairman*
PETER RENTZEPIS
ALAN SCHECHTER
V. ADRIAN PARSEGIAN, *Editor, Biophysical Journal*

BIOPHYSICAL DISCUSSIONS

 Fast Biochemical Reactions in Solutions, Membranes, and Cells

AIRLIE HOUSE, AIRLIE, VIRGINIA

2–5 APRIL 1978

BIOPHYSICAL JOURNAL

OCTOBER 1978 · VOLUME 24, NUMBER 1

THE ROCKEFELLER UNIVERSITY PRESS

COPYRIGHT © 1978 BY THE BIOPHYSICAL SOCIETY
LIBRARY OF CONGRESS CATALOGUE NUMBER 78-64444
ISBN 87470-030-2
PRINTED IN THE UNITED STATES OF AMERICA

Recommended citation:
[Author(s)]. 1978. [Title]. *Biophys. J.* **24**:[Page numbers].

Editor's Foreword

V. ADRIAN PARSEGIAN, *Physical Sciences Laboratory, National Institutes of Health, Bethesda, Maryland 20014 U.S.A.*

The interdisciplinary nature of biophysics makes it a particularly suitable area for meetings between groups who do not normally work together. We have patterned this Biophysical Discussion on the Discussions of the Faraday Society. That series has a proud history of presenting major advances and controversies to all parts of the scientific community. The unique feature of those meetings, the distribution of information to participants before the meeting, is a useful device for overcoming language barriers inherent in dialogue between disciplines.

Robert Berger's suggestion of fast biochemical reactions as a topic for the first discussion seemed most appropriate. The development of physical methods for studying fast reactions and the application of such methods to biochemical problems have gone hand in hand for more than 50 years. Indeed, much of the pioneering work in making fast reactions accessible to measurement came from biochemists. The continuous flow apparatus of Hartridge and Roughton (1923) and the accelerated and stopped-flow instruments of Chance (1940) first made it possible to measure chemical reactions with half-times in the millisecond range.

An important advance was achieved in 1954 by Gibson's introduction of the stopping syringe and the Gibson-Milne instrument. The 1954 *Discussion of the Faraday Society* on the study of fast reactions in solution provided a necessary meeting place where this emerging work could be better recognized. At that Discussion the new methods of Eigen and deMaeyer for observing chemical relaxation and of Norrish and Porter for doing flash photolysis were first introduced to the scientific community.

In 1964 Chance organized a symposium published as part of the Johnson Foundation series on rapid mixing and sampling techniques. This meeting stimulated a considerable amount of new work leading to a variety of new scientific instruments in this field.

The development of the dye laser by Bradley and others and its exploitation in several areas of perturbation chemistry, together with the advance of the mode-locked laser with its application to picosecond excitation by Rentzepis and others, set a new stage for the exploration of molecular dynamics, energy conversion, exchange, and coupling in biochemical reactions.

The time seemed perfect for a third printed discussion on the state of the art in the field of fast reaction methods. Included here are those areas where new methods may be applied to tell us how biochemical molecules and assemblages of such molecules interact. Important problems related to macromolecular structure and function involve very rapid relaxations and conformational changes of nucleic acids, proteins, and lipids. Important questions about the interaction of light with matter in biological systems involve the rapid processes after the absorption of photons. The questions of membrane structure and function involve translational and rotational diffusion of proteins and other molecules embedded in the complex lipid-containing membrane structure.

At the meeting itself, except for the keynote speakers, authors were allowed only 5 minutes to remind participants of what was in the Study Book distributed several weeks earlier. The actual discussion that followed was recorded, transcribed, and edited for presentation here. The anonymous questions inserted into the dialogue had been submitted earlier by the referees for written or oral response by the authors. "Extended" Comments (10-minute limit) were invited from all participants and took precedence over regular exchange; their prepared texts were inserted into the individual discussions. "Notes added in proof" were permitted to help authors incorporate revisions or new work arising in the 4 months between submission and the Discussion.

No one who was actually there can forget the excitement and intense spontaneity of the actual discussion, or the delight of receiving an "instant replay" from the original, unedited transcript, like these quotations:

> I have a question about this isomeier having an affair with an incepter.
>
> This extended comment is intended to illicit an extended response from Dr. Frazoose if possible.
>
> At three mezazorts the double ears should come to the cell impedence. I wonder if you had any chance to see this contribution.
>
> Now when orientation is slow, for example from fever, or slow because we have pretty large areas of moleculses of fivefullness, then the system will tend to propose an electric field.
>
> We don't know whether it goes through an intermediate or whether it represents an electron transfer or removal of one of these hydropaltts. But we do arrive at this radical incident.
>
> And call it felix as a dive board which is not indirection of the sphere will disappear.

The Organizing Committee—Robert Berger, Peter Rentzepis, and Alan Schechter—provided both help and guidance in defining the first Discussion topic, attracting participants and financial support, designing the program, chairing the meeting, and preparing the final text. Valerie Parsegian, nominally Assistant to the Editors, did most of the supporting paper work for everything from manuscript processing to arrangement of transportation to the meeting.

Financial support for this Discussion has come from many sources: The Biophysical Society, the Office of Naval Research (contract N0014 77 G 068), the Department of Energy, the Division of Research Resources (National Institutes of Health), the National Institute of Environmental Health Sciences, the National Cancer Institute, the Fogarty International Center (National Institutes of Health), and the National Institute of General Medical Sciences. The Rockefeller University Press, especially Margaret Broadbent and Judith Hoffmann, did extraordinary work in modifying its publication procedures to prepare the pre-meeting Study Book as well as this final record.

Contents

EDITOR'S FOREWORD. *V. Adrian Parsegian*

STOPPED FLOW, MIXING, METHODS, AND ANALYSIS

SOME PROBLEMS CONCERNING MIXERS AND DETECTORS FOR STOPPED FLOW KINETIC STUDIES. KEYNOTE ADDRESS. *Robert L. Berger*	2
DISCUSSION	19
THE RELATION BETWEEN CARBON MONOXIDE BINDING AND THE CONFORMATIONAL CHANGE OF HEMOGLOBIN. *Charles A. Sawicki and Quentin H. Gibson*	21
DISCUSSION	29
BICARBONATE-CHLORIDE EXCHANGE IN ERYTHROCYTE SUSPENSIONS. STOPPED-FLOW pH ELECTRODE MEASUREMENTS. *Edward D. Crandall, A. L. Obaid, and R. E. Forster*	35
DISCUSSION	42

Extended Abstracts

THE MOTION OF CYTOCHROME b_5 ON LIPID VESICLES MEASURED VIA TRIPLET ABSORBANCE ANISOTROPY. *R. H. Austin and W. Vaz*	49
SPECTRAL INTERMEDIATES IN THE ACTIVATION OF GLYCERALDEHYDE-3-PO_4 DEHYDROGENASE-CATALYZED REACTIONS. *Sidney A. Bernhard*	49
STUDIES OF THE ACTIVATION OF YEAST ENOLASE BY METALS USING A "TRANSITION STATE ANALOGUE." *John M. Brewer*	53
SCANNING MOLECULAR SIEVE CHROMATOGRAPHY OF INTERACTING PROTEIN SYSTEMS. EFFECT OF KINETIC PARAMETERS ON THE LARGE ZONE BOUNDARY PROFILES FOR LOCAL EQUILIBRATION BETWEEN MOBILE AND STATIONARY PHASES. *Paul W. Chun and Mark C. K. Yang*	56
A SOLUTION MIXER WITH 10-μs RESOLUTION. *Robert Clegg and Mack J. Fulwyler*	57
KINETIC TRANSIENTS. A WEDDING OF EMPIRICISM AND THEORY. *Alan H. Colen*	58
A METHOD FOR DETERMINING THE KINETIC TYPE OF FAST KINETIC DATA. *David C. Foyt and John S. Connolly*	60
THE EFFECT OF PRETREATMENT WITH CALCIUM AND MAGNESIUM IONS ON PHOSPHOENZYME FORMATION BY SARCOPLASMIC RETICULUM ATPASE. *Jeffrey P. Froehlich*	61
RECORDING OF FAST BIOCHEMICAL REACTIONS USING A LOGARITHMIC TIME SWEEP. *Eugene Hamori and Toshihiro Hirai*	63
FAR-ULTRAVIOLET STOPPED-FLOW CIRCULAR DICHROISM. *Jerry Luchins*	64
A SIMPLE SYSTEM FOR MIXING MISCIBLE ORGANIC SOLVENTS WITH WATER IN 10–20 ms FOR THE STUDY OF SUPEROXIDE CHEMISTRY BY STOPPED-FLOW METHODS. *Gregory J. McClune and James A. Fee*	65
FLUID MECHANICS OF RAPID MIXING. *Jerome H. Milgram*	69

Kinetics of Association and Dissociation Phenomena in Human Hemoglobin Studies in a Laser Light-Scattering Stopped-Flow Device. *Lawrence J. Parkhurst and Duane P. Flamig*   71

Extending the Wavelength Range of Fundamental Laser Sources. *R. A. Popper, T. Nowicki, W. Ruderman, and J. Ragazzo*   73

Kinetic Studies of Fast Reactions at Water-Micelle Interfaces. *Michael A. J. Rodgers, David C. Foyt, and Maria F. da Silva e Wheeler*   75

A Macroscopic Approach to Fluctuation Analysis. Solution of the Phase Problem of Relaxation Spectrometry. *W. O. Romine, C. Watkins, S. T. Christian, and M. C. Goodall*   76

An Inexpensive Microcomputer-Based Stopped-Flow Data Acquisition System. *D. G. Taylor, J. N. Demas, R. P. Taylor, and M. J. Zenkowich*   77

On the Integration of Coupled First-Order Rate Equations. *Darwin Thusius*   79

SMALL PERTURBATIONS

Picosecond Fluorometry in Primary Events of Photosynthesis. Keynote Address. *Leonid B. Rubin and Andrew B. Rubin*   84
    discussion   91

A Time-Resolved Electron Spin Resonance Study of the Oxidation of Ascorbic Acid by Hydroxyl Radical. *Richard W. Fessenden and Naresh C. Verma*   93
    discussion   101

Magnetic Relaxation Analysis of Dynamic Processes in Macromolecules in the Pico-to the Microsecond Range. *R. King, R. Maas, M. Gassner, R. K. Nanda, W. W. Conover, and O. Jardetzky*   103
    discussion   114

New Electric Field Methods in Chemical Relaxation Spectrometry. *A. Persoons and L. Hellemans*   119
    discussion   131

Response of Acetylcholine Receptors to Photoisomerizations of Bound Agonist Molecules. *Menasche M. Nass, Henry A. Lester, and Mauri E. Krouse*   135
    discussion   154

Hapten-Linked Conformational Equilibria in Immunoglobulins XRPC-24 and J-539 Observed by Chemical Relaxation. *S. Vuk-Pavlović, Y. Blatt, C. P. J. Glaudemans, D. Lancet, and I. Pecht.*   161
    discussion   170

The Rate of Entry of Dioxygen and Carbon Monoxide into Myoglobin. *Robert H. Austin and Shirley SuiLing Chan*   175
    discussion   182

SINGLE CELL OBSERVATIONS OF GAS REACTIONS AND SHAPE CHANGES IN NORMAL AND SICKLING ERYTHROCYTES. *E. Antonini, M. Brunori, B. Giardina, P. A. Benedetti, G. Bianchini, and S. Grassi* — 187
DISCUSSION — 190

NANOSECOND RELAXATION PROCESSES IN LIPOSOMES. *Mugurel G. Badea, Robert P. DeToma, and Ludwig Brand* — 197
DISCUSSION — 209

DETECTION OF HINDERED ROTATIONS OF 1,6-DIPHENYL-1,3,5-HEXATRIENE IN LIPID BILAYERS BY DIFFERENTIAL POLARIZED PHASE FLUOROMETRY. *J. R. Lakowicz and F. G. Prendergast* — 213
DISCUSSION — 227

Extended Abstracts

THE APPLICATION OF SELECTIVE EXCITATION DOUBLE MÖSSBAUER TO TIME-DEPENDENT EFFECTS IN BIOLOGICAL MATERIALS. *Bohdan Balko and Eugenie V. Mielczarek* — 233

ON MAGNETICALLY INDUCED TEMPERATURE JUMPS. *George H. Czerlinski* — 234

KINETIC AND TRANSIENT ELECTRIC DICHROISM STUDIES OF THE IREHDIAMINE-DNA COMPLEX. *N. Dattagupta, M. Hogan, and D. Crothers* — 238

OPTICAL DETECTION OF COMPRESSIBILITY DISPERSION. RELAXATION KINETICS OF GLUTAMATE DEHYDROGENASE SELF-ASSOCIATION. *Herbert R. Halvorson* — 239

CREATION OF A NONEQUILIBRIUM STATE IN SODIUM CHANNELS BY A STEP CHANGE IN ELECTRIC FIELD. *Eric Jakobsson* — 240

HIGH-FREQUENCY DIELECTRIC SPECTROSCOPY OF CONCENTRATED MEMBRANE SUSPENSIONS. *Donald S. Kirkpatrick, John E. McGinness, William D. Moorhead, Peter M. Corry, and Peter H. Procter* — 243

SUBNANOSECOND FLUORESCENCE LIFETIMES BY TIME-CORRELATED SINGLE PHOTON COUNTING USING SYNCHRONOUSLY PUMPED DYE LASER EXCITATION. *Vaughn J. Koester and Robert M. Dowben* — 245

ALLOSTERY IN AN IMMUNOGLOBULIN LIGHT-CHAIN DIMER. A CHEMICAL RELAXATION STUDY. *D. Lancet, A. Licht, and I. Pecht* — 247

THE STRUCTURE OF THE RETINYLIDENE CHROMOPHORE IN BATHORHODOPSIN. *Aaron Lewis* — 249

CONTRACTILE DEACTIVATION BY RAPID, MICROWAVE-INDUCED TEMPERATURE JUMPS. *Barry D. Lindley and Birol Kuyel* — 254

CALCULATION OF DIELECTRIC PARAMETERS FROM TIME DOMAIN SPECTROSCOPY DATA. *W. D. Moorhead* — 256

LIGHT-JUMP PERTURBATION OF CARBON MONOXIDE BINDING BY VARIOUS HEME PROTEINS. *Emilia R. Pandolfelli, Celia Bonaventura, Joseph Bonaventura, and Maurizio Brunori* — 257

MEASUREMENT OF INTERCONVERSION RATES OF BOUND SUBSTRATES OF PHOSPHORYL TRANSFER ENZYMES BY ^{31}P-NUCLEAR MAGNETIC RESONANCE. *B. D. Nageswara Rao and Mildred Cohn* — 258

DISSOCIATION RATE OF SERUM ALBUMIN-FATTY ACID COMPLEX FROM STOP-FLOW DIELECTRIC STUDY OF LIGAND EXCHANGE. *Walter Scheider*   260

TIME-RESOLVED RESONANCE RAMAN CHARACTERIZATION OF THE INTERMEDIATES OF BACTERIORHODOPSIN. *James Terner and M. A. El-Sayed*   262

STUDIES ON PROTEINS AND tRNA WITH TRANSIENT ELECTRIC BIREFRINGENCE. *Michael R. Thompson, Ray C. Williams, and Charles H. O'Neal*   264

METAL ION INTERACTIONS WITH FLUORESCENT DERIVATIVES OF NUCLEOTIDES. *J. M. Vanderkooi, C. J. Weiss, and G. Woodrow III*   266

RUPTURE DIAPHRAGMLESS APPARATUS FOR PRESSURE-JUMP RELAXATION MEASUREMENT. *Tatsuya Yasunaga and Nobuhide Tatsumoto*   267

LARGE PERTURBATIONS

PROBING ULTRAFAST BIOLOGICAL PROCESSES BY PICOSECOND SPECTROSCOPY. KEYNOTE ADDRESS. *P. M. Rentzepis*   272
    DISCUSSION   283

FAST ELECTRON TRANSFER PROCESSES IN CYTOCHROME *c* AND RELATED METALLOPROTEINS. *Michael G. Simic and Irwin A. Taub*   285
    DISCUSSION   293

EXPLORING FAST ELECTRON TRANSFER PROCESSES BY MAGNETIC FIELDS. *Klaus Schulten and Albert Weller*   295
    DISCUSSION   303

APPLICATION OF PULSE RADIOLYSIS TO THE STUDY OF PROTEINS. CHYMOTRYPSIN AND TRYPSIN. *Moshe Faraggi, Michael H. Klapper, and Leon M. Dorfman*   307
    DISCUSSION   315

FAST REACTIONS IN CARBON MONOXIDE BINDING TO HEME PROTEINS. *N. Alberding, R. H. Austin, S. S. Chan, L. Eisenstein, H. Frauenfelder, D. Good, K. Kaufmann, M. Marden, T. M. Nordlund, L. Reinisch, A. H. Reynolds, L. B. Sorensen, G. C. Wagner, and K. T. Yue*   319
    DISCUSSION   329

PHOTOINITIATED ION FORMATION FROM OCTAETHYL-PORPHYRIN AND ITS ZINC CHELATE AS A MODEL FOR ELECTRON TRANSFER IN REACTION CENTERS. *S. G. Ballard and D. Mauzerall*   335
    DISCUSSION   342

A PICOSECOND PULSE TRAIN STUDY OF EXCITON DYNAMICS IN PHOTOSYNTHETIC MEMBRANES. *N. E. Geacintov, C. E. Swenberg, A. J. Campillo, R. C. Hyer, S. L. Shapiro, and K. R. Winn*   347
    DISCUSSION   355

MODULATION OF THE PRIMARY ELECTRON TRANSFER RATE IN PHOTOSYNTHETIC REACTION CENTERS BY REDUCTION OF A SECONDARY ACCEPTOR. *M. J. Pellin, C. A. Wraight, and K. J. Kaufmann* 361
DISCUSSION 367

Extended Abstracts

THE REACTION OF "BLUE" COPPER OXIDASES WITH O_2. A PULSE RADIOLYSIS STUDY. *M. Goldberg and I. Pecht* 371

VOLTAGE-INDUCED CHANGES IN THE CONDUCTIVITY OF ERYTHROCYTE MEMBRANES. *Kazuhiko Kinosita, Jr., and Tian Yow Tsong* 373

FAST BIOCHEMICAL REACTIONS IN THIN FILMS INDUCED BY NUCLEAR FISSION FRAGMENTS. *R. D. Macfarlane* 375

NONHOMOGENOUS CHEMICAL KINETICS IN PULSED PROTON RADIOLUMINESCENCE. *John Howard Miller and Martin L. West* 376

PICOSECOND PHOTODISSOCIATION AND SUBSEQUENT RECOMBINATION PROCESSES IN CARBOXYHEMOGLOBIN, CARBOXYMYOGLOBIN, AND OXYMYOGLOBIN. *L. J. Noe, W. G. Eisert, and P. M. Rentzepis* 379

TIME-RESOLVED MAGNETIC SUSCEPTIBILITY. A NEW METHOD FOR FAST REACTIONS IN SOLUTION. *J. S. Philo* 381

CARBOXYLATION KINETICS OF HEMOGLOBIN AND MYOGLOBIN: LINEAR TRANSIENT RESPONSE TO STEP PERTURBATION BY LASER PHOTOLYSIS. *D. D. Schuresko and W. W. Webb* 382

LIST OF PARTICIPANTS 386
AUTHOR INDEX 391
SUBJECT INDEX 395

Stopped Flow, Mixing, Methods, and Analysis

ROBERT L. BERGER, *Chairman*

KEYNOTE ADDRESS

SOME PROBLEMS CONCERNING MIXERS AND DETECTORS FOR STOPPED FLOW KINETIC STUDIES

ROBERT L. BERGER, *Laboratory of Technical Development, National Heart, Lung, and Blood Institute, National Institutes of Health, Bethesda, Maryland 20014 U.S.A.*

This conference is taking place on the 55th anniversary of the introduction of rapid mixing methods to the field of biology by Hartridge and Roughton (1923a). Their specific aim was to study the reactions of hemoglobin with its ligands (Hartridge and Roughton, 1923b). The tenacity with which nature holds onto its secrets is nowhere better exemplified than with this remarkable molecule. The very nature of the reaction of hemoglobin with its ligands (O_2, CO, and NO) and its effectors (H^+, 2,3-diphosphoglycerate[phosphates], CO_2, and salt), together with its tetrameric cooperative intramolecular interaction, taxes our ingenuity and technology to explore, and we still do not understand these reactions in detail. However, methods and techniques discussed at this meeting offer the greatest promise so far for forcing this important model of molecular dynamics to yield its secrets.

During the past ten years a number of advances have been made that are only now bearing fruit, not just in hemoglobin research but also in many other areas of biology. The wide range of papers and posters presented in this Discussion bears witness to the breadth of application that has occurred as reliable commercial instruments have become available. These instruments have in general a time resolution of 1 ms, limited primarily by flow velocity and stopping time; they are useful for solutions whose viscosity is essentially that of water.

The state of the art in mixing and flow systems up to 1972 has been reviewed recently by Chance (1974). In this talk I would like to point out some of the technical problems that still exist, with the hope that this discussion will stimulate research efforts to solve them. Since the introduction of the rapid reaction apparatus and early experimental work of Hartridge and Roughton (1923a, 1924, 1926), Roughton and Millikan (1936), Millikan (1936), and Chance (1940a,b), very little quantitative work has been done on the theory of mixing and flow for the particular needs of stopped flow. Schlichting (1968), Pai (1954), Kay (1963), Hinze (1959), Davies (1972), and

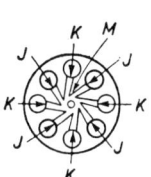

FIGURE 1 Original Hartridge-Roughton mixer.

Batchelor (1971) summarize much of the work on compressible and incompressible fluids. The fully developed turbulent flow of incompressible fluids in which eddy equilibrium has been reached and where Kolmogoroff radii can be measured has received some attention (Friedlander and Topper, 1961; Brodkey, 1975; Corrsin, 1961; and Toor, 1969). However, the mixer-flow-stop problem peculiar to the continuous, stopped, and accelerated flow systems used in the study of chemical and biochemical reactions needs to be considered in the light of a very great need to conserve solutions, increase velocity without cavitation, and stop the flow very rapidly. Smith (1973) has systemized the study of jet mixers and deals with scaled models. The problem of scaling turbulence and cavitation together is not dealt with. Work at very low temperatures (Chance, 1978) and at high viscosities (due to high protein concentration or high stabilizing medium concentration) demand efficient mixing. Moskowitz and Bowman (1966) have demonstrated the advantages of small multiple capillaries for knowing the precise time of the initiation of mixing, mixing of multiple components, and the use of a mesh to improve efficiency.

To define the problems more clearly, let me briefly remind all of you of the general

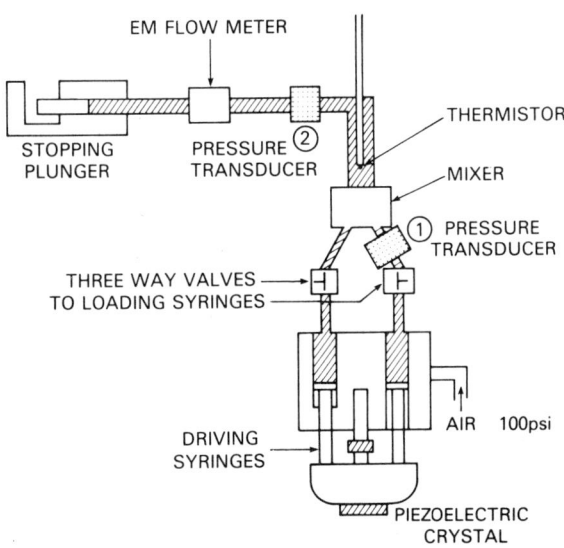

FIGURE 2 Stopped-flow method using the Gibson stopping syringe concept.

R. L. BERGER *Keynote Address*

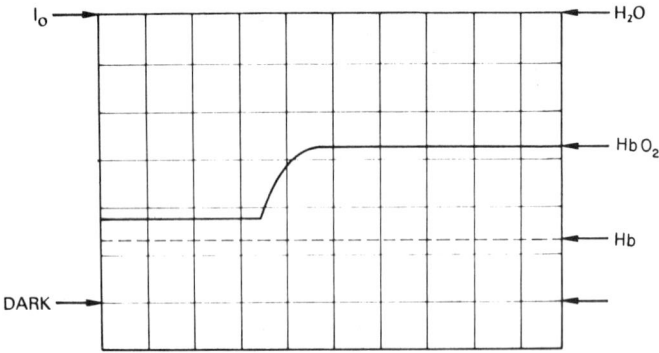

FIGURE 3 The reaction of 40 μM hemoglobin with 120 μM oxygen per iron.

scheme of a mixing system. Fig. 1 shows the first Hartridge-Roughton mixer made of brass from a Rolls-Royce carburetor. The volumes used in their first continuous flow experiment were 100 ml/point. In the mid-thirties a number of glass mixers were made and tested by Millikan (1936). These mixers together with stopped-flow brought about a reduction in volume to about 1 ml or less for the entire curve. Further improvement in optical detection, better mixers, and a greatly improved overall apparatus by Chance (1940a,b) brought the volume needed for a curve to 0.2 ml. Gibson's introduction of the stopping syringe in 1950 and a greatly improved apparatus with a double mixer (Chance, 1974) brought about the first commercial instrument. Fig. 2 shows the general stopped flow scheme, in which stopping is achieved by filling a syringe, which hits a stop. These were tested by visual observation of indicator reac-

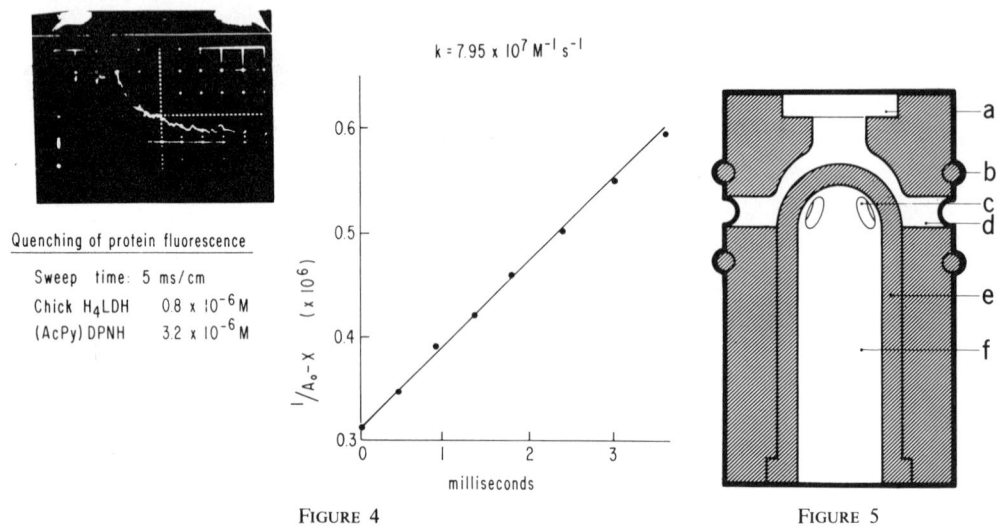

FIGURE 4 The reaction of 0.8 μM lactate dehydrogenase with 3.2 μM (AcPy) ADH (per monomer).

FIGURE 5 The ball mixer.

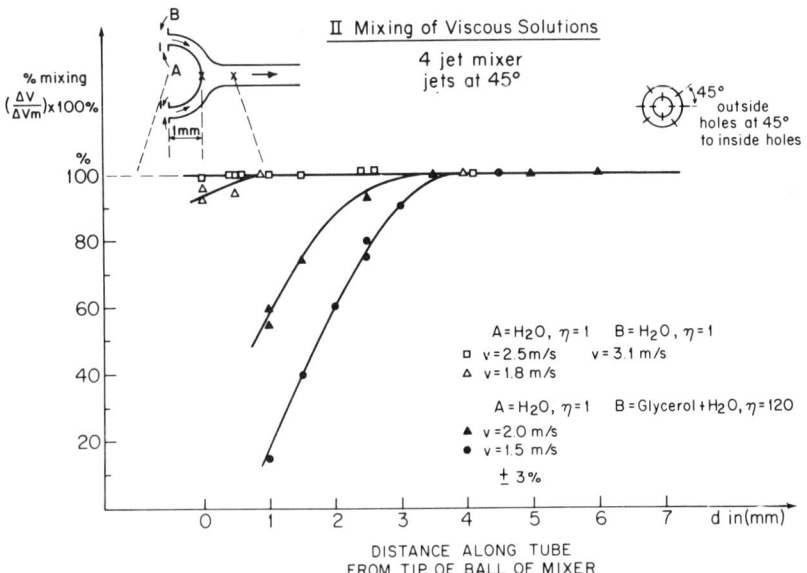

FIGURE 6 Mixing efficiency of the ball mixer. Reproduced by permission from Berger et al. (1968a). Copyright © 1968 American Institute of Physics.

tions and the use of characterized reactions. Roughton used equal titers of NaOH and HCl and a thermocouple moved along the observation tube as a test of mixing. This has proved useful, particularly when mixing glycerol and water, where the high heat of hydration can be used.

If we consider the hemoglobin reaction with oxygen at a concentration after mixing of 40 μM in heme, we see in Fig. 3 that at even 120 μM oxygen, part of the reaction is being missed. To simplify the analysis, we would like to use at least 250 μM heme, thus reducing dimerization as oxygen binds. This is particularly true for isoionic hemoglobin. Thus, if a reaction has a 1 ms half-life at 40 μM, it will move to roughly 150 μs half-life at 250 μM heme and become even faster as we go to 1 mM oxygen. If the "on" rate is of the order of 10^7 1/M-s, the half-life will be of the order of 75 μs. 10-μs resolution would be desirable. If 1 mm is the median point of closest approach to the point of mixing, a velocity of 100 m/s is required. We must clearly stop in this length of time as well, or characterize the system so that accelerated flow (Chance, 1940b) and simultaneous mixing and chemical reaction theory (Corrsin, 1961; Toor, 1969) can be used. Fig. 4 indicates the additional problem of light intensity insufficient to measure the reaction. This can likely be improved with a continuous wave laser ratioed to correct for variable intensity, if variable wavelength from 260 to 700 nM is available at a power level of 1 W. Lastly, when whole blood, high protein concentrations, etc, are mixed with plasma or buffer solutions, or low temperatures are used, severe viscous mixing problems arise. The ball mixer introduced by Berger and Bowman (1964) and extended to flow systems by Berger et al. (1968a) pushed the resolution time for mixing to less than 200 μs at flow velocities of 40 m/s. This mixer, shown in

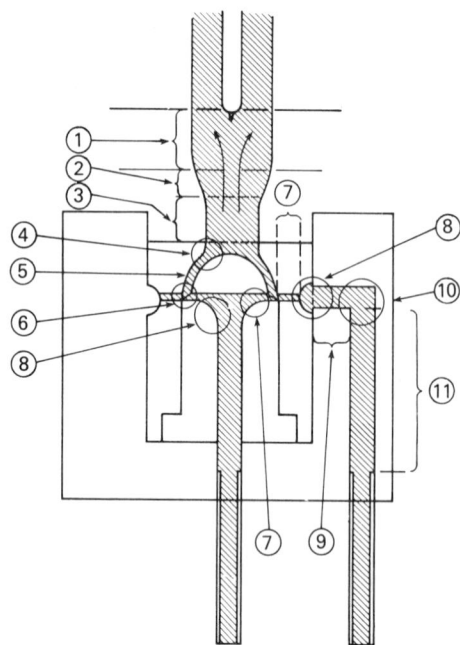

FIGURE 7 High-energy dissipation points in the micro ball mixer used in stopped flow calorimetry.

Fig. 5, is basically a simple turbulence-creating device, as mentioned by Goldstein (1938). The ball is in a flow stream and a turbulent wake is created on the downstream side of the ball, as if the stream were stationary and the ball moving. Most important, turbulence moves back against the ball as flow velocity increases. As shown by Berger et al. (1968a) in Fig. 6, optimal mixing occurs with the entry jets at a 45° angle to each other so that a close approach to a sheet of fluid flowing over the ball occurs. Balko et al.[1] have shown that energy dissipation in the form of heat in this mixer occurs as right-angle turns, rough points, etc. This is seen as the circled areas in Fig. 7. These points also act as nucleating centers for cavitation.

The development of the Parschall flume laid the foundation for the necessary shaping of the change in diameter of the passages to minimize cavitation. These principles have had to be modified in the design of the ball mixer, entrance, and exit tubes etc. shown in Fig. 7, due to the short distance available if time resolution, fluid conservation, heat losses, and cavitation minimization were to be achieved. These mixers have been made in a range of physical sizes from 1.5 mm diameter of the mixer on up (C. L. Gwen. 1977. 3860 Mt. Aladin, San Diego, Calif., and Update Instruments, Inc., Madison, Wis.). They are normally made of Kel-F (3M Company, St. Paul, Minn.) or any of a variety of stainless steels, Hastaloy C (Union Carbide Corp., New York), or Carpenter 20Cb being excellent for HCl and KCl solutions (1977, Carpen-

---

[1] Balko, B., R. L. Berger, and P. D. Bowen. 1978. High speed stopped-flow microcalorimeter. In preparation.

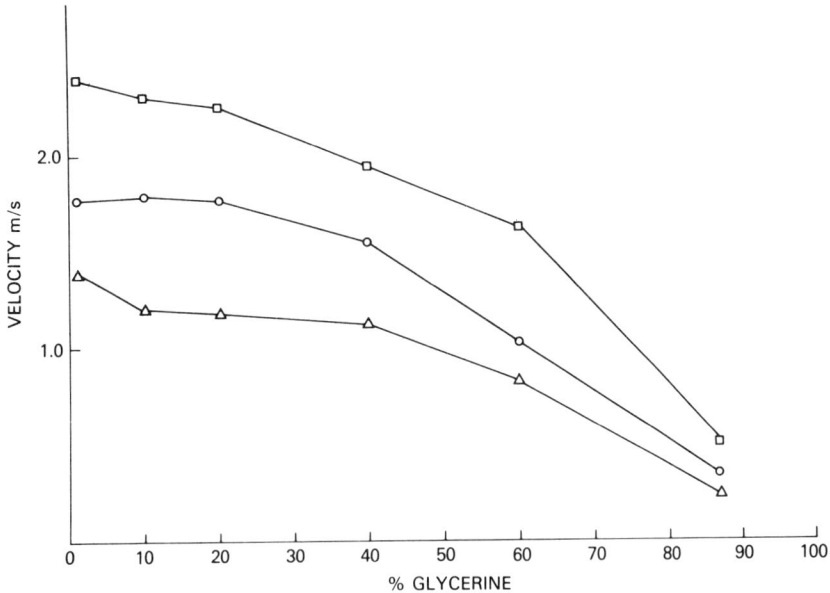

FIGURE 8 Velocity changes with driving pressures through a 3-mm diameter tube.

ter Technology Corp., Reading, Penn.). McClune and Fee (1976) report the use of this mixer as an augmenter to the Gibson-Durrum mixer (Durrum Instrument Corp., Sunnyvale, Calif.), but equal results are achieved with the ball mixer alone as shown by the mixing experiments in Fig. 6 with glycerol and water. An excellent discussion of reactions suitable for the testing of flow devices has recently been given by Tonomura et al. (1978) and should be referred to for details. Fig. 8 shows the change in velocity through a 3-mm diameter observation tube at three different driving pressures and various glycerine concentrations. Fig. 9 shows the nonlinear variation of viscosity with glycerine concentration. Fig. 10 is a plot of kinetic energy versus dissipation (temperature rise) through the observation tube.

For stopped flow at 85 m/s, a speed realized in the Berger-Science Products apparatus (Science Products Corp., Dover, N.J.) (Berger et al. 1968b), 11.8 μs are required to travel 1 mm. Minimum distance to center of the observation point is 2.5 mm; thus, a resolution time of 30 μs would be achieved if the solution can be stopped in less time and if cavitation can be avoided. The stopped valve used in the Berger-Science Products apparatus (see Chance, 1974), shown in Fig. 11, is a modification of an earlier valve (Berger et al., 1968b). It is normally closed. Flow is started by opening the valve and at the same time pressurizing an acoustic timer located in the body of the valve, the RC of which is set by a needle valve and the chamber volume. At a preset time, usually 5 ms, the valve closes in 25 μs, as measured with a phototube and light pipe. Due to cavitation, no better than 200-μs resolution at 40 m/s has been obtained (Berger et al., 1973; Berger et al. 1978).

It would appear that a 25–30 μs resolution time to the point of observation after stop

FIGURE 9

FIGURE 10

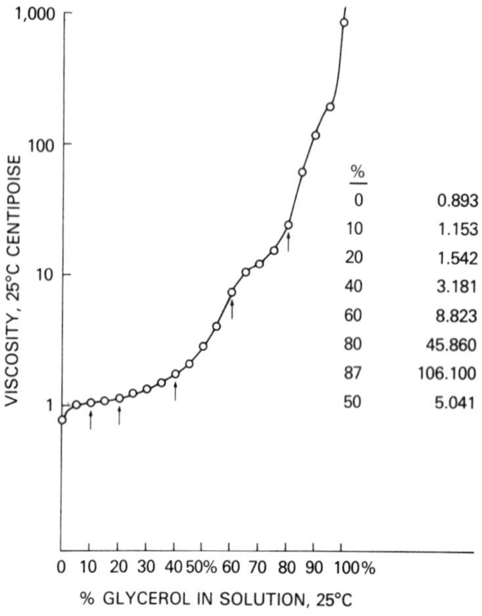

FIGURE 9 Nonlinear viscosity changes versus percentage of glycerine.
FIGURE 10 Energy dissipation versus fluid kinetic energy.

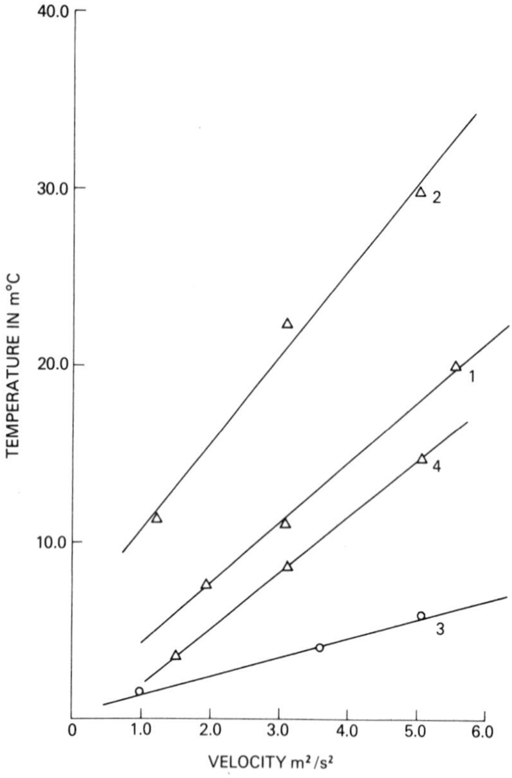

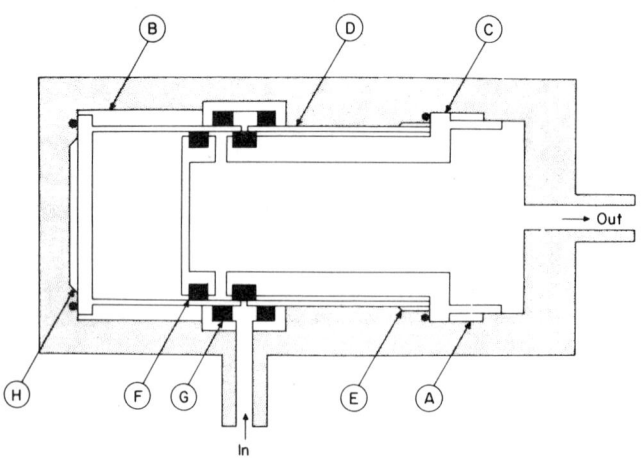

FIGURE 11 High-speed acoustic stop-valve. Reproduced by permission from page 40 of Chance (1974). Copyright © 1974 by John Wiley & Sons, Inc.

is obtainable, providing cavitation can be reduced to zero (Duffy and Staerk, 1969). One does have the problem of a pressure pulse at the front of the stop valve, which we have measured at about 1,000 psi. It appears to dissipate rapidly and, unless an air bubble or cells are involved, does not seem to affect the optical path, about 2-4 cm below the valve. There is no reason this valve cannot be closed in 10 μs but a larger pressure pulse will certainly occur. The theoretical and practical problem is the design of a highly turbulent system free of cavitation. While turbulence can be reasonably well scaled, turbulence and cavitation have offered severe problems. Nevertheless, work is proceeding on this problem with a time resolution of 10 μs the goal.

The major optical improvements made in stopped flow work since the last reviews (Schechter, 1970; Reich, 1971; Chance, 1974), are mainly in the various rapid scan spectrophotometers that have appeared. These have been discussed by Ridder and Margerum (1977). The apparatus of Kawania is described by Wightman et al. (1974); it has the ability to scan 250 nm in 1.2 ms with an optical density sensitivity of 0.005 absorbance units, and at 0.5 nm resolution 50 points can be taken. A new Tracor Northern unit offers a scan rate of 10 μs/diode with a 512-diode array (Tracor Northern, Middleton, Wis.). Optical densities to over 3 can be utilized with an intensified detector.

For single cell work, the system described by Benedetti et al. (1976) and Benedetti and Lenci (1977) is most interesting. The condenser stage is moved under computer control. Both spectra and location can be scanned as well as total shape; the latter is done independently with an infrared vidicon. Thus for stopped-flow work on single red cells, reactants could be flushed past the cell and the reactions followed inside the cells as, for example, oxygen is either added or taken away. Dr. Giardina will discuss its uses for red cell work later at this meeting (Antonini et al., 1978), with flash photolysis.

A variety of detection methods has been applied to stopped flow devices in the last several years. Of particular interest is the nuclear magnetic resonance flow system of Manuck et al. (1973); the circular dichorism system of Luchins (1977), used for the study of hemoglobin denaturation; and the combined stopped-flow laser flash photolysis system of Sawicki and Gibson (1978) and McCray and Smith[2] for the study of carbon monoxide and oxyhemoglobin. Less than 20-ms resolution times are obtained with the first two methods and about 0.2 ms with the flow-flash system. To study the hemoglobin reaction inside the red cell, a new dual-wavelength detection system has been developed (Berger et al., 1978) for use with the laser flash-flow system. It has a 1-μs response time and utilizes the near infrared region of the spectrum. It is a modification of the Optisat (Vurek, 1973) and, as shown in Fig. 12, utilizes two light-emitting diodes, one at 660 nm and one at 900 nm. Two phototransistors receive the light. Dichromatic filters separate the beams. The new fiber optic-single chip laser detector systems developed at the Bell Laboratories, (Murray Hill, N.J.) (Tien, 1977)

---

[2] McCray, J., and P. D. Smith. 1978. Kinetic binding to tetramers. Submitted for publication.

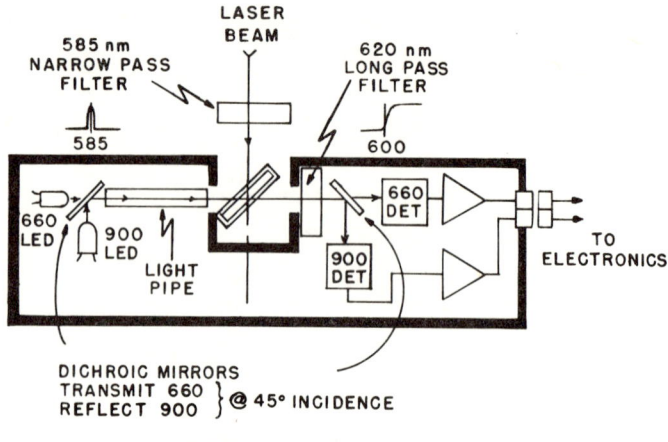

FIGURE 12 100 kHz dual-wavelength spectrometer.

for light communication lines offer fascinating opportunities for future single cell research.

Considerable effort has been expended over the years in looking at other detection methods. Some of these have been reviewed by Roughton (1963). Chance (1974) has reviewed quench flow by acid or freezing. Since these reviews, advances have been made in pH detectors (Crandall et al., 1971, 1978) and thermodetection (Balko and Berger, 1968; Balko et al. 1969).[3] Recently we have developed a thermistor probe, shown in Fig. 13 (Thermometrics, Inc., Edison, N.J.) and a modified differential bridge (Berger et al., 1974; Linear Research, San Diego, Calif.) shown in Fig. 14, which

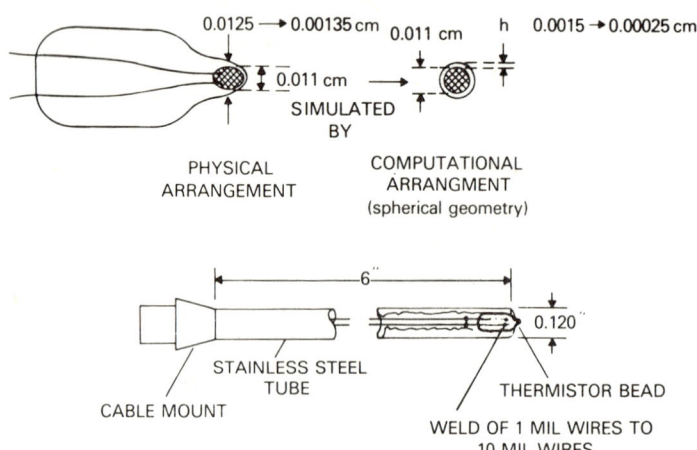

FIGURE 13 High-speed thermistor probe.

---

[3]See note 1.

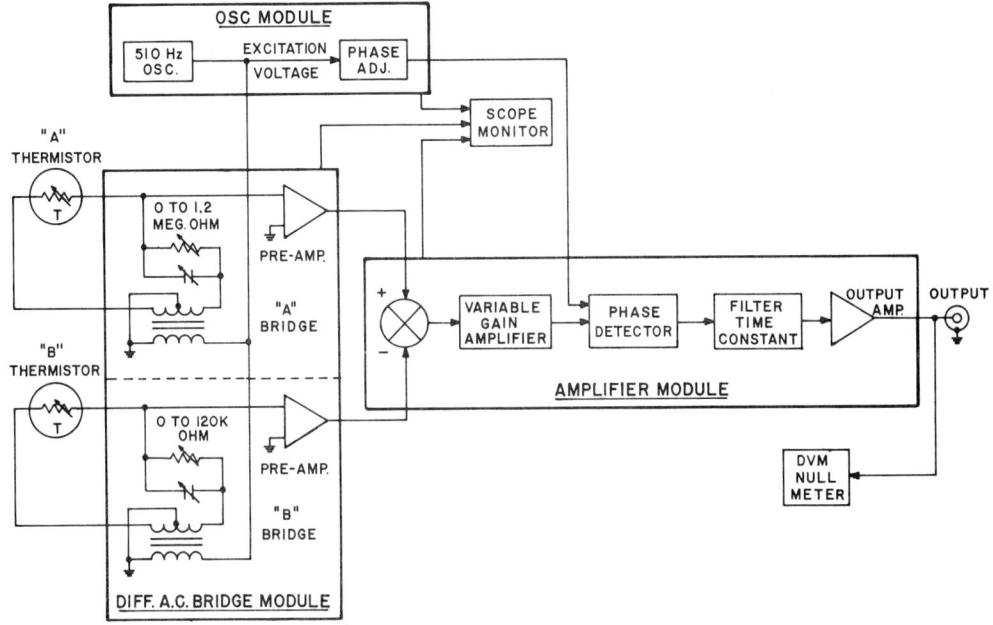

FIGURE 14 Differential 200 AC bridge.

has a response time to $1/e$ of 3 to 7 ms. The response is measured with the spring-loaded plunger shown in Fig. 15 (Berger and Balko, 1972). Fig. 16 shows a typical thermistor response to this plunge test. A detailed study of thermal losses in flow systems was carried out (Balko et al., 1978) with the result that the areas shown encircled in Fig. 7 have been smoothed, greatly decreasing heat losses. Fig. 17 shows sodium bicarbonate plus HCl (1.2 and 1.2 kcal/mol for each of the reactions given in Eqs. 1 and 2) at 28°C

$$H^+ + HCO_3^- \leftrightarrow H_2CO_3 \qquad (1)$$

$$H_2CO_3 \rightarrow CO_2 + H_2O \qquad (2)$$

Heats of flow, mixing, etc., can be kept to about 5 millidegrees or less. A number of problems have appeared. A pressure effect on the thermistor occurs when the glass coating is too thin, so that it begins to develop microcracks, or when bubbles exist in the thin glass coating of the thermistor. It is hoped this can be corrected by the evaporation of silicon dioxide onto the thermistor to hermetically seal it. The need for much faster and more sensitive thermal sensors has led to some work on thin film capacitance devices (Maserjian, 1972). Work is continuing on these devices in collaboration with the Thermometry Section of the National Bureau of Standards.

pH detectors also have proven troublesome for several reasons. In general, pH electrodes have resistances of 100–500 MΩ. Electrode and lead capacitance amount to

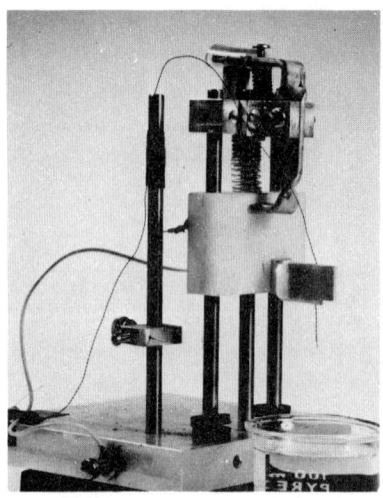

FIGURE 15 Spring-loaded guillotine for testing thermistor and pH electrodes. Reproduced by permission from Berger and Balko (1972). Copyright © 1972 Instrument Society of America.

about 300 pF. Thus, a time response of about 90 ms to $1/e$ is to be expected. A 3-mm-diameter combination electrode such an Ingold type (1977, W. Ingold AG, Urdorf-Zurich, Switzerland) made of LOT glass, a calomel reference, and a porous plug, together with a pH meter such as the one shown in Fig. 18, is presently being used. This pH meter uses a driven shield, thus reducing the lead capacitance to nearly zero. The electrode has about 150 pF capacitance and a resistance, at 25°C, of about 100–200 MΩ. We thus expect a response time of 15 ms. Fig. 19 shows the time of response to the plunge test (Berger and Balko, 1972), to be about 10 ms uncoated (*a*) and 14 ms coated (*b*). Faster times may be achievable, as suggested by Crandall (1978) at this discussion. $t_{1/2}$ of only several milliseconds using L & N electrodes (Leeds & Northrup

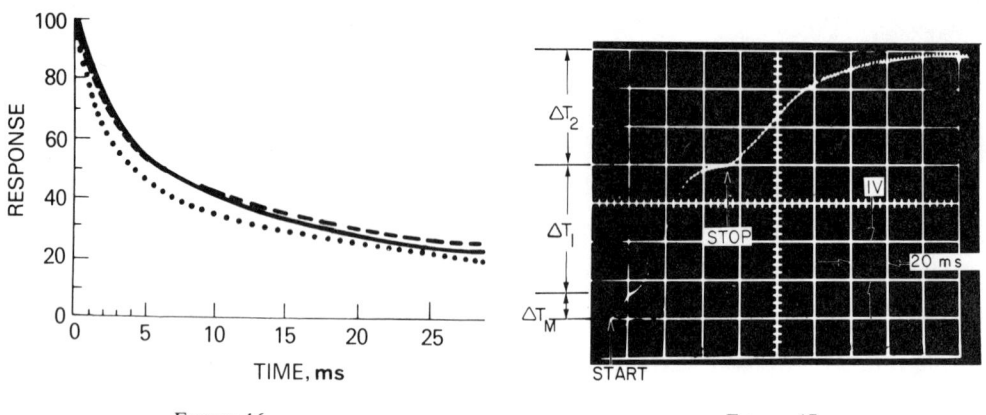

FIGURE 16

FIGURE 17

FIGURE 16 Thermistor response to plunge test.
FIGURE 17 Stopped-flow thermal reaction of 0.02 M NaHCO$_3$ with 0.01 M HCl.

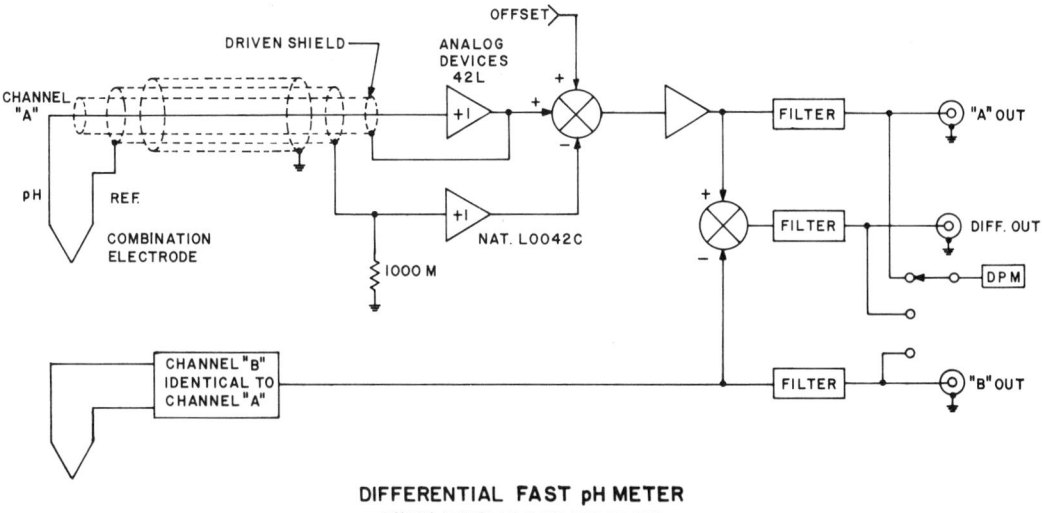

DIFFERENTIAL FAST pH METER
—SIMPLIFIED BLOCK DIAGRAM—

FIGURE 18 Differential fast pH meter.

Co., North Wales, Pa.) were estimated. The electrodes, in our hands, have a resistance of 5×10^8 Ω and a capacitance of 140 pF. Since the driven shield eliminates the cable capacitance, probe capacitance is about 20–30 pF, giving a time constant of about 10–15 ms. Plunge tests give a response time of 9 ms, as shown in Fig. 20a and b. An additional problem is the loss of sensitivity and response time of the electrode caused by protein and cell contamination. This has been solved by the use of Lycra (1976; Ethicon, Inc., Somerville, N.J.) without serious loss of response time or sensitivity. Biocompatibility appears to be excellent (Kolobow et al., 1977). Studies of pH elec-

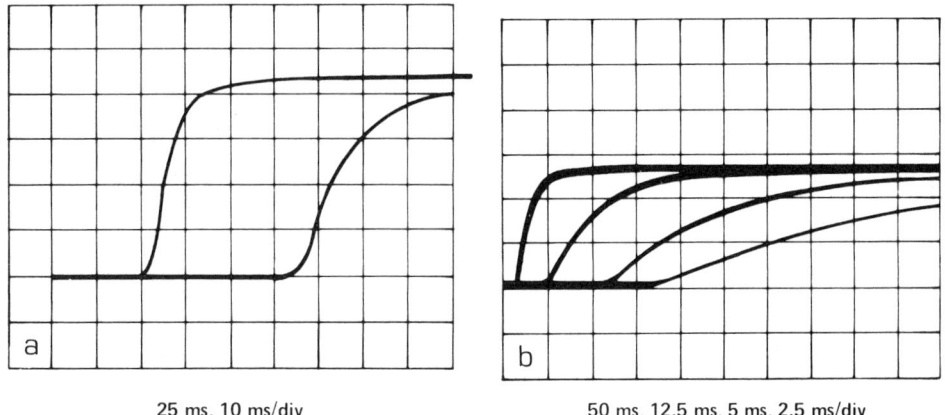

25 ms, 10 ms/div   50 ms, 12.5 ms, 5 ms, 2.5 ms/div

FIGURE 19 Response of Ingold-type lot 405 3472 3 mm combination pH electrode (Calomel ref. with porous plug) to the plunge test: *a.* uncoated $t_{1/2} = 10$ ms; *b.* coated with Lycra (Ethicon) $t_{1/2} = 14$ ms.

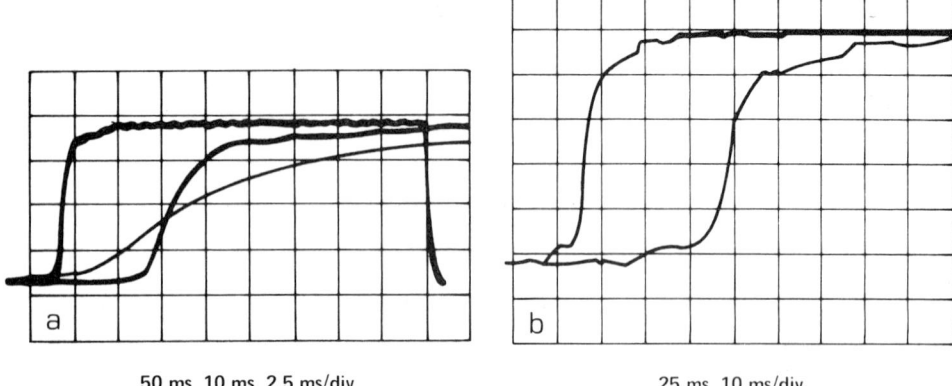

50 ms, 10 ms, 2.5 ms/div   25 ms, 10 ms/div

FIGURE 20 Plunge tests of Leeds and Northrup electrode type. Reference is silver-silver chloride with a crack-type leak. *a*. uncoated, $t_{1/2}$ = 9 ms; *b*. coated with Lycra, $t_{1/2}$ = 10 ms.

trode response times by Wikby (1972) made by electron flow measurements suggest a large number of time constants; one, accounting for about 10% of the electrode response, is as slow as 2–20 s. We do not see this. A careful investigation of this will use the plunge method developed for thermistors (Berger and Balko, 1972).

Froelich et al. (1976) has recently described a quench flow apparatus using a stepping motor and a ball-mixer system for multiple mixing experiments. Time resolutions of 1.5 ms are reported for sodium-potassium ATPase and a sacroplasmic reticulum-CaATPase. Multiple mixing can be carried out in up to three mixers in the present apparatus (Commonwealth Technology, Alexandria, Va.; Update Instrument, Inc.). An automatic sampler adds to the speed of operation. Four points on the rate curve can be obtained with each particular size tubing by changing the speed of the advance of the stepping motors. About 0.25 ml of each reagent are required per point. Application to several systems will be discussed by Froelich in a poster session at this discussion. Brahm (1977) has improved on the original Tosteson continuous flow filtration apparatus for use with red cells; he reports a time resolution of about 7 ms. Utilization of Nuclepore filters (Nuclepore Corp., Pleasanton, Calif.) pushed up into the flow path has proven very successful. It would appear that current attempts to make such an adapter for red cell volume regulation studies to the Froelich quench flow apparatus are meeting with success.[4]

Finally, I would like to say a word about the advances being made in data handling over the last several years. Stopped flow systems produce vast amounts of data and this is not a trivial problem. DeSa and Gibson (1969) worked out a minicomputer-based system for taking, correcting, calculating, and storing the data from a rapid flow apparatus. A commercial system to do this is available (OLS, Athens, Ga.), as are several systems that allow some data manipulation (Wightman et al., 1974; Mieling et

---

[4]Froelich, J. P., F. Kregenow, and R. L. Berger. In preparation.

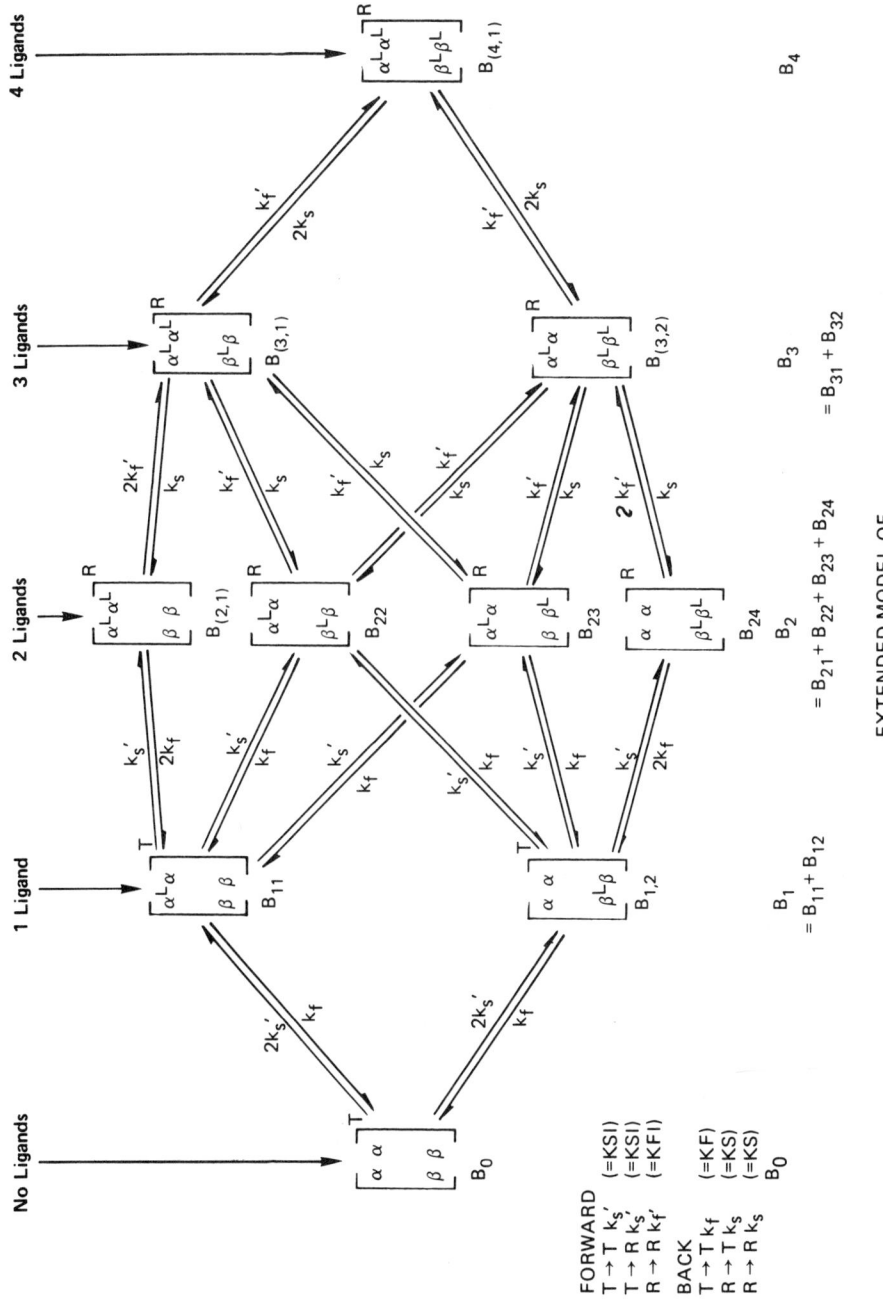

FIGURE 21 Expanded two-state finite element model of hemoglobin.

R. L. BERGER *Keynote Address*

al., 1976). These tasks are rapidly being expanded to microprocessor based systems. The apparatus described by Winslow et al. (1978), while not a stopped flow instrument, is an example of automation where the entire experiment is carried out to the point of writing out the final graph. This machine is interesting from a different standpoint; it produces an oxygen dissociation curve of hemoglobin in whole blood or hemoglobin solution to an accuracy of about 0.1% over the entire range of saturation. This affords the possibility of testing kinetic models by forcing the kinetic data to fit the much more accurate equilibrium data. In the Adair-Perutz two-state model, a large number of pathways are possible, as shown in Fig. 21. McCray and Smith[2] 1978 and Berger et al. (1978) have analyzed this model by LaPlace transforms and the finite element simulation technique (FEST) of Davids and Berger (1960, 1969, 1977). This latter technique was developed to correct heat conduction calorimeters so they could be used for kinetics (Rehak et al., 1976; Berger et al., 1975). By coupling this to MLAB (Knott and Shrager, 1972), a curve-fitting technique, FEST can be used to determine automatically the correct rate constants to make the kinetic data fit the equilibrium curve. FEST is now being used on a microprocessor-based system to correct for the response

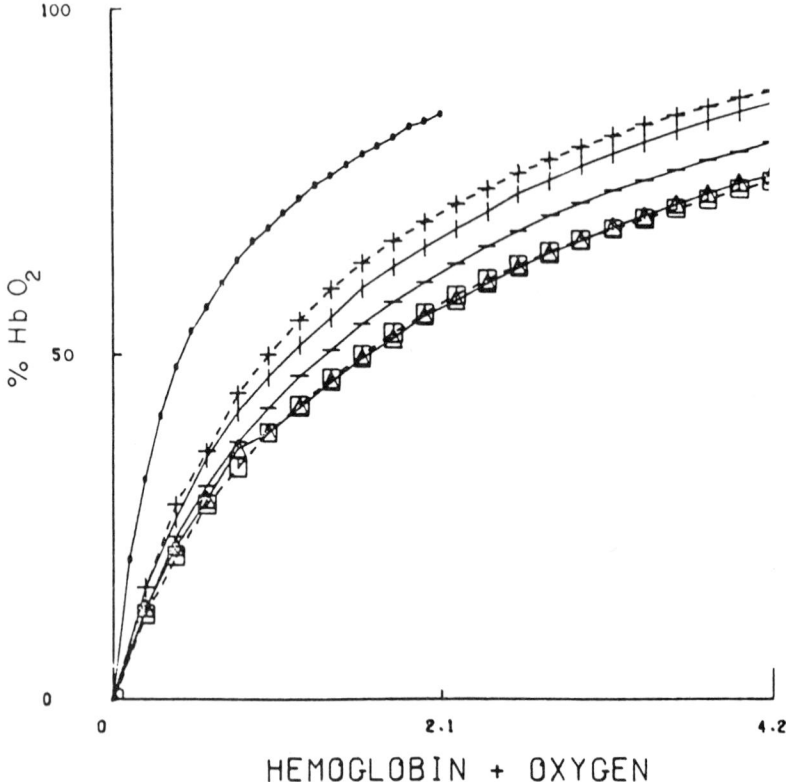

FIGURE 22 Reaction of 40 μM (in heme) hemoglobin with 120 μM oxygen as a. ● Isoionic; b. +0.02 M bis-tris, pH 7.4; c. □ 1 M/M 2,3-DPG; d. Δ 2 M/M 2,3-DPG; e. | 0.1 M PO$_4$, pH 7.4; f. — 0.3 M KCl and 1 M/M 2, 3-DPG. Abscissa in milliseconds.

time of detectors[1] and to fit kinetics and thermodynamic models to obtain rate and equilibrium constants. While many schemes are presently available (see Garfinkel et al., 1977; Berman and Weiss, 1972), the finite element simulation technique offers the advantage of allowing the investigator to remain with the physics and chemistry of the problem.

The improvement in the data obtained by use of the microprocessor to "run" the experiment, by adaptations of modern signal processing methods, and by simulation methods makes it possible to perform biological experiments on a single small sample, thus greatly improving our ability to analyze complex biochemical reactions. The rising importance of this field is indicated by the number of discussions and poster papers at this meeting and by the large increase in descriptions of stopped-flow investigations appearing in the literature.

With our ability to produce isoionic proteins (Righetti and Drysdale, 1976), we must begin to work on the effect on reactions of the various constituents of the "real" world in which the reactions run. Fig. 22 shows such effects on the kinetics of the oxygen and hemoglobin reaction (Berger et al., 1973). The discussions of Moelwyn-Hughes (1967) and Guggenheim and Turgeion (1955) on pressure, salt, and dielectric constant effects, and Schurr (1970) on diffusion and, of course, the better-known temperature and entropy problems will add greatly to our understanding of the system with which we are dealing. That this system is, in a very real sense, a solid-state system is becoming clearer as our membrane friends demonstrate that one major metabolic system after another is clearly bound to the membrane. One would hope that solid-state physicists would team up with biochemists to explore, with what I am sure will be a new generation of solid-state experimental methods, the molecular interactions on the membrane surface.

REFERENCES

ANTONINI, E., M. BRUNORI, B. GIARDINA, P. A. BENEDETTI, G. BIANCHINI, and S. GRASSI. 1978. Single cell observations of gas reactions and shape changes in normal and sickling erythrocytes. *Biophys. J.* 24:187.
BALKO, B., and R. L. BERGER. 1968. Measurement and computation of therm-junction response times in the submillisecond range. *Rev. Sci. Instrum.* 39:498.
BALKO, B., R. L. BERGER, and W. S. FRIAUF. 1969. Stopped-flow calorimetry for biochemical reactions. *Anal. Chem.* 41:1506.
BATCHELOR, G. K. 1971. Scientific Papers of Sir Geoffrey Ingram Taylor. Cambridge University Press, London, U.K.
BENEDETTI, P. A., G. BIANCHINI, and G. CHITI. 1976. Fast scanning microspectroscopy: an electrodynamic moving-condenser method. *Appl. Opt.*, 15:2554.
BENEDETTI, P. A., and F. LENCI. 1977. In vivo microspectrofluorometry of photoreceptor pigments in *Euglena gracillis, Photochem. Photobiol.* 26:315.
BERGER, R. L., and B. BALKO. 1972. Thermal sensor coatings suitable for rapid response biomedical applications. *Temp. Meas. Control Sci. Ind.* 5:2169.
BERGER, R. L., and R. L. BOWMAN. 1964. Ball mixers. *In* Rapid Mixing and Sampling Techniques in Biochemistry. B. Chance et al., editors. Academic Press, Inc., New York.
BERGER, R. L., and N. DAVIDS. 1965. General computer method of analysis of conduction and diffusion in biological systems with distributive sources. *Rev. Sci. Instrum.* 36:88.
BERGER, R. L., B. BALKO, and H. CHAPMAN. 1968a. High resolution mixer for the study of the kinetics of rapid reactions in solution. *Rev. Sci. Instrum.* 39:493.

BERGER, R. L., B. BALKO, and W. BORCHERDT. 1968b. High speed optical stopped flow apparatus. *Rev. Sci. Instrum.*, **39**:486.

BERGER, R. L., J. EVERSE, L. CARPENTER, and J. L. KAPLAN. 1973. Human isoionic hemoglobin: preparation and kinetic properties. *Anal. Lett.* **6**:125.

BERGER, R. L., W. S. FRIAUF, and H. E. CASCIO. 1974. A low-noise thermistor bridge for use in calorimetry. *Clin. Chem.* **20**:1009.

BERGER, R. L., N. DAVIDS, and E. PANEK. 1975. The design and development of a stopped-flow microcalorimeter for the study of enzyme kinetics. *J. de Calorimetrie d'anal. Therm.* **6**:1–10.

BERGER, R. L., N. DAVIDS, J. A. McCRAY, and P. D. SMITH. 1978. Finite element modeling of oxyhemoglobin reaction. *Fed. Proc.* **37**:1673.

BERMAN, M., and M. F. WEISS. 1972. SAAM Manual. Government Printing Office, Publication no. 1708.

BRAHM, J. 1977. Temperature dependent changes of chloride transport kinetics in human red cells. *J. Gen. Physiol.* **70**:283–307.

BRODKEY, R. S., editor. 1975. Turbulence in mixing operations. Academic Press, Inc., New York.

CHANCE, B. 1940a. The accelerated-flow method for rapid reactions. I. Analysis. *J. Franklin Inst.* **229**:455–613.

CHANCE, B. 1940b. Accelerated and stopped-flow apparatus. II. *J. Franklin Inst.* **220**:613, 737.

CHANCE, B. 1974. Investigation of rates and mechanisms of reactions. *Techn. Chem.* **6**:5–62.

CHANCE, B., et al., editors. 1978. Tunnelling Biological Molecules. Academic Press, Inc., New York.

CORRSIN, S. 1961. Turbulent flow. *Am. Sci.* **49**:300.

CRANDALL, E. D., and R. E. FORSTER. 1978. Studies of pH equilibration and anion exchanges in erythrocyte suspensions. Stopped-flow pH electrode measurements. *Biophys. J.* **24**:35–41.

CRANDALL, E. D., R. A. KLOCKE, and R. E. FORSTER. 1971. Hydroxyl ion movements across the human erythrocyte membrane. *J. Gen. Physiol.* **57**:666–683.

DAVIDS, N., and R. L. BERGER. 1964. A computer analysis method for thermal diffusion in biochemical systems. *Commun. Am. Comput. Mach. Assoc.*, **7**:547.

DAVIDS, N., and R. L. BERGER. 1969. Finite element simulation method for the design and data correction of calorimeters. *Curr. Mod. Biol.* **3**:169.

DAVIES, J. T. 1972. Turbulence Phenomena. Academic Press, Inc., New York.

DESA, P. J., and Q. H. GIBSON. 1969. A practical automatic data acquisition system for stopped-flow spectrophotometers. *Comput. Biomed. Res.* **2**:494.

DUFFY, D., and H. STAERK. 1969. An energetic model for rapid mixing. *Rev. Sci. Instrum.* **40**:789.

FRIEDLANDER, S. K., and TOPPER, L. editors. 1961. Turbulence. Interscience, John Wiley & Sons, Inc., New York.

FROEHLICH, J. P., J. V. SULLIVAN, and R. L. BERGER. 1976. A chemical quenching apparatus for studying rapid reactions. *Anal. Biochem.* **73**:331–341.

GARFINKEL, L., M. C. KOHN, and D. GARFINKEL. 1977. Systems analysis in enzyme kinetics, *CRC Crit. Rev. Bioeng.* **329**.

GOLDSTEIN, S. 1938. Modern Developments in Fluid Dynamics, Vol. I. The Oxford University Press, London, U.K.

GUGGENHEIM, E. A., and J. C. TURGEION. 1955. Specific interaction of ions. *Faraday* Soc. Trans. **51**:747.

HARTRIDGE, H., and F. J. W. ROUGHTON. 1923a. Method of measuring the velocity of very rapid chemical reactions. *Proc. R. Soc. (Lond.) A Math. Phys. Sci.* **104**:376–394.

HARTRIDGE, H., and F. J. W. ROUGHTON. 1923b. The velocity with which carbon monoxide displaces oxygen from combination with haemoglobin. I. *Proc. R. Soc. (Lond.) B Biol. Sci.* **94**:336–267.

HARTRIDGE, H., and F. J. W. ROUGHTON. 1924, 1926. Improvements in the apparatus for measuring the velocity of very rapid chemical reactions. *Proc. Cambridge Phil. Soc.* **22**:426–431. **23**:450–460.

HINZE, J. O. 1959. Turbulence. McGraw-Hill Book Company, New York.

KAY, J. M. 1963. Introduction to Fluid Mechanics and Heat Transfer. Cambridge University Press, London, U.K. 2nd edition.

KNOTT, G. D., and R. I. SHRAGER. 1972. On-line modelling by curve fitting computer graphics. Proceedings of the Siggraph Computer in Medicine Symposium. *Assoc. Comput. Mach. Siggraph Notes.* **6**(No. 4):138–151.

KOLOBOW, T., T. A. TOMLINSON, and J. E. PIERCE. 1977. Blood compatibility of methyl, methyl vinyl, methyl phenyl, and trifluoropropylmethylvinyl silicone rubber without silica fillers in the spiral-coiled membrane lung. *J. Biomed. Mater. Res.* **11**:471–481.

LUCHINS, J. I. 1977. Stopped-flow circular dichrosim. Ph.D. Thesis. Columbia University, New York. University Microfilms, Inc., Ann Arbor, Mich.

MANUCK, B. A., J. G. MALONEY, JR., and B. D. SYKES. 1973. Kinetics of the interaction of methyl isonitrile with hemoglobin β chains: measurement by NMR. *J. Mol. Biol.* **81**:199–205.

MASERJIAN, J., 1972, Thin-film temperature sensor. *Temperature.* **4**:2159–2167.

MCCLUNE, G. J., and J. A. FEE. 1976. Stopped-flow spectrophotometric observation of superoxide dismutation in aqueous solution. *Fed. Eur. Biochem. Soc. (FEBS) Lett.*, **67**:294–298.

MIELING, G. E., R. W. TAYLOR, L. G. HARGIS, J. ENGLISH, and H. L. PARDUE. 1976. Fully automated stopped-flow studies with a hierarchical computer controlled system. *Anal. Chem.* **48**:1686.

MILLIKAN, G. A. 1936. Photoelectric methods of measuring the velocity of rapid reactions. III. A portable micro-apparatus applicable to an extended range of reactions. *Proc. Roy. Soc. (Lond.) Math. Phys. Sci.* **A155**:277–292.

MOELWYN-HUGHES, E. A. 1967. The influence of pressure on the velocity constants of biomolecular ionic reactions in aqueous solution. *J. Phys. Chem.* **71**:4120–4121.

MOSKOWITZ, G. W., and R. L. BOWMAN. 1966. An efficient multicapillary mixer. *Science (Wash. D.C.).* **153**:428–429.

PAI, S. 1954. Fluid Dynamics of Jets. Van Nostrand Reinhold Company, New York.

REHAK, N. N., J. EVERSE, N. O. KAPLAN, and R. L. BERGER. 1976. Determination of the activity and concentration of immobilized and soluble enzymes by microcalorimetry. *Anal. Biochem.* **70**:381–386.

REICH, R. M. 1971. Instrumentation for the study of rapid reactions in solution. *Anal. Chem.* **43**:85A.

RIDDER, G. M., and D. W. MARGERUM. 1977. Simultaneous kinetic and spectral analysis with a Vidicon rapid-scanning stopped-flow spectrometer. *Anal. Chem.* **49**:2098–2108.

RIGHETTI, P. G., and J. W. DRYSDALE. 1976. Isoelectric Focussing, North Holland Publishing Co., Amsterdam.

ROUGHTON, R. J. W. 1963. Rates and mechanisms. *Tech. Org. Chem.* **8**:303–392.

ROUGHTON, F. J. W., and G. A. MILLIKAN. 1936. Photoelectric methods of measuring the velocity of rapid reactions. I. General principals and controls. *Proc. Roy. Soc. (Lond.). Math. Phys. Sci.* **A155**:258–269.

SAWICKI, C. A., and Q. M. GIBSON. 1978. The relation between carbon monoxide binding and the conformational change of hemoglobin. *Biophys. J.* **24**:21–28.

SCHECHTER, A. N. 1970. Measurement of fast biochemical reactions. *Science (Wash. D.C.).* **170**:273.

SCHLICHTING, H. 1968. Boundary Layer Theory. McGraw-Hill Book Company, New York. 6th edition.

SCHURR, M. M. 1970. The role of diffusion in enzyme kinetics. *Biophys. J.* **10**:717–727.

SMITH, M. H. 1973. Observations relating to efficiency mixing in rapid reaction. *Biophys. J.* **13**:817–821.

TIEN, P. K. 1977. Integrated optics and new wave phenomena in optical waveguides. *Rev. Mod. Phys.* **49**:361.

TONOMURA, B. H., M. HAHATANI, J. OHNISHI, Y. ITO, and K. HIROMI. 1978. Test reactions for stopped flow apparatus. *Anal. Biochem.* **84**:370–383.

TOOR, H. L. 1969. Turbulent mixing of two species with and without chemical reactions. *Ind. Eng. Chem. Fundam.* **8**:655–659.

VUREK, G. G. 1973. Oxygen saturation for extracorporeal circulation applications. *Med. Instrum. (Arlington).* **7**:262–267.

WIGHTMAN, R. M., R. L. SCOTT, C. N. REILY, and R. W. MURRAY. 1974. Computer controlled rapid scanning stop flow spectrometer. *Anal. Chem.* **46**:1492–1499.

WIKBY, A. 1972. The influence of isoproparol on the surface resistance of glass electrodes. *J. Electroanal. Chem.* **38**:441–443.

WINSLOW, R. M., J. M. MORRISSEY, R. L. BERGER, P. D. SMITH, and C. G. GIBSON. 1978. Variability of oxygen affinity of normal blood. *J. Appl. Phys.* In press.

DISCUSSION

SUTTER: At 1000–2000 psi, what precautions are necessary to keep from fracturing the windows or otherwise blowing them out?

BERGER: Considerable work has gone into the construction of our observation tubes. They are presently made by casting the circular quartz windows in Stycast 2057 resin (Emerson & Cuming, Inc., Canton, Mass.) using Catalyst #9. The mold is made of stainless steel and precision ground to the correct dimensions. This results in a square observation tube with the appropriate shaping to the circular mixer. The resin must be carefully degassed before and during pouring and setting, i.e. under vacuum. Curing is carried out at room temperature, under vacuum, for 24 h. The tubes are stored under water to prevent dehydration and shrinkage.

CZERLINSKI: In your stopping value of Borcherdt, you mentioned a pressure pulse of 1000 psi (occurring presumably at the end of your 25-μs stopping time). What is the time profile of this pressure pulse? If Δt (pressure) is short in relation to the time window for one observation point, you would not expect to observe (reversible) effects.

SMITH: The pressure pulse had a tau of 5 μs and is therefore a delta function.

KIRSCHENBAUM: Under what conditions does one have to worry about the effect of the pressure of the stopped-flow experiment on the kinetic results?

BERGER: Under conditions where the adiabatic compressibility of the sample, i.e. solvent-buffer-salt-solute, is large enough to produce a relaxation effect. In general, water can be neglected below 1,000 atm. In the range of 50–200 atm it would appear from Knoche and Wiese's work (1976) that a great many of the systems we are interested in have relaxations both to ΔP and ΔT. These often are in the microsecond range but do extend for many systems to the millisecond range. See Yasunaga and Tatsumoto's extended abstract (p. 267, this discussion). However, this will only be true if the ΔP or ΔT lasts for several milliseconds. In general the ΔP lasts for only 5–10 μs.

JAKOBSSON: With regard to time response of pH electrodes, one should be able to extrapolate to the "true" voltage across the glass from the measured one, even if the measurement is lagging, by rearranging the equation

$$dV_{meas}/dt = (1/RC)(V_{true} - V_{meas.}), \qquad (1)$$

to

$$V_{true} = RC\, dV_{meas}/dt + V_{meas}. \qquad (2)$$

If the RC time constant is determined independently, then the true voltage and hence the pH during the actual experiment (when one monitors dV_{meas}/dt and $V_{meas.}$) can be calculated from Eq. 2. This should also apply to lag in temperature measurement, replacing RC in Eq. 2 with the time constant for response of the temperature probe.

REFERENCE

KNOCHE, W., and G. WIESE. 1976. Pressure-jump relaxation technique with optical detection. *Rev. Sci. Instrum.* **47**:220–221.

THE RELATION BETWEEN CARBON MONOXIDE BINDING AND THE CONFORMATIONAL CHANGE OF HEMOGLOBIN

Charles A. Sawicki and Quentin H. Gibson, *Section of Biochemistry, Molecular and Cell Biology, Cornell University, Ithaca, New York 14853 U.S.A.*

ABSTRACT The spectral difference between normal and rapidly reacting deoxyhemoglobin (Sawicki and Gibson [1976], *J. Biol Chem.* **251**:1533-1542) is used to study the relationship between CO binding to hemoglobin and the conformational change to the rapidly reacting form in a combined flow-laser flash experiment. In both pH 7 phosphate buffer and pH 7 bis(2-hydroxy-ethyl)imino-tris (hydroxymethyl)methane buffer (bis-Tris) with 500 μM 2,3-diphosphoglycerate (DPG), the conformational change lags far behind CO binding; rapidly reacting hemoglobin is not observed until more than 10% of the hemoglobin is liganded. In pH 9 borate buffer the formation of rapidly reacting hemoglobin leads CO binding by a significant amount.

A simple two-state allosteric model (Monod et al. [1965], *J. Mol. Biol.* **12**:88-118) which assumed equivalence of the hemoglobin subunits in their reaction with CO was used to simulate the experimental results. In terms of the model, the conformational change lead observed at pH 9 suggests that significant conformational change has occurred after binding of only one CO molecule per tetramer. In the presence of phosphates good agreement between experimental results and simulations is obtained using parameter values suggested by previous experimental studies. The simulations suggest that the conformational change occurs after binding of three CO molecules.

INTRODUCTION

Ligand-linked conformational changes of the hemoglobin molecule are generally thought to be responsible for the sigmoidal binding curves observed for oxygen (1) and carbon monoxide (2). Several experimental techniques have been used to study the relationship between fractional ligation and conformational change in an attempt to test models describing cooperative ligand binding by hemoglobin. Equilibrium experiments have usually shown a proportionality between saturation and conformational change (3-6) as might be expected in view of the great difference (several hundredfold) in the affinity of hemoglobin for the first and last molecules of ligand. Unless some special requirements are imposed, this affinity difference means that intermediate species will be sparsely populated, with approximate proportionality as a necessary consequence. Such experiments have little value in testing specific models for hemoglobin cooperativity. Equilibrium experiments involving hemoglobin spin-labeled at the β93 position will not be discussed in detail because, in relation to native hemoglobin, these modified hemoglobins have a very high affinity which is relatively insensitive to phosphates (3, 7). Considerable disruption of functional properties by the spin label is not surprising in light of structural studies (8).

Kinetic techniques combining rapid mixing with other methods have provided a more powerful approach to the study of conformational changes in hemoglobin because nonequilibrium distributions of intermediates are present in these experiments. The release of a fluorescent analogue of 2,3-diphosphoglycerate (DPG) was shown to lag CO binding in flow-fluorescence experiments (9). Experiments combining flow with partial photolysis (10) showed that the formation of rapidly reacting hemoglobin lagged CO binding in pH 7 phosphate buffer.

In the present experiments, a new flow-laser flash technique is presented which makes use of the spectral difference between rapidly reacting deoxyhemoglobin and normal deoxyhemoglobin to measure the extent of conformational change at various times after mixing of deoxyhemoglobin and CO solutions. This rapidly reacting transient species has been interpreted as deoxyhemoglobin remaining in the carboxyhemoglobin conformation after the sudden removal of ligand by a laser pulse. At 20°C in pH 9 borate buffer this species relaxes to normal deoxyhemoglobin at a rate of 6,500 s^{-1} and much more quickly in the presence of phosphates at pH 7 (11). The primary advantage of this technique is that the relation between ligand binding and conformational change can be studied for unmodified hemoglobin under a wide range of conditions both in the presence and absence of phosphates. In this study the population of rapidly reacting hemoglobin present after mixing is found by direct absorbance measurements rather than by inference from the kinetics observed after flow-partial photolysis (10). Analysis of the partial flash experiments requires the assumption that the conformational distribution present after the flash reaches equilibrium before significant ligand binding occurs at the rapid rate (10, 12). This assumption is not necessary in the analysis of the present experiments.

MATERIALS AND METHODS

Hemoglobin was prepared and freed of residual CO as previously described (11). Deoxygenated stock solutions, typically 5 mM in heme, were stored at 4°C in a tonometer and used within 5 days of preparation. All experiments were carried out at 20°C. Buffers and dilute working solutions of deoxyhemoglobin and CO were prepared as discussed previously (11). The experimental geometry (11) and apparatus (13) are similar to those used in previous studies. Aside from a brief description, only differences in experimental approach will be discussed here.

A Gibson-Milnes stopped-flow apparatus (14) was used to mix deoxyhemoglobin and CO solutions. The photolysing dye laser pulse (0.2 J at 540 nm) and the observation beam enter the same window of the 2.7-mm path length optical cell. The ends of this 2 mm diameter optical cell were sealed with large windows (6 mm diameter, 1.5 mm thick) epoxied to the metal surface to allow a free path for the laser pulse and sample beam. The stopped flow was positioned so that the optical cell takes the place of the sample cell in the previously described experimental geometry (11). Experiments performed by mixing myoglobin and CO solutions and fully photolysing the mixture after 2 ms gave a mixing dead time of 1.1 ± 0.2 ms. Calculations suggest that the temperature jump associated with the absorption of laser light by the hemoglobin is <0.1°C (for a 0.2 J pulse and 20 µM hemoglobin).

Changes in the absorption of light by the mixture were detected photoelectrically and voltage changes digitized with a transient recorder (Biomation, Cupertino, Calif., model 805) and transferred to a PDP 8/E minicomputer (Digital Equipment Corp., Marlboro, Mass.) for averaging and conversion to absorbance. In the flow-flash mode, an electronic time delay started by the stopping syringe was used to fire the laser and start data collection after an adjustable delay.

Data were collected at two wavelengths. The fractional saturation of hemoglobin with CO was observed as a function of time after the stopping of flow at the isosbestic point of rapidly reacting deoxyhemoglobin and normal deoxyhemoglobin near 436 nm (11). At this wavelength the absorbance change is proportional to ligation. The formation of rapidly reacting hemoglobin was observed at the isosbestic point of deoxy and carboxyhemoglobin near 425 nm. The initial absorbance excursion at this wavelength after full photolysis is taken to be proportional to the concentration of rapidly reacting hemoglobin (11) present in the mixture just before the flash. The initial absorbance excursion in a flow-flash run was recorded as a function of the time delay. The fraction of hemoglobin in the rapidly reacting state as a function of time was obtained by dividing these absorbance excursions by the full absorbance excursion observed after the hemoglobin was fully liganded (2-s delay). A more detailed discussion of the properties of these isosbestic points and the methods used to locate them has been given previously (11).

RESULTS AND DISCUSSION

Fig. 1 presents plots of the fraction of rapidly reacting hemoglobin vs. fractional ligation for several CO concentrations in pH 9 borate and in pH 7 phosphate buffer. Within experimental error, the fraction of rapidly reacting hemoglobin depends on fractional saturation but not on CO concentration over the twofold range of concentrations studied. Under these conditions, it appears that the conformational changes

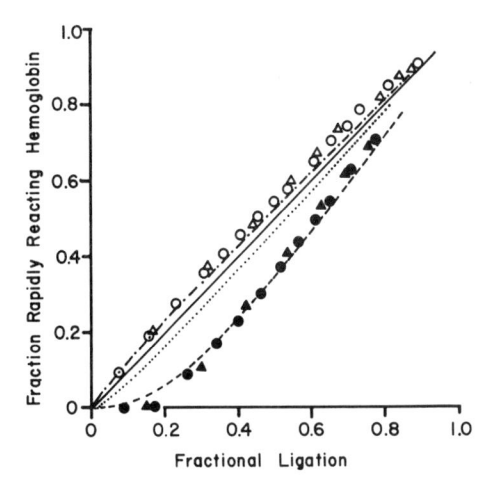

FIGURE 1 The dependence of the fraction of hemoglobin in the rapidly reacting state on fractional ligation, measured as described in the text. The solid diagonal line represents proportionality between conformational change and ligation. The open symbols correspond to experiments in pH 9 borate buffer whereas the closed symbols resulted for experiments in pH 7 phosphate buffer. The hemoglobin and CO concentrations after mixing were: o, 21.5 and 45 μM; △, 20.3 and 90 μM; ●, 18.5 and 45 μM; and ▲, 20 and 90 μM. The broken lines are the result of simulations discussed in the text.

responsible for the formation of rapidly reacting hemoglobin are quick by comparison with CO binding. Rapid mixing experiments have led to similar conclusions (15). The broken lines are the results of simulations that will be discussed later in this section.

The simplest form of two-state allosteric model (16) has proven to be a useful tool for description of a wide range of the equilibrium and kinetic properties of hemoglobin (12, 17). This model, which neglects differences between hemoglobin subunits in reactions with CO, is adopted in the analysis of the present experiments, for although there is some evidence for subunit differences in the presence of phosphates (18, 19), too little is known to justify the use of a more complex model.

Eqs. 1–3 give the adaptation of the two-state model used to simulate the results of these experiments. L and c are, respectively, the allosteric parameter and the ratio of the microscopic dissociation constants for the R and T states. Hb^R and Hb^T represent the R and T conformations of hemoglobin. Eqs. 1 and 2 give the reactions assumed to occur between hemoglobin and CO. l'_R and l'_T are, respectively, the microscopic rate constants for CO combination with R and T state hemoglobin. Similarly, l_R and l_T are the microscopic rate constants for dissociation from the R and T states. The original statement of the two-state model (16) assumes that the affinity of a hemoglobin molecule depends only on conformational state. It has been further assumed in Eqs. 1 and 2 that the rate constants depend only on conformational state (12). Eq. 3 expresses the conformational equilibrium (17) assumed to apply during the flow part of these experiments. Dissociation of hemoglobin into dimers is not considered in this form of the two-state model.

An upper limit to the fraction of carboxyhemoglobin present as dimers may be obtained from the fraction of rapidly recombining hemoglobin observed after flash photolysis (11, 20). 10 s after mixing a 90-μM CO solution with a 40-μM hemoglobin

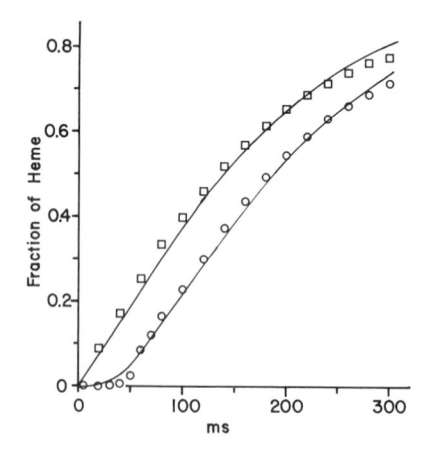

FIGURE 2 Fraction of CO-bound heme (□) and the fraction of heme in the rapidly reacting state (○) plotted against time for pH 7 phosphate buffer. The hemoglobin and CO concentrations were 18.5 and 45 μM. The solid curves are the result of a simulation using the two-state model described in the text for the parameter values: $l_R = 0.01\ S^{-1}$, $l'_R = 6.5\ \mu M^{-1} S^{-1}$, $l_T = 0.1\ S^{-1}$, $l'_T = 0.09\ \mu M^{-1} S^{-1}$, and $L = 1.4 \times 10^7$.

solution, full photolysis produced recombination kinetics at 436 nm that were 15% rapid for pH 9 borate buffer and 10% rapid for pH 7 phosphate buffer and pH 7 bis-Tris buffer with 500 μM DPG. Recombination of deoxy dimers to form tetramers is negligible on the time scale of the rapid CO binding reaction in these experiments (21). Flow-flash experiments begin with deoxyhemoglobin which is dimerized to a negligible extent (22) so that in the worst case <15% of the hemoglobin is in the form of dimers during these experiments. Dimers, which have a deoxy spectrum similar to that of rapidly reacting deoxyhemoglobin (23), form primarily from liganded R state hemoglobin so they should not have an important effect on the present experimental results.

Differential equations derived from Eqs. 1–3, with appropriate parameter values, were solved numerically with a PDP 8 computer to produce simulations for comparison with experimental data. Further discussion of the simulation process is given in the Appendix.

Fig. 2 presents flow-flash data for pH 7 phosphate buffer. The open squares give the observed fraction of heme with bound CO as a function of time,

$$\text{Hb}^R(\text{CO})_n + \text{CO} \underset{(n+1)\cdot l_R}{\overset{(4-n)\cdot l'_R}{\rightleftharpoons}} \text{Hb}^R(\text{CO})_{n+1}. \qquad 0 \leq n \leq 3, \qquad (1)$$

$$\text{Hb}^T(\text{CO})_n + \text{CO} \underset{(n+1)\cdot l_T}{\overset{(4-n)\cdot l'_T}{\rightleftharpoons}} \text{Hb}^T(\text{CO})_{n+1}. \qquad 0 \leq n \leq 3, \qquad (2)$$

$$\text{Hb}^R(\text{CO})_n \overset{Lc^n}{\rightleftharpoons} \text{Hb}^T(\text{CO})_n. \qquad 0 \leq n \leq 4. \qquad (3)$$

$$c = (l_R \cdot l'_T)/(l'_R \cdot l_T).$$

and the open circles give the observed fraction of heme in R state molecules as a function of time. The solid curves are simulations using the parameter values listed in the legend. The dissociation rate constants were obtained by assuming that the dissociation rates observed for HbCO and Hb(CO)$_4$ (24) reflect the properties of the T and R states, respectively.

The value used for l'_T is reasonable in relation to measurements of the second-order rate constant for CO binding as a function of fractional ligation (25, 26). l'_R was determined from partial photolysis experiments as previously described (12). Accurate determination of equilibrium curves for CO binding to hemoglobin has not been possible due to the high binding affinity (2), so that the allosteric parameter L, which is ligand independent, is taken from a laser photolysis study in which both equilibrium and kinetic data were obtained simultaneously for partially oxygenated hemoglobin solutions (13). The spectra and relaxational properties of rapidly reacting deoxyhemoglobin formed by photolysis of oxy and carboxyhemoglobin are very similar (27), in agreement with two-state model predictions of ligand independence.

The broken lines in Fig. 1 present two-state model simulations using parameter values taken from the legend to Fig. 2. The dotted line is derived for equilibrium con-

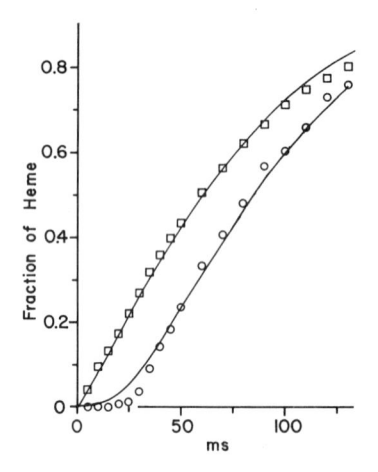

FIGURE 3 Fraction of CO-bound heme (□) and fraction of heme in R state molecules (o) plotted against time for pH 7 bis-Tris buffer with 500 μM DPG. The hemoglobin and CO concentrations after mixing were 20.3 and 90 μM. The solid curves are the result of a simulation using the two-state model described in the text for the parameter values: $l_R = 0.01$ S^{-1}, $l'_R = 6.5$ μM^{-1}S^{-1}, $l_T = 0.1$ S^{-1}, $l'_T = 0.1$ μM^{-1}S^{-1}, and $L = 3 \times 10^7$.

ditions corresponding to Eqs. 1–3, whereas the dashed curve results from the simulation of the flow-laser flash experiment presented in Fig. 2. These simulations suggest that it should be difficult under equilibrium conditions to discriminate experimentally between a spectral change linked to ligation and one linked to quaternary structural changes. Experimental studies have generally been consistent with these conclusions (3–6, 28). In terms of the two-state model, the large lag in the formation of R state hemoglobin in the present experiments results because of the slow dissociation rate of CO. Thus redistribution of ligand to form more high affinity R state molecules after initial statistical binding to T state molecules is a very slow process by comparison with the time scale of these experiments.

Fig. 3 presents flow-flash data for hemoglobin in pH 7 bis-Tris buffer with 500 μM DPG. The solid curves are the result of a simulation with parameter values identical to those in the legend for Fig. 2 but l'_T was taken to be 0.1 μM^{-1}S^{-1} and L was taken to be 3×10^7 from laser-flash experiments similar to those for hemoglobin in pH 7 phosphate buffer (13).[1] In the presence of phosphates the simulations agree reasonably well with the experimental data; however, systematic deviations may be seen in Figs. 2 and 3. In particular, the population of rapidly reacting hemoglobin develops more slowly at short times than predicted by the model. It should be noted that the agreement between simulation and experiment is better in Fig. 1 than in Fig. 2 only because the explicit time dependence is removed in a plot of fractional conformational change vs. fractional ligation.

Fig. 4 presents flow-flash results for hemoglobin in pH 9 borate buffer. In this case

---

[1] Sawicki, C. A., and Q. H. Gibson. 1977. Unpublished experiments.

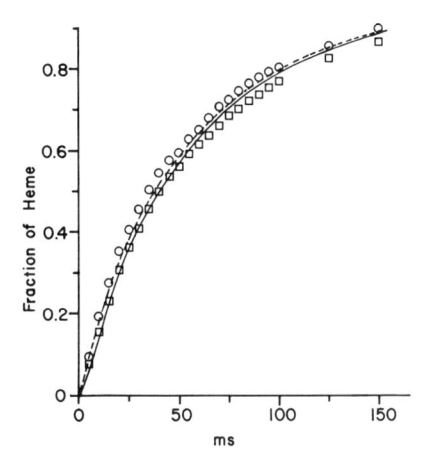

FIGURE 4 Fraction of CO-bound heme (□) and fraction of heme in R state molecules (o) plotted against time for pH 9 borate buffer. The hemoglobin and CO concentrations after mixing were 21.5 and 45 μM. The curves are the result of a simulation using the two-state model described in the text for the parameter values: $l_R = 0.01$ S^{-1}, $l'_R = 11 \mu$M^{-1}S^{-1}, $l_T = 0.1$ S^{-1}, $l'_T = 0.1$ μM^{-1} S^{-1}, and $L = 550$. The dashed curve gives the calculated fraction of heme in R state molecules, whereas the solid curve presents the calculated fraction of CO-bound heme.

the conformational change leads ligand binding by a significant amount. Relatively little is known about the parameters of the two-state model that apply at pH 9. l'_R was determined by partial photolysis (12). L was taken to be 550. Other parameters were arbitrarily set to the values given in the legend to Fig. 3. Further simulation suggests that in terms of the simple two-state model the conformational change will only lead ligand binding when significant conformational change occurs after the binding of one CO molecule. The alternately dashed and dotted line in Fig. 1 plots the simulations of Fig. 4 as: fraction of hemoglobin in the R state vs. fractional ligation.

To summarize, flow-laser flash experiments produce much larger effects than those seen in equilibrium experiments, and so permit a more demanding test of models describing cooperativity in hemoglobin. The conformational changes responsible for the formation of rapidly reacting hemoglobin after mixing of deoxyhemoglobin and CO appear to occur much faster than CO binding under the conditions employed here. In pH 7 phosphate buffer and in pH 7 bis-Tris buffer with 500 μM DPG, the formation of rapidly reacting hemoglobin lags far behind ligand binding. In these cases simulations using the two-state model with parameter values suggested by independent experiments are in reasonable agreement with the experimental observations. These simulations suggest that hemoglobin, under these conditions, is essentially completely switched to the R state after binding three CO molecules, in contrast to the situation at pH 9, where it appears that significant conformational change occurs after binding of one CO molecule.

This research was supported by National Institutes of Health Grant GM-14276-12.

Received for publication 19 November 1977.

REFERENCES

1. ROUGHTON, F. J. W., and R. L. J. LYSTER. 1965. Some combination of the Scholander-Roughton syringe capillary and van Slyke's gasometric techniques, and their use in special haemoglobin problems. *Hvalradets Skrifter.* **48**:185–198.
2. JOELS, N., and L. G. C. E. PUGH. 1958. The carbon monoxide dissociation curve of human blood. *J. Physiol.* **142**:63–77.
3. OGAWA, S., and H. M. MCCONNELL. 1967. Spin-label study of hemoglobin conformations in solution. *Proc. Natl. Acad. Sci. U.S.A.* **58**:19–26.
4. SIMON, R. S., and C. R. CANTOR. 1969. Measurement of ligand-induced conformational changes in hemoglobin by circular dichroism. *Proc. Natl. Acad. Sci. U.S.A.* **244**:205–212.
5. OGATA, R. T., and H. M. MCCONNELL. 1972. Mechanism of cooperative oxygen binding to hemoglobin. *Proc. Natl. Acad. Sci. U. S. A.* **69**:335–339.
6. OGATA, R. T., and H. M. MCCONNELL. 1971. The binding of a spin-labeled triphosphate to hemoglobin. *Cold Spring Harbor Symp. Quant. Biol.* **36**:325–336.
7. COLEMAN, P. F. 1977. A study of conformational changes in two β-93 modified hemoglobin A's using a triphosphate spin label. *Biochemistry.* **16**:345–351.
8. MOFFAT, J. K. 1971. Spin-labelled haemoglobins: a structural interpretation of electron paramagnetic resonance spectra based on x-ray analysis. *J. Mol. Biol.* **55**:135–146.
9. MACQUARRIE, R., and Q. H. GIBSON. 1972. Ligand binding and release of an analogue of 2,3-diphosphoglycerate from human hemoglobin. *J. Biol. Chem.* **247**:5686–5694.
10. GIBSON, Q. H., and L. J. PARKHURST. 1968. Kinetic evidence for a tetrameric functional unit in hemoglobin. *J. Biol. Chem.* **243**:5521–5524.
11. SAWICKI, C. A., and Q. H. GIBSON. 1976. Quaternary conformational changes in human hemoglobin studied by laser photolysis of carboxyhemoglobin. *J. Biol. Chem.* **251**:1533–1542.
12. HOPFIELD, J. J., R. G. SHULMAN, and S. OGAWA. 1971. An allosteric model of hemoglobin. I. Kinetics. *J. Mol. Biol.* **61**:425–443.
13. SAWICKI, C. A., and Q. H. GIBSON. 1977. Properties of the T state of human oxyhemoglobin studied by laser photolysis. *J. Biol. Chem.* **252**:7538–7547.
14. GIBSON, Q. H., and L. MILNES. 1964. Apparatus for rapid and sensitive spectrophotometry. *Biochem. J.* **91**:161–171.
15. ANTONINI, E., and M. BRUNORI. 1971. Haemoglobin and Myoglobin in Their Reactions with Ligands. North-Holland Publishing Co., Amsterdam. 203.
16. MONOD, J., J. WYMAN, and J. CHANGEUX. 1965. On the nature of allosteric transitions: a plausible model. *J. Mol. Biol.* **12**:88–118.
17. SHULMAN, R. G., J. J. HOPFIELD, and S. OGAWA. 1975. Allosteric interpretation of haemoglobin properties. *Q. Rev. Biophys.* **8**:325–420.
18. GRAY, R. D., and Q. H. GIBSON. 1971. The binding of carbon monoxide to α and β chains in tetrameric mammalian haemoglobins. *J. Biol. Chem.* **246**:5176–5178.
19. JOHNSON, M.E., and C. HO. 1974. Effects of ligands and organic phosphates on functional properties of human adult hemoglobin. *Biochemistry.* **13**:3653–3661.
20. EDELSTEIN, S. J., and Q. H. GIBSON. 1971. Probes of Structure and Function of Macromolecules and Membranes. Vol. II. Academic Press, Inc., New York. 417–429.
21. IP, S. H. C., and G. K. ACKERS. 1977. Thermodynamic studies on subunit assembly in human hemoglobin. *J. Biol. Chem.* **252**:82–87.
22. EDELSTEIN, S. J., M. J. REHMAR, J. S. OLSON, and Q. H. GIBSON. Functional aspects of the subunit association-dissociation equilibria of hemoglobin. *J. Biol. Chem.* **245**:4372–4381.
23. KELLETT, G. L., and H. GUTFREUND. 1970. Reactions of haemoglobin dimers after ligand dissociation. *Nature (Lond.).* **227**:921–925.
24. SHARMA, V. S., M. R. SCHMIDT, and H. M. RANNEY. 1976. Dissociation of CO from carboxyhemoglobin. *J. Biol. Chem.* **251**:4267–4272.
25. ANTONINI, E., E. CHIANCONE, and M. BRUNORI. 1967. Studies on the relations between molecular and functional properties of hemoglobin. *J. Biol. Chem.* **242**:4360–4366.
26. HOPFIELD, J. J., S. OGAWA, and R. G. SHULMAN. 1972. The rate of carbon monoxide binding to hemoglobin Kansas. *Biochem. Biophys. Res. Commun.* **49**:1480–1484.

27. SAWICKI, C. A., and Q. H. GIBSON. 1977. Quaternary conformational changes in human oxyhemoglobin studied by laser photolysis. *J. Biol. Chem.* **252**:5783–5788.
28. HUESTIS, W. H., and M. A. RAFTERY. 1972. ^{31}P-NMR studies of the release of diphospholyceric acid on carbon monoxide binding to hemoglobin. *Biochem. Biophys. Res. Commun.* **49**:428–433.
29. PENNINGTON, R. H. 1970. Introductory Computer Methods and Numerical Analysis. Macmillan & Co., Ltd. London. 2nd edition. 475.

APPENDIX

Simulations

The two-state model used is specified by Eqs. 1–3. Kinetic simulations (Figs. 1–4) were produced by numerically solving the corresponding differential equations, incorporating the parameter values listed earlier, using a second order Runge-Kutta algorithm (29). The program keeps account of the population of the 10 R and T state species R_n and T_n, where n is the number of bound CO molecules. In these experiments the CO concentration was kept relatively low so that very little ligand binding (<2%) occurs during the flow dead time. Thus the initial conditions when flow is stopped are $T_0 = Hb_0/4$ (Hb_0 is the total heme concentration) whereas all other species are initially unpopulated. The time increments used to produce the simulated curves were equal to or less than one-fifth of the time separation of the experimental points.

The dependence of the fraction of hemoglobin in the R state on fractional ligation under equilibrium conditions (dotted line in Fig. 1) was calculated by solution of the equilibrium expressions corresponding to Eqs. 1–3 for a set of CO concentrations.

DISCUSSION

OGAWA: In the curve-fitting of the allosteric model, how much can the parameters be allowed to vary in these measurements? The real unknown parameter is l_T (off rate) and all other parameters are essentially known. (*a*) It is known that K_1 (first oxygen binding constant) is pH-dependent. If this pH dependence (l'_T and l_T should be pH-sensitive) does not fit the present results, the consequence is extremely interesting. It could be that Bohr proton release is quaternary linked and the salt bridge breakage is not ligand linked. (*b*) Similar points apply when DPG is present. The so-called R-T difference of the deoxyheme absorption spectra comes from the alpha subunit heme only, although ligand binding properties of the alpha and beta subunits may not differ appreciably. Since the Cornell group has been strong on alpha-beta difference in the past, I am curious to know whether measurements at other wavelengths can deviate from the present analysis.

SAWICKI: In response to your first question, we have not tried to determine l_T and l'_T in these experiments. One of the aims of this work was to use values for the allosteric constant L and the rate constants taken from the literature rather than allowing these parameters to vary rather arbitrarily to obtain a best fit to our data. As you point out, the greatest uncertainty is in l_T; however, in the case of pH 7 phosphate buffer this rate constant can be obtained from the stopped flow experiments of Sharma et al.(24).

For simulation of the pH 9 data, we have arbitrarily assumed that l'_T and l_T have values similar to those at pH 7, because it appears that significant conformational change occurs after binding of one molecule of carbon monoxide, so that it may be difficult to separate the properties of the R and T states. For example, the binding of the first ligand to a hemoglobin molecule may occur with a slow T rate while the dissociation of ligand from $Hb(CO)_1$, occurs with a rate characteristic of the R state.

With regard to your second question, it is only practical from the point of view of analysis to perform experiments at an isobestic wavelength like 425 nm. Other experiments with CO suggest that chain differences are relatively small compared with those of oxygen, where differences of a factor of 10 or so appear to exist.

C. BONAVENTURA: The rate of CO binding at pH 9.1 in borate is decelerating, yet from equilibrium studies we know that cooperative interactions occur. How do you fit the kinetic behavior into a simple two-state model?

SAWICKI: From our point of view this is a real but second-order sort of problem. Under all of the conditions we studied in Figs. 2, 3, and 4, the agreement between model simulations and binding data becomes less satisfactory at longer times. The meaning of this misfit is unclear.

LESTER: I don't understand Dr. Bonaventura's question. Are you wondering about the deviation of the binding kinetics from first order at pH 9?

C. BONAVENTURA: No, we are not. The binding shows a gradual decrease in rate as more ligand is bound. The binding rate slows down.

MCCRAY: Just for clarification, is the equilibrium curve cooperative at pH 9?

SAWICKI: Yes, it is. We also see the conformational change spectrum and large changes in the carbon monoxide binding rate.

BALLOU: The concentration of CO is only twofold in excess of the heme. This would explain deviations from first-order behavior in Fig. 4.

SAWICKI: The effect Dr. Bonaventura refers to is also seen when the ligand concentration is much greater than the heme concentration.

FERRONE: I would like to get back to the question of inequivalence again. I don't know to what extent you have discriminated against alpha and beta inequivalence in the experiments that you alluded to a moment ago, but I think that from the data you present one might actually be able to get further limitations on the degree of equivalence in the following sense: If in fact the alpha chains are the only ones contributing to the spectral change that you are observing, and if you can fit the progress of the alpha chain with parameters derived from an average of alpha and beta, that is L and c obtained from say, equilibrium experiments, and if you knew how tight that fit was, you could get a limitation on how close the properties of the alpha chains agree with the average of alpha and beta properties.

SAWICKI: No, that isn't right. In terms of the two-state model we observe a spectral difference between T deoxy and R deoxy hemoglobin after full photolysis. Spectral changes in the alpha chains are used as a marker of conformational state. It doesn't matter where the ligand is before photolysis. Only the conformational state is important. The fact that the absorbance difference comes mostly from the alpha chains, however, suggests something isn't quite right with the simple two-state model.

FERRONE: That is not quite true, in that if the ligand binds to the alpha chains, then there is no spectral change to be observed. That is, the spectral change only occurs in the deoxy form so that there is some spectral discrimination that way.

SAWICKI: That isn't true. What we do is take all of the ligand off with a laser pulse so we are looking at an absorbance change that corresponds to the difference between deoxy R and deoxy T hemoglobin. If the simple two-state model is valid, it doesn't matter where the ligand is before photolysis.

SCHUSTER: You haven't taken into consideration in this model the difference in the binding of phosphate between oxy and deoxyhemoglobin.

SAWICKI: Not explicitly, but this is included in L.

PRENDERGAST: We have been interested in two-state models for another purpose, but it intrigued us that at short times Figs. 2 and 3 do not show a very good agreement between your model simulations and your experimental data. Would you care to comment on that at all?

SAWICKI: That is true, but considering the simplicity of the model, I think that the fit is surprisingly good. Actually I was expecting a more dramatic disagreement, because the model parameters are taken from other independent experiments. In addition, several unusual effects, seemingly in contradiction with the simple two-state model, have been observed for carboxyhemoglobin, for example, the data presented by Dan Schuresko and Watt Webb in the poster session at this meeting and the flow-flash quantum efficiency experiment I presented at the 1978 joint American Physical Society/Biophysical Society meeting. Although this experiment doesn't relate directly to functional properties, it suggests that other states are present.

ROMINE: Have you observed the change of the initial time lag before going to the rapidly reacting state as a function of pH between 7.0 and 9.0?

SAWICKI: No, but we did experiments in pH 7 bis-Tris buffer. In that case it appears that the conformational change occurs after binding of two ligands. The development of R state hemoglobin lags only very slightly behind carbon monoxide binding in this case.

MCCRAY: With respect to the data at pH 9, what evidence do you have that the T → R rate of conformational change is not affecting the results?

SAWICKI: As shown in the Fig. 1, changing CO concentration doesn't change the relationship between fractional conformational change and fractional ligation. This suggests that conformational change is quick compared with ligand binding.

FERRONE: John Hopfield and I have actually measured that rate at the level of threefold ligation for pH 7 and find that in fact it is rapid—approximately 2×10^3/s at threefold ligation.

SAWICKI: That is an interesting experiment, because you see the results of Hopfield and Ferrone contradict our results, because they would say that the conformational state after binding three ligands is about 30% T while the values for L and c we used predict only about 3% T when you have three ligands.

SUTTER: Why do you use a second-order Runge-Kutta method rather than a fourth-order method?

SAWICKI: It is mostly a matter of having a small computer. Rounding off errors and space considerations lead to the use of the simplest methods. The second-order Runge-Kutta method works fine so we use it.

Maybe I should say something else in relation to Frank Ferrone's comment. We did partial photolysis experiments that disagree with their numbers. This may have something to do with the validity of the two-state model. The conformational change rate constants given by Ferrone and Hopfield imply that under the proper conditions of CO concentration the rebinding reaction after low levels of partial photolysis (5% should be biphasic because all of the deoxy sites on Hb(CO)$_3$ are R state immediately after laser pulse, but conformational equilibrium with about 30% of these species in the T state is approached with a rate of 3,300 s^{-1}. The biphasicity expected is very slight, amounting to about a 30% change in the observed binding rate. We looked carefully and didn't see this biphasicity. Another interesting thing is the binding rate they found for the R state. They remove one ligand from Hb(CO)$_4$, using a slightly different procedure from ours, and get a binding rate of about 3 μM^{-1}s^{-1}, while other experiments, ours included, give a rate of about 6 M^{-1}s^{-1}. These differences need to be investigated.

FERRONE: I think one of the difficulties we had was in a proper assignment of the total free CO. In 10% experiments we got binding rates that gave us higher rates such as 6 and 7 rather than 3 μM^{-1}s^{-1}. We weren't primarily interested in the binding rates in those experiments, so those remain not as hard numbers as the others.

HOFRICHTER: As a point of information, are there any data on the rate of disphosphoglycerate dissociation from hemoglobin at pH 7? If this dissociation is slow, then T → R quartenary structural change could, in part, be rate limited by the necessity for DPG to be released. The kinetic inhibition of the T → R transition could then work together with the thermodynamic criteria that you included in the allosteric model in generating the delay in the T → R transition.

SAWICKI: I don't know the rate of DPG release but again in Fig. 1 the relation between fractional conformational change and fractional ligation is the same for the range of CO concentrations studied.

HOFRICHTER: Is anything known about the speed of the R → T conversion?

SAWICKI: In the case of deoxyhemoglobin we have a fairly good idea of these rates. In the absence of phosphates at 20°C the rate is about 6,000 s^{-1}, while in the presence of phosphates the rates lie in the range 50,000–100,000 s^{-1}.

LUCHINS: Do you have any feel as to how far along in the shifting of the pairs of chains your spectral marker on the alpha chain's heme appears? To what extent does your deoxy conformation, the appearance of which is signaled by that marker, correspond to what a crystallographer would call deoxy?

SAWICKI: The unstable, rapidly reacting species (R deoxyhemoglobin) presumably has structural features in common with liganded R-state hemoglobin, which shows similar CO binding rates, but the exact nature of these similarities isn't clear.

BERGER: Have you been able to look at the effect of the dielectric constant of the medium on your results?

SAWICKI: No, we haven't looked at this yet.

MCCRAY: One point that has bothered me is the breakup of R-state tetramers to dimers during CO binding. As CO molecules are bound and the T-state tetramers change to R-state tetramers, the concentration of R-state tetramers is initially very small. How fast do these tetramers break up?

SAWICKI: First, the effect of dissociation is not very serious because deoxy dimers have a spectrum similar to R deoxy hemoglobin. The tetramer dimer dissociation constant is relatively small so that under equilibrium conditions, after CO binding, only about 10–15% of the heme exists as dimers. Also, other experiments indicate that the formation of dimers is relatively slow, so that at the longest times given in Figs. 2, 3 and 4 less than half of the final equilibrium population of dimers has formed.

SCHUSTER: Several experiments conflict with the two-state model, while your experiments and, for example, equilibrium experiments can be well fit into the two-state model. Doesn't this suggest that these types of experiments may be inadequate to distinguish between different possible models?

SAWICKI: Distinguishing between models is often a difficult problem, particularly in the case of a cooperative system. Because of the large population of slowly reacting intermediates present in our flow-laser flash experiments, they can't be compared with equilibrium experiments. Although a number of doubts have been raised about the strict validity of the two-state model, it still seems to offer a reasonable description of the major functional change in the process of binding ligand to hemoglobin.

BICARBONATE-CHLORIDE EXCHANGE IN ERYTHROCYTE SUSPENSIONS

Stopped-Flow pH Electrode Measurements

Edward D. Crandall, A. L. Obaid, and R. E. Forster,
Departments of Physiology and Medicine, University of Pennsylvania, Philadelphia, Pennsylvania 19104 U.S.A.

ABSTRACT A pH-sensitive glass electrode was used in a temperature-controlled stopped-flow rapid reaction apparatus to determine rates of pH equilibration in red cell suspensions. The apparatus requires less than 2 ml of reactants. The electrode is insensitive to pressure and flow variations, and has a response time of <5 ms. A 20% suspension of washed fresh human erythrocytes in saline at pH 7.7 containing $NaHCO_3$ and extracellular carbonic anhydrase is mixed with an equal volume of 30 mM phosphate buffer at pH 6.7. Within a few milliseconds after mixing, extracellular HCO_3^- reacts with H^+ to form CO_2, which enters the red cells and rehydrates to form HCO_3^-, producing an electrochemical potential gradient for HCO_3^- from inside to outside the cells. HCO_3^- then leaves the cells in exchange for Cl^-, and extracellular pH increases as the HCO_3^- flowing out of the cells reacts with H^+. Flux of HCO_3^- is calculated from the dpH/dt during HCO_3^--Cl^- exchange, and a velocity constant is computed from the flux and the calculated intracellular and extracellular [HCO_3^-]. The activation energy for the exchange process is 18.6 kcal/mol between 5°C and 17°C (transition temperature), and 11.4 kcal/mol from 17°C to 40°C. The activation energies and transition temperature are not significantly altered in the presence of a potent anion exchange inhibitor (SITS), although the fluxes are markedly decreased. These findings suggest that the rate-limiting step in red cell anion exchange changes at 17°C, either because of an alteration in the nature of the transport site or because of a transition in the physical state of membrane lipids affecting protein-lipid interactions.

INTRODUCTION

Anion exchanges across the erythrocyte membrane are of particular physiological significance because of the importance of rapid HCO_3^--Cl^- exchange for CO_2 transport in lung and tissue capillaries (1). Recent studies (2, 3) have suggested that a specialized transport system facilitates Cl^--Cl^- self-exchange across the membrane, based on the observation of saturation kinetics, competitive inhibition, high activation energy at low temperature, and the behavior of exchange flux as a function of pH. The exchange may be obligatory and electrically neutral, since membrane conductance appears to be much lower than the calculated exchange permeabilities would suggest (4). This has led to speculation that both exchange and net translocation pathways for anions exist in the red cell membrane, and that both pathways may share common sites (5, 6).

These data have been made more convincing by the demonstration that specific sites in the red cell membrane are likely to be involved in anion exchange processes (7). Much of the kinetic data has been obtained at low temperatures, but Brahm (3) has recently shown that similar phenomena occur at 37°C. Gunn (6) has proposed a model for anion exchanges that may explain the experimental data for divalent and monovalent anions, and includes net translocation as a probabilistic event occurring through the exchange pathway.

Red cell membrane HCO_3^--Cl^- exchange, because of its critical role in gas exchange, has been studied by a number of workers over the last 50 years (8–11). Since the half-time of this process may be of the same order of magnitude as the transit time of red cells through capillary beds during exercise, the exchange might limit the amount of CO_2 that can be transported between blood and tissue or between blood and alveolar gas. Recently, a new approach was developed (12) to study HCO_3^--Cl^- exchange in red cells (or any closed permeable vesicle) over a wide range of physiological conditions. The method is based on observing the rate of pH change in the extracellular fluid of a suspension under conditions where the HCO_3^--Cl^- exchange is rate-limiting. A large quantity of H^+ is added to a red cell suspension that contains some HCO_3^- and a high concentration of carbonic anhydrase in its extracellular fluid. As a result, some H^+ combines with HCO_3^- in the extracellular fluid rapidly to form CO_2 (Fig. 1). This dissolved CO_2 quickly enters the red cells and rehydrates to form H^+ and HCO_3^-, setting up an electrochemical potential gradient for HCO_3^- from inside to outside the cells. As a result of these rapid readjustments in intra- and extracellular [HCO_3^-], HCO_3^- flows out of the cells in exchange for Cl^-. As HCO_3^- enters the extracellu-

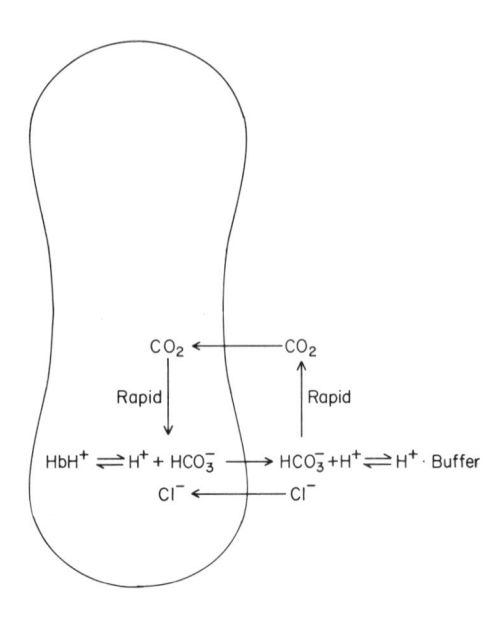

FIGURE 1 Schematic diagram of the Jacobs-Stewart cycle in the presence of extracellular carbonic anhydrase.

lar fluid, however, it rapidly combines with H^+ there to form CO_2 and raise extracellular pH. The CO_2 re-enters the cell and rehydrates to form HCO_3^- and H^+. The net result of a complete cycle of one bicarbonate ion leaving the cell and CO_2 entering to reform HCO_3^- is the transfer of one H^+ and one Cl^- (with the necessary osmotic flow of water) from outside to inside the cell (Fig. 1). This process has been called the Jacobs-Stewart cycle (13,14). dpH/dt in the extracellular fluid after the initial rapid readjustments of intra- and extracellular $[HCO_3^-]$ is a measure of the rate of exchange of HCO_3^- for Cl^- across the red cell membrane.

In the present study, we have used the stopped-flow rapid reaction apparatus to measure pH changes in human erythrocyte suspensions after pulsing the extracellular fluid with acid at temperatures ranging from 5° to 40°C. The flux of HCO_3^- in the Jacobs-Stewart cycle was calculated from the measured dpH/dt, along with a velocity constant based on computed intracellular and extracellular $[HCO_3^-]$. We have shown that there is a transition at 17°C from an activation energy of 18.6 kcal/mol below to 11.4 kcal/mol above 17°C. SITS, a potent anion exchange inhibitor (4-acetamide-4'-isothiocyanostilbene-2,2'-disulfonic acid), does not change the activation energies or the transition temperature, although the fluxes are markedly decreased. The data agree with recently reported measurements of Cl^--Cl^- self-exchange (3), and suggest different rate-determining steps for anion exchange across the red cell membrane above and below the transition temperature. The change in activation energy with temperature may be mediated by a change in the nature of the transport site of the membrane, or by alterations in protein-lipid interactions due to changes in the physical state of lipids in the cell membrane.

METHODS AND MATERIALS

Apparatus

The stopped-flow rapid reaction apparatus used in these experiments has been described previously (15) and is shown schematically in Fig. 2. In the apparatus, equal volumes of a red cell suspension A and phosphate-buffered saline solution B are forced through a four-jet mixer (0.004 ml) into a 0.1-ml measuring chamber. A pH-sensitive glass electrode (117145 Leeds & Northrop Co., North Wales, Pa.) is used to follow the pH of the mixture as a function of time, both before and after flow stops. The reference electrode liquid junction is a KCl-saturated cotton wick bridging a snug-fitting Teflon plug, and is pressure- and flow-insensitive (16). The voltage across the electrodes is amplified (Transidyne General Corp., Ann Arbor, Mich., MPA-6 with its own power source MPS-15) and monitored on a storage oscilloscope screen (5103N, Tektronix, Inc., Beaverton, Ore.). A measure of flow velocity is simultaneously monitored on the oscilloscope screen by recording voltage output from a magnet-in-coil device mounted on the stopped-flow apparatus drive block. The entire apparatus is water-jacketed, and the experiments reported here were carried out between 5°C and 40°C.

The response time of the electrode system has been estimated to be <5 ms, using a ramp change in pH due to carbonic acid dehydration as a test reaction. The lag time of the apparatus (elapsed time between mixing and reaching the glass electrode) is less than 20 ms at the linear flow rates used in these experiments (25–50 cm/s). Further details of the characteristics of the rapid-reaction apparatus are available in the literature (15–17).

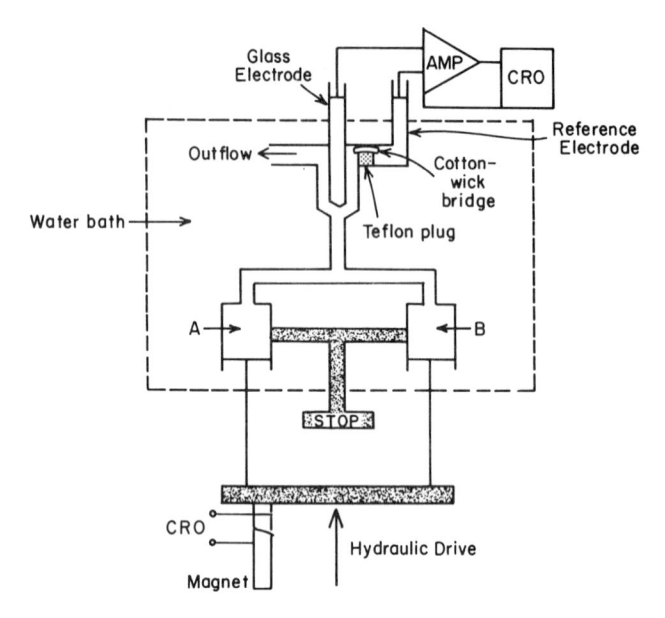

FIGURE 2 Schematic diagram of the stopped-flow rapid reaction apparatus with the pH-sensitive glass electrode as the measuring device.

Solutions and Procedure

Fresh heparinized human blood was centrifuged for 10 min at 3,000 g. The cells were resuspended in 10 times their volume of 146.5 mM NaCl, 3.5 mM KCl, and recentrifuged, this procedure being repeated three times. The washed cells were then resuspended in 146.5 mM NaCl, 3.5 mM KCl to about 20% hematocrit to form suspension A. NaOH was added to the suspending medium to reach a final pH_A of 7.7. Carbonic anhydrase (bovine carbonate hydrolase, Sigma Chemical Co., St. Louis, Mo.) was added to the suspension to a concentration of 80,000 Wilbur-Anderson U/100 ml of suspension. Freshly-prepared $NaHCO_3$ was then added to a final concentration of 4.4 mM and the suspension maintained in a closed tonometer thereafter. The phosphate-buffered solution B consisted of 112.5 mM NaCl, 15 mM Na_2HPO_4, and 15 mM KH_2PO_4 with pH_B of 6.7. All suspensions and solutions were prepared at room temperature.

Hematocrit (Hct) of suspension A was measured in standard Wintrobe tubes. pH was determined anaerobically in a Radiometer BMS3 Mk 2 blood gas machine (Radiometer Co., Copenhagen, Denmark). Supernatant hemoglobin concentration in suspension A was measured spectrophotometrically at 541 nm in a Perkin-Elmer Coleman 124 spectrophotometer (Perkin-Elmer Corp., Norwalk, Conn.). Intracellular pH was determined from red cell lysates produced by freezing and thawing packed cells separated by centrifugation. A mixture of equal volumes of suspension A and solution B was collected in a syringe after the mixture had passed through the rapid reaction apparatus. After centrifugation, the supernatant from the mixture was titrated anaerobically with a microburette (S-1100A, Roger Gilmont Instruments, Inc., Great Neck, N.Y.) and a water-jacketed titration chamber. The buffer capacity β of the extracellular fluid of the mixture as a function of pH (around 6.7) was obtained by differentiation of the titration curve.

For the experiments in which cells were treated with an anion exchange inhibitor, the red cells were washed three times as described above and resuspended to 10% hematocrit in a solution containing 146.5 mM NaCl, 20 mM Tris buffer (pH 7.4), 0.2% ethanol and 0.11 mM

SITS (PolyScience Corp., Niles, Ill.). This suspension was incubated at 37°C for 10 min with constant agitation. The cells were then rewashed three times with 146.5 mM NaCl, 3.5 mM KCl, resuspended to 20% hematocrit, and treated further as described above.

Computations

The flux ϕ of HCO_3^- out of the red cells per unit of membrane surface area was determined from the initial dpH/dt observed in the mixture after stopping flow in the rapid reaction apparatus: $\phi = \beta(dpH/dt)(1 - Hct)/(Hct \times A/V)$, where V = volume per cell and A = surface area per cell. The intracellular and extracellular $[HCO_3^-]$ at the time of stopping flow were computed from the measured extracellular pH at that time ("plateau" pH) with the equilibrium constant for carbonic acid and the assumptions that $[CO_2]$ is the same intra- and extracellularly and that total CO_2 content remains constant. The equations, given previously (12), were solved on a PDP-10 digital computer (Digital Equipment Corp., Marlboro, Mass.) using an iterative procedure. A velocity constant for HCO_3^--Cl^- exchange was then calculated: $k = \phi/([HCO_3^-]_i - [HCO_3^-]_0)$.

RESULTS

A typical experimental record is shown in Fig. 3. The upper tracing represents the pH of the fluid in the measuring chamber as a function of time. The lower trace indicates where flow of reactants starts and stops. Each trace was swept across the screen several times. Before flow starts, solution B (pH 6.75) is in the measuring chamber. During flow, the "plateau" pH is that of the mixture during flow, about 20 ms after mixing (pH 6.77). This is the extracellular pH after the rapid movement of HCO_3^- into the cells (described above), but before significant HCO_3^--Cl^- exchange has had time to occur (see Fig. 1). After flow stops, the pH of the mixture in the measuring chamber rises towards its final equilibrated value (pH 7.14), as the Jacobs-Stewart cycle effects the transfer of H^+-equivalents from outside to inside the erythrocytes.

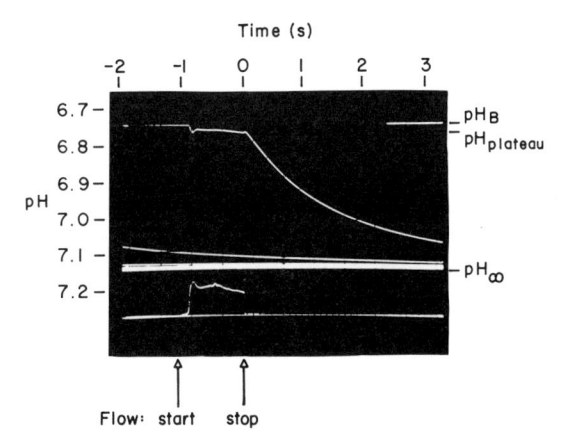

FIGURE 3 Oscilloscope tracing of the change of pH with time in a mixture of equal volumes of solution B and suspension A. In this experiment, suspension A was at pH 7.7 with hematocrit of 16.3%, total CO_2 content of 4.4 mM, and carbonic anhydrase concentration of 800 Wilbur-Anderson U/ml. The times of starting and stopping flow in the rapid reaction apparatus are indicated. Further details of this record, solution B, and suspension A are given in the text.

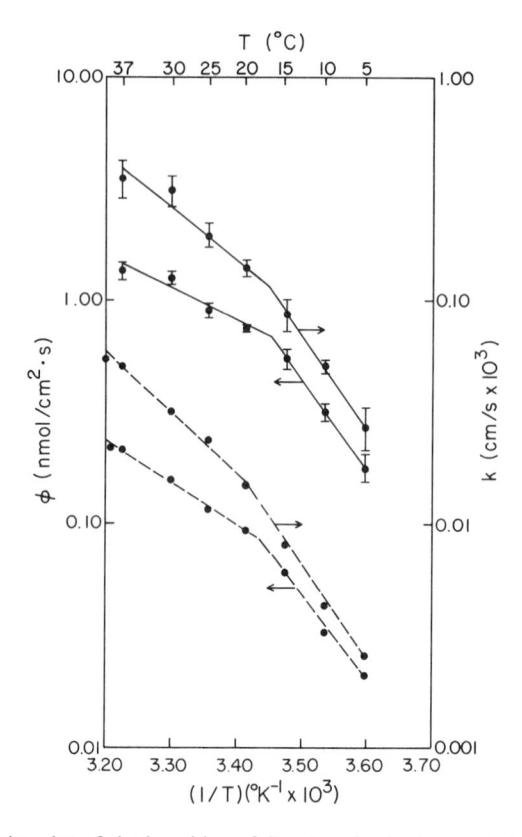

FIGURE 4 Arrhenius plot of the logarithm of flux ϕ and velocity constant k vs. $1/T$ for bicarbonate-chloride exchange. Data for untreated cells are represented by the points on the solid lines (means ± SE, $3 \leq n \leq 8$), and those for cells exposed to SITS are given by the points on the broken lines (means, $n = 2$). The activation energies quoted in the text are based on the velocity constant data.

Fig. 4 shows the Arrhenius plot of the logarithm of both ϕ and k versus $1/T$. The activation energy of the exchange process changes from 18.6 kcal/mol (low temperature) to 11.4 kcal/mol (high temperature) at the transition temperature of 15°C. The Arrhenius plot for SITS-treated cells is also shown in Fig. 4. Although the fluxes are markedly decreased in the presence of SITS, the activation energies and the transition temperature are not significantly changed.

DISCUSSION

Although the observation that activation energy for the exchange of bicarbonate for chloride ions across the red cell membrane is a function of temperature has been reported earlier (12), the present data show that this relationship can be characterized by just two values for activation energy, with the change occurring at about 17°C. Similar behavior has been measured for red cell glucose transport (18) and for viscosity of suspensions of red cell membrane fragments (19), both showing transition temperatures in the vicinity of 17°C. It was suggested (19) that the change in activation energy is due to an alteration in the physical state of membrane lipids.

Previous studies on Cl^--Cl^- and other anion exchange processes across the red cell membrane had failed to show a change in activation energy with temperature (20), primarily due to the inability to follow rapid fluxes at higher temperatures by isotope techniques. The rapid reaction method reported here, capable of determining fluxes over a wide range of temperature, revealed the transition phenomenon. Very recently, Brahm (3) used a modification of a continuous flow rapid reaction apparatus to study Cl^--Cl^- self-exchange across the red cell membrane. It was found that the activation energy is a function of temperature, being 30 kcal/mol below and 20 kcal/mol above 15°C. Given that we are studying HCO_3^--Cl^- exchange rather than Cl^--Cl^- self-exchange, and allowing for the differences in the experimental conditions, the similarities in the data are notable.

A change in activation energy for anion exchanges at a transition temperature implies that different processes are limiting the rate of the exchanges below and above this temperature. One possible explanation is that the membrane transport mechanism is not the same in the cold as it is at higher temperatures. For example, a carrier-mediated mechanism might limit the rate of exchange at low temperature, while a diffusion process determines the rate at higher temperature (12). This possibility was made less likely by the demonstration (3) that at both low and high temperature, a saturation phenomenon and a maximum flux at $7 < pH < 8$ characterize Cl^--Cl^- exchange. It is more likely that the change in activation energy is mediated by alterations involving the membrane proteins that may participate in anion exchanges (7). These alterations at the transition temperature could include modifications within a given pathway, a switch of transport from one saturable pathway to another, or changes in the physical state of membrane lipids. The latter possibility, however, would not explain the transition temperature of 25°C for Br^--Br^- self-exchange (3) compared to that for Cl^--Cl^- or HCO_3^--Cl^- exchanges. Brahm (3) has suggested a maximum turnover number at a specific site in the membrane as the factor that determines the rate-limiting step for anion exchanges.

SITS-treated cells, although flux was uniformly reduced by 90%, showed transition temperature and activation energies not significantly different from those for untreated cells. This agrees with data reported for cells exposed to another anion exchange inhibitor, DIDS (4,4'-diisothiocyano-2,2'-stilbene disulfonic acid) (3). These data are interpreted as being consistent with unchanged transport pathway and kinetics at a given temperature, but a diminished number of sites available for the exchange. Further investigation is needed to elucidate the specific mechanism underlying the change in activation energy for transport processes across the erythrocyte membrane.

Research support was provided by U.S. Public Health Service grant HL19737, AHA 75-992, and Research Career Development Award HL00134 (to E.D.C.).

Received for publication 1 December 1977.

REFERENCES

1. CRANDALL, E. D., and R. E. FORSTER. 1973. Rapid ion exchanges across the red cell membrane. *Adv. Chem. Ser.* **118**:65–87.
2. GUNN, R. B., M. DALMARK, D. C. TOSTESON, and J. O. WIETH. 1973. Characteristics of chloride transport in human red blood cells. *J. Gen. Physiol.* **70**:185–206.
3. BRAHM, J. 1977. Temperature-dependent changes of chloride transport kinetics in human red cells. *J. Gen. Physiol.* **70**:283–306.
4. LASSEN, U. V., L. PAPE, B. VESTERGAARD-BOGIND, and O. BENGTSON. 1974. Calcium-related hyperpolarization of the *Amphiuma* red cell membrane following micropuncture. *J. Membr. Biol.* **18**:125–144.
5. KNAUF, P. A., G. F. FUHRMANN, S. ROTHSTEIN, and A. ROTHSTEIN. 1977. The relationship between anion exchange and net anion flow across the human red blood cell membrane. *J. Gen. Physiol.* **69**:363–386.
6. GUNN, R. B. 1977. A titratable lock-carrier model for anion transport in red blood cells. *Proc. Int. Union Physiol. Sci.* **12**:122.
7. ROTHSTEIN, A., Z. I. CABANTCHICK, and P. KNAUF. 1976. Mechanism of anion transport in red blood cells: role of membrane proteins. *Fed. Proc.* **35**:3–10.
8. DIRKEN, M. N. J., and H. W. MOOK. 1931. The rate of gas exchange between blood cells and serum. *J. Physiol. (Lond.).* **73**:349–360.
9. Luckner, H. 1948. Die temperaturabhängigkeit des anionen-austausches roter blutkörperchen. *Pflügers Arch. Eur. J. Physiol.* **250**:303–311.
10. HEMINGWAY, A., C. J. HEMINGWAY, and F. J. W. ROUGHTON. 1970. The rate of chloride shift of respiration studied with a rapid filtration method. *Respir. Physiol.* **10**:1–9.
11. KLOCKE, R. A. 1976. Rate of bicarbonate-chloride exchange in human red cells at 37°C. *J. Appl. Physiol.* **40**:707–714.
12. CHOW, E. I., E. D. CRANDALL, and R. E. FORSTER. 1976. Kinetics of bicarbonate-chloride exchange across the human red blood cell membrane. *J. Gen. Physiol.* **68**:633–652.
13. JACOBS, M. H., and D. R. STEWART. 1942. The role of carbonic anhydrase in certain ionic exchanges involving the erythrocyte. *J. Gen. Physiol.* **25**:539–552.
14. FORSTER, R. E., and E. D. CRANDALL. 1975. Time course of exchanges between red cells and extracellular fluid during CO_2 uptake. *J. Appl. Physiol.* **38**:710–718.
15. CRANDALL, E. D., R. A. KLOCKE, and R. E. FORSTER. 1971. Hydroxyl ion movement across the human erythrocyte membrane. *J. Gen. Physiol.* **57**:664–683.
16. CRANDALL, E. D., and J. DELONG. 1976. A pressure- and flow-insensitive reference electrode liquid junction. *J. Appl. Physiol.* **41**:125–128.
17. GROS, G., R. E. FORSTER, and L. LIN. 1976. The carbamate reaction of glycylglycine, plasma and tissue extracts evaluated by a pH stopped flow apparatus. *J. Biol. Chem.* **251**:4398–4407.
18. LACKO, L., B. WITTE, and P. GECK. 1973. The temperature dependence of the exchange transport of glucose in human erythrocytes. *J. Cell. Physiol.* **82**:213–218.
19. ZIMMER, G., and H. SCHIRMER. 1974. Viscosity changes of erythrocyte membrane and membrane lipids at transition temperature. *Biochim. Biophys. Acta.* **345**:314–320.
20. DALMARK, M., and J. O. WIETH. 1972. Temperature dependence of chloride, bromide, iodide, thiocyanate and salicylate transport in human red cells. *J. Physiol. (Lond.).* **224**:583–610.

DISCUSSION

BECKER: Have you measured the phase transition temperature and activation energies in red cells having different cholesterol content?

CRANDALL: Not yet. That is one of the things we have on the drawing board. We have a way to make the cholesterol-enriched and cholesterol-depleted cells, and we are about to do that. That is one of the variations we have not done yet, and it should be interesting.

BECKER: Regarding your finding of the transition temperature of 17°C for chloride ion exchange and Brahm's finding that bromide ion exchange has a phase transition temperature at 25°C, I would like to suggest that Cl and bromide transport may be located in two patches of membrane lipid, each patch having a different fatty acid composition, and therefore characterized by different transition temperatures.

CRANDALL: When we first found our data on bicarbonate-chloride ion exchange, it seemed likely to us that a transport pathway was interacting with membrane lipids as the membrane lipids change phase in some way or other. However, Brahm's finding (1977) that bromide has a transition temperature at 25°C makes it less likely. It is always possible that the bromide pathway and the chloride pathway are in different portions of the lipid membranes, but that is difficult for me to believe. I suspect that bromide-bromide and chloride-chloride use the same pathway, and that the bicarbonate-chloride exchange is also using the same transport pathway.

BECKER: Recent studies from our laboratory indicate such a situation exists in endoplasmic reticulum membranes from rat liver. We studied the membrane-bound mixed function monoxygenase enzyme system by assaying different enzyme activities as a function of temperature. Benzo(α)pyrene hydroxylase exhibited a T_c of 29°C while p-nitroanisole o-demethylation had T_c of 24°C, significantly different.

CRANDALL: I think that is interesting data and may indicate the same sort of difference that chloride and bromide exchange in the red cells exhibits.

BECKER: Feeding rats a fat-free diet caused the membrane fatty acid composition to become more saturated, which in turn caused a significant increase in the transition temperature of the benzo(α)pyrene hydroxylase.

CRANDALL: I think this is a very interesting finding, and I hope we can produce a similar experiment.

BARISAS: (a) Do your data, especially those for untreated cells, actually require a biphasic Arrhenius plot? It would seem that the data in Fig. 4 could just as well be fitted with a curved Arrhenius plot. This would imply, instead of a transition at 17°C, a $\Delta C^{\ddagger}_p$ of -400 cal deg^{-1} mol^{-1}. (b) In this connection, does Brandt's differential scanning calorimeter work on erythrocyte membranes show any evidence of reversible transitions at 17°C?

CRANDALL: (a) We originally thought we had a continuously changing E_A with temperature (Chow et al., 1976). Many further experiments, however, have convinced us that we are really dealing with a single transition temperature with constant E_A above and below that temperature. This is further reinforced by the data by Brahm (1977). (b) I'm not sure I understand this question. I don't think reversible transitions have been looked for, in the sense that varying the temperature of a given population of cells around T_c and then studying possible alterations in transport have not been done.

ROMINE: In reference to the change in flux, $\Delta\phi/\Delta(1/T)$, at 17°C, is it not possible that this reflects a change in the diffusion rate of molecular CO_2 across the erythrocyte membrane with the change from liquid to solid phase? Has any work been done on identifying the role of CO_2 diffusion? The data would be completely consistent, especially in view of the lack of SITS-treated cells having changed Arrhenius plots, with the CO_2 being the actual energy barrier measured.

CRANDALL: Experiments on CO_2 movements across the red cell membrane at different temperatures above T_c have shown that there is no measurable resistance to the diffusion of this gas across the membrane. Since gases in general diffuse very rapidly across cell membranes, it seems reasonable to assume that CO_2 diffusion could not be the rate-limiting step even below T_c, although we know of no direct measurements at that temperature.

COLEN: Returning to the question of Dr. Barisas about whether you could be seeing a heat capacity change, I would like to make this comment. Apparent heat capacity changes can be generated by coupled reactions. Thus either a shift in rate-determining step or simply a shift in a coupled equilibrium can produce an apparent ΔC_P for a process. All that is required is a difference in ΔH value for the steps involved. One can easily either obtain curved behavior for such a system or even sharply bending behavior for a cooperative process.

R. P. TAYLOR: It looks like 90% of the sites are blocked by your SITS inhibitor. Is it possible that is you work at high concentration you could block the rest?

CRANDALL: Yes.

JAIN: If the K_m changes with temperature, then at certain temperatures you might be on the saturating portion of the curve; isn't this a problem? As you change temperatures, what you're really doing is moving down the saturation curve. This could have several possible interpretations, not all based on a highly cooperative lipid system. In 1973–1974 we reported on the effect of temperature on K_m of galactose transport system in *E. coli* (Sullivan et al. 1974. *Biochim. Biophys. Acta.* **352**:288).

CRANDALL: Yes, it's a potential problem. However, the K_m for Cl^-–Cl^- exchange varies only from 28 to 65 mM over 0–38°C (3). In our work we always have chloride concentrations above the K_m but bicarbonate concentration is always well below the chloride K_m. Since we are dealing with two ionic species, a single K_m is not defined directly. Since, however, chloride concentration is always very high and bicarbonate concentration is always very low, we do not think it likely alterations in K_m with temperature can explain our results.

JAIN: If the K_m is changing continuously or even abruptly with temperature, I don't know what it is due to in your plots. It is the sharpness of the change that bothers me. A broad transition could be due to smaller cooperative units (a realistic possibility) or due to a large number of phases coexiting in membranes. These have different significance. I don't have anything against the role of lipids to account for the breaks in Arrhenius plots; however, this is by no means a proof of the role of lipids.

MIZUKAMI: Could you comment some more on your pH studies, especially in relation to your transport site and lipid protein interactions?

CRANDALL: I could. It's not in this paper, but what we found is that as we decrease extracellular pH from 6.8, the bicarbonate-chloride exchange flux falls, reaches a minimum about pH 5, and then starts to increase again. Above pH 5, activation energies do not change, but below pH 5 they do. Perhaps we're looking at a different transport process below pH 5.

Above pH 5, what we think is happening due to increasing extracellular H^+ is similar to what SITS is doing. That is to say, the hydrogen ions interact with transport sites at the outside surface of the membrane, thereby inactivating a certain fraction of the sites. It's a much poorer inhibitor than SITS is, obviously, but that's what we think is happening. The transport sites that are not inactivated continue to exchange bicarbonate for chloride with the same kinetics as at pH 6.8.

MIZUKAMI: Do you think that carbonic anhydrase has anything to do with the change of the rates?

CRANDALL: We don't think so. Carbonic anhydrase remains active over a wide range of pH and temperature.

BROWN: Have you taken into account the fact that chloride is a potent inhibitor of carbonic anhydrase and that most of the enzyme inside the cell is believed to be bound to the membrane and is known to be largely inactive?

CRANDALL: We know that is true for human carbonic anhydrase B, but C probably is quite active under the conditions of our experiments. We do not believe the intracellular reactions could ever be rate-limiting unless inhibited.

J. BONAVENTURA: What is the nature of the anion exchange inhibitors SITS and DIDS?

CRANDALL: They are disulfonic acids.

BONAVENTURA: What is the mechanism of inhibition?

CRANDALL: They are thought to bind to the transport site in band 3 protein of the red cell membrane.

J. BONAVENTURA: Do they bind irreversibly to the red blood cell membrane and are they transportable across the membrane?

CRANDALL: They probably bind mostly irreversibly and do not get inside.

CHUN: Please comment on the interaction between carbonic anhydrase and hemoglobin, since carbonic anhydrase is the second largest enzyme protein found in the erythrocyte. I wonder whether the interaction between carbonic anhydrase and hemoglobin modulates the exchange properties of HCO_3-anion and Cl-anion.

CRANDALL: I don't know any data at all on that point. In any case, red cell carbonic anhydrase greatly catalyzes the CO_2 reactions.

PARKHURST: Have you made any measurements in deoxygenated cells? If there is coupling of the exchange with chloride ion or CO_2 binding to hemoglobin, there might be an effect.

CRANDALL: We have. We find that whether the hemoglobin is oxygenated or deoxygenated doesn't make any difference to this exchange. If saturation of hemoglobin changes during the experiment, then we have a problem because protons will be released or bound due to the Bohr effect, but we make sure that the two fluids have the same oxygen tension.

BROWN: Does the system run backwards? If you raise the pH, does chloride flow out of the cell?

CRANDALL: We believe it does. We've done a couple of experiments in the reverse direction by going to the alkaline pH, and it works. We get data of the sort we expect, but we haven't done enough experiments to give you results now.

BREWER: You have no trouble with breakage of the glass electrode?

CRANDALL: We've had no trouble at all with that. These 3-mm electrodes with pH-sensitive tips are very sturdy, and we have been very, very pleased with them.

ROMINE: You state that there is no effect on exchange in the change from oxygenated to deoxygenated Hb. Can this be reconciled with the fact that deoxygenated Hb has an "oxy-labile" proton acceptor that is physiologically very important to maintenance of ionic balance? It certainly causes a significant change in the ionic concentrations within the erythrocyte and should show as part of the flux equations a temperature-sensitive association-disassociation to the oxy-labile site.

CRANDALL: You are right that an effect would be expected if hemoglobin saturation changes during the experiment, which we made sure it didn't. We would not make the assertion you've attributed to me.

BERGER: In response to the question of fragility of the electrodes, Fig. 20 of my paper shows the response of the Leeds & Northrup electrodes, reference and glass together, to the plunge test I described earlier. Fig. 20*a* shows the uncoated electrodes and Fig. 20*b* shows the electrode coated with Lycra.

CRANDALL: We have similar data to Dr. Berger's, although we use the hydrochloric acid-bicarbonate test reaction system in stopped-flow, and we find that by extrapolating back the linear portion we've got a 5-ms response. We've never bothered pushing it further than that, and I think you can get it down below that.

SUTTER: You have a stopped flow system essentially without a stop syringe. If you take out the pH-detecting electrodes, and replace them with an optical detection system where you don't have the built-in 1.6-ms time constant of the pH electrode, how good would your apparatus be, ultimately?

CRANDALL: We do have a stop signal from a magnet-in-coil device mounted on the driving block. We think the rapid mixing apparatus itself is extremely efficient.

SUTTER: Correct. But using it as is, without the stop syringe, and removing this 1.6-ms time constant from your pH-sensitive electrode, and using some other detection device, how fast do you think it could be?

CRANDALL: Well, we should certainly be able to get down below the millisecond range. Beyond that I don't know offhand how far down it could go.

JAKOBSSON: It seems to me that to the extent that the pH electrode really is an RC circuit, you can calculate how far off you are by putting a real step change on it, so that maybe the response time on the pH electrode shouldn't be such a limitation on knowing where you are.

CRANDALL: I agree. I don't think necessarily that this 1.6 ms I found by a different test system is in the electrode; it might just have been in the electronics by which we measured it. So the electrode may be faster than I said, but we haven't pushed it.

BERGER: We've measured the impedance of the electrode, and it's about 50 MΩ. It's a fairly low-resistance glass, actually. And the capacitance, if you balance out the cable capacities, is only about 15 or 20 pF. You can calculate what the RC is. Thus I don't think we're anywhere near the diffusion of protons across the glass membrane.

GOSS: Have you looked at the effect on sickle cells?

CRANDALL: We haven't. It's something, again, we'd like to do, but it's just a matter of doing things sort of in order, but thank you for the suggestion.

EXTENDED ABSTRACTS

THE MOTION OF CYTOCHROME b_5 ON LIPID VESICLES MEASURED VIA TRIPLET ABSORBANCE ANISOTROPY

R. H. AUSTIN AND W. VAZ, *Max Planck Institute for Biophysical Chemistry, Göttingen, West Germany*

We have replaced the iron atom in the heme group with the chemically similar atom rhodium (1). Unlike iron-protoporphyrin IX, rhodium-protoporphyrin IX has a reasonable triplet yield easily measured via extinction coefficient changes in the Soret region. However, like iron-protoporphyrin, the rhodium-protoporphyrin can axially ligate to the protein, thus making it an ideal probe to study the protein motion via anisotropy decay of the triplet state.

We have inserted our rhodium-protoporphyrin IX into the intrinsic membrane protein, cytochrome b_5, and used the anisotropy decay of the triplet state excited by the pulse from a nitrogen laser-driven dye laser to measure the motion of the cytochrome b_5 molecule on the membrane surface. Possible values for orientation of the porphyrin plane relative to the membrane surface and the rotational mobility of the protein in the lipid are found.

REFERENCES

1. HANSON, L., M. GOUTERMAN, and J. HANSON. 1973. The crystal and molecular structure and luminescence of bis(dimethylamine)etio(I)porphinatorhodium(III) chloride dihydrate. *J. Am. Chem. Soc.* **95**: 4822.

SPECTRAL INTERMEDIATES IN THE ACTIVATION OF GLYCERALDEHYDE-3-PO$_4$-DEHYDROGENASE-CATALYZED REACTIONS

SIDNEY A. BERNHARD, *Institute of Molecular Biology, University of Oregon, Eugene, Oregon 97403 U. S. A.*

In the presence of the effector, NAD$^⊕$, glyceraldehyde-3-PO$_4$-dehydrogenase (GPDH) reacts rapidly with the substrate (or product) analogue, β-(2-furyl)acryloyl phosphate

TABLE I
λ_m of Various Furylacryloyl Derivatives (FA-X)

X	λ_m
	(nm)
OH	310
O$^\ominus$	292
OCH$_3$	309
NHCH$_3$	301
H	322
SCH$_2$O	337
SCH$_2$E (apo)	346

(FAP) to form an acyl-enzyme (FAE) (Eq. 1) (1). The electronic character of the covalently-attached ligand to the furylacryloyl chromophore contributes substantially to the energy of the first electronic transition (Table I).

Although slower in rate than the enzymic reaction with true substrate by about one or two orders of magnitude, the otherwise kinetically stable reagent (FAP) reacts rapidly and uniquely with the enzyme at the SH of Cys 149 and demonstrates kinetic biphasicity in the resultant spectral perturbations (1) (Fig. 1). The rate of the fast phase of reaction depends on the concentrations of both FAP and NAD$^\oplus$. The faster rate saturates in [NAD$^\oplus$] at concentrations consistent with the known equilibrium dissociation constants for E[NAD$^\oplus$] ($\sim$1 μM) (2). The faster rate saturates only at high concentrations of FAP ($\sim$10 mM). The slower phase velocity is independent of both [FAP] and [NAD$^\oplus$], provided that substrate and effector are in substantial excess over enzyme sites. The amplitude of the slow phase is [NAD$^\oplus$]-dependent, reaching a saturation value at high [NAD$^\oplus$] ($\sim$5 mM).

A partial resolution of these complex phenomena is derivable from studies of the equilibrium and kinetic properties of the isolated acyl-enzyme (FAE). Equilibrium studies (3) indicate that although the dissociation of NAD$^\oplus$ from the enzyme is small ($K_d \sim 1$ μM); the dissociation of NAD$^\oplus$ from the acyl-enzyme is much greater ($K_d \sim 50$ μM). Hence at intermediate [NAD$^\oplus$] ($\sim$10 μM) the chemical reaction is approximated by Eq. 2:

$$E(NAD^\oplus) + FAP \rightleftharpoons EFA + NAD^\oplus + P_i. \qquad (2)$$

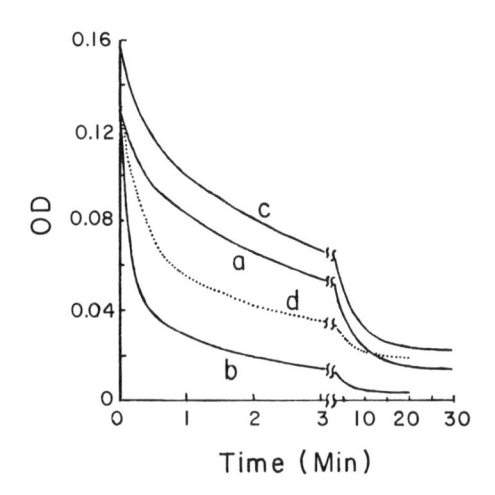

FIGURE 1 Kinetics of deacylation of FA-GPDH. In every case the reaction was started with a small volume of acceptor and monitored at 360 nm. Protein and NAD$^\oplus$ were preincubated. Concentration conditions are as follows: (a) 2.4 µM protein with 2.08 FA groups per mole enzyme, 25 µM NAD$^\oplus$, and 0.5 mM arsenate; (b) 2.4 µM protein with 2.08 FA groups per mole enzyme, 250 µM NAD$^\oplus$ and 0.5 mM arsenate; (c) 4.1 µM protein with 2.08 FA and two carboxymethyl groups per mole enzyme, 65 µM NAD$^\oplus$, and 0.5 mM arsenate; (d) 3.1 µM protein with 1.83 FA groups per mole enzyme, 131 µM NAD$^\oplus$, and 0.5 mM phosphate (some of the curves have been shifted along the vertical scale for clarity).

At higher [NAD$^\oplus$], the binding of NAD$^\oplus$ to the acyl (FA)-enzyme can be detected by its effect on the FAE absorption spectrum (3) (Table I). [NAD$^\oplus$] influences (increases) the amplitude, but not the velocity, of the slow step of acylation, suggesting that this slow phase is a consequence of the reaction of the sequence of Eq. 3.

$$\text{FAE} \underset{}{\overset{\text{SLOW}}{\rightleftharpoons}} \text{FA*E} \underset{}{\overset{\text{FAST}}{\rightleftharpoons}} \text{FA*E (NAD}^\oplus\text{)}. \qquad (3)$$

These speculations can be tested directly by studying the kinetics of interaction of the acyl-enzyme with the effector [NAD$^\oplus$] in the absence of any acyl acceptor (HPO$_4^{2-}$, HAsO$_4^{2-}$, or NADH). The kinetics of spectral perturbation upon addition of NAD$^\oplus$ to the FA enzyme show biphasic behavior. The fast phase of reaction requires stopped-flow techniques for its quantitative measurement, whereas the slow phase reaction has a half-life of the order of 1 min. The amplitudes for each step are wavelength-dependent (Fig. 2). The overall equilibrium of Eq. 3 has been investigated spectrophotometrically as well (Fig. 2). At equilibrium, the reaction shows only two spectrally discernible species (Table I) with a single isobestic point at 360 nm (3). Indeed, the acylation of GPDH by FAP exhibits monophasic pseudo-first-order kinetics at this wavelength (1). From the wavelength dependence of stopped-flow experiments on the interaction of apo-FA enzyme with NAD$^\oplus$, the kinetic isobestic point wavelengths can be determined for each of the two readily separable steps of acyl enzyme perturbation. These isobestic points differ from one another (Fig. 2) and from the equilibrium isobestic point. The spectually identifiable species in the interaction of

Extended Abstracts

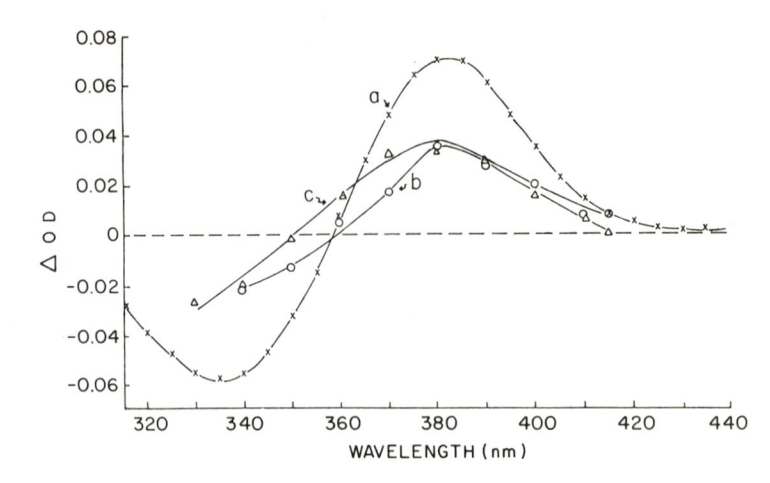

FIGURE 2 (a) Total ΔOD FAE + 1 mM NAD⊕ vs. FAE, after equilibrium (X). (b) ΔOD after the fast step after NAD⊕ addition (o—o). (c) ΔOD after the slow step; OD after slow step − OD after fast step (△).

NAD⊕ with acyl (FA) enzyme at equilibrium are summarized in Table II. The qualitative and quantitative kinetic data allow an estimation of the number and chemical bonding properties of addition transient acyl-enzyme intermediates. These analyses identify three spectral components; EFA (λ_m = 340 nm) EFA* (λ_m = 350 ± 10 nm), and EFA*(NAD⁺) (λ_m = 360 nm). The equilibrium apo acyl-enzyme (λ_m = 346 nm) is a mixture of EFA and EFA*. The fast component of reaction is the reversible formation of EFA*(NAD⊕) from EFA*, and the slow NAD⁺-independent rate is the rate of the isomerization reaction (Eq. 2). "Red-shifts" in the longest wavelength absorption bands of the furylacryloyl chromophore are associated with increased polarity of the carbonyl (4) [(+)C—O(−)] and hence with greater susceptibility to nucleophilic attack, for example by HPO_4^{2-}. This enhanced nucleophilicity is inherent in the conformational isomerization of the acyl enzyme to a state (EFA*), which is stabilized by the noncovalent interaction with the effector (NAD⊕), only the red-shifted acyl-enzyme (EFA*) is catalytically competent towards nucleophilic attack.

The conformational isomerization (Eq. 3) limits the overall turnover of the enzyme.

TABLE II
IDENTIFICATION OF BOUND DINUCLEOTIDE TO VARIOUS SITES OF THE DIACYL ENZYME TETRAMER (FA_2E_4)

Dinucleotide	Site	Diagnostic
NAD⊕	Free SH	Racker band[6] λ_m = 360 nm
NAD⊕	FASCH₂E	λ_m shifts from 340 to 360 nm
NADH	Free SH	Spectral shift at 275 nm
		Fluorescence of NADH quenched
NADH	FASCH₂E	λ_m shifts from 240 to 328 nm

There is a stoichiometric limitation of one FA group per two subunits (1).

Hence, the addition of HPO_4^{2-}, AsO_4^{2-}, or NADH[5] to FA enzyme at suboptimal concentrations shows a common slow phase of reaction velocity with a specific rate or approximately 0.01 s^{-1}, as in the slow phase of spectral perturbation. A common limiting rate of turnover has been observed in the saturated rates with glyceraldehyde-3-PO$_4$ or 1,3 diphosphoglycerate (2) presumably due to a corresponding conformational change in the intermediate 3-phosphoglyceroyl enzyme. In this case, the specific rate is approximately 10^3-fold faster, thus indicating a synergistic effect of acyl-structure and protein tertiary structure.

The work was supported by grants from the Public Health Service (GM 10451-12) and the National Science Foundation (BMS 75-23297).

REFERENCES

1. MALHOTRA, O. P., and S. A. BERNHARD. 1968. *J. Biol. Chem.* **243**:1243-1252.
2. SEYDOUX, F., S. A. BERNHARD, O. PFENNINGER, M. PAYNE, and O. P. MALHOTRA. 1974. *Biochemistry.* **12**:4290-4300.
3. MALHOTRA, O. P., and S. A. BERNHARD. 1968. *Proc. Natl. Acad. Sci. U.S.A.* **70**:2077-2081.
4. CHARNEY, E., and S. A. BERNHARD. 1966. *J. Am. Chem. Soc.* **89**:2726-2733.
5. SCHWENDIMANN, B., D. INGBAR, and S. A. BERNHARD. 1976. *J. Mol. Biol.* **108**: 123-138.
6. RACKER, E., and I. KRIMSKY. 1952. *J. Biol. Chem.* **198**:731-736.

STUDIES OF THE ACTIVATION OF YEAST ENOLASE BY METALS USING A "TRANSITION STATE ANALOGUE"

JOHN M. BREWER, *Department of Biochemistry, University of Georgia, Athens, Georgia 30602 U.S.A.*

Yeast enolase, a dimeric protein, binds up to 2 mol of "conformational" Mg, which enables up to 2 mol of substrate or competitive inhibitor to bind, which in turn enables more Mg to bind. The latter Mg produces the actual catalysis. A putative "transition state analogue," aminoenolpyruvic acid-2-phosphate (AEP), synthesized by Spring and Wold, exhibits a large 295-nm difference spectrum upon binding to enolase with Mg present, permitting the monitoring of binding of AEP, conformational Mg, "catalytic" Mg, and other metals. Spectrophotometric titrations and stopped-flow measurements have led to some tentative conclusions:

(*a*) The strength of catalytic Mg and AEP binding is interdependent, consistent with an ordered sequence of addition. With saturating AEP, about two-thirds of the 295 nm absorbance change occurs on addition of "conformational" Mg, the rest on adding catalytic Mg (Fig. 1).

(*b*) The nonactivating metals Ca, Hg, and Ba do not give this 295-nm absorbance change though at least Ca binds at the same sites as Mg, with a similar affinity and effect on the protein. Ni (a weak activator), Mn (intermediate), and Mg (best) give

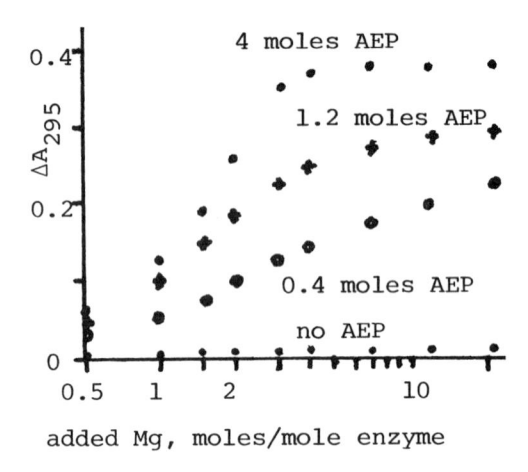

FIGURE 1

the full change—the reactions appear to be all or none (Table I). This suggests that the rate of the reaction is determined by the catalytic metal, a suggestion supported by stopped-flow experiments in which enzyme with Mg, Ca, or Ni is mixed with substrate and additional Ca, Ni, or Mg (Table II). These observations suggest that the conformational metal plays a key role in substrate activation, rather than passively providing only a binding site.

(c) Calorimetric measurements of the enthalpy of mixing the competitive inhibitor 3-phosphoglyceric acid with the Ca-enzyme show that the inhibitor does bind. In addition, stopped-flow measurements of the effect of the substrate and AEP on the 295-nm absorbance change produced on adding excess EDTA to the Ca enzyme or the Mg enzyme show that the substrate slows the rate of the reaction from 0.9/s to 0.7/s while AEP reduces it a factor of two, to 0.45/s (Fig. 2). This is consistent with the suggestion that AEP is a transition state analogue and also shows that the different metals behave identically as far as enabling substrate or AEP to bind. The difference between activating and nonactivating metals is that the nonactivating ones do not cause the change in AEP absorbance.

TABLE I

Metal	Activating?	ΔOD_{295}: moles metal/mole enzyme		
		1	2	100
Mg	Yes	0.089	0.189	0.284
Mn	Yes	0.110	0.214	0.298
Ni	Yes	0.098	0.186	0.290
Ca	No	0.014	0.018	0.006
Ba	No	0.009	0.016	0.009
Cu	No	0.012	0.021	—
Hg	No	—	0.001	—

TABLE II

Moles metal/ moles enzyme (initial)	Moles/mole added with substrate	Initial activity
		%
2 Mg	2 Mg	1.16 (100)
2 Mg	2 Ni	0.42 (36)
2 Ni	2 Mg	0.62 (54)
2 Mg	2 Ca	0.38 (32)
2 Ca	2 Mg	0.20 (17)
2 Ni	2 Ni	0.16 (14)

These observations also suggest that the "conformational" metal interacts directly with the substrate in the transition state. This suggestion is supported by stopped-flow experiments in which the AEP is displaced with excess substrate: the rate of the reaction is 0.08/s, lower than that produced by addition of EDTA or excess Mg (to the Ca enzyme) (0.14/s) or Ca (to the Mg enzyme) (0.16/s). And the observed rates of apparent metal loss or displacement in the presence of AEP are much lower than the rates of metal loss in the absence of AEP, monitored with chlorophosphonazo III, a chromophoric metal-chelating agent (2/s).

(d) The subunits of enolase dissociated in the absence of magnesium are inactive because the substrate (AEP) binding site has been affected. Stopped-flow measurements indicate that this site is restored only after subunit association (Fig. 3).

This work was supported by National Science Foundation grant PCM76-21378.

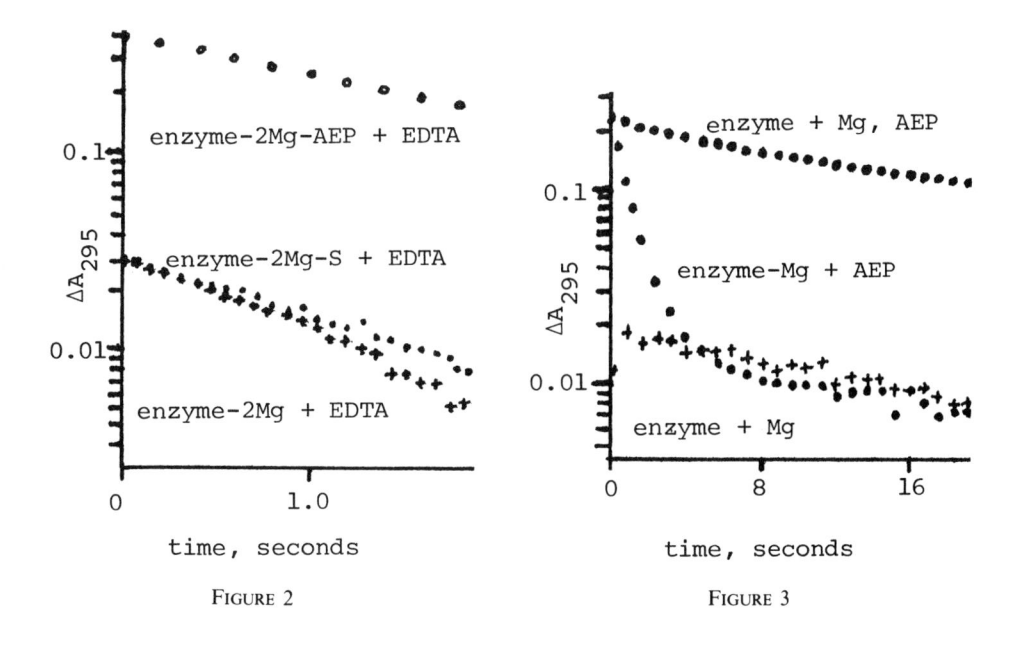

FIGURE 2

FIGURE 3

SCANNING MOLECULAR SIEVE CHROMATOGRAPHY OF INTERACTING PROTEIN SYSTEMS

Effect of Kinetic Parameters on the Large Zone Boundary Profiles for Local Equilibration Between Mobile and Stationary Phases

Paul W. Chun and Mark C. K. Yang, *Department of Biochemistry and Molecular Biology, College of Medicine, University of Florida, Gainesville, Florida 32610 U. S. A.*

Large zone reaction boundary profiles for molecular sieve chromatography as affected by kinetic parameters have been simulated for local equilibration between the mobile and stationary phases. Our studies of monomer-dimer and monomer-tetramer systems indicate that in a slowly equilibrating system, the kinetic controls operating between the mobile and stationary phases contribute most significantly to the overall boundary profile. In a rapidly equilibrating system, however, the kinetic parameters k_{ij} and k_{ji} operating in the mobile phase are the principal determinants of the reaction boundary, while the kinetic effects of k_{ii} and k_{jj} between the mobile and stationary phase are minimal.

The solute partition cross-section and axial dispersion operator (1) of the boundary profiles of such an interacting protein system under kinetic control have been simulated by (*a*) time discretization, (*b*) the finite difference approximation of Fick's second law (2) and (*c*) the kinetic expression for the concentration of monomer, $C^*_{1(t)}$, as a function of equilibration time (3). Here, the expression for idealized chromatographic continuity is (4):

$$(\partial C_i^*/\partial t) = L_i(\partial^2 C^*/\partial X^2) - (F/\xi_i)(\partial C/\partial X)$$
$$- \sum_{j \neq i} k_{ij}(C_i^*)^j + \sum_{j \neq i} k_{ji}(C_j^*)^i - \sum_i k_{ii} C_i^* + \sum_j k_{jj} C_i^{**}, \quad (1)$$

where k_{ij} is the chemical reaction rate from species i to j. C_i^* is the concentration of species i. The $-\sum_{j \neq i} k_{ij}(C_i^*)^j + \sum_{j \neq i} k_{ji}(C_j^*)^i$ term represents a local equilibration in the mobile phase and $-\sum_i k_{ii} C_i^* + \sum_j k_{jj} C_i^{**}$ represents an equilibration between the mobile and stationary phases. F is the elution flow rate, ξ_i is the solute partial cross-section, and L_i the axial dispersion coefficient.

In considering a monomer-*n*-mer system, the kinetic effect varies with the distribution of the concentration of monomer along the column. Due to kinetic effects, the column equilibrium coefficient for a given self-associating species in such a system is not always constant. Hence, the apparent equilibrium constant will vary as a function of the distance coordinate, i.e., $dK/dX \neq 0$ (5).

The general kinetic expressions for the concentration of monomer as a function of equilibration time, $C^*_{1(t)}$, in the mobile phase for various types of self-associating systems become:

(I) $n = 1$, Isomerization: $dC^*_1/\{[k_{ln} + k_{nl}]C^*_1 - k_{nl}C^*_T\} = -dt$,

(II) $n = 2$, Monomer-dimer system:

$$dC^*_1/\{[k_{ln}(C^*_1)^2 + k_{nl}C^*_1] - k_{nl}C^*_T\} = -dt,$$

(III) $n = 3$, Monomer-trimer system:

$$dC^*_1/\{[k_{ln}(C^*_1)^3 + k_{nl}C^*_1] - k_{ln}C^*_T\} = -dt,$$

(IV) $n = 4$, Monomer-tetramer system: $dC^*_1/\{k_{ln}(C^*_1)^4 + k_{nl}C^*_T\} = -dt$. (2)

Our simulation studies in monomer-dimer and monomer-tetramer systems have indicated that in a slowly equilibrating system, the kinetic rate constants k_{ii} and k_{jj} between the mobile and stationary phase contribute significantly to the overall gradient boundary profile. In contrast, in the rapidly equilibrating system, the kinetic parameters k_{ij} and k_{ji} in the mobile phase are the principal determinants of the reaction boundary. Although the kinetic effects of k_{ii} and k_{jj} may be noted, their contribution to the overall boundary profile is minimal.

This work was supported by National Science Foundation grant PCM 76-04367 and in part by the University of Florida computer center.

REFERENCES

1. ACKERS, G. 1967. *J. Biol. Chem.* **242**:3026, 3237.
2. COX. 1969. *Arch. Biochem. Biophys.* **12**:106.
3. CHUN, P. W., and M. C. K. YANG. 1978. *Biophys. Chem.* In press.
4. CHUN, P. W., and M. C. K. YANG. 1977. *Fed. Proc.* **36**: 3073.
5. CHUN, P. W., and YOON. 1977. *Biopolymers.* **16**:2579.

A SOLUTION MIXER WITH 10-μs RESOLUTION

ROBERT CLEGG, *Max Planck Institute for Biophysical Chemistry,*
Department of Molecular Biology, 3400 Göttingen, West Germany, and
MACK J. FULWYLER, *Becton Dickinson Electronics Laboratory,*
Mountain View, California 94043 U. S. A.

An apparatus has been developed which mixes two solutions within a time limit of 10–100 μs. The technique is based upon the mixing properties of a fluid flowing around a sphere. The two solutions are juxtaposed and constrained to a very thin layer as they flow around the sphere. The time resolution of the method depends upon the ability to reduce the mixing dimensions as much as possible. Optical detection of the progress of the reaction is accomplished by measuring the fluorescence of the liquid jet at different positions along the emerging stream.

KINETIC TRANSIENTS

A Wedding of Empiricism and Theory

Alan H. Colen, *Veterans Administration Hospital, Kansas City, Missouri 64128, and The University of Kansas School of Medicine, Lawrence, Kansas 66044 U. S. A.*

The purpose of the study of transient-state kinetics is to separate events in time and thus contribute to a detailed understanding of underlying mechanism. We have recently developed some simple but novel methods to achieve this, including the rapid-flow calorimetric determination of the kinetics of oxime formation (1) as a model for imine formation in enzymatic reactions, the stopped-flow ultraviolet spectrophotometric measurement of hydrogen exchange in nucleotides and of the effect of binding by dehydrogenases on that process (2), and the observation of transient features of enzymatic reactions over a period of minutes by using the cryoenzymological approach of Douzou and Fink (3). Thus, either by mixing and observing rapidly or by slowing the reaction sufficiently we can observe and characterize complex behavior ordinarily inaccessible to steady-state or relaxation kinetics techniques.

Many qualitative empirical conclusions can be obtained from studies of kinetic transients: "lags" give information about the kinetics of formation of precursor complexes; the existence of a "burst" suggests that there is a slow step late in the mechanism, possibly involving the formation of tight product complexes; time-dependent spectral shifts often help identify reaction intermediates; and preincubation effects (or lack thereof) give clues about slow steps and obligatory pathways in biochemical reactions. All of these phenomena can yield detailed information about the relative rates of formation and breakdown of complexes with effectors.

Quantitative mathematical and theoretical treatment of biochemical transients and related fast reactions is required, however, to give flow methods their full power as tools for the determination of mechanism. Although detailed treatments are available and many transient kinetics studies have been performed, some powerful but simple empirical methods for obtaining and handling data to yield a quantitative understanding of the processes underlying observed transients far from equilibrium have not been fully exploited. Two such approaches are developed and illustrated here with applications to the study of the glutamate dehydrogenase-catalyzed reaction, relating empirical observation to mechanistic description.

The first method is the study of the initial velocities of transients. These may be measured either directly for initially linear time-courses or by extrapolation of nonlinear experimental time-course curves to an empirically determined experimental time zero. Once obtained, these velocities are usually susceptible to analysis by standard kinetics methods, with the steady-state formalism of enzyme kinetics often being the most powerful. The advantage of the measurement of initial transient velocities is that one thereby can isolate the very initial reaction steps of complicated mechanisms, selectively avoiding the effects of reverse reaction steps and later complex transient behavior.

In the oxidative deamination of L-glutamate by $NADP^+$ and glutamate dehydrogenase, there is an initial transient burst of reactivity, producing reduced nicotinamide absorbance of enzyme-NADH product complexes in the 340-nm region. Lineweaver-Burk plots of initial transient velocities (4) are linear in both substrate and coenzyme concentration, obeying a concentration dependence of the form:

$$e/v = 0_0 + [0_1/(NADP^+)] + [0_2/(\text{L-Glu})] + [0_{12}/(NADP^+)(\text{L-Glu})]. \quad (1)$$

Since no preincubation effects are observed, and dissociation constants for binary enzyme-$NADP^+$ and enzyme L-glutamate complexes (which can be calculated from the 0_i values obtained by fitting Eq. 1) agree with those measured independently at equilibrium, it can be concluded that both binary complexes are formed and equilibrate rapidly with coenzyme and substrate on the stopped-flow time scale. One may also calculate the concentration of material tied up in ternary enzyme-$NADP^+$-L-glutamate complexes and an apparent heterotropic cooperativity of formation of such complexes. Using the pH dependence of the 0 values, it can be shown that a ternary complex also equilibrates rapidly, with dissociation constants equal to the limiting Michaelis constants determined experimentally (5).

Not all initial velocity experiments can be handled so simply by steady-state techniques. In some cases, especially when preincubation effects are observed, it is necessary to solve the initial velocity rate equations in a quasi-equilibrium or quasi-steady-state approximation coupled to a relatively slow transient process. Such is the case for the product inhibition by α-ketoglutarate of the burst described above. In this case, the catalytic reaction is inhibited by a tight dead-end enzyme-$NADP^+$-α-ketoglutarate complex, which forms and breaks down in times commensurate with the duration of the transient burst (6).

The second, more commonly used approach to kinetic transients is the measurement of the apparent first-order rate constants that characterize the transient phenomena. It has not been generally recognized, however, that such data, even far from equilibrium, may be treated by using the powerful formalism already developed for relaxation kinetics. This is true not only for systems approaching equilibrium but also for a first-order transient approach to a steady state. This latter application is particularly valuable to the biochemical kineticist. With this approach, it has been demonstrated quantitatively that the increase of the apparent first-order rate constant for the L-glutamate deamination burst in the presence of the product inhibitor NH_4^+ is caused by the formation of a new stable intermediate in rapid equilibrium with a product complex of enzyme, NADH, and α-ketoglutarate, previously characterized at equilibrium (7).

Using these powerful theoretical tools, coupling the results of transient kinetics with those obtained at equilibrium, and extending them over a range of solvents and temperatures, one can construct a detailed picture of biochemical mechanism, complete with both free energy and enthalpy characterization of important reaction intermediates.

This work was supported in part by grants from the National Science Foundation (PCM75-17107) and from the General Medicine Institute of the National Institutes of Health (GM15188).

REFERENCES

1. FISHER, H. F., D. C. STICKEL, A. BROWN, and D. CERRETTI. 1977. Determination of the thermodynamic parameters of individual steps of pyruvate-oxime formation by rapid, continuous flow microcalorimetry. *J. Am. Chem. Soc.* **99**:8180–8182.
2. CROSS, D. G., A. BROWN, and H. F. FISHER. 1976. Hydrogen exchange at the amide group of reduced pyridine nucleotides and the inhibition of that reaction by dehydrogenases. *J. Biol. Chem.* **251**:1785–1788.
3. FINK, A. L. 1976. Cryoenzymology: the use of sub-zero temperatures and fluid solutions in the study of enzyme mechanisms. *J. Theor. Biol.* **61**:419–445.
4. COLEN, A. H., R. A. PROUGH, and H. F. FISHER. 1972. The mechanism of glutamate dehydrogenase reaction. IV. Evidence for random and rapid binding of substrate and coenzyme in the burst phase. *J. Biol. Chem.* **247**:7905–7909.
5. COLEN, A. H., R. R. WILKINSON, and H. F. FISHER. 1977. The transient-state kinetics of L-glutamate dehydrogenase: pH-dependence of the burst rate parameters. *Biochim. Biophys. Acta.* **481**:377–383.
6. COLEN, A. H. 1978. Transient-state kinetics of L-glutamate dehydrogenase: mechanism of α-ketoglutarate inhibition in the burst phase. *Biochemistry.* **17**:528–533.
7. BROWN, A., A. H. COLEN, and H. F. FISHER. 1978. Effect of ammonia on the glutamate dehydrogenase catalyzed oxidative deamination of L-glutamate: the production of an ammonia-containing intermediate in the "burst" phase. *Biochemistry.* In press.

A METHOD FOR DETERMINING THE KINETIC TYPE OF FAST KINETIC DATA

DAVID C. FOYT AND JOHN S. CONNOLLY, *Center for Fast Kinetics Research, University of Texas, Austin, Texas 78712 U. S. A.*

Several well-established methods exist for the treatment of data obtained by fast kinetic techniques, whereby the appropriate kinetic parameters (e.g., rate constants) may be extracted. Typically, these methods involve an initial assumption as to the appropriate type of kinetics, which is then employed to fit the corresponding kinetic equations to the data. Particularly in the case of pulse radiolysis and flash photolysis, the frequent occurrence of mixed-order processes (whether independent or competing, growth or decay), of simultaneous detection of more than one species, and of residual base-line concentrations often make it difficult to choose the correct initial assumption for data analysis.

Furthermore, a given data set can frequently be fit with comparable success by more than one type of kinetics, making it necessary to vary the experimental conditions to distinguish among the possible analyses. The systematic application of several different data reduction techniques to several sets of data becomes tedious, and the sheer volume of results thus obtained may actually obscure rather than clarify the correct solution to the problem.

We present a data reduction technique applicable without any initial assumptions regarding the kinetics involved, which produces a graphical output from whose shape the type of kinetic analysis appropriate to the data can, in most cases, be determined by simple visual examination. This technique is intended as a preliminary to the fitting and extraction of kinetic parameters. Let $V(t)$ be the data trace proportional to concentration, and let $d/dt\,[\log(V(t_0)/V(t)0]$ be plotted vs. $V(t)$. The resulting transform is linear in the special case of competing first- and second-order processes with the intercept and slope giving the first- and second order-rate constants, respectively (1, 2). For more complicated kinetics this transform may be sigmoid or otherwise curved, but it is generally quite distinctive.

Computation of the transform described involves the difficult problem of evaluating the derivative of experimental data containing noise. Acceptable results were obtained by first applying a simple linear filter to the function to be differentiated. Then a least-squares fit of the first eight Chebyshev polynomials was obtained and differentiated analytically.

Results are presented for synthetic data of various types, and the limiting signal-to-noise ratio for successful analysis is examined. The method is also applied to experimental data and the usefulness of families of such curves, corresponding to different experimental conditions, is illustrated.

REFERENCES

1. LINSCHITZ, H., and K. SARKANEN. 1958. *J. Am. Chem. Soc.* **80**:4826.
2. PEKKARINEN, L., and H. LINSCHITZ. 1960. *J. Am. Chem. Soc.* **82**:2407.

THE EFFECT OF PRETREATMENT WITH CALCIUM AND MAGNESIUM IONS ON PHOSPHOENZYME FORMATION BY SARCOPLASMIC RETICULUM ATPase

JEFFREY P. FROEHLICH, *National Institute on Aging, National Institutes of Health, Baltimore, Maryland 20014 U. S. A.*

It has previously been shown (1, 2) that pretreatment of $(Na^+ + K^+)$ATPase with K^+ (10–20 mM) before the addition of Na^+ and ATP slows the rate of phosphoenzyme formation and reduces the early phosphate "burst." These effects have been explained as resulting from the slow conversion of E_2, a form of the enzyme stabilized by high K^+ to E_1, a form stabilized by Na^+ and rapidly phosphorylated by ATP. Having Na^+ present initially with K^+ overcomes the effect resulting from pretreatment with K^+ alone by a mechanism postulated to involve Na^+-induced conversion of E_2 to E_1. To determine whether analogous effects occur in the reactions catalyzed by sarcoplasmic reticulum ATPase, the kinetics of ^{32}P incorporation were investigated after pretreatment with high levels of Mg^{2+}. Rapid mixing was carried out with a quench-flow apparatus (3), which mixes the enzyme and substrate solutions at a volume ratio of 1 to 20. This arrangement was used to bring about the dissociation of Mg^{2+} from the enzyme and to minimize possible competitive interactions between the ligands. If the enzyme is pretreated with 1 mM Mg^{2+} and 1 mM EGTA to remove tightly bound Ca^{2+}, phosphoenzyme formation resulting from the addition of 10 μM ATP and 50 μM Ca^{2+} is rapid ($t_{1/2}$ = 15 ms) and displays a transient overshoot. Pretreatment of the enzyme with 20 mM Mg^{2+} and 1 mM EGTA reduced the rate of phosphorylation ($t_{1/2}$ = 28 ms) and eliminated the overshoot without affecting the steady-state level of phosphoenzyme. A small amount of Ca^{2+} (~ 10 μM) included in the enzyme medium with 20 mM Mg^{2+} was able to restore completely the rapid rate of phosphorylation obtained when 1 mM Mg^{2+} is present initially. Excluding both Ca^{2+} and Mg^{2+} from the enzyme medium resulted in a high rate of phosphorylation ($t_{1/2}$ = 13 ms) and no overshoot. These findings are interpreted with the aid of the following diagram:

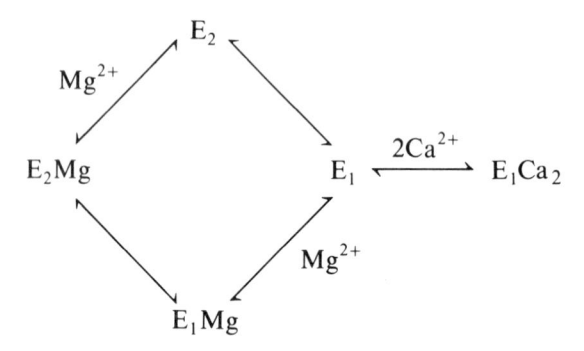

$$\begin{array}{ccc}
 & E_2 & \\
Mg^{2+} \nearrow & & \searrow \\
E_2Mg & & E_1 \xrightleftharpoons{2Ca^{2+}} E_1Ca_2 \\
 & \nwarrow \quad \nearrow Mg^{2+} & \\
 & E_1Mg &
\end{array}$$

where, by analogy to the events described in $(Na^+ + K^+)$ATPase, the unliganded enzyme exists in two separate conformational states characterized by having either high affinity for Ca^{2+} (E_1) or low affinity for Mg^{2+} (E_2). In the absence of ligands, E_1, the form of the enzyme rapidly phosphorylated by ATP, is favored. High Mg^{2+} drives the equilibrium in favor of E_2; upon addition of Ca^{2+} and ATP, E_2 is converted to E_1 at a rate slower than the rates of substrate binding and phosphorylation.

REFERENCES

1. MÄRDH, S. 1975. *Biochim. Biophys. Acta.* **391**:448–463.
2. LOWE, A. G., and J. W. SMART. 1977. *Biochim. Biophys. Acta.* **481**:695–705.
3. FROEHLICH, J. P., J. V. SULLIVAN, and R. L. BERGER. 1976. *Anal. Biochem.* **73**:331–341.

RECORDING OF FAST BIOCHEMICAL REACTIONS USING A LOGARITHMIC TIME SWEEP

EUGENE HAMORI AND TOSHIHIRO HIRAI, *Department of Biochemistry, Tulane Medical Center, New Orleans, Louisiana 70112 U. S. A.*

Kinetic data are conventionally obtained by recording concentration changes on a linear time scale. Our work is concerned with the application of a time base (oscilloscope time sweep) which moves not linearly but logarithmically with the elapsed time of observation.

When using such a time base the units of the horizontal scale (x) are related to the elapsed time of the observation (t) as $x = \ln(t)$. Since in this type of recording the fast initial changes of the reaction are spread out by the rapidly moving time coordinate and the slower changes are compressed by the decelerating time base, practically the entire time-course of the reaction can be accommodated on a small record with an even apportionment of the abscissa to the time domains involved (nanoseconds, microseconds, seconds, etc.). In a logarithmic recording a first-order reaction represented by the function $y = \exp(-kt)$ will be transformed to a sigmoidal curve $y = \exp[-k\exp(x)]$ shown in Fig. 1. If the data are stored in a digitized form, the derivative of this curve can be readily computed, $y' = -k\exp[x - k\exp(x)]$, an asymmetrical bell-shaped curve whose peak is at $x = -\ln(k)$ (Fig. 2). Thus the numerical value of the rate constant can be simply read off from the peak position of the curve. The rate equation of a second-order reaction, $1/y = 1/y_0 + kt$, would appear as $y = y_0/[y_0 k \exp(x) + 1]$ in the x-y system. The derivative of this curve, $y' = -y_0^2 k \exp(x)/[y_0 k \exp(x) + 1]^2$, is a symmetrical bell-shaped curve whose peak is at $x = -\ln(y_0 k)$ (Fig. 3). Thus logarithmic recording allows an immediate visual distinction between the first- and second-order reactions and the establishment of the rate constants without additional replotting. (A zeroth order reaction, $y = kt$, will appear as an exponential curve, $y = k[\exp(x)]$, on the logarithmic record, easily distinguishable from the traces of the first- or second-order reactions.) Complex reactions will appear as multiple-step sigmoidal curves. The derivatives of these will have several peaks related to changes occurring at various time domains. If the complex reaction is the result of a minor perturbation of an equilibrium (as in a relaxation experiment), the peak positions of the curve will indicate the individual relaxation

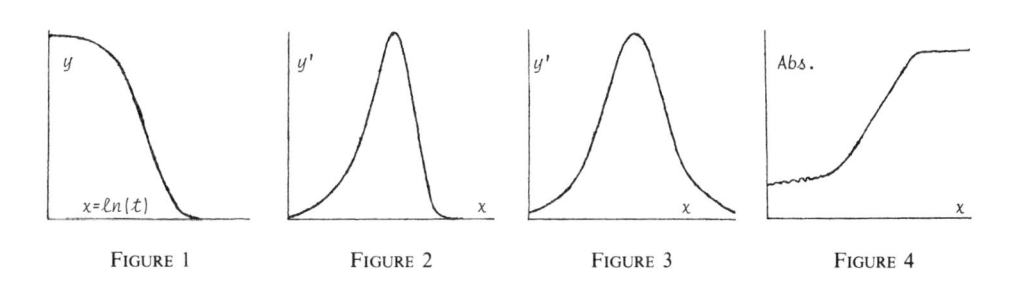

FIGURE 1   FIGURE 2   FIGURE 3   FIGURE 4

FIGURE 5 FIGURE 6 FIGURE 7 FIGURE 8

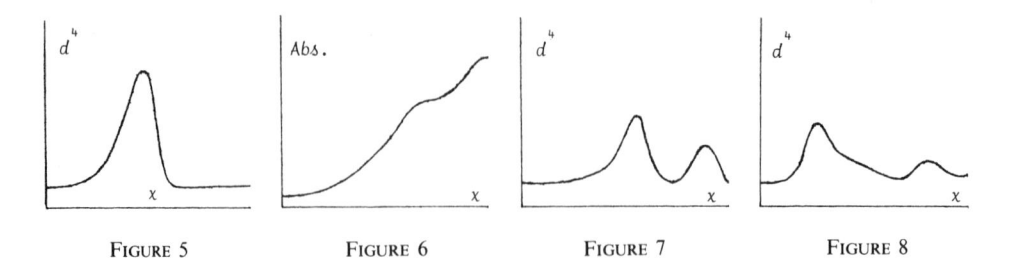

times involved, provided they are sufficiently separated in time (if not, a shifting of the peak positions will occur; however, this can be corrected by a simple iterative computer program).

Our instrumentation consisted of a Durrum stopped-flow apparatus (Durrum Instrument Corp., Sunnyvale, Calif.) and a Tektronix digital processing oscilloscope (DPO) coupled to a programmable desk calculator (TEK 31 Tektronix, Inc., Beaverton, Ore.). The logarithmic time base was generated by feeding an adjustable ramp function into a four-decade logarithmic amplifier (Solid State Electronics Corp., Sepulveda, Calif., 3076) and connecting its output to the external-volts output of the DPO's time base (7B70). The direct record of the reaction captured by the DPO had to be smoothed first in the calculator to eliminate the excessive noise which hindered the subsequent differentiation of the curve. Fig. 4 is the record of the pseudo-first-order reaction between Fe^{+++} and SCN^- before curve smoothing. Fig. 5 is the derivate of the smoothed version of the same curve raised to the fourth power. Figs. 6 and 7 are corresponding curves for the reaction between chymotrypsin and *p*-nitrophenyl acetate. Fig. 8 represents the amylose/iodine reaction. The latter figures indicate the presence of two distinct processes in these complex reactions, respectively.

Our work suggests that the logarithmic recording of kinetic results combined with the processing of the data in a DPO can be a very useful tool in the fast-reaction kinetic studies of biochemical reactions.

This work was supported by National Institutes of Health grant GM 20008.

FAR-ULTRAVIOLET STOPPED-FLOW CIRCULAR DICHROISM

JERRY LUCHINS, *Department of Biological Sciences, Columbia University, New York 10027 U. S. A.*

To follow directly the secondary structure changes involved in rapid protein-folding processes, we have designed a stopped-flow circular dichroism (CD) instrument capable of millisecond-range time resolution in the far-ultraviolet region. A stabilized Xe light source, piezo-optical birefringence modulator and phase-sensitive, heterodyning lock-in amplification techniques are utilized in conjunction with a special stopped-flow observation chamber to produce and detect the small, rapidly varying signals. We present initial results of the application of the instrument to a complicated protein subunit folding and assembly reaction system involving reorganization on the secondary, tertiary, and quaternary structural levels. These results demonstrate the instrument's utility for separating out the kinetics and, thus, for elucidating the interplay of the structural changes at those various levels.

To illustrate instrumental capabilities, the acid denaturation of ferrihemoglobin at three pH values and a pH-jump renaturation of denatured heme-free α globin are presented, as monitored by CD at 222-nm. The fastest of these reactions, exhibiting a 42-ms reaction half-time for a total CD change of 88 millidegrees (17.2×10^3-deg-cm^{2}dmol^{-1} mean residue ellipticity), was detected with a signal-to-noise ratio of 5 to 1.

At 222 nm, with an effective OD between 0.5 and 1.5 and a smoothing time constant of 4 ms, the noise level does not exceed approximately 17 millidegrees with current components and design. The noise is sharply reduced by increasing wavelength by even a few nanometers. Modifications to reduce the noise at 222 nm by an order of magnitude are outlined.

With a total accessible wavelength range from ~200 nm to ~800 nm, the instrument should prove a useful tool in kinetic investigation of a wide array of reactions involving altered optical activity and modification of chiral centers.

A SIMPLE SYSTEM FOR MIXING MISCIBLE ORGANIC SOLVENTS WITH WATER IN 10-20 ms FOR THE STUDY OF SUPEROXIDE CHEMISTRY BY STOPPED-FLOW METHODS

GREGORY J. MCCLUNE AND JAMES A. FEE, *Biophysics Research Division and Department of Biological Chemistry, The University of Michigan, Ann Arbor, Michigan 48109 U. S. A.*

We describe a simple device capable of mixing dimethyl sulfoxide (DMSO) and aqueous solutions for spectrophotometric observation of superoxide (O_2^-) chemistry with common stopped-flow methodology (1). Studies on superoxide and particularly on its dismutation catalyzed by so-called superoxide dismutases have been forced to rely on the expensive pulse radiolysis technique or poorly defined chemical or biochemical techniques to produce superoxide (2). Pulse radiolysis can generate superoxide concentrations up to 300 μM, while chemical methods produce steady-state concentrations 1,000 times lower. A method for dissolving high concentrations (>100 mM) of commercially available potassium superoxide in DMSO has been

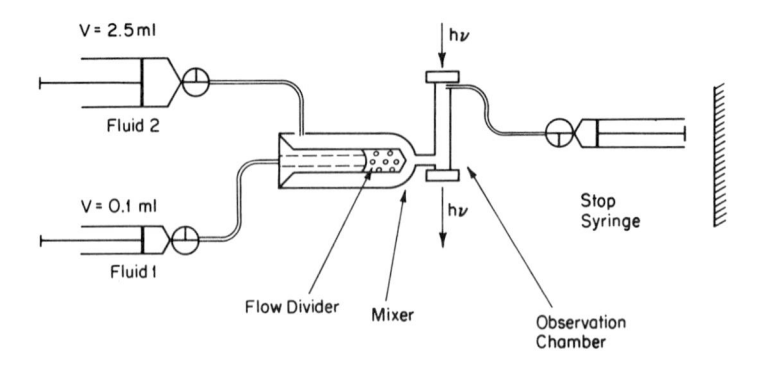

FIGURE 1 Schematic representation of stopped-flow apparatus with organic-aquo solution mixer (not drawn to scale).

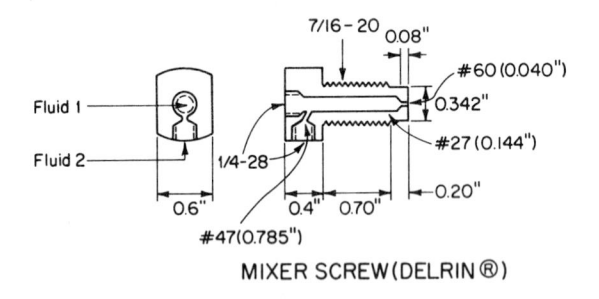

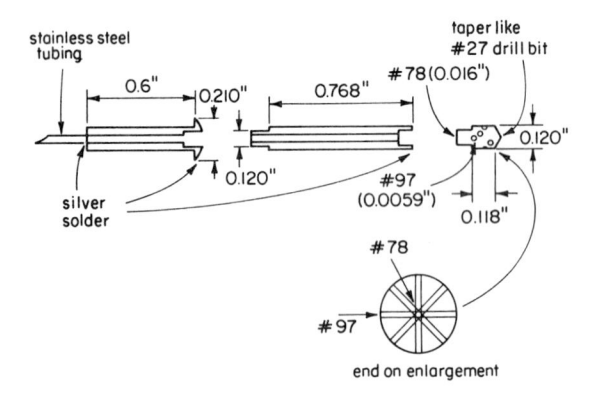

MIXER SCREW (DELRIN®)

MIXER NOZZLE (STAINLESS STEEL)

FIGURE 2 Details of mixer screw and flow divider. The flow divider is inserted into hole 1 of the mixer screw and secured as described in Ref. 8. Fluid 2 enters hole 2 by a stainless steel connector similar to that described by Ballou (8). Holes ~0.003″ can be placed in the stainless tip with a laser (Indiana Wire and Die Co., Fort Wayne, Ind.). The tip of the flow divider can also be constructed from Delrin and screwed into the central part of the flow divider. Holes ~0.006″ can be drilled through the Delrin tip.

MILLED MIXER (KEL-F)

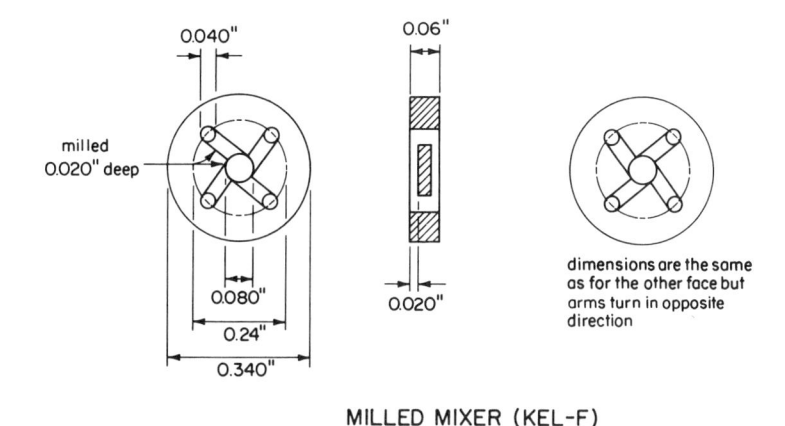

FIGURE 3 Detail of four-jet tangential mixer.

described (3, cf.4), making stopped-flow experiments with high superoxide concentrations feasible, provided mixing can be achieved in the time range of 10 ms.

Although DMSO and water are miscible, they have different refractive indices, a small volume contraction occurs on mixing, and there is a considerable heat of mixing (5). It is not easy to mix them rapidly. Our first attempts to mix equal volumes of DMSO and water using a variety of mixing chambers and different flow velocities revealed a propensity for these two liquids to layer on one another, forming a structure like a rolled blanket, which caused severe optical absorption artifacts and dissipated only over a period of minutes. Since useful kinetic data can usually be obtained only if the half-time for mixing, considered as a first-order process, is less than one-tenth of the half-time of the chemical reaction to be studied, it was clear that an alternative approach had to be taken.

The method we have developed to accomplish rapid, homogeneous mixing of DMSO and aqueous solutions involves (*a*) a ratio of aqueous to DMSO solution volume of at least 25:1, (*b*) a flow divider to separate the DMSO stream into eight or more parts and to force them into the aqueous solution stream normal to the net flow direction of the resultant mixture, (*c*) turbulence in the wake of the flow divider tip (6), and (*d*) a tangential mixer that further mixes the mixture produced by the flow divider. The mixing process is generally considered to involve the physical breakup of the solutions into small packets by turbulence. These packets then mix to homogeneity by diffusion (7). The flow divider, which aids in the physical mixing of the solutions, will ideally have a large number of very small holes. The flow divider is required to prevent the occurrence of the layering phenomenon mentioned above. Further break up of the DMSO solution packets is accomplished by the turbulence produced in the remainder of the mixer, in which final diffusional mixing is also occurring.

Since this mixing device is quite simple to build and may have a general applicability in chemistry, we include detailed plans for its construction. Fig. 1 shows an overall

Quartz flats

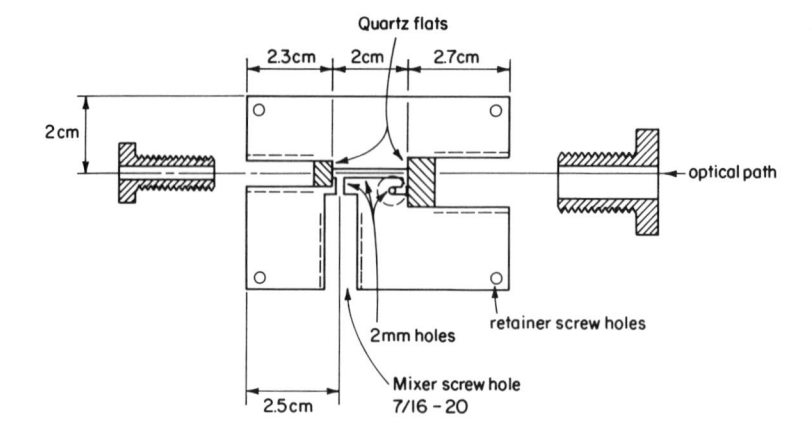

MIXING BLOCK (BLACK KEL-F 7x5x3cm)

FIGURE 4 Detail of mixing block. Fluid exit part is out of the plane and is directed to a stop syringe by stainless steel connectors similar to the plastic devices described by Ballou (8).

schematic, indicating the liquid flow pattern from left to right. Fluid 1, from the small syringe, is forced into the flow divider (Fig. 2). Fluid 2 enters a chamber around the flow divider, where it encounters fluid 1 in the form of a group of fine jets emanating from the tip of the flow divider. There follow one or two four-jet tangential mixers only 0.06" in thickness (Fig. 3), separated by a thin spacer having a 1-mm hole at its center. In each of these the heterogeneous fluid is divided into four streams on

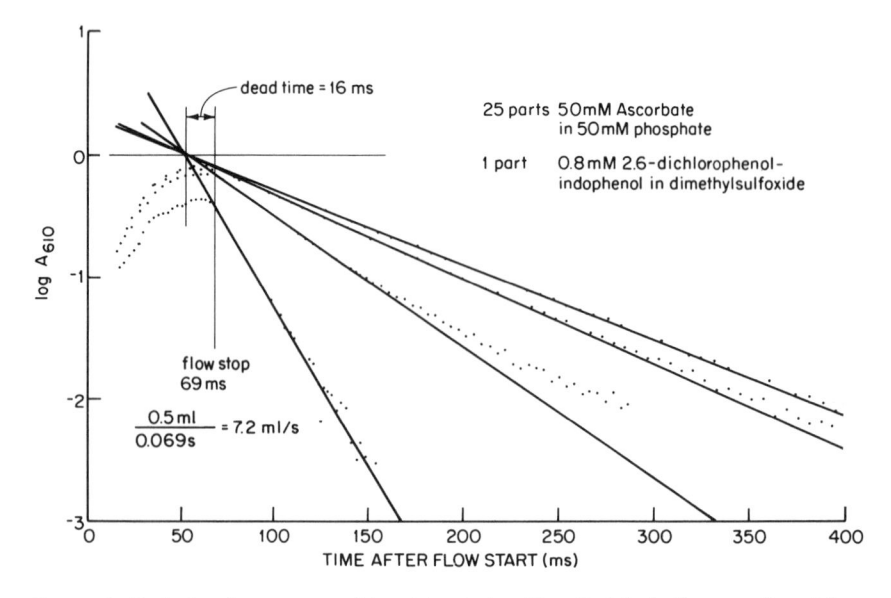

FIGURE 5 Typical performance test of the mixing device. The pH of the buffer was adjusted from 6.4 through 8.4 to obtain the different rates of reaction.

one side and re-mixed on the other side. Finally the mixed solution enters the observation chamber, indicated by the passage of light, and into the stop-syringe. Fig. 4 shows the plastic block in which the various parts are assembled.

Methods for testing stopped-flow mixers have been described (1, 8). One involves observation of optical absorption changes of acid-base indicators upon changing the pH in the mixing process. The protonation and deprotonation reaction of acid-base indicators are very fast and therefore, any absorbance change observed after flow stop is due to the kinetics of mixing and not of the chemical reaction (7). In tests of our mixing system, we observe less than 2% of the expected absorbance change to occur after flow stop. Another test method involves the extrapolation of absorbance versus time data from an observed first-order reaction back to the known initial absorbance at the point of mixing to obtain a measure of the dead time of the mixer. For our mixer this varies from 10 to 20 ms, though it may be reduced by increasing the flow rate of the reacting solutions above that used for this test (Fig. 5), which was about 7 ml/s (i.e., a maximal linear flow velocity of about 9 m/s).

This mixing method has been applied to the study of spontaneous and protein-catalyzed superoxide dismutation (9, 10) and the reactions of O_2^- with Fe (II) and Fe(III) chelated by ethylenediaminetetraacetic acid (EDTA) (11).

This work was supported in part by U. S. Public Health Service grant 21519 and a National Science Foundation Graduate Fellowship to G. J. McC.

REFERENCES

1. GIBSON, Q. H. 1969. *Methods Enzymol.* Vol. 16.
2. BORS, W., M. SARAN, E. LENGFELDER, R. SPOTTL, and C. MICHEL. 1974. *Curr. Top. Radiat. Res.* **9**: 247.
3. VALENTINE, J. S., and A. B. CURTIS. 1975. *J. Am. Chem. Soc.* **97**:224.
4. MARKLUND, S. 1976. *J. Biol. Chem.* **251**:7504.
5. FOX, M. F. 1975. *J. Chem. Soc. Faraday Trans. I.* **21**:1407.
6. BERGER, R. L., B. BALKO, and H. F. CHAPMAN. 1968. *Rev. Sci. Instrum.* **39**:493–498.
7. BRODKEY, R. S., editor. 1975. Turbulence in Mixing Operations. Academic Press, Inc., New York.
8. BALLOU, D. 1971. Ph. D. Dissertation, University of Michigan, University Microfilms, Ann Arbor, Mich.
9. McCLUNE, G. J., and J. A. FEE. 1976. *Fed. Eur. Biochem. Soc. (FEBS) Lett.* **67**:294.
10. FEE, J. A., and G. J. McCLUNE. 1978. Mechanisms of Oxidizing enzymes. Proceedings of an International Symposium on the Mechanisms of Ozidizing Enzymes, La Paz, Baja, Mexico, Dec. 5–7, 1977. T. P. Singer and P. N. Ondarza, editors. Elsevier-North Holland, Amsterdam. 273.
11. McCLUNE, G. J., J. A. FEE, G. A. M. McCLUSKEY, and J. T. GROVES. 1977. *J. Am. Chem. Soc.* **99**: 5229.

FLUID MECHANICS OF RAPID MIXING

JEROME H. MILGRAM, *Massachusetts Institute of Technology, Cambridge, Massachusetts 02139 U. S. A.*

Theoretical and experimental studies and analysis of rapid mixing processes will result in an increased understanding of the mechanics that may lead to the development of improved devices. For example, suppose the order of magnitude of the mixing time that can be achieved is examined. The diffusion length l_d is given by $l_d \sim \sqrt{Dt}$ where D is the diffusion coefficient and t is the mixing time. The length scale, η, of the microscale eddies is given by $\eta \sim (\nu^3/\epsilon)^{1/4}$, where ν is the kinematic viscosity of the fluid and ϵ is the rate of dissipation of the energy of the turbulence (per unit time and mass). An efficient mixing apparatus will dissipate a substantial fraction of the turbulence energy in the mixing time, so $\epsilon \sim E/t$, where E is the initial energy of the turbulence. Sufficient mixing requires that the order of magnitude of η be at least as small as that of l_d, so the minimum order of magnitude of the mixing time is found by equating η to l_d: $t \sim \nu^3/D^2E$. We can write $E = u^2$, where u is the characteristic speed of the turbulence, so that $t \sim \nu^3/D^2u^2$. Now suppose we choose typical values of $D = 10^{-5}$ cm^2/s and $\nu = 10^{-2}$ cm^2/s. Then, $t \sim 10^4/u^2$ cm^2/s. Thus, for a characteristic turbulence flow speed, u, of 10 m/s, $t \simeq 10^{-2}$ s, and for $u = 100$ m/s, $t \simeq 10^{-4}$ s.

This simple analysis yields an important relationship between the order of magnitude of the mixing time and the flow speed. It can be used to compare the mixing times for a given apparatus run at different speeds or with materials of different kinematic viscosities or diffusion coefficients. Similar analyses can be used for obtaining limits on orders of magnitude of other important factors, such as cavitation and pressure forces. These results are helpful in relating gross design features to performance. However, such analyses do not provide a detailed comparison of various mixer types or yield much information on design details. Those matters are best studied with experiments. Most flow apparatuses are too small to allow convenient use of hot film anemometry, the most useful technique for measuring turbulence in liquids. However, it is possible to properly scale both Reynolds and cavitation numbers in scaled-up apparatuses and make the turbulence measurements in them.

For our experiments, a Johnson Research Foundation Model B mixing apparatus design is scaled up by a factor of six, giving a cross-sectional size for the observation section of 6 cm × 6 cm. To scale the Reynolds number, the flow speed is reduced by a factor of six. Then to scale the cavitation number simultaneously, the pressure (referenced to the vapor pressure) is reduced by a factor of 36. These scaled parameters result in flow speeds and pressures easily attained in the scaled-up apparatus. Most of the measurements of the flow field, including mixing effects, are made with hot film anemometry. However, the device is also suitable for measurements by laser anemometry, and these are being contemplated for those functions of the turbulence that can be measured by single-component anemometry. The degree of mixing is being measured by colorimetry on a pH-sensitive indicator.

The author of this paper did not attend the actual Discussion. The present text was submitted and circulated to participants before the meeting.

KINETICS OF ASSOCIATION AND DISSOCIATION PHENOMENA IN HUMAN HEMOGLOBIN STUDIED IN A LASER LIGHT-SCATTERING STOPPED-FLOW DEVICE

LAWRENCE J. PARKHURST AND DUANE P. FLAMIG, *Department of Chemistry, University of Nebraska, Lincoln, Nebraska 68588 U. S. A.*

One of the central problems in hemoglobin studies is that of assigning kinetic properties to known aggregated states of the protein. To attack this problem, as well as study the kinetics of association, dissociation, and denaturation, we have constructed a dual-beam stopped-flow device that employs laser (Ar-ion, He-Cd, He-Ne) excitation for kinetic light-scattering measurements. Absorbance and fluorescence changes after rapid pH and concentration jumps can also be measured in the same apparatus. The apparatus has also been used in the flow-flash mode where flash-photolysis measurements at known times after flow permitted a determination of the rate constant for the dissociation of tetrameric hemoglobin. Light-scattering changes as a function of protein concentration after pH jumps can be treated as relaxation phenomena and allow us to determine $K_{4,2}$, the tetramer-dimer equilibrium constant, over the range pH 7–10.9. These determinations are not only orders of magnitude more rapid than those by ultracentrifugation, but are equally or much more precise. We have found that the tetramer to dimer reaction, pH 10.3–11.6, in HbCO is remarkable in that the rate constant increases 100-fold over this interval and reaches a plateau at ca. pH 11.4, well before the onset of denaturation (1). HbCO treated with carboxypeptidase-A, (CPA), which removes βHis(146) and βTyr(145), showed greatly increased rate constants over the interval pH 10.3–11.5 compared to native HbCO, whereas rates for hemoglobin treated with carboxypeptidase-B, which removes αArg(141), were identical. Fig. 1 shows clear evidence of the importance of the β chain COOH-terminus in maintaining the integrity of the tetramer for liganded hemoglobin. The effects of various SH reagents on the tetramer-to-dimer process were studied. Some, such as PMB, greatly increased the rate, whereas N-ethyl-maleimide gave a modified hemoglobin (NES) with a decreased dissociation rate (Fig. 1). From the light-scattering amplitudes, both the enzymatically and SH-treated hemoglobins were shown to be more dissociated at pH 7 than native protein. A detailed analysis of the pH kinetic profiles for the native and modified hemoglobins shows that a simple charged-sphere model cannot account for the phenomenon. Various two-state models that assume rapid protonic and conformational equilibria between two forms of the tetramer cannot explain the results. A minimum mechanism for the process must include a pH-dependent conformational change occurring on the same time scale as the tetramer-dimer dissociation. Similar dissociation studies for deoxy-hemoglobin reveal an exceedingly complex problem. First, our kinetic light-scattering studies show that the protein is much more disso-

FIGURE 1

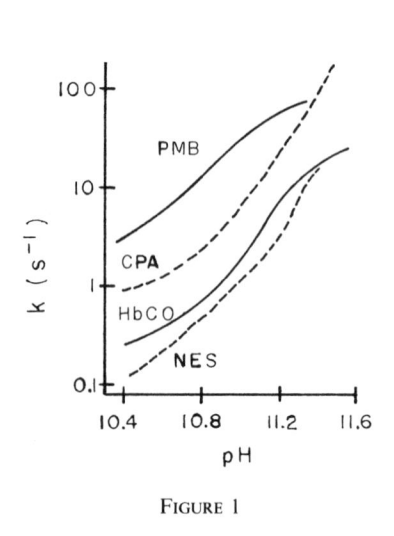

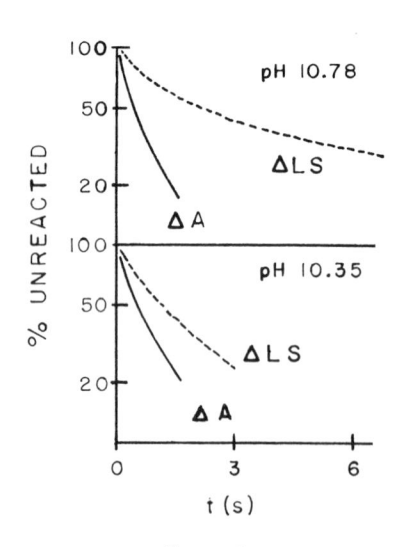

FIGURE 2

ciated at pH values greater than 10 than had been previously assumed. The reaction is at least biphasic in light-scattering, the rate does not increase smoothly with pH, and for some pH values, the rate appears to decrease with increasing total protein concentration. Furthermore, the absorbance changes (ΔA) after a rapid increase in pH precede the light-scattering changes (ΔLS) and display at least triphasic kinetics, with a very rapid change contributing about 50% to the total amplitude (rapid phase not shown in Fig. 2). In the case of HbCO, absorbance changes also occur, but at 287, 408, and 420 nm, they follow rather than precede the light-scattering changes and must therefore correspond to conformational changes in the dimers. Unlike the behavior of HbCO, a description of Hb requires at least five species. Other investigators following only absorbance changes have been entirely misled as to the time-course and the complexity of the process. Fitting the combined absorbance and light-scattering results by various models required the development of minimization procedures that used numerical integration of both the differential equations for the chemical species and the partial differential equations for the rate constants. Eigenvalues and eigenvectors of the relaxation matrix were useful in providing initial parameter values for the minimization routine.

This work was supported by National Institutes of Health grant HL 15284-06 and the Research Council of the University of Nebraska.

REFERENCES

1. FLAMIG, D. P., and L. J. PARKHURST. 1977. *Proc. Natl. Acad. Sci. U. S. A.* **74:**3814.

EXTENDING THE WAVELENGTH RANGE OF FUNDAMENTAL LASER SOURCES

R. A. POPPER, T. NOWICKI, W. RUDERMAN, AND J. RAGAZZO,
Interactive Radiation, Inc., Northvale, New Jersey 07647 U. S. A.

The ability to extend the wavelength range of fundamental laser sources provides the opportunity to match a light source to a specific chemical or biological interaction wavelength. The most commonly used technique to alter laser wavelengths involves mixing two optical waves in a nonlinear crystal to produce the sum frequency. Energy is efficiently transferred from the two input waves to the output wave only when a proper phase-matching condition is maintained. Special cases arise when the two input waves have the same optical frequency (i.e., are components of the same wave) or have in themselves a harmonic relationship with a common fundamental source. The first instance results in an output wavelength equal to exactly one-half that of the fundamental input and is called second harmonic generation (SHG). In the second instance, combining the fundamental and second harmonic of a laser will result in third harmonic generation (THG). THG is a two-stage process, a doubling stage followed by a mixing stage.

Phase matching is accomplished by utilizing the birefringence of the nonlinear crystal to overcome the refractive index dispersion from one wavelength to another. The indices are precisely tuned to the values desired for a specific set of wavelengths either by altering the optical propagation direction with respect to the crystal optic axis (angle matching) or by changing the temperature of the crystal and utilizing the subsequent index change (temperature matching). Whichever method is used, the physical properties of individual crystal species impose a range limitation on phase matching that requires that several crystals be used for broad wavelength coverage.

Laser sources used in the study of fast reactions must be characterized by short pulse widths. In addition, to achieve maximum efficiency in harmonic generating and mixing applications, the laser should have a narrow line width, high peak power, and low beam divergence. Spectral brightness (W/cm^2 sterad Å) is a single parameter relating the three variables. For efficient harmonic generation and mixing, the laser must exhibit a high spectral brightness. Among the more common and flexible lasers that exhibit high spectral brightness are Nd:YAG, ruby, N$_2$-pumped dye lasers and Nd:YAG-pumped dye lasers. Techniques used to generate short-pulse outputs include Q-switching and mode-locking.

A wavelength region of high interest to biologists, chemists, physicists, and spectroscopists is the ultraviolet (UV), since this is a region of high interaction of light with molecules. Several techniques are employed to extend the wavelength range of tunable or fixed-frequency laser sources to the ultraviolet. SHG is the most direct approach. In those cases where the desired wavelength regions fall within the umbrella of 90° phase-matching of KDP homologues (see figure), an extremely efficient tem-

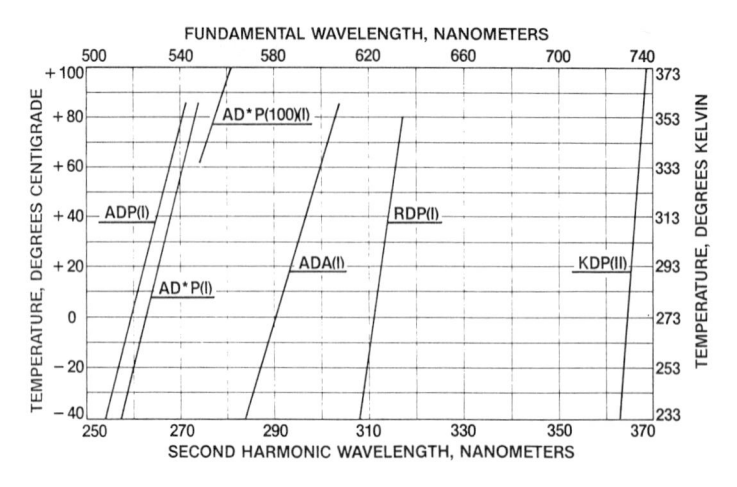

FIGURE 1

perature-tuned SHG approach may be used. Where this is not possible less efficient critical phase matching (angle-tuned) SHG techniques are available.

Mixing techniques frequently offer a great improvement in efficiency over conventional SHG approaches. For example, using a temperature-tuned ammonium dihydrogen arsenate crystal (ADA) to mix the output of a dye laser pumped by the second harmonic of a Nd:YAG laser with that same second harmonic has resulted in highly efficient conversion from 285.1 nm to 302.8 nm. The nonlinear crystal potassium pentaborate (KPB) makes it possible to obtain intense tunable UV radiation down to 196.6 nm by mixing the output of the fourth harmonic (266 nm) of a Nd:YAG laser and a near-infrared dye laser pumped by the second harmonic (532 nm) of the same YAG laser. Five ns pulses with peak powers of 40 kW at 10 pulse/s at 196.6 nm have been achieved.

Parametric conversion (difference-frequency) techniques are used for generating continuously variable wavelengths longer than those of the pump pulses. Continuously tunable radiation from 660 to 2,750 nm can be obtained by parametric conversion of 532 nm radiation (the second harmonic of a mode-locked Nd:YAG laser) in lithium iodate. Pulse widths as short as 2–3 ps with pulse energies of 50 μJ are achievable. In addition, the fourth harmonic (266 nm) of Q-switched or mode-locked Nd:YAG lasers can pump an ammonium dihydrogen phosphate (ADP) parametric amplifier to produce nanosecond or picosecond pulses at visible wavelengths. A synchronous mode-locked tunable dye laser using a frequency-doubled mode-locked Nd-YAG laser pump can be sum and difference mixed with the 1,064-nm and 532-nm pump pulses in lithium iodate, KDP, and ADP. This results in tunable picosecond pulse generation in the ultraviolet from 270 to 432 nm and in the infrared from 1,130 to 5,600 nm.

KINETIC STUDIES OF FAST REACTIONS AT WATER-MICELLE INTERFACES

MICHAEL A. J. RODGERS, DAVID C. FOYT, AND
MARIA F. DA SILVA E WHEELER, *Center for Fast Kinetics Research, University of Texas at Austin, Texas 78712 U. S. A.*

Micellar dispersions have frequently been used to model the changes in diffusional and reactivity properties of substrates associated with biological supramolecular assemblies such as cell membranes. Our approach has been to use electron beam and laser generation of short-lived transients with detection by time-resolved spectroscopy to observe the effects of micellar phases on simple physicochemical processes such as collisional quenching and electron transfer.

When molecules of pyrene (or other aromatic fluorescent probes) solubilized by the lipid region of surfactant micelles are excited to the S_1 state by an ultraviolet laser pulse, the decay rate of the fluorescence is critically dependent on the nature and concentration of counter-ions in the diffuse double layer. For some ions (e.g., Cs^+, Ag^+) the rate law for decay is exponential; for other ions (Cu^{++}, dimethyl viologen) a complex decay is observed that becomes exponential at long times. The initial rate of the early component is dependent on the quencher concentration; the exponential part is concentration independent.

A model that leads to an understanding of this phenomenon has been developed, based on quenching interactions occurring at interfacial collisions. Using the notion of quencher ions distributing themselves among micelle double layers according to a Poisson function, a mathematical analysis leads to the expression $N^*(t) = N^*(0) \cdot \exp\{<r>e^{-t/\tau_1} - <r> - t/\tau_0\}$, where $N^*(t)$ and $N^*(0)$ refer to the concentration of pyrene S_1 states at times t and zero after the excitation pulse, $<r>$ is the mean value of the occupancy index, and τ_0 and τ_1 are the S_1 natural lifetimes for pyrene in micelles associated with zero and one quencher, respectively. This expression describes the observed complex cases adequately. The simple exponential rates can be understood in terms of weaker micelle-ion binding, causing a nonstatic ion population about micelles that leads to a time-averaged decay rate being observed.

This variability in micelle-ion association strength is supported by electron pulse radiolysis measurements of the rate of reaction of hydrated electrons with e.g., dimethyl viologen cations. When the concentration of surfactant molecules is raised above the critical micelle concentration, the rate constant for electron capture falls, consistent with the idea that such cations are associated with micelles in a quasistatic distribution, at least for several microseconds. Details of the concentration dependence lead to consideration of the primary salt effect associated with micellar reactions.

Dr. da Silva e Wheeler's present address is: Department of Physics, Manchester University, Manchester M13 9PL, England.

A MACROSCOPIC APPROACH TO FLUCTUATION ANALYSIS

Solution of the Phase Problem of Relaxation Spectrometry

W. O. Romine, C. Watkins, S. T. Christian, and M. C. Goodall,
Department of Chemistry and School of Medicine, University of Alabama in Birmingham, University Station, Birmingham, Alabama 35294 U. S. A.

One of us (1) has demonstrated that by multivariate spectral analysis of the statistical mechanical noise and fluctuations about equilibrium, one may obtain all relevant rate parameters (including Onsager coefficients) from the power spectral matrix and its inverse at $f = 0$. The proof is based on the fluctuation-dissipation theorem of Callen and co-workers (2, 3) and in canonical systems (systems of chemical interest), the fundamental equation is

$$\Omega \equiv L\Phi = 4 <\delta\vec{\alpha}\,\delta\vec{\alpha}^*> \{G(0)\}^{-1} = 4 \int_0^\infty G(f)\,df\{G(0)\}^{-1}, \quad (1)$$

where Ω is the reduced matrix of rate coefficients, defined in terms of an exactly determined set of progress variables (4), L is the Onsager flow matrix, Φ is the matrix of second partials of free energy with respect to the extensive thermochemical parameters, $G(f)$ is the spectral matrix, and $<\delta\vec{\alpha}\,\delta\vec{\alpha}^*>$ is the matrix of second moments of the extensive parameters.

The above equation is also macroscopically applicable to a relaxation process of Langevin form

$$\frac{d\delta\vec{\alpha}}{dt} + L\Phi\delta\vec{\alpha} = \vec{\chi}(t); \quad <\vec{\chi}(t)> = 0 \quad (2)$$

where $\vec{\chi}(t)$ is a random force induced by a Gaussian white noise pertubation of some intensive thermochemical parameter (T, P, E, M, etc.). It can be proven that solution of the spectral equation then reduces to Eq. 1, if the pertubations are sufficiently small as to allow the approximation of linearity.

Experimentally, this allows one to solve, using simple algebraic matrix methods, all rate parameters within a system. Since time series averaging in the frequency domain is employed, one can use Wiener and subsequent stochastic analysis theory (5, 6) to draw kinetic noise from extraneous noise through the use of spectral averaging. In fact, there is (in principle) no limitation (to the Heisenberg limit) to the kinetic information that can be extracted. Thus, it might be possible even to measure the intrinsic statistical mechanical fluctuations in a variety of systems. Some possible approaches are discussed.

REFERENCES

1. ROMINE, W. O. 1976. *J. Chem. Phys.* **64**:2350–2358.
2. CALLEN, H. B., and T. A. WELTON. 1951. *Phys. Rev.* **83**:34–40.
3. CALLEN, H. B., and R. F. GREEN. 1952. *Phys. Rev.* **86**:702–710.
4. KIRKWOOD, J., and I. OPPENHEIM. 1961. Chemical Thermodynamics. McGraw-Hill Book Company, New York.
5. WIENER, N. 1949. Extrapolation, Interpolation, and Smoothing of Stationary Time Series. The M.I.T. Press, Cambridge, Mass.
6. JENKINS, G., and D. WATTS. 1968. Spectral Analysis and Its Applications. Holden-Day, Inc., San Francisco, Calif.

AN INEXPENSIVE MICROCOMPUTER-BASED STOPPED-FLOW DATA ACQUISITION SYSTEM

D. G. TAYLOR, J. N. DEMAS, R. P. TAYLOR, AND M. J. ZENKOWICH,
Departments of Chemistry and Biochemistry, University of Virginia, Charlottesville, Virginia 22901 U. S. A.

ABSTRACT A low-cost (<$2,500) microcomputer-controlled data acquisition system for use with a stopped-flow instrument is described. Data acquisition, reduction, signal averaging, kinetic modeling, and plotting are performed under software control. Applications to biological and inorganic systems are presented.

INTRODUCTION

Kinetics studies typically generate enormous amounts of raw data that traditionally have been laboriously hand-digitized and reduced. The described microcomputer-controlled data acquisition system improves accuracy, eliminates the labor-intensive steps, and permits data acquisition, reduction, and plotting of final results in seconds without the operator ever having to handle raw data.

METHODS

Instrumentation

The stopped flow instrument has been described previously (1). In all experiments reported concentrations are before mixing.

The computer system consists of an MITS Altair 8800 microcomputer (Altair Corp., Chicago, Ill.) using an Intel 8080A microprocessor (Intel Corp., Santa Clara, Calif.), 9,216 bytes of semiconductor random access memory, 2,048 bytes of permanent read-only memory (PROM), and an ASR33 Teletype interface (Processor Technology 3P+S serial card). The data acquisition control and clock circuitry plugs directly into the Altair bus; it uses ~50 7400 type small and medium-scale integration integrated circuits and was wire-wrap constructed on a Vector 8800V prototyping card (Vector General, Inc., Woodland Hills, Calif.). The separate analog card contains a Hybrid Systems ADC-550-10E-G 10 bit, 27-μs analog-to-digital converter (ADC) (Hybrid Systems Corp., Burlington, Mass.) and two 371I-10 10-bit digital-to-analog convertors (DAC's), which communicate with the control card through multiconductor cables. The real-time clock controls the conversion rate of the ADC and is selectable in a 1, 2, 4, 5, 8, 10, 16 sequence from 40 μs to 160 s per data point.

Analog signals for the ADC and the DAC's are run through coaxial cables to an external control center. The analog input voltage signal is buffered with operational amplifiers with selectable normal or inverted levels (1 or 2 gain), and variable DC offset ranges (± 10, 1, and 0.1 V). A Datel IC-SHM-1 sample-and-hold amplifier (Datel, Inc., Holly Hill, Fla.) between the buffer amplifiers and ADC eliminates digitization errors for rapidly changing signals. An LM311 comparator trigger circuit provides normal oscilloscope triggering functions (± 10 V level control, rising or falling edge, internal or external triggering, and an light-emitting diode status indicator). The DAC outputs are converted to 0–10.23 V to drive a display oscilloscope and x-y plotter.

SOFTWARE

The PROM stores machine language utility (768 bytes) and data acquisition (256 bytes) routines. PROM permits immediate loading of paper tape programs including the MITS 8 K BASIC Interpreter upon power up. The acquisition software, accessed through a BASIC subroutine call, permits selection of the real-time clock interval, execution of the experiment, and examination and plotting of stored data or calculated results.

The acquisition software instructs the interface control card from commands entered on the Teletype and the computer's switch register. After a keyboard-issued arming command and manual operation of the stopped-flow instrument, flow stoppage triggers collection of 101 "decay" points and 101 "infinity" points separated by a 200 time interval delay. Other collection schemes are readily implemented. After acquisition, stored data may be visually evaluated for data quality, and the best region to fit selected from the computer generated oscilloscope display or an x-y plot. Upon return to BASIC, on-line data reduction is performed by least-squares programs in BASIC. As configured, approximately 100–120 lines of

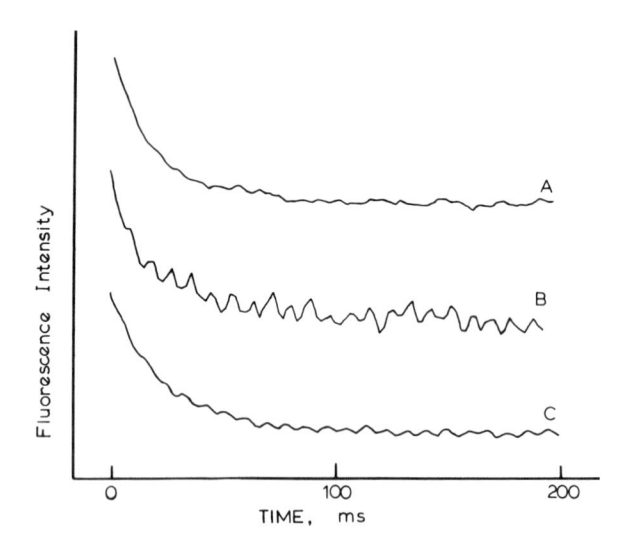

FIGURE 1 A. Tryptophan fluorescence monitored on the mixing of equal volumes of IA-BSA (ca. 1.5×10^{-5} M) in 0.04 M NaClO$_4$ at pH 3.5 and 0.05 M phosphate (pH 8.0). Increasing fluorescence intensity is down. On-line computer-calculated refolding rate constant is 28.8 ± 0.4 s^{-1} ($r = 0.97$). B. The same as A with the initial protein concentration reduced to 5×10^{-6} M to increase the noise level. Refolding rate constant is 24.8 ± 0.6 s^{-1} ($r = 0.85$). C. Results of recording and on-line ensemble averaging 16 transients of B. Refolding rate constant is 32.3 ± 0.5 s^{-1} ($r = 0.98$).

BASIC programming are accommodated. A first-order rate constant error estimates are typically computed in <5 s for 101 data points.

RESULTS

The kinetics of iodoacetamide-blocked bovine serum albumin (IA-BSA) refolding have recently been characterized by stopped-flow monitoring of the protein's tryptophan fluorescence after a rapid pH jump (1). We reexamined this system using our microcomputer-interfaced stopped-flow instrument.

Fig. 1 shows computer-acquired and -plotted data taken with good signal-to-noise ratio (S/N) (1A), poor S/N (1B), and with computer-controlled ensemble averaging of the noisy data (1C). The computed constants are in excellent agreement with the previously reported value of 32 s^{-1} (1).

We also examined the reactions of 0.02 M NaHCO$_3$ with 0.01 N HCl (3.4 × 10^{-5} M bromophenol blue indicator) and of 10^{-3} M Fe(NO$_3$)$_3$ in 0.1 M H$_2$SO$_4$ with 1.0 M KSCN. The NaHCO$_3$ reaction yielded a rate constant of 21.1 s^{-1} (24°C), which compares well with the literature value of 19.0 ± 2.0 s^{-1} (24.1°C) (2).

For the Fe^{3+} system, four replicate runs (24°C) yielded a lifetime of 57.5 ± 0.5 ms, which compares well with an estimated value of 45 ms calculated for a slightly different ionic strength (3). The total time to acquire all four transients, display and verify their quality on the oscilloscope, compute the rate constants, and plot the observed and calculated best-fit curves on an *x-y* recorder was under 20 min.

We gratefully acknowledge support by the National Science Foundation (MPS 74-17916 and PCM 75-22703).

REFERENCES

1. TAYLOR, R. P., V. CHAU, M. ZENKOWICH, and L. LEAKE. 1977. Stopped-flow studies of the NEF transition in serum albumin. *Biophys. Chem.* In press.
2. BERGER, R. L., and L. C. STODDART. 1965. Combined calorimeter and spectrophotometer for observing biological reactions. *Rev. Sci. Instrum.* 36:78–84.
3. BELOW, J. F., JR., R. E. CONNICK, and C. P. COPPEL. 1958. Kinetics of the formation of the ferric thiocyanate complex. *J. Am. Chem. Soc.* 81:2961–2967.

ON THE INTEGRATION OF COUPLED FIRST-ORDER RATE EQUATIONS

DARWIN THUSIUS, *Instituto de Biofísica, Centro de Ciências da Saúde, Rio de Janeiro, Brasil and Laboratoire d'Enzymologie Physico-chimique et Moléculaire, Orsay, France*

Differential rate equations for many chemical reaction mechanisms are inherently linear or can be linearized through judicious choice of experimental conditions.

The author of this paper did not attend the actual Discussion. The present text was submitted and circulated to participants before the meeting.

Examples in biochemistry are: ligand binding to a macromolecule followed by a series of conformational changes under pseudo-first-order conditions; reversible denaturation of a protein or nucleic acid induced by a rapid change in pH; "single turnover" enzyme catalysis; pre-steady-state enzyme catalysis. In these cases the reaction mechanism can usually be written as a system of coupled first- or pseudo-first-order reactions involving n components related through a single mass conservation relationship. The time dependence of the component concentrations is given by

$$X_i = \overline{X}_i + \sum_{k=1}^{n-1} C_{ik} \exp(-\lambda_k t), \quad i = 1, 2 \ldots n, \tag{1}$$

where $\overline{X}_i$ denotes final equilibrium concentrations, the λ_k are macroscopic rate constants, and the C_{ik} are integration constants which depend on initial concentrations (X_i^0). Eq. 1 holds for sequential, branched, and cyclic reaction mechanisms.

The n differential rate equations are defined by the "coefficient matrix" $\mathbf{A}$ (1). It proves convenient to reduce $\mathbf{A}$ to an $(n-1) \times (n-1)$ matrix $\mathbf{A}'$ by eliminating one of the rate equations with mass conservation. The macroscopic rate constants are then the $n-1$ roots of the secular equation,

$$|\mathbf{A}' - \lambda \mathbf{I}| = 0 \tag{2}$$

The present communication describes a computational method that considerably simplifies the derivation of the analytical expressions of Eq. 1. The procedure involves: construction of the transpose of the $n \times n$ secular determinant, $|\mathbf{A}^T - \lambda \mathbf{I}|$, by examination of the kinetic model; reduction to $|\mathbf{A}' - \lambda \mathbf{I}|$ by examination of $|\mathbf{A}^T - \lambda \mathbf{I}|$; derivation of the $\overline{X}_i$ and C_{ik} by systematic evaluation of subdeterminants of $|\mathbf{A}^T - \lambda \mathbf{I}|$.

The first step in the calculations is to write the kinetic mechanism as a set of coupled first-order reactions, with the elementary steps and microscopic rate constants defined as

$$X_i \underset{k_{ji}}{\overset{k_{ij}}{\rightleftarrows}} X_j \quad i \neq j; i, j = 1, 2 \ldots n. \tag{3}$$

The species to be eliminated by mass conservation is written X_n; otherwise the numbering of components is arbitrary. Pseudo-first-order reactions are denoted by primed rate constants, k'_{ij}. Steps resulting in formation of a product other than an X_i must be irreversible, or the product must be present in sufficient excess to allow the reverse reaction to be written as a pseudo-first-order process.

The determinant $|\mathbf{A}^T - \lambda \mathbf{I}|$ is constructed in essentially two steps: (a) the typical off-diagonal element in the ith row and jth column is written as $-k_{ij}$; (b) the diagonal element of row i is formed by taking the sum of all off-diagonal elements of row i, multiplying by -1, and subtracting λ. Then the reduced secular determinant is formed by: (a) subtracting each element in the last row of $|\mathbf{A}^T - \lambda \mathbf{I}|$ from all remaining elements of the same column; (b) striking out the last column and last row of the new determinant, followed by transposing to $|\mathbf{A}' - \lambda \mathbf{I}|$.

Finally, the constants $\bar{X}_i$ and C_{ik} are derived from initial concentrations and elements of determinant $|\mathbf{A}^T - \lambda \mathbf{I}|$. It can be shown that[1]

$$\bar{X}_i = B_i^{(0)} \Big/ \prod_l \lambda_l; \quad C_{ik} = \frac{B_i^{(k)}}{\lambda_k \prod_{l \neq k} (\lambda_k - \lambda_l)}, \tag{4}$$

where the index l runs from 1 to $n - 1$. The coefficients $B_i^{(0)}, B_i^{(l)}, \ldots B_i^{(n-1)}$ are explicit functions of specific rate constants, initial concentrations and the roots of Eq. 2. The superscript ($k = 0, 1, 2 \ldots n - 1$) indicates that $B_i^{(k)}$ is evaluated using λ_k. By definition we set $\lambda_0 = 0$. Conveniently, the $B_i^{(k)}$ for a given component X_i are formally identical, differing only in the definition of λ_k:

$$B_i^{(k)} = \sum_{j=1}^{n} (-1)^{1+j} \Delta_{ij}^{(k)} x_j^0, \tag{5}$$

where $\Delta_{ij}^{(k)}$ is the subdeterminant obtained by striking out the ith row and jth column of $|\mathbf{A}^T - \lambda_k^1 \mathbf{I}|$. Often, only one X_j^0 will have a finite value at the moment the reaction is initiated. The computational labor in deriving expressions for $\bar{X}_i$ and C_{ik} is then reduced to evaluating a single subdeterminant for each component.

The time dependence of products that do not feed back into the sequence of formal first-order reactions is calculated by direct integration of the precursor concentration. If product P is formed irreversibly from component X_i, we may use the following general relation:

$$P = \frac{(k_{ij} B_i^{(0)}) t}{\prod_l \lambda_l} + \sum_{k=1}^{n-1} \frac{k_{ij} B_i^{(k)} (1 - \exp(-\lambda_k t))}{\lambda_k^2 \prod_{l \neq k} (\lambda_k - \lambda_l)}. \tag{6}$$

Eq. 6 is always valid in the early stages of reaction, where the rate of $P \to X_i$ transformation can be ignored.

As an example of a practical calculation using the above method, consider the derivation of the integrated rate equation for product appearance in a two-product, two-substrate enzyme reaction, such as

$$\mathrm{NAD} \searrow \underset{k_{21}}{\overset{k'_{12}}{\rightleftarrows}} X_1 \quad \mathrm{S} \searrow \underset{k_{32}}{\overset{k'_{23}}{\rightleftarrows}} X_2 \quad \overset{k_{34}}{\longrightarrow} X_3 \quad \overset{\mathrm{P} \nearrow}{\longrightarrow} X_4 \quad \overset{\mathrm{NADH} \nearrow}{\underset{k_{41}}{\longrightarrow}} X_1,$$

where $X_1 = \mathrm{E}$, $X_2 = \mathrm{E\text{-}NAD}$, $X_3 = \mathrm{E\text{-}NAD\text{-}S}$, and $X_4 = \mathrm{E\text{-}NADH}$. From inspection of the above scheme we construct the 4×4 determinant $|\mathbf{A}^T - \lambda \mathbf{I}|$,

$$\begin{vmatrix} k'_{12} - \lambda & -k'_{12} & 0 & 0 \\ -k_{21} & k_{21} + k'_{23} - \lambda & -k'_{23} & 0 \\ 0 & -k_{32} & k_{32} + k_{34} - \lambda & -k_{34} \\ -k_{41} & 0 & 0 & k_{41} - \lambda \end{vmatrix}$$

---

[1] D. Thusius. To be published.

which is reduced by examination to $|\mathbf{A}' - \lambda \mathbf{I}|$:

$$\begin{vmatrix} k'_{12} + k_{41} - \lambda & k_{41} - k_{21} & k_{41} \\ -k'_{12} & k_{21} + k'_{23} - \lambda & -k_{32} \\ 0 & -k'_{23} & k_{32} + k_{34} - \lambda \end{vmatrix}$$

When applied to enzyme reactions, Eq. 6 gives the pre-steady-state (exponential terms) and steady-state (linear term) appearance of products. If the early phase production of NADH is required, we set $i = 4$ and derive the analytical expression for $B_4^{(k)}$ from initial concentrations and the subdeterminants $\Delta_{4j}^{(k)}$ of $|\mathbf{A}^T - \lambda \mathbf{I}|$. When the reaction is initiated by adding a mixture of substrate and NAD to enzyme, the initial conditions are defined by: $X_2^0 = X_3^0 = X_4^0 = 0$; $X_1^0 = X_{\text{total}}$. Thus, the integration coefficients for NADH production in this case are completely determined with $\Delta_{41}^{(k)}$. It is readily shown that,

$$B_4^{(k)} = -\Delta_{41}^{(k)} X_1^0 = k'_{12} k'_{23} k_{34} X_{\text{total}}. \tag{7}$$

Substitution of the above relation into Eq. 6 gives the expression for initial appearance of reduced coenzyme:

$$\frac{(\text{NADH})}{X_{\text{total}}} = k'_{12} k'_{23} k_{34} k_{41} \left[\frac{t}{\lambda_1 \lambda_2 \lambda_3} + \frac{1 - \exp(-\lambda_1 t)}{\lambda_1^2 (\lambda_1 - \lambda_2)(\lambda_1 - \lambda_3)} \right.$$

$$\left. + \frac{1 - \exp(-\lambda_2 t)}{\lambda_2^2 (\lambda_2 - \lambda_1)(\lambda_2 - \lambda_3)} + \frac{1 - \exp(-\lambda_3 t)}{\lambda_3^2 (\lambda_3 - \lambda_1)(\lambda_3 - \lambda_2)} \right].$$

The macroscopic rate constants are given implicitly as the roots of the secular polynomial (Eq. 2). If the roots have very different values, one can generally factorize $|A' - \lambda I|$ to obtain the λ_k in terms of the k_{ij}.[1] At the same time, the limiting conditions $\lambda_k \gg \lambda_l$ lead to considerable simplification of the denominators in the pre-exponential terms.

REFERENCES

1. FROST, A. A., and R. G. PEARSON. 1960. Kinetics and Mechanism. John Wiley & Sons, Inc., New York. 2nd edition. 173–177.

Small

Perturbations

"/"

ALAN SCHECHTER, *Chairman*

KEYNOTE ADDRESS

PICOSECOND FLUOROMETRY IN PRIMARY EVENTS OF PHOTOSYNTHESIS

LEONID B. RUBIN, *Department of Quantum Radiophysics, Faculty of Physics,* and
ANDREW B. RUBIN, *Department of Biophysics, Faculty of Biology Moscow State University, Moscow, USSR*

ABSTRACT Many laboratories in different countries are involved in the study of the mechanism of conversion of light energy into chemical energy, namely photosynthesis. As is evident from the literature, the initial phases of photosynthesis, which determine the character of this process, proceed at time intervals of 10^{-8} and 10^{-13} s. They are associated with absorption of light quanta and energy transfer from the molecules of light-harvesting antenna (LHA) chlorophyll and accessory pigments to the reaction centers (RC), where the key reaction of photosynthesis occurs: photo-induced charge separation. Evidently it is of importance to study experimentally the process that occurs within the 10^{-8}–10^{-13} s time domain.

In my lecture, I will try to discuss how picosecond fluorometry can help us to understand the mechanism and nature of primary events in photosynthesis. But before presenting any results, I want to acquaint you with our group in Moscow University working in the picosecond fluorometry region: Professor Andrew Rubin, Dr. V. Paschenko, and Dr. A. Kononenko. Our group consists of biologists and physicists who help us do experimental work with a wide variety of biological systems. Chloroplasts, isolated photosystems I and II (PI and PII), as well as reaction centers (RC) from photosynthetic bacteria were studied with a picosecond spectroscopy setup with high sensitivity and time resolution.[1]

One of the problems that can be solved by picosecond fluorometry techniques is the energy migration between light harvesting antennae (LHA) and RC. We can describe this process in terms of energy transfer from donor to acceptor, where the RC plays the role of acceptor of energy or quencher of electronic excitation of LHA. If one excites the LHA by a picosecond light pulse, the measurements of the fluorescence decay of the LHA will give information about the energy migration from LHA to the

---

[1]Dr. Rubin did not supply references with the text of his talk.—*Editor.*

RC. It is well known that the maximum of fluorescence of the LHA of PI is at 735 nm and that of PII is at 690 nm. This makes it possible to study separately the processes of energy migration in both photosystems of the green plants.

Fig. 1 presents the principal scheme of our picosecond fluorometry apparatus. It consists of a picosecond ruby or Nd^{3+} laser, a Pockels cell selecting a single picosecond pulse, and a streak camera with a time resolution of 10^{-11} s and an amplification of 5×10^6. In all of our experiments the energy of the picosecond pulse did not exceed 5.10^{12} photons cm^{-2}; much higher energies damage the chloroplast.

The fluorescence decay of pea chloroplasts has three components with a lifetime (τ) of 80, 300, and 4,500 ps, ascribed to PI, PII, and monomeric chlorophyll not involved in photosynthesis, respectively. The only reason that the lifetimes of PI and PII are much shorter than the lifetimes in solutions of chlorophyll is the very efficient energy migration between the LHA and RC. This fast transfer makes it possible to evaluate the constant of energy migration: $K_M \approx 1/\tau$.

Our measurements resulted in the following values of $K_{MPSI} = 1.25 \times 10^{10}$ s^{-1}, $K_{MPSII} = 3.3 \times 10^9$ s^{-1}. It is not difficult to show clearly that if one knows the value of the constant of energy migration K_M and spectral properties of LHA and RC, one can apply the formula that describes the process of inductive-resonance energy migration and evaluate the distance R between the donor (LHA) and acceptor (RC) of energy. Our calculations indicate that $R \approx 15$ Å.

According to the well-known hypothesis suggested by Duysens, the RC can quench the electronic excitation of the LHA only in the reduced form; in the oxidized form it is "closed," and as a consequence there should be an increase in the fluorescence lifetime τ and quantum yield. Taking into account the high quantum yield of primary steps of photosynthesis, $\sim 100\%$, one would expect that when the RC is oxidized, the molecules of chlorophyll in LHA would emit fluorescence with a lifetime and quantum yield close to those observed in pigment solution.

In another series of experiments we studied the dependence of fluorescence lifetimes of PI and PII on the degree of oxidation of the RCs induced by potassium ferricyanide

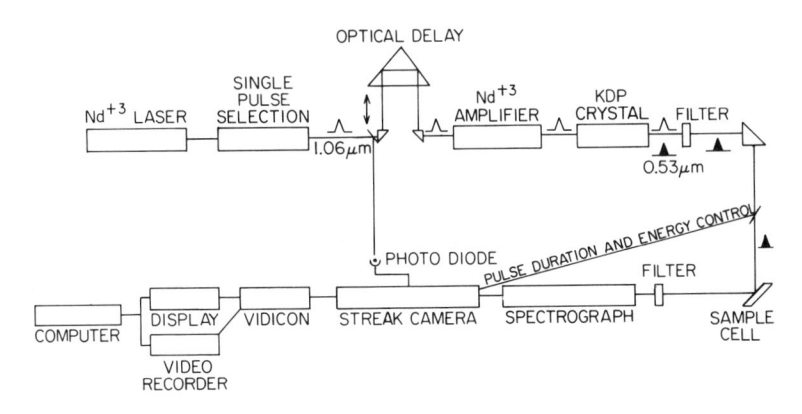

FIGURE 1 Picosecond fluorometry apparatus.

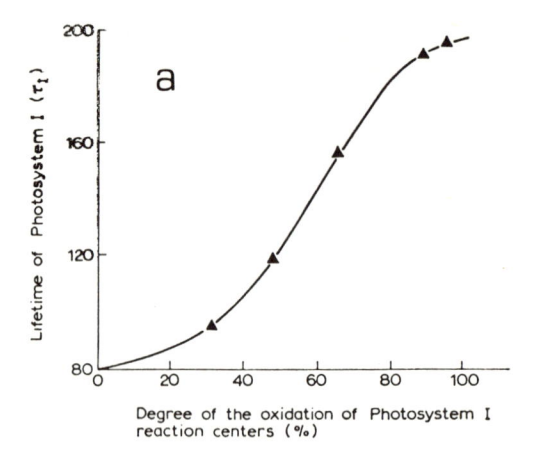

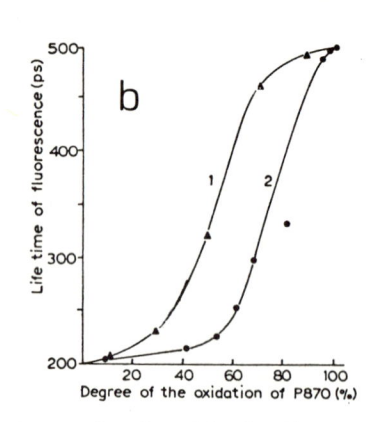

FIGURE 2 *a*. Dependence of the fluorescence lifetime (in picoseconds) of PI light-absorbing pigments in sub-chloroplast particles on the degree of RC oxidation. *b*. Lifetime of the fluorescence emitted by chromatophores from *R. sphaeroides*, 1760-I, as a function of a redox state of BChl P870 of photosynthetic RCs. λ_{exc} = 694.3 nm; λ_{meas} = 730–1,000 nm. The redox state of P870 was monitored by electron spin resonance measurements. P870 was oxidized either chemically by adding potassium ferricyanide (1) or photochemically with background illuminations from a He-Ne laser (2).

and by exposing chloroplasts to saturating intensities of actinic light and blocking PII by DEMU treatment. In all cases, we found only a two- to threefold increase in the maximum values as a result of any of the above-mentioned treatments (Fig. 2 *a*, *b*). Thus the lifetime of LHA of PI increased from 80 to 20 ps, PII from 300 to 600 ps, and in bacterial chromophores from 300 to 550 ps.

The relatively small observed rise of τ for PSI and LHA indicates that when the RC's are in a closed state and cannot donate electrons to the primary acceptor, the quenching of the singlet excited electron state of LHA still occurs considerably efficiently. The conversion of electron excitation into heat seems improbable, for under "normal" conditions this would lead to considerably lower quantum yields of the primary events of photosynthesis.

It is well known that the oxidation of RCs causes a strong red shift of its absorption band from 700 to 830 nm in the case of RC of PSI and from 830 nm to 1,250 nm in the bacterium *Rhodopseudomonas sphaeroides*. These changes in the absorption spectra must decrease the value of integral overlapping at least by 20–30 times and accordingly increase the τ of LHA by 20–30-fold. But our experiments indicate that the efficiency of energy migration from LHA to RC$^+$ does not decrease by 20–30 times, but only by a factor of 2 to 3. To explain this result, we propose that oxidation of the RC causes the conformational changes in photosynthetic apparatus, decreasing the distance between the RC$^+$ and LHA by a factor of 1.3 to 1.4. This is evident if one considers the sixth-power dependence of the K_M on the distance between energy donor and acceptor. It can compensate by the decrease in the value of the overlap integral.

We also studied the temperature dependence of the lifetime of the LHA of PI and

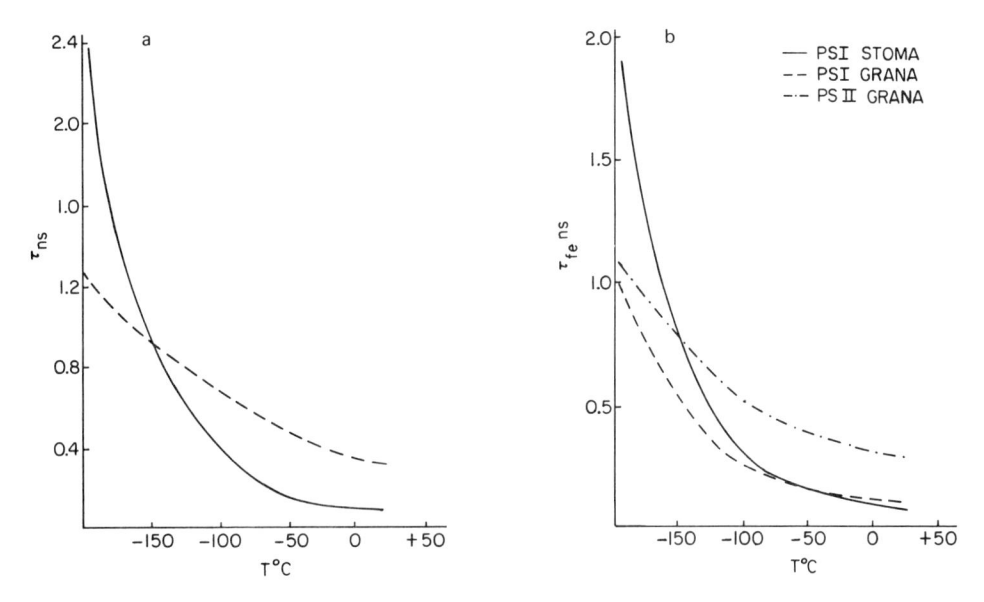

FIGURE 3 *a.* Dependence of the lifetime (τ) of the fluorescence of pea chloroplasts on temperature. Spectral region of the fluorescence measurement: λ = 730 nm (———); λ = 680 nm (- - -). *b.* Dependence of the lifetime (τ) of the fluorescence of PI (stroma), PI (grana), and PII (grana) of pea chloroplasts on temperature.

PII. These results are shown in Fig. 3 *a* and *b*. It was strange for us to find a dependence of τ on temperature because in a pure solution of chlorophyll the dependence of τ on temperature is very slight. But according to our results, the τ of PSI from the stroma part of pea chloroplasts increased from 80 to 2,000 ps and the change in the τ value occurs only below −70°C. These experiments certainly indicate that at temperatures below −70°C the efficiency of energy migration between LHA and RC decreases, although the fluorescence and absorption spectra of LHA and RC do not show any change. We propose that the main reason for the strong increase in the value of the fluorescence lifetime is that the temperature-dependent conformational changes increase the distance between LHA and RC. I want to draw your attention to the fact that the dependence of τ of PI from stroma and grana parts of the pea chloroplasts are different. This means that the structural organization of these photosystems cannot be identical. Therefore, if one studies the nature of PI, one must know whether the sample contains PI particles from the stroma or from the grana part of the chloroplasts.

According to our experimental results, the energy migration from LHA to the RC is provided by inductive-resonance mechanism and it takes about 100–300 ps for the energy of light quanta absorbed by the LHA to be trapped by the RC. It is clear that measurements of the LHA fluorescence lifetime cannot give information as to what happens to the energy of the electron excitation in the RC. To study this problem by picosecond fluorometry, one must provide the direct excitation of the RC's pigments so that its fluorescence would not be masked by a more intense fluorescence from LHA. These experiments must be done with isolated RC without any antenna chlorophyll or carotenoid molecules.

Unfortunately, because I came to the USA in January 1978, I cannot present the most recent results of B. Gulayev on the primary step of energy conversion in the RC of PI isolated from pea chloroplasts. The details of the electron transfer reactions in the RC, isolated from purple bacteria, have recently been revealed by Rentzepis and co-workers, using picosecond absorption spectroscopy. It was shown that within 10 ps after light absorption the first step of charge separation occurs: electron transfer from the bacteriochlorophyll dimer $(BChl)_2$ to the bacteriopheophytin (BPh); the second step

$$\text{light quanta} \rightarrow (BChl)_2 \, BPh \, X \xrightarrow[\text{I}]{10 \text{ ps}} (BChl)_2^+ BPh^- \, X \xrightarrow[\text{II}]{150 \text{ ps}} (BChl)_2^+ BPh \, X^-,$$

includes the process of electron transfer from BPh^- to the primary acceptor quinone X.

We have also studied the fluorescence kinetic decay from RCs isolated from *Rhodopseudomonas sphaeroides*, strain 1760-I, after excitation with a picosecond laser pulse. It appeared that the fluorescence decay consists of two components with $\tau_1 = 15 \pm 7$ ps and $\tau_2 = 150$ ps (Fig. 4). When the redox potential of the medium is increased to a point that $(BChl)_2$ becomes fully oxidized, the amplitude of the short-lived component diminished, while the lifetime of the long-lived one increased up to 700 ps. A detailed analysis shows that the τ_1 of 15 ± 7 ps corresponds to the fluorescence of $(BChl)_2^+$ and a τ of 250 ps to the BPh.

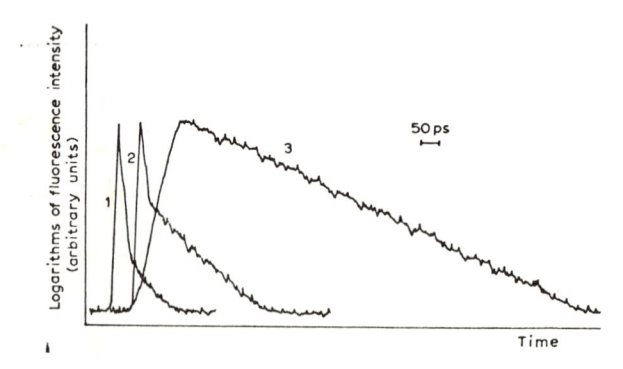

FIGURE 4 Decay kinetics of the fluorescence from photosynthetic RC perparations made from *R. sphaeroides*, strain 1760-I. $\lambda_{exc} = 694$ nm; $\lambda_{meas} = 850$–1,000 nm. 1, 100% reduction of P870; 2, 40–50% oxidation of P870, 3, 100% oxidation of P870. The redox state of P870 was posed by adding sodium ascorbate and potassium ferricyanide and monitored by electron spin resonance measurements (after V. Paschenko, 1977).

These results strongly indicate that picosecond fluorometry as well as absorption spectroscopy can give information about the energy distribution and electron transfer in the RC.

According to our measurements, the electronic excitation of (BChl)$_2$ is released within 15 ps in the electron transfer to the BPh. In that case BPh cannot emit fluorescence, since all its π-bonding orbitals are occupied. But the ruby laser radiation 6,943 Å can be directly absorbed by BPh molecules and the excited molecules of BPh emit fluorescence with $\tau_2 = 250$ ps, much shorter than the 2,000-ps lifetime of the fluorescence of BPh in solution. It is reasonable to ascribe such a strong decrease in τ to the electron transfer from BPh to the primary acceptor X. This means that BPh can be the donor of the electron to the primary acceptor X.

Summarizing our results, we come to the conclusion that picosecond fluorometry gives us approximately the same information about the primary steps of electron transfer in the RC as picosecond absorption spectroscopy. I want to draw your attention to the fact that the electron transfer from (BChl) to BPh occurs between the excited electron levels in both molecules and only on the next step of electron transfer from BPh to primary acceptor X will an electron be localized on the ground level X.

Among the problems regarding the picosecond events occurring in the RC, so far little is known about the mechanisms governing the fast electron transfer between components of the RC. To answer this question, we studied the dependence of the fluorescence lifetime of (BChl)$_2$ and BPh on temperature that gives us information about the dependence of the electron transfer from (BChl)$_2$ to BPh and to X (Fig. 5a, b). It is clear that the τ of (BChl)$_2$ and BPh fluorescence depends very little on the temperature. That means that the constants of electron transfer within the RC's components are also independent of temperature in the region from 0°C to -186°C. It is interesting to mention that at temperatures of about -70 to -90°C, the τ of (BChl)$_2$ and BPh fluorescence doubled.

These results lead us to the conclusion that the electron transfer within the RC is due to a tunneling mechanism. Without going into details beyond the scope of my lecture, I want to stress that all the proposed tunneling mechanisms invoke the coupling of the electron-nuclear interaction, which dissipates enough energy to prevent back electron tunneling, providing a high efficiency of charge separation.

I want to draw your attention to the fact that electron transfer between the cytochrome and RC also proceeds by a tunneling mechanism, but the dependence on the temperature in the +300 to -70°C region of that step of photosynthesis is quite different from those occurring in the RC itself. That means that the mechanism and nature of processes preventing the back electron transfer reaction may not be identical.

We hypothesize that the charge separation with the RC is identical to the appearance of the dipoles within a very strong electric field. According to measurements provided recently by J. Lee and P. M. Rentzepis, the value of the electric field can exceed 10^5 V·cm^{-1} in the 20 Å area. Such a strong electric field will induce the polariza-

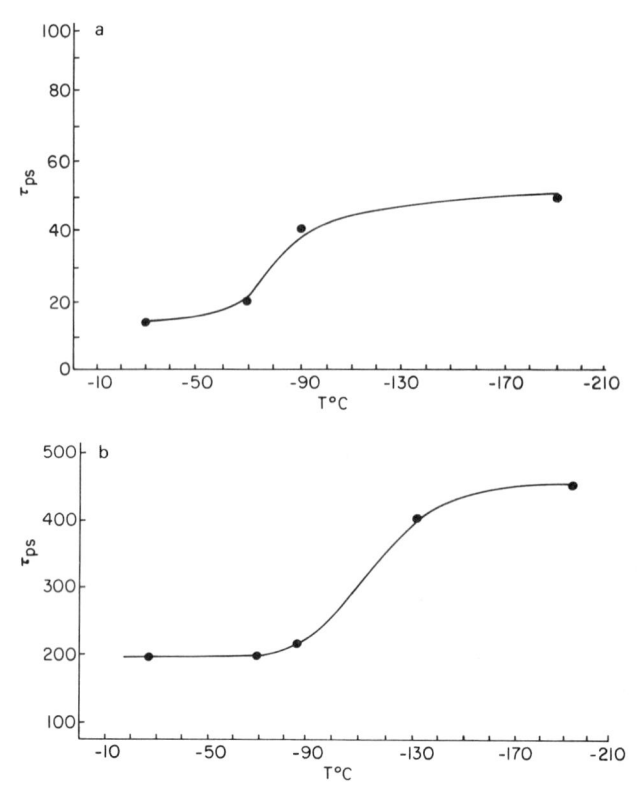

FIGURE 5 Dependence of the lifetime (τ) of the fluorescence of *R. sphaeroides* RCs on the temperature. *a*, fast component; *b*, slow component.

tion of the surrounding media-protein matrix. In its turn, the polarization of protein can change the efficiency of the back electron tunneling.

According to this hypothesis, the local electric fields within the RC may provide excellent feedback that regulates the efficiency of charge separation and stabilization. That means that the changes in the redox state of the components in the RC must influence the constant of the electron transfer. In this case I want to remind you of results that show that oxidation of the $(BChl)_2$ increases the time of the electron transfer from BPh to primary acceptor X from 250 ps to up to 700 ps. The paper by M. J. Pellin, C. A. Wraight, and K. J. Kaufmann on page 361 of this Discussion also gives evidence that the reduction of the secondary acceptor changes the constant of the electron transfer between BPh and the primary acceptor.

From this point of view the RC is a unique photoactive enzyme complex, not merely a set of a separate redox reactions. Light-induced electron-conformational changes proceed in a definite sequence and lead eventually to the release of an electron, while the whole complex relaxes through several steps to the original state, thus completing its working cycle. It is possible that the extra energy of ~0.5 eV released for such a cycle is not lost as heat dissipation, but is stored in some form of membrane-protein conformation.

DISCUSSION

GEACINTOV: You have reported a series of lifetime measurements and it is now known that the intensity of the excitation pulses plays a very important role in the measured decay times. What was the intensity of your pulses in your measurements, and are you sure that nonlinear effects were unimportant in your fluorescence decay measurements?

RUBIN: It is an important question. About two or three years ago it was shown that the lifetimes of PI and PII depend on the light intensity. Light intensities of more than 10^{13} or 10^{14} quanta/cm^2 generate nonlinear effects. In our experiment we use the light energy of a single pulse, 5.10^{12} photons/cm^2.

About two years ago Shapiro published some results saying that everybody who used the high intensity pulse got incorrect results—everybody who used picosecond for fluorescence or photosynthesis, PI as well as PII. Now I know that maybe even you or Shapiro, or both together, have proved that the lifetime of PI does not depend on the light intensity, but it is not so.

GEACINTOV: At low temperatures only! At 77°K, the decay time of the PI emission does not depend on the intensity of the picosecond pulses (at least up to 10^{15} quanta/cm^{-2} pulse^{-1}), while that of PII varies strongly with intensity.

RUBIN: But we measured the dependence of the lifetime of the intensity up to the 10^{16} quanta/cm^2. We measured the decrease in lifetime in the region where there is no linear response of the decreasing of lifetime.

GEACINTOV: If your lifetimes were indeed measured with low-intensity pulses, then there seems to be a difference between your measurements and those of other workers using more conventional phase fluorometry and single-photon counting techniques, which show that the decay is longer than 30–200 ps. In fact the values given for the decay times are ~400–800 ps.

RUBIN: 400 ps. You are talking about which lifetime?

GEACINTOV: Lifetime of PI at room temperature.

RUBIN: No. The lifetime of PI at room temperature is 80 ps. But when you lower the temperature the lifetime of PI can be increased up to 2,500 ps.

GEACINTOV: In fact it is difficult to resolve photosystem I emission at room temperature because of its intense fluorescence tail above 700 nm.

RUBIN: We can separate the fluorescence of PI and PII because of their different region of fluorescence. And I tell you that one cannot compare the results of the streak camera measurements of fluorescence decay and the data obtained by phase fluorometry. They are quite different. Phase fluorometry gives you some average lifetime; if there is nonexponential decay it is difficult to get a direct answer.

GEACINTOV: However, there are also single-photon decay curves which show the comparably longer decay times.

RUBIN: What can I say? You can see our results; you can believe your results and I can believe my results. It is up to you.

WELLER: There have been measurements of only the first step of the electron transfer from the bacterial chlorophyll dimer to the bacterial pheophytin. It has been shown that there is a magnetic field effect on the production of triplets, when the electron goes back to the bacterial chlorophyll (page 88). This magnetic field effect seems to show that the electron transfer occurring in less than 10 ps must be a tunneling process. Otherwise the exchange interaction between the chlorophyll radical cation and the pheophytin radical anion would be much too strong to allow for a magnetic field effect. And this is another indication of the importance of electron tunneling in this system.

RENTZEPIS: Since Professor Weller mentioned tunneling, may I say that we have recently measured the rate of transfer from pheophytin to Q as a function of temperature from 300 to 4°K? These experiments indicate that the rate is temperature-independent throughout this range. Utilizing the Hopfield-Jortner theories, we estimate the distance between pheophytin and Q to be 9–13 Å. Similarly the radical ion pair information is also temperature-independent within this range, and we estimate a similar distance or slightly smaller between the $BChl^+$ and BPh^- (K. Peters, P. Avouris, and P. M. Rentzepis. 1978. *Biophys. J.* **23**:207–217.).

FOYT: I have a comment. The appeal to proton tunneling assumes the standard model of a potential barrier to nuclear motion, the barrier being due to the electronic potential surface in the Born-Oppenheimer approximation. Two alternatives to tunneling should also be considered. (*a*). The Born-Oppenheimer approximation may not be valid. There may be a second electronic potential surface near the ground state surface, so that the reaction is not confined to a single surface and the electronic and nuclear motions are strongly coupled. (*b*) Even within the Born-Oppenheimer approximation, the assumption of a single reaction coordinate passing through the saddle point of the potential surface may be a poor approximation in the present case. To be strictly correct, it is necessary to calculate the reaction probability by performing a Feynman path integral over all possible reaction paths. In simple cases, a qualitatively correct picture is obtained by considering only the single most important path, i.e. the path of least action, which typically passes through the saddle point. However, if the potential surface is strongly shaped, then other quite different paths may make significant contributions. Either of these two alternatives could produce an anomalous Arrhenius plot without the necessity of involing proton tunneling.

RENTZEPIS: Your first point is correct if such a state exists. We have no evidence of a state between the first singlet and ground test other than the "prelumi," which are considered in the scheme presented. The second point is a bit philosophical. Yes, we would like to calculate the probability of *all* possible reaction paths but that is impossible in practice. With regard to a possible "strangely shaped" potential surface, you suggest that other paths might make a contribution. Possible, but what are they? Any suggestions? Even if some unknown "anomalous Arrhenius" is operative, how do you explain the $K_H:K_D$ ratio of F? Similarly the constant rate of *H* and *D* at low temperatures would necessitate such an anomalous Arrhenius process to be equivalent to proton tunneling.

A TIME-RESOLVED ELECTRON SPIN RESONANCE STUDY OF THE OXIDATION OF ASCORBIC ACID BY HYDROXYL RADICAL

RICHARD W. FESSENDEN AND NARESH C. VERMA, *Department of Chemistry and Radiation Laboratories, Carnegie-Mellon University, Pittsburgh, Pennsylvania 15213, and The University of Notre Dame, Notre Dame, Indiana 46556 U.S.A.*

ABSTRACT Time-resolved electron spin resonance (ESR) spectroscopy for the study of radicals produced by pulse radiolysis is illustrated by a study of the oxidation of ascorbic acid by OH radical in aqueous solution. In basic solution, the direct oxidation product, the ascorbate mono-anion radical, is formed within less than 2 μs of the radiolysis pulse. In acid solutions (pH 3–4.5, N_2O:saturated) three radicals are initially formed, the ascorbate mono-anion radical, an OH adduct seen also in steady-state ESR experiments, and an OH adduct at C2 with the main spin density at C3 of the ring. The first OH adduct decays with an initial half-life of about 100 μs, probably by bimolecular reaction. The second OH adduct, which shows one hyperfine splitting with a^H = 24.4 ± 0.3 G and g = 2.0031 ± 0.0002, decays with a half-life of about 10 μs. On this same time scale the concentration of the ascorbate radical approximately doubles. It is concluded that the adduct at C2, but not the other adduct, loses water rapidly to form the ascorbate radical.

INTRODUCTION

Electron spin resonance (ESR) experiments with microsecond time resolution to study radicals produced by pulse radiolysis have been carried out in two laboratories (1–5). Much of the initial thrust has been directed toward understanding chemically induced dynamic electron polarization (CIDEP) a process by which chemical reactions either producing or destroying radicals can be spin-dependent and thereby modify the population of the various spin states and the intensities of the ESR transitions. In addition, however, several kinetic and mechanistic applications have been reported (6–8). To illustrate the potential of such experiments, this paper will present data from time-resolved ESR experiments which help clarify the complex chemistry involved in the oxidation of ascorbic acid by hydroxyl radical. A brief preliminary report of some observations on this system was included in an earlier paper reviewing *in situ* radiolysis-ESR experiments (9).

The oxidation of ascorbic acid by OH has been studied by both steady-state radiolysis-ESR experiments and pulse radiolysis with optical spectrophotometric and conductometric detection. The ESR experiments (10) showed that the ascorbate mono-anion radical formed by the net loss of an electron from ascorbate was present over the pH range 0–13 and a second radical also was formed in acid solution (pH < 5). The most detailed optical study is that by Schöneshöfer (11), who studied changes in the optical spectra as a function of pH for acid solutions. He proposed a very complex reaction scheme. Although some effects are seen by both spectrophotometric and ESR methods, the present data show that some of the details of that proposed mechanism are incorrect. The earlier steady-state ESR work (10) showed clearly that the ascorbate anion radical has a pK of −0.45, in contrast to the value of 3 concluded from the optical study (11).

The oxidation of ascorbic acid by OH is potentially complex because of the possibility of addition to either end of the double bond of either of the tautomeric forms.

The radical found in neutral and basic solutions can result either from loss of water

from an OH adduct or by direct electron transfer oxidation. The various intermediates either have similar absorption spectra with peaks at 360 nm or have little absorption in the accessible region (>300 nm). The ESR method can be used to advantage in such a situation because of the better resolution of different spectra.

THE TIME-RESOLVED ESR METHOD

ESR experiments with a time resolution of better than 1 μs have been demonstrated (3, 5). Radicals are produced by pulse radiolysis with 3 MeV electrons with pulse periods of 0.1–1 μs. Direct detection of ESR absorption is preferred because quantitative analysis of the resulting curves is much easier (3). In contrast to optical techniques, fast ESR spectroscopy has a number of potential complications, which must be kept in mind in analyzing curves of absorption vs. time after the radiolysis pulse. The strength of an ESR absorption does not necessarily reflect the concentration of radicals producing it. In dynamic experiments the chemical reactions forming or destroying radicals can be spin-selective, perturbing the relative spin populations upon which the absorption intensities depend. This latter effect, as mentioned above, is called CIDEP. In the usual form the low-field lines of a spectrum appear weak or even in emission and high-field lines are in enhanced absorption. The strength of the effect is proportional to the rate of the radical-radical encounters and so the amplitude of a high-field line will depend more than linearly on the radical concentration. When observing radical disappearance with such a line, the apparent half-life will be somewhat shorter than the correct chemical value. A related effect involves the fact that in any radical transformation, such as addition to a double bond, the product radical will initially possess the spin populations of the reactant. This effect is particularly pronounced in radicals formed from H atoms, because the latter can develop strongly perturbed populations. Another important consideration is that the ESR signal does not appear immediately after formation of the radicals. The rise time of the ESR signal is related to the spin relaxation time of the radical, the initial population difference for the particular transition, and the microwave power level. A final effect occurs at high microwave power levels, where overshoot and damped oscillations on the rise of the ESR signal are possible. This transient ESR response disappears with the spin relaxation time. A quantitative treatment of all of these effects has been given in a previous paper (3).

In spite of all these potential problems, the observation of ESR signals in such pulse experiments provides very useful information, which often can be analyzed quantitatively to determine rate constants. The rates of reaction of H atoms with several alcohols (7) and of phenyl radicals with isopropyl alcohol (8) have been determined in this way. An important conclusion is that for times long with respect to the spin relaxation time the signals reflect mainly the chemical kinetics. In the work to be reported here, the relaxation times are fairly short (1–2 μs) so that the transient effects and the effects of the initial spin populations are minor. Only one of the radicals appears to be affected by CIDEP.

Implicit in the above discussion is the idea that the spectrometer is set on a given ESR line and a time profile of the absorption recorded. To provide sufficient signal-to-noise ratio, a number of curves must be recorded for signal averaging. A similar number of curves are taken at a nearby position to provide a base line with no ESR adsorption. The difference of the two accumulations is the true time profile. Because the ESR lines usually are narrow with respect to the total extent of a spectrum (0.1 vs. 100 G), it is difficult to search for lines of new radicals. A continuous scan using a boxcar integrator is possible, but this mode has not proved as sensitive as recording time dependences. The time-resolved experiments are clearly less sensitive than the steady-state ones.

EXPERIMENTAL DETAILS

The ESR spectrometer was as described previously (3). Absorption was detected directly. Radiolysis was with 2.8 meV electrons. Each pulse produced about 3×10^{-5} M total radicals. To provide for signal averaging, the pulses were repeated at 100 s^{-1} repetition rate and formation-decay curves for about 5,000 pulses were digitized with a Biomation 8100 transient recorder (Biomation, Cupertino, Calif.) and processed in a PDP8 minicomputer (Digital Equipment Corp., Marlboro, Mass.). The computer also provided a field-frequency lock to maintain a given position in the spectrum throughout an accumulation. The microwave

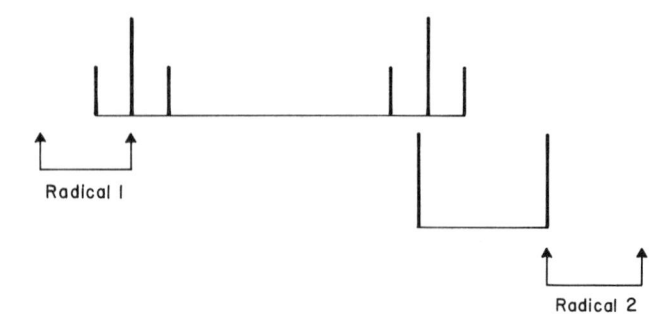

FIGURE 1 A schematic representation of the ESR spectra of the first and second radicals observed in acid, N$_2$O-saturated ascorbic acid solutions at about pH 4 (10). The arrows mark the positions of signal and base-line accumulations for the principal radical (low-field side) and the second radical (high-field side). The doublet splitting for radical 1 is 1.76 G. The lines for radical 3 are centered at fields 15.7 G above and 8.7 G below the center of this spectrum.

power level was about 1 mW and the spectrometer time constant 1 μs. Sample solution flowed through the flat aqueous cell at about 0.5 cm^3/s so that each volume element (20 μl) received only a few irradiation pulses. Increases in the flow rate had no significant effect on the data obtained.

The solutions were prepared in water which had been distilled and passed as vapor with oxygen through a silica oven. Oxygen was removed by bubbling with N$_2$O. The ascorbic acid was obtained from Calbiochem (San Diego, Calif.) and was the same as that used for the steady-state experiments (10). The pH was changed by addition of Baker analyzed KOH or HClO$_4$ (J. T. Baker Chemical Co., Phillipsburg, N.J.).

RESULTS AND DISCUSSION

The ESR spectrum of N$_2$O-saturated solutions of ascorbic acid at pH 4, as observed in continuous radiolysis experiments (10), is shown schematically in Fig. 1. The first (or principal) radical shows a doublet of triplets and is present with unchanged line positions from pH 1 to 13. (This radical starts to protonate at pH < 1 and the lines then shift.) The second radical shows a doublet and its lines start to shift below about pH 3.

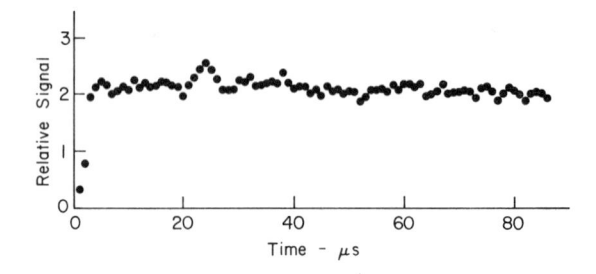

FIGURE 2 Time dependence of the ESR absorption signal of the first radical in basic (pH 11), N$_2$O-saturated solutions of ascorbic acid (5 mM). The time scale is 1 μs/point. The time origin is the end of the 1-μs irradiation pulse. The 2-μs delay in response after the pulse is the result of switching the signal amplifier off during the irradiation period to prevent interference. The data represents the average of about 5,000 pulses.

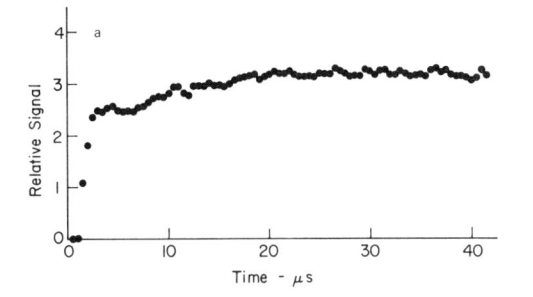

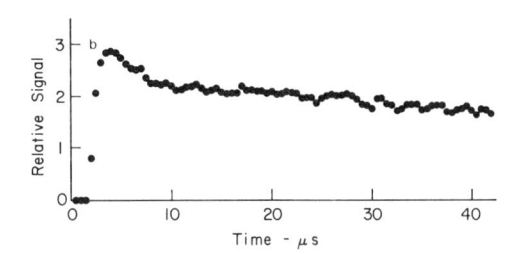

FIGURE 3 Time dependence of the first (a) and second (b) radicals in an N_2O-saturated solution of 5 mM ascorbic acid at pH 3.9. The time scale is 0.5 μs/point.

The pulse experiments described here show the time profiles taken at the positions indicated.

A time profile for the first radical at pH 11 is given in Fig. 2. This curve shows that the transient effects associated with the rise of the ESR signal are not a problem because of a relatively short relaxation time for the radical. Based on the observed rise time, the relaxation time cannot be longer than about 2 μs. The lack of decay is as expected from the known disappearance rate constant (11). Because of the slow reaction, no effects of CIDEP should be present. A curve (not shown) for the central line of the upper triplet of this radical (the only one present at pH 11) is essentially identical to that shown, confirming this expectation.

Data taken at pH 3.9 and 3.5 are shown in Figs. 3 and 4 for the lines of the first and second radicals. In each case the signal of the second radical rises within 3–4 μs to a level which then decays with a half-life of about 100 μs. Such a decay is typical of radical-radical reaction at the concentration levels used. (The slight overshoot on the curve at pH 3.9 is a result of transient ESR response as described above. At a lower power level this overshoot is absent.) The signal from the first radical shows a growth on two time scales. In each case there is a fast growth, as observed at pH 11, followed by a slower growth. At pH 3.9 (see Fig. 3) 65% of the signal (extrapolated to the end of the pulse) appears rapidly, with the remaining 35% growing with a half-life of about

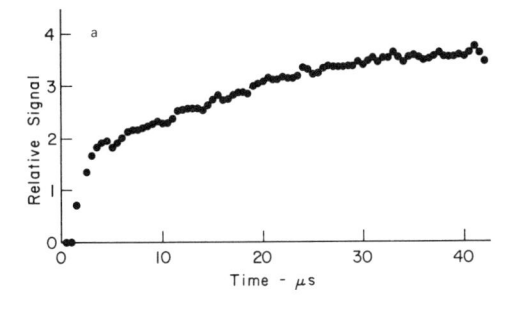

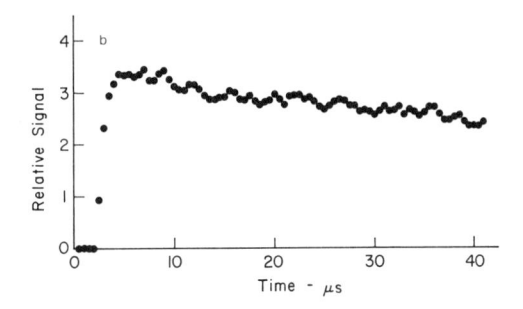

FIGURE 4 Time dependence of the first (a) and second (b) radicals in ascorbic acid at pH 3.5. Other conditions were as in Fig. 3.

8 μs. At pH 3.5 (see Fig. 4) the faster portion represents 40% and the rest grows with about 13-μs half-life. The division between the two portions does not continue to change with lower pH, as curves (not shown) for pH 3.2 give 50% for the fast portion. (The difference between 40 and 50% represents the experimental error of the determination.) The smaller slow portion at pH 3.9 is probably the result of a change in mechanism, as a considerable amount of ascorbate anion (pK = 4.1) is present. At pH 4.5 the slow portion is not readily discernable. The estimate of the half-period of the slower component is rather crude, because the level shown at 40 μs is not a true plateau, as some decay occurs at longer times. A curve taken at twice the time scale showed 20% decay from 30 to 80 μs. Corrections for this effect give 15 μs for the lifetime at pH 3.5.

The longer period of decay of the second radical (~100 μs) as compared with that for the growth for the first radical, as discussed above, shows that the decay of the second radical probably does not lead to formation of the first radical. Thus a third radical must be present. The most likely species is one formed by addition of OH at position C2 of the ring with the main spin density at C3. This radical would have one large hyperfine splitting of >20 G from the single hydrogen atom on C4. A search for such a species was carried out by running accumulations at successive magnetic fields in the appropriate regions of the spectrum. In fact, lines were found as shown by data for pH 2.9 in Fig. 5. Although it is not certain that this spectrum does not contain other lines, the structure matches the expected radical. In contrast to the other radicals, this radical has broad lines—about 2 G full width at half-amplitude. The parameters are a^H = 24.4 ± 0.0002 G and g = 2.0031 ± 0.0002. The accuracy of these values is limited because of the weakness of the spectrum and the width of the lines. This radical is also present at pH 3.2 and 3.9 but is barely detectable at pH 4.5. It is not present when 0.1 M KBr is added so that oxidation of the ascorbic acid is by Br_2^- and only the first radical is formed as previously described (11). These latter results also show that the third radical is not formed from the direct yield of H atoms in water. The emission observed for the low-field line is a result of CIDEP, which would be expected for a radi-

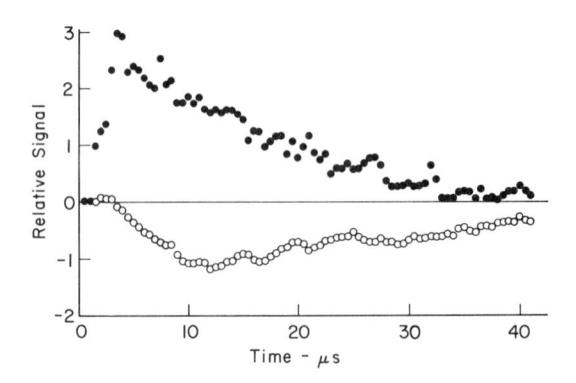

FIGURE 5 Time dependences found for the high- and low-field lines of the third radical in N_2O-saturated 5 mM solutions of ascorbic acid at pH 2.95. The low-field line (negative going) shows emission as a result of CIDEP produced by fast radical-radical reaction.

cal with a sizable hyperfine splitting. The high-field line must represent enhanced absorption and so the apparent half-life (~ 10 μs) will be shorter than the actual chemical value. The approximate values does, however, correspond to that for the slower growth of the first radical. It is very likely that the third radical converts to the first radical with such a period.

On the basis of these results, the mechanism for reaction of OH with ascorbic acid in acid solutions can be written as follows:

$$\begin{bmatrix} \text{R-C(O)-C=O structure with H-C=C-OH OH below} \\ \updownarrow \\ \text{R-C(O)-C-OH structure with H-C=C-O OH} \end{bmatrix} \xrightarrow{+OH} \begin{bmatrix} (\text{Intermediate?}) \longrightarrow \text{radical (1)} \quad \tau \sim 15 \mu s \\ \text{radical (2)} \\ \text{radical (3)} \end{bmatrix}$$

where R represents HOCH$_2$CHOH-. The curve of Fig. 4a shows that about 50% of the radical 1 is formed in a few microseconds and that the other 50% is formed with a half-period of about 15 μs. Conversion of radical 3 to 1 is the most likely path of this reaction. Radical 3 is clearly identified as the result of addition of OH at C2. This radical can be produced from either of the two forms of ascorbic acid because of an expected fast equilibration between radicals protonated at the oxygen atoms on C1 and C3. Radical 2 represents one of the two remaining adducts. Although it is not certain that this form is responsible for the spectrum observed in steady-state experiments, the adduct at C1 should be the least stable as a result of three oxygen substituents on that carbon atom. It is likely that this form is produced and loses water rapidly to form 1.

This mechanism differs in a number of ways from the very detailed one given by Schöneshöfer (11). His mechanism involved only OH adducts at C1 and C3. Changes seen on the 25-μs time scale were attributed to loss of water from an OH adduct at C3. On the basis of the present results, this reaction is clearly loss of water from 3. We cannot comment on many of the other features of his mechanism except to note that the steady-state ESR experiments clearly show pK for 1 and 2 to be -0.45 and 2.0, respectively.

These results on ascorbic acid are relatively qualitative but illustrate very well the power of the ESR method in resolving the separate radicals in a system where the optical spectra do not do so. It is to be emphasized that the qualitative nature of these experiments results from the difficulty in carrying them out because of the relatively long accumulation times necessary. The detection of radical 3 was the most difficult experiment carried out to date and all portions of the equipment had to be functioning at peak consistency. With the present data in hand it is possible that more quantitative experiments could be contemplated.

This work was supported in part by the Division of Basic Energy Sciences of the Department of Energy. This is document NDRL-1833 from the Notre Dame Radiation Laboratory.

Received for publication 1 December 1977.

REFERENCES

1. SMALLER, B., J. R. REMKO, and E. C. AVERY. 1968. Electron paramagnetic resonance studies of transient free radicals produced by pulse radiolysis. *J. Chem. Phys.* **48**:5174–5181.
2. FESSENDEN, R. W. 1973. Time-resolved ESR spectroscopy. I. A kinetic treatment of signal enhancements. *J. Chem. Phys.* **58**:2489–2500.
3. VERMA, N. C., and R. W. FESSENDEN. 1976. Time-resolved ESR spectroscopy. IV. Detailed measurement of the ESR time profile. *J. Chem. Phys.* **65**:2139–2155.
4. TRIFUNAC, A. D., and M. C. THURNAUER. 1975. Chemically induced dynamic electron polarization. Pulse radiolysis of aqueous solutions of alcohols. *J. Chem. Phys.* **62**:4889–4895.
5. TRIFUNAC, A. D., K. W. JOHNSON, B. E. CLIFFT, and R. H. LOWERS. 1975. Submicrosecond studies in pulse radiolysis by time-resolved EPR spectroscopy. *Chem. Phys. Lett.* **35**:566–568.
6. SMALLER, B., E. C. AVERY, and J. R. REMKO. 1971. EPR pulse radiolysis studies of the hydrogen atom in aqueous solution. I. Reactivity of the hydrogen atom. *J. Chem. Phys.* **55**:2414–2418.
7. FESSENDEN, R. W., and N. C. VERMA. 1977. Studies of the reactions of hydrogen atoms by time-resolved ESR spectroscopy. *Discuss. Faraday Soc.* **63**:104–111.
8. MADHAVAN, V., R. H. SCHULER, and R. W. FESSENDEN. 1978. The absolute rate constants for reactions of phenyl radicals. *J. Am. Chem. Soc.* In press.
9. FESSENDEN, R. W. 1975. Steady-state and time-resolved ESR studies of radiolytically produced radicals. *In* Fast Processes in Radiation Chemistry and Biology, G. E. Adams, E. M. Fielden and B. D. Michael, editors. The Institute of Physics and John Wiley & Sons, Inc., New York. 60–75.
10. LAROFF, G. P., R. W. FESSENDEN, and R. H. SCHULER. 1972. The electron spin resonance spectra of radical intermediates in the oxidation of ascorbic acid and related substances. *J. Am. Chem. Soc.* **94**:9062–9073.
11. SCHÖNESHÖFER, M. 1972. Pulsradiolytische untersuchung zur oxidation der ascorbinsaüre durch OH-radikale und halogen-radikal-komplexe in wässriger lösung. *Z. Naturforschung B Teil Anorg. Chem. Org. Chem. Biochem. Biophys. Biol.* **27**:649–659.

DISCUSSION

CZERLINSKI: Is one control point on one side of your signal line sufficient, or should you use two points, symmetrical to your signal line? Your base line could be inclined or you could be observing the line of an undiscovered radical intermediate (or side product).

FESSENDEN: That is a good point. Because of the way we did the experiment, we see only changes that are a function of time *and* of the difference between the two magnetic fields selected for the accumulations. No sloping base line is possible unless it represents a resonance line produced by the irradiation. If both fields are positioned away from lines, the result is a signal less than 5% of the amplitudes shown. Certainly there is a chance of undetected lines in the spectrum. We have explored the spectrum by varying the positions of the magnetic fields used for the two accumulations to verify that the position taken as the base line is not on any resonance line. The lines are sufficiently narrow that the base line has little contribution from the tail of the resonance line under study.

PERSOONS: What is the rate of equilibration of the two tautomers of ascorbic acid? Because these tautomerizations are acid-base catalyzed, the equilibrium may be slow.

FESSENDEN: I have no information on the rate of interconversion between the tautomers of ascorbic acid. The hydroxyl radical will react with each form according to its concentration and rate constant. As long as equilibration takes place between solution preparation and reaction, the interconversion will not affect these results.

BERGER: I think that this is a very intriguing method; I just don't understand the hardware. I wonder if you could show the configuration of the cavity. The ability to detect ESR signals in 20 ms is a most important advance and a careful elucidation of the experimental apparatus is very important to us.

FESSENDEN: There are two major considerations in a time-resolved ESR spectrometer. Because the time dependence of the ESR signal corresponds to a high frequency, it is not necessary to use field modulation to put the signal at a frequency where the detector has little noise. In the description given in ref. 3, the signal from the microwave detector diode is amplified by an AC-coupled video amplifier. Thus the signal observed corresponds to changes in ESR absorption. In the amplifier used by us, several methods (such as a gated amplifier) were taken to reduce the effect caused by the irradiating beam.

The second consideration is the automatic frequency control system. In most spectrometers this function is accomplished by frequency modulation of the klystron. There will be signals at the detector at the frequency of this modulation and they may interfere with observation of the transient signals. In my spectrometer the automatic frequency control (AFC) function is accomplished without frequency modulation of the klystron. It now appears that if the AFC signal at the detector can be kept small enough, it will not interfere. Since it is not synchronous with the irradiation, the averaging will remove it. Thus, the use of a commercial spectrometer with minimum modification may be possible.

The cavity is relatively simple since no field modulation is necessary. It is constructed of a brass wave guide with suitable posts for the irradiation beam pipe and for the cell. For laser photolysis experiments, the usual multipurpose cavity could be used.

The cavity-loaded Q is about 1,500. The response time limit posed by this Q is about 60 ns. The observed response time of our spectrometer of about 0.3 μs is limited by the electronics.

MAGNETIC RELAXATION ANALYSIS OF DYNAMIC PROCESSES IN MACROMOLECULES IN THE PICO- TO MICROSECOND RANGE

R. KING, R. MAAS, M. GASSNER, R. K. NANDA, W. W. CONOVER,
AND O. JARDETZKY, *Stanford Magnetic Resonance Laboratory, Stanford University, Stanford, California 94305 U.S.A.*

ABSTRACT A formalism based on the theory of Markov processes and suitable for the analysis of multiple internal motions in macromolecules is presented. Computer calculations of specific motional models for ^{13}C nuclear magnetic resonance (NMR) relaxation, treated as special cases of the proposed formalism, demonstrate the potential of this approach for discriminating between different motional models on the basis of NMR relaxation data.

INTRODUCTION

It has been known for some time (1, 2) that nuclear magnetic resonance (NMR) relaxation data on macromolecular systems contain, at least in principle, a wealth of information on dynamic processes, including internal motions, association and dissociation rates for complexes, rates of energy transfer, etc. In recent years, experimental relaxation measurements on various macromolecules, using various nuclei (^{1}H, ^{13}C, ^{31}P, ^{9}F), have begun appearing in the literature in ever increasing numbers (3–10). In all cases reported so far the measurements were interpreted by assuming a specific relaxation model, a correlation time for the overall motion of the macromolecule, and one degree of internal motional freedom. The most widely used model is that calculated by Woessner (11, 12) for anisotropic diffusion or a methyl group rotation on a rigid sphere and the extension of this model to treat sequential rotations in a hydrocarbon chain (specifically that of a membrane lipid) developed by Levine et al. (13, 14). Other models dealing with correlated motions and a limited degree of internal rotation have also been treated (15–18). Interpretation of relaxation data in terms of specific models is undoubtedly instructive if a suitable model can be found and if the required calculation is tractable. The procedure is, however, not without its dangers. In the absence of a general theory which permits the comparison of different models, the inapplicability of an available model may not be easily recognized. Nor can the possibility be tested that the data may be compatible with more than one model. Yet no general theoretical effort has been made to deal with these problems.

We have recently pointed out that a general formalism for the analysis of NMR relaxation measurements on systems with multiple degrees of freedom can be based on the theory of Markov processes and have derived a general equation for the analysis of data on systems undergoing an arbitrary number of motions of an arbitrary nature (19). In this communication we report a comparison between different relaxation models calculated within the framework of this formalism.

THEORETICAL

The calculation of relaxation parameters is in essence a calculation of spectral density functions $J_{F_j}(\omega)$ defined as

$$F_{F_j}(\omega) = \int_{-\infty}^{\infty} e^{i\omega\tau} \overline{F_j(t)F_j^*(t + \tau)}d\tau \qquad (1)$$

for a dynamic variable F_j, with an autocorrelation function $\overline{F_j(t)F_j^*(t + \tau)}$ (1, 2). F_j may be, as is usual, taken to be the time-dependent part of an appropriate interaction Hamiltonian (1) or used to denote the total Hamiltonian for a given relaxation mechanism.

As already pointed out (19), the sole additional assumption necessary to develop a general formalism for the analysis of relaxation phenomena in systems with multiple degrees of freedom is that any motion contributing to relaxation is a Markov process. The simple case of rotational diffusion or exchange as a Markov process contributing to magnetic relaxation has been discussed by Kubo (20) and others (21, 22). Our formulation may be regarded as a generalization of these earlier treatments. It should be emphasized that the assumption of a general Markov process is far less restrictive than the common use of Markov statistics describing specific models. The assumption in its general form implies no more than that a probability can be assigned to each step in the process and that the system has no memory. The basic relationships that can be derived from it cover a wide range of specific models, and provide a general framework for systematic comparisons among them. As previously shown (19) for a system undergoing N independent motions, $J_{F_j}(\omega)$ is given by:

$$J_{F_j}(\omega) = \sum_{l_1 \ldots l_N} \frac{2|\langle F_j, \prod_{k=1}^{N} {}^k\phi_{lk}^* \rangle|^2 \sum_{k=1}^{N} \lambda_{lk}^k}{\left\{\sum_{k=1}^{N} \lambda_{lk}^k\right\}^2 + \omega^2}, \qquad (2)$$

where λ_n, ϕ_n are eigenvalues and ϕ_n, ϕ_n^* the eigenfunctions of the transition operator Ω so that

$$\Omega\phi_n = \lambda_n\phi_n, \qquad (3)$$

and similarly for Ω^*, with $\phi_n, \phi_n^* = 1$. The general properties of the transition operator Ω are discussed elsewhere (19, 23). Of more immediate importance is the physical significance of λ and φ. If F_j is a dynamic variable, ϕ_n are the amplitude parameters and λ_n the rate parameters of the motion, $\lambda_n = 1/\tau_n$, where τ_n is the correlation time for the nth individual motion. Eq. 2 thus allows us to calculate amplitudes and frequencies of any combination of internal motions or other processes affecting relaxation from a given set of relaxation data or, conversely, to predict a set of relaxation parameters from any assumed combination of processes. Although the equation strictly applies to independent processes, the condition of independence is not as restrictive as it may appear at first. If two or more processes are weakly coupled, it is possible to take this into account using standard perturbation theory, assuming

$$\Omega = \Omega_1 + \Omega_2 + \epsilon\Omega_{12}, \text{ where } \epsilon \text{ is small}, \quad (4)$$

and obtaining the solution in terms of Bohr expansions for eigenfunctions and eigenvectors (24).

If, on the other hand, the processes are strongly coupled, Ω for the combined process will not be diagonalizable in the separate components Ω_1 and Ω_2. Strongly coupled processes will therefore appear in this formalism as a single process, which may nevertheless be a reasonable approximation of physical reality. It is worth noting that Eq. 2 may be applied in one of two ways. Taking internal motions as an example, if the character of each motion is precisely known, the eigenvector ϕ_n can be derived from the geometric constraints of the motion, leaving λ_n as the sole unknown. In this form, Eq. 2 is a straightforward generalization of existing relaxation theory. Alternatively, the formalism defined by Eq. 3 and contained in Eq. 2 allows both the eigenvectors ϕ_n and eigenvalues λ_n to be treated as unknowns. Given a sufficient number of experimental measurements, Eq. 2 can thus be solved for both the amplitudes and the frequencies of each motion. If n motions are involved, $2^n + n - 1$ measurements will be necessary to obtain a self-consistent solution of Eq. 2. Considering that a single set of ϕ_n and λ_n should account for the measured values of all relaxation parameters (T_1, T_2, $T_{1\rho}$, NOE) at all frequencies and for all nuclei, the problem is of manageable size, especially in systems in which cross-correlation and spin-diffusion effects may be neglected or accounted for. The calculation of eigenfunctions and eigenvalues to interpret relaxation data will be a practical necessity in many cases of interest in macromolecules. In proteins where one may encounter overall rotation (diffusion), domain rotation, segmental flexibility, side chain motions, and relaxation by ligand binding, one is not likely to be able to rigorously specify the range of each motion for each observable group. This use of Eq. 2 will be discussed separately. The discussion in this paper is confined to the simpler problem of comparing different motional models, given that the relaxation mechanism is known.

MODEL CALCULATIONS

A crucial question for deciding whether an analysis of the proposed type can lead to physically interesting insight is whether it is possible to distinguish between different kinds of motions (and combinations of motions), given a set of relaxation measurements. To answer this question we have carried out a series of computer calculations of relaxation rates according to Eq. 2, assuming a specific relaxation mechanism and varying (a) the amplitude, (b) the frequency of each motion, and (c) the number of motions contributing to relaxation. Our initial choice of motional models was determined by what we believe to be the simplest reasonable approximations to the actual internal motions occurring in proteins. T_1 T_2 and NOE for $^{13}C - {}^1H$ and $^1H - {}^1H$ dipolar interactions (at 100 and 360 MHz ^1H frequency) have been calculated from standard expressions (1). Calculations for $^{13}C - {}^1H$ relaxation will serve as an illustration here. Application to proton relaxation will be covered elsewhere.[1]

For $^{13}C - {}^1H$ relaxation:

$$\frac{1}{T_1} = K_1 R^{-6} \{ J(\omega_H - \omega_C) + 3J(\omega_c) + 6J(\omega_H + \omega_C) \}$$

$$\frac{1}{\pi T_2} = K_2 R^{-6} \{ 4J(0) + 3J(\omega_C) + J(\omega_H - \omega_C) + 6J(\omega_H) + 6J(\omega_H + \omega_C) \}$$

$$\text{NOE} = 1 + \frac{6J(\omega_H + \omega_C) - J(\omega_H - \omega_C)}{J(\omega_H - \omega_C) + 3J(\omega_C) + 6J(\omega_H + \omega_C)} \cdot \frac{\omega_H}{\omega_C}$$

where $R = 1.09$ Å is the $^{13}C - {}^1H$ distance, K_1 and K_2 are proportionality constants, and $J(\omega)$ the spectral density functions.

The internal motions assumed in the calculation of $J(\omega)$ have been classified as follows and used singly or in combination. (i) Three-state Woessner rotation: This well-known model is typical of a methyl group in a side chain, jumping between its three low energy states. (ii) Two-state wobble: A jump between two different sites of varying amplitude, typical of a tetrahedral side-chain carbon in which two of the three low-energy states have a high-energy barrier due to steric hindrance, a *cis-trans* isomerization of proline, or a segment of a polypeptide chain constrained on a hinge. (iii) Three-state wobble: A restricted jump between three sites of varying amplitude, as may occur where segmental flexibility is allowed.

Eight specific models have been compared thus far. (a) Isotropic diffusion; (b) semi-anisotropic diffusion (ellipsoid of revolution), which is formally equivalent to (c); (c) isotropic diffusion with internal continuous Woessner rotation; (d) isotropic diffusion with internal three-state Woessner rotation; (e) isotropic diffusion with internal two-state wobble; (f) isotropic diffusion with internal three-state wobble on a cone; (g) isotropic diffusion with an internal four-state (or *n*-state) wobble; (h) isotropic diffusion with internal three-state Woessner rotation plus a two-state wobble. To illustrate the calculation we use case f. Calculations for cases a–c have been published in detail by previous workers and are included here only for purposes of comparison. Calculations for cases d and e are equally simple. For case f we have

---

[1] King, R., and O. Jardetzky. Manuscript in preparation.

$$J(\omega) = \frac{2}{5} \sum_{n=-2}^{2} \left\{ \frac{\frac{1}{\tau_0} |<\tilde{Y}_n | \phi_1^{(w)} \cdot \phi_1^{(j)}>|^2}{\omega^2 + \left(\frac{1}{\tau_0}\right)^2} \right.$$

$$+ \frac{\left(\frac{1}{\tau_0} + 2\mu\right) |<\tilde{Y}_n | \phi_2^{(w)} \cdot \phi_1^{(j)}>|^2}{\omega^2 + \left(\frac{1}{\tau_0} + 2\mu\right)^2}$$

$$+ \frac{\left(\frac{1}{\tau_0} + 3\eta\right) |<\tilde{Y}_n | \phi_1^{(w)} \cdot \phi_2^{(j)}>|^2}{\omega^2 + \left(\frac{1}{\tau_0} + 3\eta\right)^2}$$

$$+ \frac{\left(\frac{1}{\tau_0} + 2\mu + 3\eta\right) |<\tilde{Y}_n | \phi_2^{(w)} \cdot \phi_2^{(j)}>|^2}{\omega^2 + \left(\frac{1}{\tau_0} + 2\mu + 3\eta\right)^2}$$

$$+ \frac{\left(\frac{1}{\tau_0} + 3\eta\right) |<\tilde{Y}_n | \phi_1^{(w)} \cdot \phi_3^{(j)}>|^2}{\omega^2 + \left(\frac{1}{\tau_0} + 3\eta\right)^2}$$

$$\left. + \frac{\left(\frac{1}{\tau_0} + 2\mu + 3\eta\right) |<\tilde{Y}_n | \phi_2^{(w)} \cdot \phi_3^{(j)}>|^2}{\omega^2 + \left(\frac{1}{\tau_0} + 2\mu + 3\eta\right)^2} \right\}$$

with the inner products:

$$<\tilde{Y}_n | \phi_1^{(w)} \cdot \phi_1^{(j)}> = \frac{1}{6} \sum_{i=1}^{2} \sum_{j=1}^{3} \tilde{Y}_n(\theta_{ij}, \varphi_{ij})$$

$$<\tilde{Y}_n | \phi_1^{(w)} \cdot \phi_2^{(j)}> = \frac{1}{6} \sum_{i=1}^{2} \frac{3}{2} \{\tilde{Y}_n(\theta_{i2}, \varphi_{i2}) - \tilde{Y}_n(\theta_{i3}, \varphi_{i3}')\}$$

$$<\tilde{Y}_n | \phi_1^{(w)} \cdot \phi_3^{(j)}> = \frac{1}{6} \sum_{i=1}^{2} \frac{1}{2} \{\tilde{Y}_n(\theta_{i2}, \varphi_{i2}) + \tilde{Y}_n(\theta_{i3}, \varphi_{i3}) - \tilde{Y}_n(\theta_{i1}, \varphi_{i1})\}$$

$$<\tilde{Y}_n | \phi_2^{(w)} \cdot \phi_1^{(j)}> = \frac{1}{6} \sum_{j=1}^{3} \{\tilde{Y}_n(\theta_{1j}, \varphi_{1j}) - \tilde{Y}_n(\theta_{2j}, \varphi_{2j})\}$$

$$<\tilde{Y}_n | \phi_2^{(w)} \cdot \phi_2^{(j)}> = \frac{1}{6} \cdot \frac{3}{2} \{\tilde{Y}_n(\theta_{12}, \varphi_{12}) - \tilde{Y}_n(\theta_{13}, \varphi_{13})$$
$$- \tilde{Y}_n(\theta_{22}, \varphi_{22}) + \tilde{Y}_n(\theta_{23}, \varphi_{23})\}$$

$$\langle \tilde{Y}_n | \phi_2^{(w)} \cdot \phi_3^{(j)} \rangle = \frac{1}{6} \cdot \frac{1}{2} \{ -2\tilde{Y}_n(\theta_{11}, \varphi_{11}) + \tilde{Y}_n(\theta_{12}, \varphi_{12})$$
$$+ \tilde{Y}_n(\theta_{13}, \varphi_{13}) + 2\tilde{Y}_n(\theta_{21}, \varphi_{21}) - \tilde{Y}_n(\theta_{22}, \varphi_{22}) - \tilde{Y}_n(\theta_{23}, \varphi_{23}) \}$$

where $\lambda_0 = 1/\tau_0$ is the rate of isotropic diffusion, μ the wobble (w) rate, η the jump (j) rate, and $\tilde{Y}_n$ (for a C – H vector) the nth second-order normalized spherical harmonic in the coordinate system $\tilde{Y}_n = \tilde{Y}_n \exp(\text{in } \varphi)$ describing overall rotational diffusion

$$\tilde{Y}_0 = \left(\frac{5}{4}\right)^{1/2} (1 - 3\cos^2\theta),$$

$$\tilde{Y}_{1,-1} = \left(\frac{15}{2}\right)^{1/2} \cos\theta \sin\theta,$$

$$\tilde{Y}_{2,-2} = -\left(\frac{15}{8}\right)^{1/2} \sin^2\theta.$$

$\tilde{\mathbf{Y}}_n$ is the vector obtained by computing $\tilde{Y}_n$ for the C – H vector at each of the wobble and jump states and $\langle \ \rangle$ is the inner product weighted by the equilibrium probabilities of the states in the Markov processes.

The transition matrices, the equilibrium probability vector ρ, and the eigenvalues and eigenvectors for the internal motions are given by:

Two-state wobble

$$\begin{array}{cc} & \begin{array}{cc} 1 & 2 \end{array} \\ \begin{array}{c} 1 \\ 2 \end{array} & \begin{bmatrix} -\mu_{12} & \mu_{12} \\ \mu_{21} & -\mu_{21} \end{bmatrix} \end{array} \quad \begin{array}{c} \rho \\ \begin{bmatrix} \mu_{21}/\mu_{12} + \mu_{21} \\ \mu_{12}/\mu_{12} + \mu_{21} \end{bmatrix} \end{array} \quad \begin{array}{c} \lambda \\ 0 \\ \mu_{12} + \mu_{21} \end{array} \quad \begin{array}{c} \varphi \\ \sqrt{\frac{\mu_{12}}{\mu_{21}}} \ -\sqrt{\frac{\mu_{21}}{\mu_{12}}} \end{array}$$

θ and φ arbitrary

Three-state Woessner

$$\begin{array}{c} \begin{array}{ccc} 1 & 2 & 3 \end{array} \\ \begin{array}{c} 1 \\ 2 \\ 3 \end{array} \begin{bmatrix} -2\eta & \eta & \eta \\ \eta & -2\eta & \eta \\ \eta & \eta & -2\eta \end{bmatrix} \end{array} \begin{bmatrix} 1/3 \\ 1/3 \\ 1/3 \end{bmatrix} \begin{array}{ccc} \rho & \lambda & \\ 0 & 1 & 1 & 1 \\ 3\eta & 0 & \sqrt{3}/2 & -\sqrt{3}/2 \\ 3\eta & -\sqrt{2} & (\sqrt{2})^{-1} & (\sqrt{2})^{-1} \end{array}$$

$$\begin{array}{cccc} 1 & 2 & 3 & \\ 71° & 71° & 71° & \theta \\ 0 & 120° & 240° & \varphi \end{array}$$

with θ and φ obtainable by straightforward combination. The results of calculations for different motional models can be compared on plots of pairs of predicted relaxation

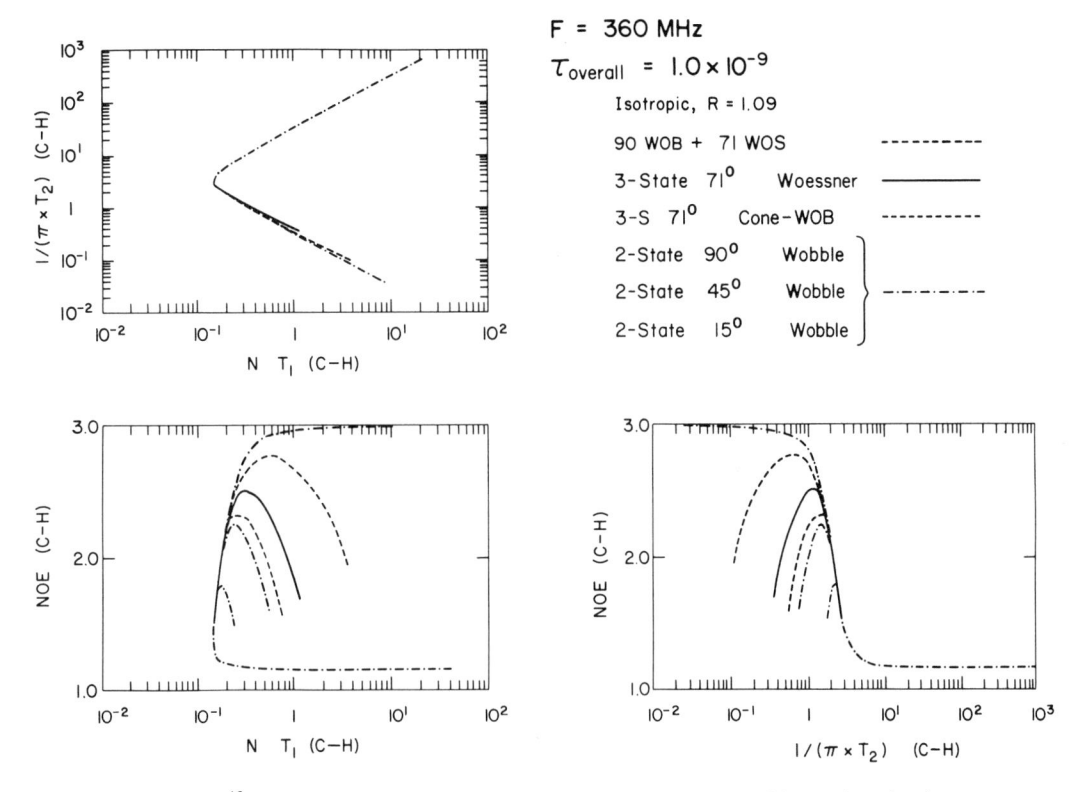

FIGURE 1 ^{13}C NMR relaxation phase diagrams for different models of internal motion in a macromolecule with a rotational diffusion constant $\tau = 10^{-9}$.

parameters, i.e., T_1 vs. $1/\pi T_2$, T_1 vs. NOE, NOE vs. $1/\pi T_2$, etc. In each plot as shown in Figs. 1–3, τ_0 is constant, $\tau_\perp$ and $\tau_\parallel$ for internal motion are variable. We find this presentation particularly useful and refer to it as a "relaxation phase diagram." It clearly defines the extent to which different combinations of assumed motions predict different combinations of observed relaxation parameters. This is important, because a unique interpretation of the data is not possible if different models account for them equally well. Furthermore, the phase diagrams emphasize that for a meaningful interpretation several relaxation parameters at several spectrometer frequencies must be accounted for by the same model. Inconsistencies between predicted and observed values for even one of the relaxation measurements should be taken as prima facie evidence that the chosen model does not apply.

Several general features emerge from the examination of diagrams such as those shown in Figs. 1–3.

(a) If the overall rotational correlation time is relatively low, discrimination between different models is best when the correlation times for internal motions are short. It becomes progressively more difficult as the correlation times for internal motions increase.

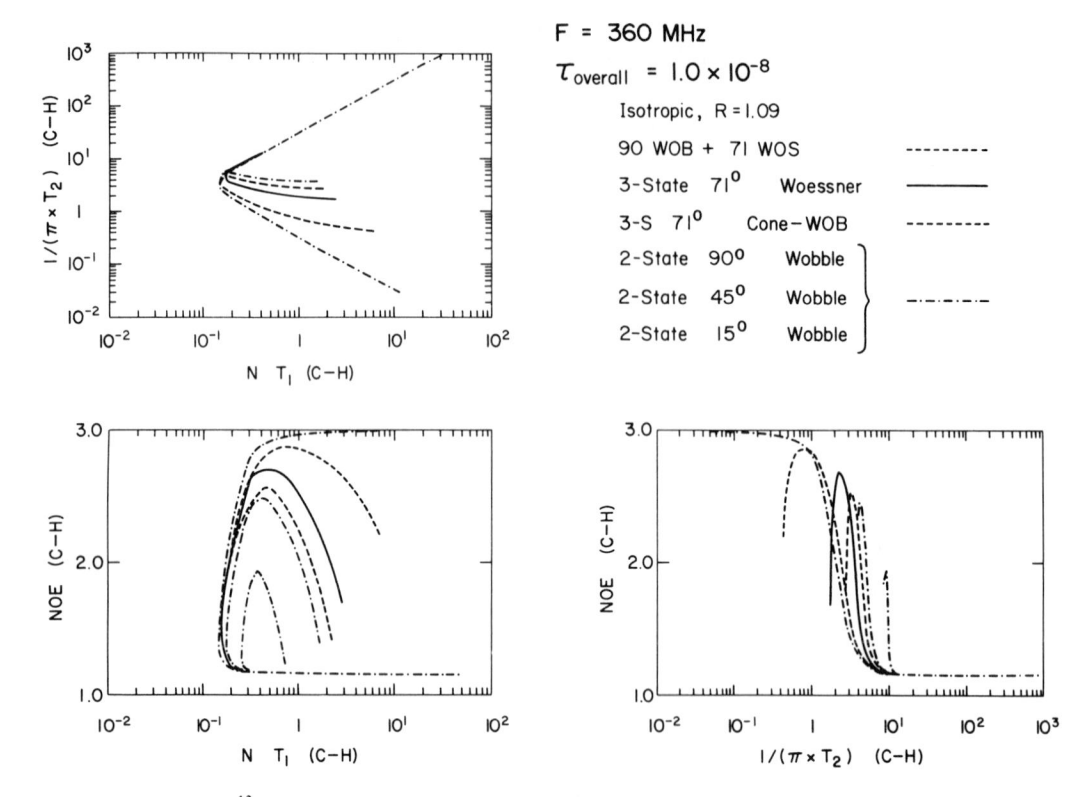

FIGURE 2 ^{13}C NMR relaxation phase diagrams for different models of internal motion in a macromolecule with a rotational diffusion constant $\tau = 10^{-8}$.

(b) For a given internal correlation time, motions of different amplitude can be distinguished if the amplitudes are relatively small. At large amplitudes, different motions become relatively indistinguishable.

(c) Discrimination between the effects of different motions is possible when either their frequencies or their amplitudes are very different. If a slower motion is superimposed in a rapid motion, discrimination from a model with a single correlation time intermediate between the two is generally possible. If two rapid motions of large amplitude are superimposed, no distinction can be made from a model with a single short correlation time. In the extreme narrowing limit, some discrimination between different motional models is still possible by considering the NOE, but not from T_1 and T_2 alone.

(d) If the overall rotational correlation is short, contributions of internal motions become undetectable or indistinguishable from each other.

(e) Motions of low amplitude at a given τ are readily discriminated by their T_2 values, but their frequencies can only be determined by a comparison of NOE and T_1 or NOE and T_2 data.

On this basis it is possible to define the general requirements for the interpretation of

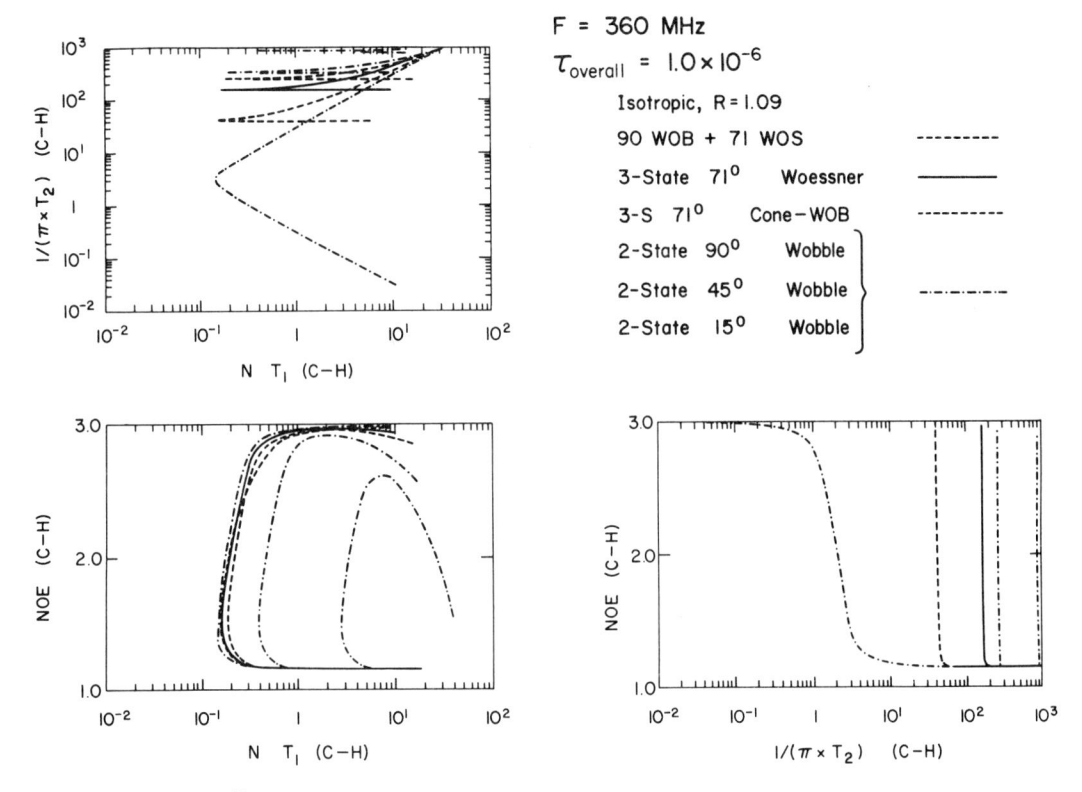

FIGURE 3 ^{13}C NMR relaxation phase diagrams for different models of internal motion in a macromolecule with a rotational diffusion constant $\tau = 10^{-6}$.

relaxation data on systems with multiple motions: (a) Data for all three, T_1, T_2, and NOE, measurements are necessary to provide a check of internal consistency; (b) the dominant relaxation mechanism must be established; (c) if the system has an overall motion corresponding to the rotation of a coordinate system originating in the center of mass, the frequency, amplitude, and symmetry of this motion should be determined by an independent method. (d) The experimental values for T_1, T_2, and NOE can be checked against a variety of motional models in the diagrams plotted for the appropriate parameters of the overall motion. (e) For existing models the values of τ_0 and τ_1 can be read off a conventional T_1 vs. τ plot for the chosen model. Fig. 4 is a previously unpublished sample plot for our case (f).

In some cases it will be found that the measurements correspond to a single specific model. In others, several models will fit the data. It might be noted that the probability of finding some model that fits all the data is very high. The probability of finding more than one model is high for those cases where the motional amplitudes are large and the overall τ is short. This is clearly a case in which a unique interpretation of relaxation data is impossible in principle. This is not to say, however, that the data provide no information about the system. It should be possible to define a class of

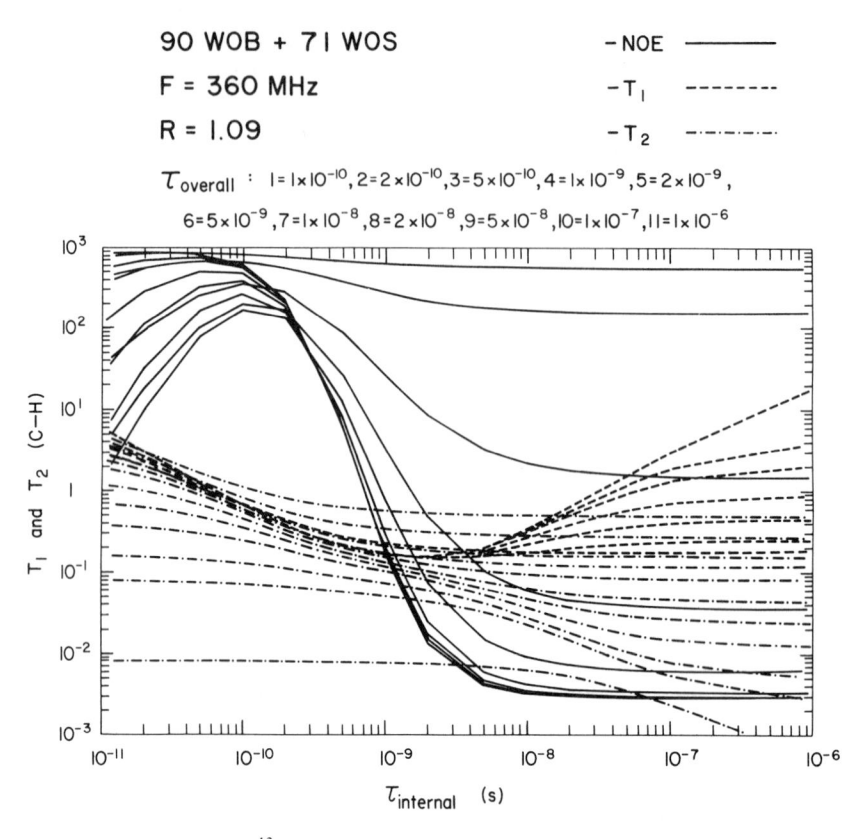

FIGURE 4 Dependence of ^{13}C NMR relaxation parameters on both the overall diffusional (τ_c) and the internal (τ_i) correlation times for the 90° Woessner model of internal motion (cf. text).

models compatible with the data and to set outer limits to the amplitudes and frequencies of these motions. Thus, for example, it might be possible to conclude that a given set of measurements cannot be accounted for by low amplitude motions of any frequency but is compatible with any number of high amplitude motions in a certain frequency range. Another set of measurements may not be compatible with any high amplitude motion, but requires a low amplitude motion in addition to the overall rotation for its explanation. Even if the amplitudes and frequencies can be specified only as allowed ranges, for some systems this provides considerably more insight into their internal dynamics than is possible by any other means.

If the system under study is a folded linear polymer, further discrimination between models in a given allowed class may be possible by (a) determining the overall motion of the macromolecule and its symmetry by light scattering or other techniques and using the parameters as knowns in Eq. 2, common to all observed groups; and (b) comparing allowed motions for neighboring segments. Most of the models will require that at least some of the motions of adjacent segments be correlated. Thus, if diffusional parameters and the sequence of the polymer chain are known, it should be possible to narrow down the choice of allowed models to one or, at most, to very few.

This research was supported by National Science Foundation Grant PCM75-02814 and National Institutes of Health Grant RR00711.

Received for publication 1 December 1977.

REFERENCES

1. ABRAGAM, A. 1961. The Principles of Nuclear Magnetism, Oxford University Press, New York.
2. JARDETZKY, O. 1964. The study of specific molecular interactions by nuclear magnetic relaxation measurements. *Adv. Chem. Phys.* **7**:499.
3. OLDFIELD, E., R. S. NORTON, and A. ALLERHAND. 1975. Studies of individual carbon sites of proteins in solution natural abundance carbon 13 nuclear magnetic resonance spectroscopy. *J. Biol. Chem.* **250**:6368.
4. OPELLA, S. J., D. J. NELSON, and O. JARDETZKY. 1976. An approach to the quantitative study of internal motions in proteins by measurements of longitudinal relaxation times and nuclear Overhauser enhancements in proton decoupled carbon-13 NMR spectra. *Am. Chem. Soc. Symp. Ser.* **34**:397.
5. DESLAURIERS, R., E. RALSTON, and R. L. SOMORJAI. 1977. Conformational studies of angiotensin-II. Least-squares fit of carbon 13 nuclear magnetic resonance relaxation times to extended and folded conformations. *J. Mol. Biol.* **113**:697.
6. JONES, W. C., T. M. ROTHGEB, and F. R. N. GURD. 1976. Nuclear magnetic resonance studies of sperm whale myoglobin specifically enriched with ^{13}C in the methionine methyl groups. *J. Biol. Chem.* **251**:7452.
7. TORCHIA, D. A., and D. L. VANDERHART. 1976. ^{13}C magnetic resonance evidence for anisotropic molecular motion in collagin fibrils. *J. Mol. Biol.* **104**:315.
8. HULL, W. W., and B. D. SYKES. 1975. Fluorotyrosine alkalene phosphatase: internal mobility of individual tyrosines and the role of chemical shift anisotropy as a ^{19}F nuclear spin relaxation mechanism in proteins. *J. Mol. Biol.* **98**:121.
9. WUTHRICH, K. 1975. NMR in Biological Research: Peptides and Proteins. American Elsevier Publishing Co., Inc., New York.
10. JARDETZKY, O., K. AKASAKA, D. VOGEL, S. MORRIS, and K. C. HOLMES. 1978. Unusual segmental flexibility in a region of tobacco mosaic virus coat protein. *Nature (Lond.).* In press.
11. WOESSNER, D. E. 1962. Nuclear spin relaxation in ellipsoids undergoing rotational Brownian motion. *J. Chem. Phys.* **37**:647.
12. WOESSNER, D. E., B. S. SNOWENDEN, JR., and G. H. MEYER. 1969. Nuclear spin-lattice relaxation in axially symmetric ellipsoids with internal motion. *J. Chem. Phys.* **50**:719.
13. LEVINE, Y. K., P. PARTINGTON, and G. C. K. ROBERTS. 1973. Calculation of dipolar nuclear magnetic relaxation times in molecules with multiple internal rotations. I. Isotropic over-all motion of the molecule. *Mol. Phys.* **25**:497.
14. LEVINE, Y. K., N. J. M. BIRDSALL, A. G. LEE, J. C. METCALFE, P. PARTINGTON, and G. C. K. ROBERTS. 1974. Calculation of dipolar nuclear magnetic relaxation times in molecules with multiple internal rotations. II. Theoretical results for anisotropic overall motion of the molecule, and comparison with ^{13}C relaxation. *J. Chem. Phys.* **60**:2890.
15. HUNTRESS, W. T. 1970. The study of anisotropic rotation of molecules in liquids by NMR quadrupolar relaxation. *Adv. Magn. Res.* **4**:1.
16. WALLACH, D. 1967. Effect of internal rotation on angular correlation functions. *J. Chem. Phys.* **47**:5258.
17. VAN PUTTE, K. 1970. Spin-lattice relaxation in anhydrous sodium and lithium soaps between 23° and 180°C. *J. Magn. Res.* **2**:23.
18. NOACK, F. 1971. Nuclear magnetic relaxation spectroscopy. *In* NMR: Basic Principles and Progress. P. Diehl, et al., editors. **3**:83.

19. KING, R., and O. JARDETZKY. 1978. A general formulism for the analysis of NMR relaxation measurements on systems with multiple degrees of freedom. *Chem. Phys. Lett.* **55**:15.
20. KUBO, R. 1962. In Fluctuation, Relaxation and Resonance in Magnetic Systems, J. ter Haar, editor. Oliver and Boyd Ltd., Edinburgh. 23.
21. ANDERSON, P. W. 1954. A mathematical model for the naming of spectral lines by exchange or motion. *J. Phys. Soc. Jpn.* **9**:316.
22. JOHNSON, C. S. 1965. Chemical rate processes and magnetic resonance. *Adv. Magn. Res.* **1**:33.
23. BHARUCHA-REID, A. T. 1960. Elements of the Theory of Markov Processes and their Applications. McGraw Hill Book Company, New York.
24. MESSIAH, A. 1961. Quantum Mechanics. North Holland Publishing Co. Amsterdam.

DISCUSSION

NAGESWARA RAO: Could you please elaborate on some of the terms in your theory? For example, what criteria do you use to determine whether two or more motional processes are weakly or strongly coupled? This puzzles me, especially because you have invoked independent motions in writing Eq. 2 in your paper. It would be especially helpful if you could provide some illustrative examples of weakly coupled and strongly coupled motions within the framework of your independent motion model.

JARDETZKY: If you are using the method with prejudice, that is, using a specific model for every motion, you can use mathematical criteria for deciding whether they are weakly or strongly coupled. You can see whether you can best fit the data by one or another specific motional models with either weak or strong coupling on the transition operator. If you are using the method without prejudice, you have to do this by trial and error, and you have to see, for example, whether you can account for the data by, let us say, two motions, or whether you require a model of three motions, two of which are weakly coupled.

NAGESWARA RAO: You see, in your Eq. 2 the eigenvalues seem to be τ_n^{-1}. Now, if I just use my elementary knowledge of quantum mechanics in considering whether the two states are coupled, I would consider the magnitude of the off-diagonal element and compare it to the difference between the two diagonal elements. What is this off-diagonal element in your case?

JARDETZKY: If you are using the method without prejudice, I suppose you would call it a degree of coupling.

NAGESWARA RAO: Well, let's go back to the quantum mechanics that you invoked as the analogy. In quantum mechanics we talk of an interaction causing this off-diagonal element. Since you didn't pick an example, let me pick one. One could be an overall reorientation. The other could be a segmental motion. The two uncoupled motions may be represented by the correlation times for the two motions. What would you consider an off-diagonal interaction that would couple an overall motion to the segmental motion?

KING: If the overall motion were discrete and made transitions between several states, and if while that overall motion was changing among those states, the molecule made a segmental motion so that the motions were stochastically coupled and not statistically independent, then you would have an off-diagonal interaction term.

NAGESWARA RAO: I don't think you are in a position really to write down the quantum mechanical states of these molecules, especially for the kind of molecules you are thinking of.

Thus for the description of your orientational degrees of freedom, you will be compelled to use a classical formulation using terms like angles and angular velocities. When you talk of angular velocity fluctuations or angle fluctuations, you'll be dealing with something like a friction tensor. So the question is, how would you couple the two friction tensors, if you can define the two in the first place?

JARDETZKY: I think you are right insofar as there is very little hope of analyzing these complex motions in the kind of detail that you suggest. The analogy with the point of mechanical analysis is not bad, but it is not perfect. This formulation does have the additional factor built into it that you can in fact derive the eigenvectors or at least the corresponding probability amplitudes as well as the eigenfunctions from experimental data. And if you can do that, then you can devise an expression defining the degree of coupling. You can also try specific detailed models, to see whether they are consistent with the kind of numbers that you get from experiments. You can do another thing: you can test simplified statistical mechanical theories for these different motions. Some of this has already been done by polymer chemists and some will have to be done for specific situations in proteins.

NAGESWARA RAO: You see I am appealing to you because you have done all these calculations. Is there some way one can guess this strong coupling or weak coupling offhand, or is there some rule of thumb you use for this purpose? That's what I'm trying to learn from this discussion.

JARDETZKY: We have not done nearly enough in terms of the calculations on coupled motions to give an answer to your question. I think you can use intuition. The most likely case is a combination of segmental and side-chain motions. If you can approximately define the states for these motions from statistical and mechanical arguments, you're likely to make an intelligent guess. I am not kidding myself that these are going to be more than intelligent guesses. I think that's something that needs to be explored further.

NAGESWARA RAO: Now for my second question: Your analysis primarily considers isotropic overall motion and only the intramolecular dipolar interactions. Could you comment on how you would modify the theory for (a) anisotropic overall reorientation (Have you tried to use the more accurate model of Hubbard rather than Woessner's model you used in your paper?), (b) effects of chemical shift anisotropy especially for ^{19}F, ^{31}P, and ^{13}C at high magnetic fields and (c) contributions from intermolecular dipolar interactions? It seems all these three factors would cause problems in the ability to extract meaningful motional information.

JARDETZKY: To answer your last question first. We are not trying to claim that you can always extract meaningful motional information. We are trying to define the conditions under which you can and the conditions under which you cannot extract meaningful motional information from relaxation data. I think on the basis of what we have already done it's very clear that there are many conditions under which it will not just be difficult, it will be impossible to extract any information. And I think a case in point is rapid motion in small molecules. Many people, for example, assume a single correlation time for something like a nucleotide, as if it were rotating as a rigid sphere, which it is not, and calculate distances. Well, this kind of analysis shows you that that's nonsense, because if there are internal motions in this kind of small molecule, you have to use a combination of correlation times. The choice of the combination of correlation times will affect the magnitude of the calculated distance. But in the extreme narrowing limit, different motional models are indistinguishable unless the distances are known a priori; the best you can hope to do is determine a set of distances and correlation times compatible with the relaxation data.

The theory needs no modification to take into account anisotropic motion, chemical shift anisotropy, or intermolecular interactions. The theory is written in terms of spectral density functions, and to consider different relaxation mechanisms, one simply must use the appropriate coefficients. Anisotropic motion results in additional terms.

NAGESWARA RAO: I'm afraid that considering anisotropic diffusion as opposed to isotropic diffusion would introduce more complications than you said. Not only must you introduce two more elements of the diffusion tensor, but also you will be required to introduce angles that specify the orientation of your dipolar vector with reference to the principal axes of the diffusion tensor, which are, in general, not known. And you have to do this for every pairwise dipolar interaction in your Hamiltonian.

JARDETZKY: That is correct.

TORCHIA: Your observation that different protons have different T_1 values doesn't necessarily imply that spin diffusion is not taking place. Although spin diffusion is not sufficiently strong to make all T_1 values the same, you cannot be sure that it does not have some influence on the proton T_1 values.

JARDETZKY: That is right. I think you have to realize with all of these measurements what is meant first of all by the 10% experimental error. You will not be able to say that there is no 10% contribution of spin diffusion. I think you can differentiate between T_1's and T_2's, so spin diffusion does not dominate. Otherwise it would be the predominant term in both cases. Maybe it is a 20% contribution but we deliberately don't worry about that.

TORCHIA: That brings me to another point. If you have reasonably large experimental uncertainties that are unavoidable because of the complexity of the molecules, I am not sure how you can discriminate between specific models for motion. Available theory permits analysis of spin relaxation in terms of moderately complex models of molecular motions, and I very much doubt that one will be able to go beyond this level of analysis with your general theory, because of uncertainties in the data.

JARDETZKY: I think in the analyses of the relaxation data you are in exactly the same boat as you were with crystal structure determination. You look at your diffraction diagram just to see how far you can analyze it, whether you can get a 14, 6, 2, or 1 Å resolution. Our framework doesn't guarantee that you will be able to reach the equivalent of 1.8 Å resolution. You may, in fact, have to do all kinds of labeling and employ all kinds of tricks to obtain sufficient data. The framework can guide you and allow you to define to what extent you can and to what extent you cannot extract information from the relaxation data.

BADEA: Suppose we are in a favorable time domain for distinguishing among different models of motion in the sense described in your paper. How different are the anisotropy correlation times obtained by fitting the acceptable models? Would you compare the accuracy of NMR measurements with other physical techniques capable of monitoring molecular motion in that particular time domain?

JARDETZKY: I think that the accuracy of the correlation times you get is probably within a factor of two. That is a clinical impression. We have written programs to see to what extent we can measure the goodness of fit. This work is not yet finished.

BADEA: You are offering two or more explanations.

JARDETZKY: That is right, two or three or four. I see what you are saying. However, you can introduce certain quite rigorous simplifications. You can, for example, derive the terms for overall motion from depolarized light-scattering experiments and you can use the appropriate exponentials as knowns. Then you can fit the unknown exponentials. If you are dealing with two or more resonances from the same side chains, you will have at least one additional exponential in the motion common to both or all. So you can do internal consistency calculations to narrow down your choice of models. Still, in many cases you probably will not be left with a single model; you will be left with a choice of a finite number of models.

NEW ELECTRIC FIELD METHODS IN CHEMICAL RELAXATION SPECTROMETRY

A. PERSOONS AND L. HELLEMANS, *Laboratory of Chemical and Biological Dynamics, Department of Chemistry, K.U. Leuven, Leuven, Belgium*

ABSTRACT New stationary relaxation methods for the investigation of ionic and dipolar equilibria are presented. The methods are based on the measurement of nonlinearities in conductance and permittivity under high electric field conditions. The chemical contributions to the nonlinear effects are discussed in their static as well as their dynamic behavior. A sampling of experimental results shows the potential and range of possible applications of the new techniques. It is shown that these methods will become useful in the study of nonlinear responses to perturbation, in view of the general applicability of the experimental principles involved.

INTRODUCTION

Electrostatic interactions between ions or dipolar molecules are fundamental to the molecular mechanisms of some major phenomena in physiology. The theoretical description of these electrostatic effects and their experimental investigation—a brilliant chapter of classic physical chemistry—are mostly restricted to the range where current densities are linearly related to the applied electric field. However, some important processes in physiology, e.g., membrane permeability changes, are unconditionally linked to the proper function of supramolecular structures whose boundaries are the site of very intense electric fields. At such high field strengths field-dependent increments in the electric admittance appear. Their amplitude and frequency behavior are related to saturation effects and molecular mobilities, respectively. In chemically reactive systems, nonlinear field effects arise as a consequence of the coupling between chemical processes and the electric field. The physical and chemical phenomena due to electrostatic interactions can therefore be characterized completely—in their energetics as well as dynamics—from the nonlinear response of the system to which an intense electric field is applied. The understanding of the observed effects needs a theoretical description on a molecular level. The current lack of theoretical developments, in particular of the dynamic aspects, may be attributed to mathematical complexity, but even more so to the absence of relevant experimental data that would instigate the design of adequate models.

It is clear that advances in our understanding of electric field effects and of their dynamics will come from the development of new experimental techniques. In the recent past we developed methods that allow the investigation of field effects and their

---

L. Hellemans is a research associate of the Belgian Research Council (bevoegdverklaard navorser N.F.W.O.).

frequency dependence. Applied to chemically reactive systems, our methods yield information on rate processes that involve ionic and/or dipolar species. These stationary relaxation techniques can be applied to the kinetic investigation of nonlinear responses upon any perturbation of a reactive system, as a result of the generality of the experimental principles involved.

Chemical Systems and Electric Fields

Any system whose energy depends on the electric field density will oppose the change in energy resulting from the application of an electric field. Generally, the electric field acts upon translational and rotational properties of the species present and induces such an anisotropy in the system as will reduce the field density; these effects are linear only at the lowest field strengths. However, in chemical reactive systems the gain in electrostatic stability can be balanced by a corresponding reduction in free energy. This is revealed by a shift in chemical equilibrium composition toward the state that interacts more strongly with the electric field. The chemical effects account for the nonlinear response of a reactive system subjected to high field conditions and are well known for ionic as well as for dipolar equilibria. In this section we illustrate the theoretical approach to these effects and show their relation with chemical dynamics.

As first noted by Wien, the conductance of a solution of a weak electrolyte increases upon the application of a high electric field (1). The effect is known as the second Wien effect (the first Wien effect being the conductance increase with field in solutions of strong electrolytes) or the field-dissociation effect because it is related to increased dissociation.

A system containing free charges never reaches equilibrium in the presence of an electric field. Therefore, a thermodynamic approach to the field-dissociation effect faces many problems. A successful description was first given by Onsager (2) from an electrodiffusion treatment of the effect. Neglecting shielding by the ionic atmosphere, as well as the hydrodynamic interactions between ions, Onsager obtained for the relative increase of dissociation of an ion-pair into free ions under field conditions:

$$K(E)/K(E = 0) = 1 + 2\beta q + \tfrac{1}{3}(2\beta q)^2 + \tfrac{1}{18}(2\beta q)^3 + \ldots \qquad (1)$$

The parameter $q(= -e_1 e_2/8\pi\epsilon_0 \epsilon kT$, symbols with their usual meaning) represents the distance within which the mutual electrostatic interaction of the two ions with charges e_1 and e_2 becomes higher than their thermal interaction with the surrounding medium. This distance is known as the Bjerrum distance and may be regarded as the radius of the association sphere, i.e., the sphere within which the two ions are considered as bound. The parameter $2\beta(= |(e_1 u_1 - e_2 u_2)E|/(u_1 + u_2)kT$, in which E is the external field, is a reciprocal distance which becomes independent of the mechanical mobilities u for symmetrical electrolytes. In the symmetrical case $(2\beta)^{-1}$ represents the distance at which two oppositely charged ions form a dipole which gained an electrostatic stabilization energy of $kT\cos\theta$, θ being the angle between the dipole moment and the field direction.

The mathematical treatment by Onsager is very difficult in some of the crucial steps of the derivation of Eq. 1. It would therefore be worthwhile to have a simple picture of the physical process responsible for the facilitation of dissociation of an ion-pair in field conditions. Such a picture can be made for a symmetrical electrolyte where apparently no parameters of transport theory appear in Onsager's result. We consider in this case an ion-pair as any configuration of two oppositely charged ions within the Bjerrum-association sphere. The configuration with the two ions in contact—or solvent-separated—will have the highest probability. The configuration of two ions at the Bjerrum distance has a fleeting existence and may be considered as the transition state for the dissociation process; this configuration has equal probability of collapsing into the ion-pair state as of separating into free ions. It is now useful to consider the facilitated dissociation under field conditions as a consequence of the reduction in electrostatic work needed to separate the "Bjerrum dipole" into free ions. An alternative point of view would be to consider the increase in free ions under field conditions to be the result of a distortion of the Bjerrum sphere.

We can now write for the standard free energy change of dissociation with and without field, assuming a fast rotation of the Bjerrum dipole toward the most favorable orientation in the field:

$$\Delta G°(E) = \Delta G°(E = 0) - \mu_{Bj} E N_0. \qquad (2)$$

Eq. 2 may directly be transformed into an expression for the dissociation constant under field conditions:

$$K(E) = K(E = 0) \exp(\mu_{Bj} E / kT). \qquad (3)$$

Writing μ_{Bj} as $e_0 q$, we obtain, by expanding the exponential and retaining only the first term:

$$K(E)/K(E = 0) = 1 + 2\beta q. \qquad (4)$$

In the linear approximation, this is the result of Onsager's theory. Although the picture presented may seem to be very crude—especially in neglecting the nonequilibrium aspects of the field-dissociation effect—we nevertheless gain some insight in the dynamic aspects. This picture introduces the hydrodynamic interactions within the association sphere. We may conjecture that for electric fields of short duration, or high frequency AC fields, the field-dissociation effect will disappear not as a result of the relaxation of the association-dissociation equilibrium but as a consequence of the finiteness of orientational mobility within the association sphere.

Unlike ionic equilibria, dipolar systems may reach a new equilibrium under field conditions. This situation allows a thermodynamic treatment. However, considering the dynamic features of the response of reactive dipolar systems, care should be taken of the possible coupling of orientational and chemical effects. It is, therefore, worthwhile to give a short illustration of the physical aspects of the interaction of dipoles with an electric field. Generally, orientational polarization is achieved by the electric field exerting a torque on any dipolar species present. The energy U of the polar particle decreases with the decrease of the angle θ between dipole moment and directing field F according to $U = -\mu F \cos\theta$. A consequence of chemical importance is the diffusion of polar species toward the region of highest field strength such as is encountered in systems with inherent anisotropy, e.g. membranes (3). For a mole of dipoles, one obtains for the electrostatic energy, neglecting contributions of induced polarization:

$$\overline{U} = -N_0\mu F L(\mu F/kT), \tag{5}$$

L being the Langevin function. In the linear approximation:

$$\overline{U} = -N_0\mu^2 F^2/3\,kT. \tag{6}$$

At high field strength the effect is susceptible to saturation when the polar particles become fully aligned with the field. The fact that orientational polarization becomes increasingly difficult is reflected in a decrease of the permittivity $\epsilon_0 \cdot \epsilon$ with field strength.

The dipolar contribution to polarization vanishes when the field frequency approaches the inverse of the orientational relaxation time τ_{or}, a quantity depending on particle size and solvent viscosity. Of course this is true under high field conditions as well as in the absence of an external high field; it then represents a classic case of dielectric relaxation.

Now we shall direct our attention toward the influence of high electric fields on chemical systems in dynamic equilibrium. When such a system consists of species with different polar character, the more polar ones will become stabilized most, requiring the readjustment of the true equilibrium composition under high field. Inasmuch as the stability gain is related to energy, the field-induced equilibrium shift must depend on the square of the field strength, as indicated by Eq. 6. This defines the effect as a nonlinear phenomenon. Similarly, Eq. 6 shows that the amplitude of the shift varies with $\Sigma\,\nu_i\mu_i^2$, in which ν_i are the stoichiometric coefficients of the reaction partners defined as usual. In a formal way we write a van't Hoff-type relation for the field dependence of the equilibrium constant K:

$$d\ln K/dE = \Delta M/RT \tag{7}$$

where ΔM is the molar change of electric moment, a quantity depending on $\Sigma\,\nu_i\mu_i^2$ and the electric field E.

In the foregoing we have deliberately omitted conditioning the validity of the result in terms of the ratio of chemical to orientational relaxation times. Schwarz (4) has shown that for cases where $\tau_{chem}/\tau_{or} \ll 1$ pseudo-orientational polarization occurs through chemical reaction, even at low field conditions, without, however, affecting the bulk equilibrium composition. Without regard to the value of the ratio τ_{chem}/τ_{or} we can state that a chemical mode of incremental polarization will be induced by the action of the field, when it consists of a low-amplitude, high-frequency field superposed on a steady high field, as in the customary experimental approach (5). (It is imperative to have L significantly larger than 0 by the DC component of the field when $\tau_{\text{chem}}/\tau_{\text{or}} \ll 1$.) The polarization is revealed by a permittivity increment which will relax at frequencies dictated by the rate of the chemical process by which the polarization is effected.

The amplitude of the field-induced permittivity increments is dependent on the magnitude of the equilibrium shift ($d \ln K/dE$) as well as on the molar change of electric moment (ΔM), which translates the change of concentration C_i to permittivity. From this we conclude that the increment is proportional to the following quantities:

$$\Delta \epsilon_{\text{chem}}(\omega) \simeq \Gamma(\Sigma \nu_i \mu_i^2)^2 E^2 \phi(\omega). \tag{8}$$

Γ is the well-known reaction capacity ($1/\Gamma = \Sigma \nu_i^2/\overline{C}_i$) and $\phi(\omega)$ the pertinent relaxation function. The dispersion of the nonlinear effect thus carries information regarding the reaction rates of the field-perturbed equilibrium and other equilibria coupled to it. The increment has the property of any accurately measured virial coefficient which is to provide a wealth of information, in this case of structural (μ_i) and thermodynamic ($\overline{C}_i$) nature.

EXPERIMENTAL METHODS

The nonlinear contributions to the electric properties of a chemical reactive system subjected to high electric fields can in principle be measured from any electric circuit in which the system is used as a circuit element. This is the central idea for the methods we developed to measure the nonlinearities in electrolyte solutions and dipolar systems. After some investigation it became obvious that the sample solutions—subjected to changing electric fields—could be used directly as the active element in a modulator circuit whose performance is critically dependent on the nonlinear behavior of the active element.

In the past the field-dissociation effect has been measured from the conductance increase upon the application of a high field (6). To increase the sensitivity of these measurements, we developed a stationary method that uses repetitive high frequency, high amplitude pulses to modulate the conductance of the sample solution. In this way the sample becomes a circuit

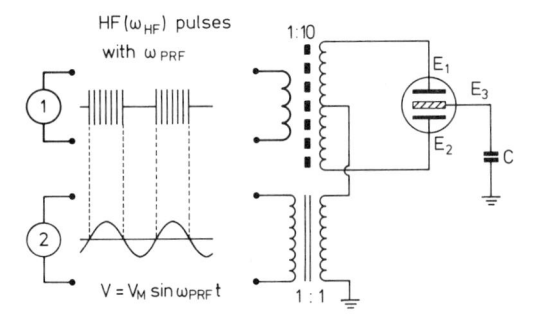

FIGURE 1 Schematic circuit for the measurement of the field-dissociation effect and its dispersion. Generator 1 is the square-wave (1 Hz–100 kHz) modulated high frequency (100 kHz–2 MHz) power oscillator (0.1–50 kV$_{\text{peak-to-peak}}$) driving the step-up (1:10) ferrite-core transformer. The ferrite core contains an adjustable air gap for tuning. Generator 2 is the low voltage sinusoidal signal generator running in synchrony with the square-wave modulating signal.

element whose nonlinear behavior can be determined from an analysis of the harmonics generated.

The principle of the method will be discussed in reference to Fig. 1 which shows a schematic of the circuit used. To the high voltage electrodes (E_1 and E_2) of the sample cell, a high-frequency voltage of high amplitude is applied as a square-wave[1] modulated signal. This high voltage is obtained from a step-up ferrite-core transformer driven from a modulated power oscillator. Depending on experimental conditions, several sample cells of a widely different design—e.g. high pressure cells—were developed. The spacing between high-voltage electrodes and the center electrode (E_2) is usually 1 mm, and electric fields of peak-to-peak amplitude of about 1–10 kV cm^{-1} are used. The center electrode of the sample cell is always short circuited for AC signals through capacitor C. The centertap terminal of the high voltage transformer is connected to a low-voltage transformer driven from a sinusoidal generator synchronously locked on the square-wave signal used to modulate the high voltage. As a consequence of the field-dissociation effect, the sample resistance will be lowered when subjected to the high-frequency high-amplitude field. At this point it should be clearly stated that this technique uses high-frequency alternating fields, whereas Onsager's theory is derived for fields of constant amplitude. In the field-on period, the capacitor C is charged through the lowered cell resistance. Because of the higher cell resistance in the field-off period, the charge on capacitor C cannot flow back completely during this period. Therefore, under stationary conditions, a net amount of charge will reside on capacitor C and can be measured as a DC voltage. It is evident that the DC signal is proportional to the amplitude of the measuring signal (V_M) and also to the relative conductance change of the sample solution. As noted in the preceeding section, an influence by the high frequency of the perturbing field pulses is also to be expected. A mathematical analysis—retaining only the square-wave envelope of the field pulses and neglecting their internal structure—of the properties of the circuit used gave for the DC voltage on capacitor C (7, 8):

$$V_{DC} = (\Delta\sigma/\sigma)(V_M/\pi) = g(\Delta\sigma/\sigma)_{Ons}(V_M/\pi). \qquad (9)$$

g is a correction factor taking into account the deviations of the experimental value of $\Delta\sigma/\sigma$ from the value $(\Delta\sigma/\sigma)_{Ons}$ calculated according to Onsager's theory.

In the foregoing discussion the response of the electrolyte solution upon perturbation was implicitly assumed to be instantaneous. This is correct as long as the pulse duration is relatively long—at low pulse repetition frequencies—and the ionic equilibrium is completely in phase with the perturbation. However, the ionic equilibrium change will progressively lag behind the perturbation as the pulse repetition frequency approaches the reciprocal relaxation time of the ionization equilibrium. This lag in response results in a smaller conductance increment lowering the modulation efficiency. The experimental consequence is a decrease of the DC voltage on capacitor C with increasing pulse repetition frequency. Ultimately, at pulse repetition frequencies much higher than the reciprocal relaxation time of the ionic equilibrium, the DC signal on capacitor C will vanish altogether. A complete analysis of the time-dependent properties of the modulator circuit yields for the DC signal as a function of pulse repetition frequency—again neglecting the high frequency structure within each pulse (7, 8):

$$V_{DC} = (\Delta\sigma V_M/\sigma\pi)\, 1/(1 + \omega^2\tau^2) = V_{DC\,max}/(1 + \omega^2\tau^2). \qquad (10)$$

This equation, derived under the usual assumption of small perturbations, is recognized as a

---

[1] In a new design sinusoidal modulation is used, which is realized more easily from an experimental point of view.

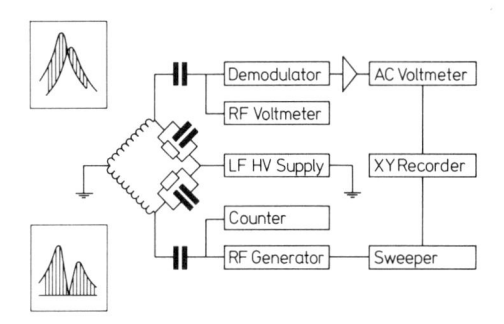

FIGURE 2 Block diagram of the apparatus. *Insets:* origin and shape of difference signals. HV, high voltage, RF, radiofrequency, and LF = 85 Hz.

classic dispersion equation. τ is the relaxation time of the ionic dissociation equilibrium, which can be expressed in the usual way as a function of the relevant kinetic constants.

A useful corollary of the technique described is the possibility of a conductance measurement from the rate of appearance or disappearance of the DC signal on the capacitor which depends, apart from the capacitance, on the cell resistance.

Nonlinear dielectric behavior appears as a change of permittivity of the system when subjected to high electric fields. Contrary to the field-dissociation effect, observable at rather low fields (~ 1 kV cm^{-1}) in low polar media, the field-induced dielectric increment is generally measurable at very high field strength (>50 kV cm^{-1}) apart from some special cases. It is also of importance to notice that as a consequence of the dependence on the field squared, the system cannot be perturbed in the limit of zero field.

Usually the field-induced dielectric increment is measured statically from the frequency shift of an oscillator, whose tank circuit includes a capacitor filled with the dielectric sample, upon the application of a constant or pulsed high electric field. We have developed a stationary perturbation technique which allows the accurate measurement of the increment of both real and imaginary parts of the permittivity in the frequency domain (9). To the dielectric losses may be added any field-induced change of conductance, either in or out of phase. A block diagram of the circuit is given in Fig. 2. As in older techniques the sample is subjected to a radio frequency (RF) low amplitude field, which is superposed on the high field. A new feature is that the sample solution is contained in a capacitor forming a parallel resonant network, which is excited externally by a RF generator. The resonant frequency of the network can be varied over a wide range (0.1–100 MHz) by choosing a coil with appropriate inductance. A second new feature is that the high field is alternating at low frequency (85 Hz).

A careful design of the sample cell—either stainless steel or gold-plated electrodes at a distance of about 0.3 mm—allows field strengths over 200 kV cm^{-1} to be sustained easily in nonconducting samples for more than 1 min.

As a consequence of nonlinear effects, the periodic high field affects the resonance characteristics of the circuit. This is reflected by the amplitude modulation of the resonant voltage at 2×85 Hz, illustrating the principle of harmonic generation by a nonlinear element. By sweeping the RF frequency over the resonance bandwidth, a signal proportional to the difference between resonance curves at peak high field and without field is detected. The signal is a measure of the modulation depth at each frequency. The sign of the difference signal is obtained from the phase relation between signal and perturbing high field. The amplitude and asymmetry of the recorded signal yield information on the changes of the complex permittivity. The amplitude modulation is very small, and we therefore consider the amount of modulation as the derivative of the resonant voltage $V(f)$. In general the three parameters characterizing the resonant circuit may change upon the application of the high field: the resonance voltage V_0, resonance frequency f_0, and the circuit quality Q. The amount of amplitude modulation at each frequency therefore can be written as:

$$\Delta V(f) = (\delta V/\delta f_0)\Delta f_0 + (\delta V/\delta V_0)\Delta V_0 + (\delta V/\delta Q)\Delta Q. \quad (11)$$

Because ΔQ can be expressed in terms of Δf_0 and ΔV_0, it is clear that the measurement of the modulation depth over the resonance bandwidth—around frequency f_0—will yield all information on the relative changes of resonant frequency and voltage, provided the static properties of the resonance circuit are known. The changes in resonant voltage and frequency with the electric field are related to the RF frequency-dependent quantities $\Delta\epsilon'(\omega)$ and $\Delta\tan\delta(\omega)$, the incremental real part of the permittivity and the increase of dielectric loss, respectively:

$$\Delta\epsilon'(\omega)/\epsilon = 2\gamma\Delta f_0/f_0 \quad \text{and} \quad \Delta\tan\delta(\omega) = (\gamma/Q)[(\Delta V_0/V_0) - F(\Delta f_0/f_0)]. \quad (12)$$

γ is a calibration factor and F a correction factor. The change in the imaginary part of the permittivity, $\Delta\epsilon''$, is contained in $\Delta\tan\delta(\omega)$; that quantity may contain a contribution from the field-induced increment in conductance, $\Delta\sigma/\epsilon_0\epsilon\omega$.

For systems of ionic equilibria, changes of $\Delta\sigma$ with the frequency of the high voltage can be observed as well. The superposed RF voltage then only acts as a probing field.

The resonant circuit is arranged as a symmetrical bridge to have maximum sensitivity, a minimum of correction factors, and an exclusion of nonlinearities from the high voltage power supply.

RESULTS AND DISCUSSION

Rather than present a complete kinetic analysis of a given system, we wish to show some typical applications of the techniques.

In Fig. 3 measurements of the field-dissociation effect in solutions of tetrabu-

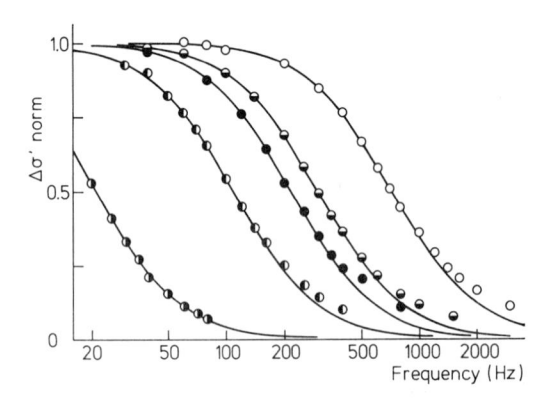

FIGURE 3 Dispersion of the field-dissociation effect in solutions of TBAP at 25°C under various conditions of pressure and solvent composition measured by the field modulation method. Solutions in benzene: ◐, 1.16×10^{-4} M; ◓, 3.80×10^{-4} M; ◒, 7.63×10^{-4} M; ●, 7.68×10^{-4} M and $P = 650$ atm ($\omega_{HF}/2\pi = 135$ kHz). Solutions in chlorobenzene/benzene (30/70 by volume): ○, 1.12×10^{-4} M ($\omega_{HF}/2\pi = 80$ kHz). The effective field strength is always about 3 kV cm^{-1}. The solid lines are Debye dispersions.

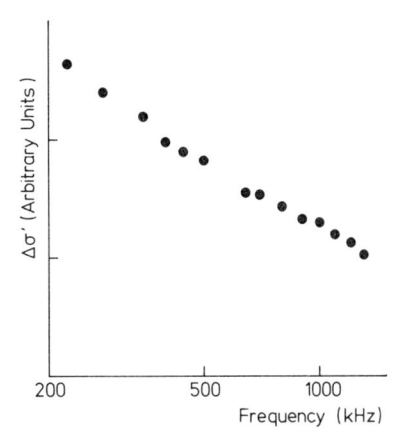

FIGURE 4 Dispersion of the field-dissociation effect in a solution of 0.95 × 10^{-4} M TBAP in chlorobenzene/benzene (20/80 by volume) at 25°C. Conductance increments are plotted as a function of the high frequency ($\omega_{HF}/2\pi$) at a modulation rate ($\omega_{PRF}/2\pi$) of 60 Hz and $E = 6$ kV cm^{-1}.

tylammonium picrate (TBAP) are presented. The relaxation behavior of the field effect can be described with good accuracy as a simple Debye dispersion process. Specific information on ionic dissociation rates, on processes related to ion-pairing and ionic aggregation, on solution phenomena, etc. is obtained from the dependence of relaxation time on concentration and experimental conditions. The relatively simple and noncritical design of cells for the field modulation method allows relaxation measurements over a broad range of temperature and pressure. The influence of permittivity on ionic processes, as opposed to the more specific ion-solvent interaction, is also open for a detailed investigation. The examples presented are all in relatively nonpolar media. However, the technique is also applicable to polar media provided the interference of joule heating, due to high field conditions, can be avoided or minimized.

In Fig. 4 we present a very interesting effect which remains, however, unexplained theoretically: under the experimental conditions of the field modulation method, the field-dissociation effect decreases when the high frequency (ω_{HF}) in the high field pulses is increased. However, the effect does not interfere with the chemical relaxation. As stated earlier we presume that the effect is related to hydrodynamic interactions of charged species in electrostatic contact. This unexpected high frequency dependence of the field-dissociation effect reflects omissions in the theoretical description of the dynamics of ionic processes.

The experimental results presented in Fig. 5 provide a link between the field modulation method and the resonance method. Indeed, the dispersion of the field-dissociation effect in a benzene solution of TBAP has been recorded with both techniques. Either one of the data sets fits rather well to a theoretical Debye dispersion curve. The relaxation times derived for the ionization process differ only slightly. This increases the reliability of both methods because identical results are obtained with distinct methods. We like to stress that with the field modulation method the in-phase component of the

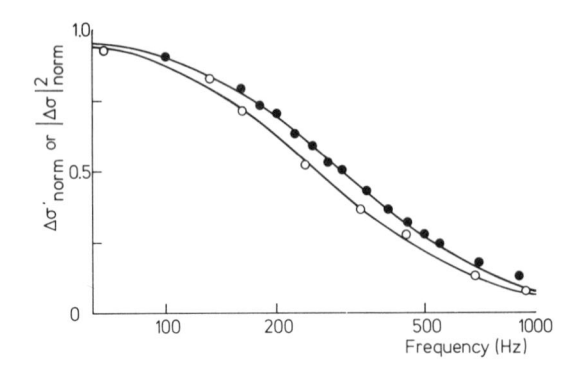

FIGURE 5 Dispersion of the field-dissociation effect in a solution of 7.83×10^{-4} M TBAP in benzene at 25°C. ●, Normalized values of the real part of the conductivity increment plotted as a function of the pulse repetition frequency ($\omega_{PRF}/2\pi$, field modulation method). E (effective) = 3 kV cm^{-1} and $\omega_{HF}/2\pi$ = 135 kHz; ○, normalized values of the square of the modulus of the conductivity increment plotted as a function of the double of the high voltage frequency (resonance method). E = 3.1 kV cm^{-1}, and resonant frequency = 123 kHz. The solid lines are Debye dispersions.

conductance increment ($\Delta\sigma'$) is measured, whereas the resonance method, being phase-insensitive, registers the modulus ($|\Delta\sigma|$) of the effect. At higher field strength and in chemical reactive systems containing dipolar species, the nonlinear parameters $\Delta\epsilon'$ and $\Delta\tan\delta$ depend most often on the square of the field, although nearly complete saturation has been observed for polypeptide solutions. Both parameters turn out to be either positive or negative depending on which contribution to the overall effect is more important: Langevin saturation and its orientational relaxation or chemical modes of permittivity increment and the related chemical relaxation.

In many instances of inter- and intramolecular processes, such effects have been observed, e.g., the self-association of *n*-butanol, of ion-pairs, and of ϵ-caprolactam (9), the proton transfer reaction between amines and phenols, and ring inversion in trans-1,2-dichlorocyclohexane. A typical analysis of conformational change by internal rotation is shown in Fig. 6. The frequency behavior of the measured quantities closely follows an ideal Debye-type relaxation. The *trans* ⇌ *gauche* equilibrium for the ethane derivative can be characterized from the ratio of chemical increment to Langevin component. The dipole moment of the polar conformer is determined. Rate and activation parameters are obtained from the relaxation times and their temperature dependence. The values agree with spectroscopic and ultrasonic absorption measurements (10). Finally, it should be noted that the conductance of the sample may not exceed such a value that heating by the high field affects the resonance parameters significantly during the RF scan.

CONCLUSION

The relaxation methods presented are suitable for the investigation of the kinetics of ionic and dipolar reactions in any medium. The frequency domain covered is very

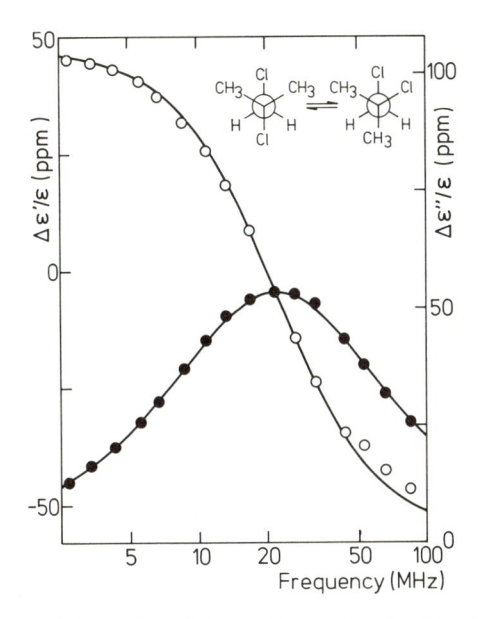

FIGURE 6 Dispersion and absorption of the nonlinear dielectric effect in a solution of 1,2-dichloro-2-methylpropane in benzene. $X = 0.112$ and $\epsilon = 2.88$ at 7.5°C and $E = 138$ kV cm^{-1}. The real ($\Delta\epsilon'$, ○) and imaginary ($\Delta\epsilon''$, ●) parts of the increment are plotted as a function of the RF frequency. The solid lines are Debye curves.

broad and can be increased with existing technology, e.g., with high power oscillators (radio-transmitters) or very high frequency (gigahertz-domain) generators. The stationary measurement is valuable because of high accuracy and sensitivity, but restricts the applicability of the methods. Indeed, joule heating in conducting media has always been the limitation for stationary high field methods. A combination of pulse techniques with our modulation methods, and trading off some accuracy against potential applications in conducting media, would perhaps reduce this limitation substantially. Another advantage of the electric field methods lies in the detection techniques that make them suitable for the study of nonhomogeneous systems. It is therefore apparent that these methods have a great potential for the investigation of chemical phenomena at interfaces and phase boundaries, which are abundant structures in biological systems.

An important feature of the methods described is how the relaxation behavior of a chemical system is studied. All relevant information is obtained from the intermodulation products generated from the interaction of perturbing and detecting signals. This experimental principle is easily generalized and allows the conception of a new class of techniques to investigate fast reactions. For example, a method for the study of dynamic aspects of photoconductivity is being developed. The method is based on light-induced conductance modulation. The analogy with the field modulation method is evident.

From the biophysicist's point of view the importance of these new techniques lies in the ubiquity of electric fields in biological systems. Important physiological processes invoke the intervention of phenomena induced by, or coupled to, electric field effects. Nonlinear electric behavior of membranes is often discussed in terms of the field-dissociation effect, while even a model of nervous excitation has been developed based on this effect (11). Experimental verification or refutation of such models is often impeded by the scarcity of experimental data on field effects in nonhomogeneous media. There is also a lack of theoretical work on field effects on ionic or dipolar processes in or near phase boundaries marked by a discontinuity in electric admittance.

To explain electrically induced conformational changes in polyelectrolytes and in membranes, polarization effects are often invoked. The field-dependent ionization or conformational transition of molecular groups at the membrane surface can cause a field-dependent surface charge density. These effects may be at the origin of the modulation of permeability by the electric field.

It is also important to note that structural transitions in biopolymers induced by electric fields may be the mechanism for translating electric signals in more enduring chemical states. Especially when the field-induced phenomena are cooperative, this translation mechanism may be of prime importance in biological information storage. Whatever the molecular basis of memory, this exchange mechanism between an electrical signal and a chemical state is an attractive hypothesis.

To assess the value of the hypothesis made on the intervention of electric field effects in biological processes, one of the main experimental criteria will be the time range in which the effects are operative. It is, therefore, promising that the relaxation methods described in this paper allow an accurate and sensitive measurement of the dynamics of these field effects. Moreover, our techniques can in principle with relatively minor technological changes be adapted to the direct study of biological systems.

The authors are much indebted to L. De Maeyer for numerous suggestions and inspiring discussions.

Financial support from the Belgian Research Council (F.K.F.O. grants 10.040 and 2.0051.77 N), the Belgian government (Programmatie van het Wetenschapsbeleid-conventie 76/81-II.4), and from the Catholic University Leuven (Derde Cyclus project OT/III/19) is acknowledged.

Received for publication 29 November 1977.

REFERENCES

1. Eckstrom, H. C., and C. Schmelzer. 1939. The Wien effect: deviations of electrolytic solutions from Ohm's law under high field strengths. *Chem. Rev.* **27**:367.
2. Onsager, L. 1934. Deviations from Ohm's law in weak electrolytes. *J. Chem. Phys.* **2**:599.
3. Lewis, G. N., and M. Randall. 1961. Thermodynamics. McGraw-Hill Book Co., New York. 2nd edition. 502.
4. Schwarz, G. 1967. On dielectric relaxation due to chemical rate processes. *J. Phys. Chem.* **71**:4042.
5. Bergmann, K., M. Eigen, and L. De Maeyer. 1963. Dielektrische Absorption als Folge Chemischer Relaxation, *Ber. Bunsenge. Phys. Chem.* **67**:819.
6. Mead, D. J., and R. M. Fuoss. 1939. Dependence of conductance on field strength. I. Tetrabutylammonium picrate in diphenyl ether at 50°. *J. Am. Chem. Soc.* **61**:2047 and 3989.

7. PERSOONS, A. 1974. Field dissociation effect and chemical relaxation in electrolyte solutions of low polarity. *J. Phys. Chem.* **78**:1210.
8. NAUWELAERS, F., L. HELLEMANS, and A. PERSOONS. 1976. Field dissociation effect, chemical relaxation, and conductance of tetrabutylammonium picrate ion pairs in diphenyl ether. *J. Phys. Chem.* **80**: 767.
9. HELLEMANS, L., and L. DE MAEYER. 1975. Absorption and dispersion of the field induced dielectric increment in caprolactam-cyclohexane solutions. *J. Chem. Phys.* **63**:3490.
10. WYN-JONES, E., and W. J. ORVILLE-THOMAS. 1968. Molecular acoustic and spectroscopic studies. Pt. 1. Rotational isomerism in some halogenated hydrocarbons. *Trans. Faraday Soc.* **64**:2907.
11. BASS, L., and W. J. MOORE. 1968. A model of nervous excitation based on the Wien dissociation effect. *In* Structural Chemistry and Molecular Biology. A. Rich and N. Davidson, editors. W. H. Freeman and Company, San Francisco. 356.

DISCUSSION

SCHEIDER: This extended comment is intended to elicit an extended response from Dr. Persoons, if possible, directed to the question of the problems and possibilities of studying *dipolar* chemical coupling in proteins, and to find out what experience you have had with this. For a long time we have been looking for a system in which to see this kind of coupling. We came up against the role of the relative time constants or relative rates of the chemical and orientation events. I can't miss the opportunity to put a plug in here for the *Biophysical Journal* and myself, because in 1965 we published an article there in which we identified the importance of the ratio, $\tau_{chem}/\tau_{orientation}$, and of the compound relaxation rate, which is the sum of the reciprocals of the time constants for chemical and rotational relaxation.

The basic fact, in a heuristic sense, is that whether or not thermodynamics is applicable, you can't induce a chemical change by an electric field unless you have preferential orientation of the molecule in which you are trying to induce the chemistry with respect to the direction of the field. If you accept the fact that you have to orient before you can induce chemistry, then the question of the relative rates is important. In the case of small signals, if the chemistry is too fast—to be specific, if the rotation time of proteins is of the order of 1 μs; it is "slow" if it is a great deal slower than 1 μs—you may see a change but you cannot distinguish it from any other induced electronic polarization unless you use other methods, such as optical detection. If you are working strictly with the electrical signal then you can't distinguish the two processes. If the chemistry is slow compared to the rotation, then for what amounts effectively to the length of the experiment (one time constant of rotation), you have no chemical activity. Another way of saying this is that in the dispersion region for molecular rotation you don't see any of the chemistry. The system of choice therefore is one in which the ratio $\tau_{chem}/\tau_{orient}$ is of magnitude unity. Now the question is: To what extent can you escape these problems by using large fields? How long can you reasonably subject the protein to large enough fields when you have to use water as a solvent and cannot avoid heating due to ionic current?

PERSOONS: We can perhaps analyze the response of a system to an electric field as a function of the ratio of chemical relaxation time to orientational relaxation time. Each time you apply an electric field to a system, it will tend to oppose the increase of the potential energy. When the orientation is much faster than the chemical process, the system will oppose the increase in potential energy by orientation, at least at small field strength. Chemical relaxation, however, will be observed at high field strengths due to a net change in concentration of dipolar species. In conclusion, at low fields we observe no chemistry and at high fields we see chemical relaxation.

This is what we actually presented. Now when the orientation is slow or hindered, because we deal with rather large molecules or biopolymers, then the system will tend to oppose the effect of an electric field not by orientation but by chemical transformation. The picture I am painting is the following: if you have a very large polypeptide and helical parts of it with the dipole moment are in an unfavorable direction in the field, the helical parts will disappear and reappear again in another part of the polypeptide but now favorably oriented in the field. This is a chemical mode of orientation and although we don't have a net change in bulk concentration of dipolar species, we see this orientation as a chemical orientation effect in dielectric absorption. Taking this picture of a helix disappearing and again reappearing in a more favorable direction in the field, you can understand that it is possible to see chemical effects even at low field strengths. If we use high fields, the polar helical parts in the biopolymer will tend not only to orient in the field but also to grow at the expense of the polar coiled parts. However, here the frequency of the high field comes in and this frequency should be low enough to allow for orientation. So when you are working at high fields at a rather low frequency, you see not only the chemical mode of orientation but also the chemical effect of dipolar shifts in the equilibrium between polar helical parts and nonpolar coiled parts of the polypeptide. When the orientational relaxation time and the chemical relaxation time are of the same order (we haven't done yet such experiments, but from theoretical considerations I think that it would be possible to analyze the system), one obtains information on both orientational and chemical processes from measuring the system response both at low *and* high field strengths. Only at high fields is a real equilibrium shift towards the more polar state possible.

SWENBERG: I have three questions: (*a*) What is the reason for modulating the high speed pulses? Is this an experimental necessity? (*b*) What is the physical interpretation of your chemical relaxation time in the experiments you are reporting? Is this the recombination rate? (*c*) In Fig. 3 you report the effects of pressure on one of the samples. I would like you to clarify why the relaxation time increases.

PERSOONS: We use a high-frequency pulsed field for technical reasons so we can isolate our power electronics from the measuring cell through the use of a transformer. This is because we are measuring rather small DC voltages—of the order of about 50–100 mV—so we don't want to inject any DC signal from the electronics into our measuring cell. Moreover, by using high-frequency, high-voltage fields we have no trouble—at least I haven't seen any—with polarization effects. Now your second question, on the physical interpretation of the chemical relaxation time: The chemical relaxation time is simply the characteristic time needed for a chemical equilibrium to adapt to new equilibrium conditions upon a small perturbation of an intensive variable determining the equilibrium. That is the most general idea behind chemical relaxation. Concerning your third question: we have presented in Fig. 3 some dispersion measurements at 1 atm and also a measurement at increased pressure, to illustrate the possibilities. We are currently measuring pressure effects in ionic dissociation in low polar media. Here we have important electrostriction effects around the ions, so high pressure is increasing the dissociation rate constant of an ion pair, as seen in the relaxation time. This enables us to measure activation volumes of ionic dissociation and recombination processes, yielding information on electrostriction and structure of the solvent molecules around ions and ion pairs. Of course, depending on the actual mechanism of the formation of charges in the media studied, the chemical relaxation time may either increase or decrease with pressure.

GEACINTOV: My comment deals with the Onsager equation (Eq. 1), which appears in your paper, and the intuitive interpretation you give to it by invoking a thermodynamic relationship (Eq. 2). Onsager's original result, shown in Eq. 1, is given in terms of ratios of probabilities and not of equilibrium constants, as is implied in this paper. The ratios of probabilities of dissociation in the presence and absence of an electric field are generally not the same as the ratios of the equilibrium constants. Thus, the agreement of the first two terms in Eqs. 1 and 4 appears to be fortuitous. This is not surprising, since the model of an ion pair described by a "Bjerrum dipole" whose free energy, given by Eq. 2, is not very realistic; it is unlikely that an ion pair should first align itself in the electric field as implied in Eq. 2 and then undergo dissociation.

PERSOONS: I was expecting such a question, because we made a simple physical picture for the field dissociation effect on a thermodynamic basis. Of course a system containing free charges never can be an equilibrium system in the presence of an electric field. I agree with you. However, I am sorry to say that Onsager gives the dissociation equilibrium constant under field conditions in his paper. Eq. 1 of our paper is exactly Eq. 1 in Onsager's paper (ref. 2) on the field dissociation effect. What Onsager did in fact was to calculate the rate constant for dissociation of an ion pair in the presence of an electric field and, because the rate constant for recombination is independent of the field, this calculation gives also the effect on the dissociation equilibrium constant. Our model may be looked at in the same way. When a Bjerrum dipole—that is, the dipole formed by two oppositely charged ions at the Bjerrum distance—orients favorably in the field, we say that the free energy for dissociation is lowered. This is the same as stating that the probability for the two ions to become free by dissociation is somewhat enhanced. Of course, we should take into account the distribution between contact ion pairs and Bjerrum dipoles, the radial distribution of the Bjerrum dipoles with respect to the field, and their rotational Brownian motion, but this would detract from the simplicity of the model. We only meant this model to show the importance of the coupling between dipolar orientation of ion pairs and their dissociation in an electric field.

CZERLINSKI: In view of the prior question I would like to be a bit more specific. In Eq. 9 of your paper you se a correction factor g. What numbers do you actually find for this correction factor? In Eq. 12 you use a calibration factor γ and a correction factor F. I would like to know what these factors consist of and how large they are.

PERSOONS: The g factor in Eq. 9 takes into account the effect of the frequency of the square wave-modulated high-frequency field on the field dissociation effect. Since we do not as yet understand this effect, it is difficult to give an answer. We are investigating this effect; in this context it is important to realize that Onsager made the theory of the second Wien effect for DC fields and we are using AC fields. Nobody has made as yet a theory for the frequency behavior of the Wien effect. Experimentally we see that there is an effect of frequency, which we suppose to be linked to the rotational motion of two ions within the ion pair sphere, the sphere as defined from the Bjerrum concept. Even with the Bjerrum concept there are some difficulties, because if you calculate the Bjerrum distance in a low-polar medium, for example benzene, you find a distance of about 70 Å. According to the Bjerrum concept two univalent ions at such a distance are an ion pair! But then gradually the difference between free ions and ion pairs disappears. As for the correction factors F and γ in Eq. 12 of our paper, F is a small correction factor taking into account the dependence of tan δ on capacitance changes and depends on the symmetry of the cell design. Usually $F\Delta\rho_0/\rho_0$ is less than 1% of $\Delta U_0/V_0$ and can be neglected. Most people don't put it down there. It really is a small correction which can be neglected for practical purposes. γ is a calibration factor measuring the ratio of the total equivalent capacitance to field dependence, or active capacitance, and is of the order of 1.5 at a dielectric constant of 2.

SCHELLY: From our investigation of the Destriau effect in ligand solutions, it seems that the electrical double layers build up at the electrodes in about 0.5–0.8 µs. Thus, at modulation frequencies below ~3 MHz, the double layers should contribute to the cell impedance. I wonder whether you made any relevant observations.

PERSOONS: We have surveyed the possibilities of double layers. So what we have done is to make a form of radial cells using steel electrodes, platinum electrodes, stainless steel electrodes, gold-plated electrodes, and platinum electrodes. The results are independent of the family of the material of the electrodes, of the cell design and of cell positioning. We have always obtained the same distortions and the same signals. So I don't think there is an interference of double layers. We used a Mylar sheet, which in fact is also the center of the electrode. The Mylar sheet isolates the metal plates. I think we use 10 different cell designs and the main result is that the signals we obtained are independent of the cells. We have no evidence for the contribution of double layers to the cell impedance.

SCHELLY: Using a single square-pulse electric field jump apparatus, we recently were able to observe the Destriau effect for the first time in liquid solutions. The effect is bascially a high-field electroluminescence evidence by the emission of a light pulse from the sample of *both* the leading and trailing edges of the perturbation pulse. The effect can be understood in terms of the dynamics of build-up and collapse of an electric double layer in the cathode, and it is essential that light emission takes place only during the variation of the field, i.e. $\pm \partial E/\partial t \neq 0$. I wonder whether you observed light emission from your sample cell? If your modulation frequency is higher than ~2 MHz, you should be able to see continuous pulsed light emission with a "DC level."

PERSOONS: We have not looked for light emission.

REFERENCES

SCHEIDER, W. 1965. Dielectric relaxation of molecules with fluctuating dipole moment. *Biophys. J.* **5:** 617–628.

SCHELLY, Z. A. 1978. *Proc. Natl. Acad. Sci. U. S. A.* In press.

RESPONSE OF ACETYLCHOLINE RECEPTORS TO PHOTOISOMERIZATIONS OF BOUND AGONIST MOLECULES

MENASCHE M. NASS, HENRY A. LESTER, AND MAURI E. KROUSE,
Division of Biology, California Institute of Technology, Pasadena, California 91125 U.S.A.

ABSTRACT In these experiments, agonist-induced conductance is measured while a sudden perturbation is produced at the agonist-receptor binding site. A voltage-clamped *Electrophorus* electroplaque is exposed to *trans*-Bis-Q, a potent agonist. Some channels are open; these receptors have bound agonist molecules. A light flash isomerizes 3–35% of the *trans*-Bis-Q molecules to their *cis* form, a far poorer agonist. This causes a rapid decrease of membrane conductance (phase 1), followed by a slower increase (phase 2). Phase 1 has the amplitude and wavelength dependence expected if the channel closes within 100 μs after a single bound *trans*-Bis-Q is isomerized, and if the photochemistry of bound Bis-Q resembles that in solution. Therefore, the receptor channel responds rapidly, and with a hundred-fold greater closing rate, after this change in the structure of a bound ligand. Phase 2 (the conductance increase) seems to represent the relaxation back toward equilibrium after phase 1, because (*a*) phase 2 has the same time constant (1–5 ms) as a voltage- or concentration-jump relaxation under identical conditions; and (*b*) phase 2 is smaller if the flash has led to a net decrease in [*trans*-Bis-Q]. Still slower signals follow: phase 3, a decrease of conductance (time constant 5–10 ms); and phase 4, an equal and opposite increase (several seconds). Phase 3 is abolished by curare and does not depend on the history of the membrane voltage. We consider several mechanisms for phases 3 and 4.

INTRODUCTION

Electrophysiological experiments have yielded formal descriptions of opening and closing kinetics for acetylcholine (ACh) receptor channels (1, 2). However, we still lack a molecular description of these rate-limiting steps in channel activation. In experiments described here, we probe these events more directly by manipulating the molecular structure of the drug-receptor complex while monitoring, with electrophysiological techniques, the population of open receptor channels. The perturbations are achieved by photoisomerizing agonist molecules that are reversibly bound to receptors.

The investigations employ 3,3-*bis*[α-(trimethylammonium)methyl]azobenzene (Bis-Q) (3, 4). The *trans* isomer of Bis-Q is a potent agonist at *Electrophorus* electroplaques but few, if any, receptor channels open in the presence of *cis*-Bis-Q. When exposed to light of wavelengths between 300 and 500 nm, Bis-Q molecules in solution undergo photoisomerization. For the present study, it is important to note that isomerization proceeds in both directions (*cis* → *trans* and *trans* → *cis*) throughout this range; however, the relative probabilities of isomerization depend on wavelength. After prolonged exposure to 320 nm (ultraviolet) light, most Bis-Q molecules are in the *cis* state while 420 nm light produces a predominantly *trans* solution.

We do not yet have access to kinetic measurements on the isomerization of Bis-Q after absorption of a photon. However, in azobenzene and stilbene derivatives, such photochemical isomerizations generally occur in much less than 1 μs at physiological temperatures; there is no evidence for long-lived intermediate states (5). Thus, these structural changes occur instantaneously on the time scale of electrophysiological signals measured in this study (100 μs to several seconds). Even though a photochemical event initiates the relaxations described here, we emphasize that these measurements reflect general properties of the interaction between agonists and the nicotinic receptor. A preliminary report has been published (6).

METHODS

Fig. 1 gives the general plan of the apparatus. Experiments are controlled and analyzed by a minicomputer.

Voltage Clamp

The electrophysiological arrangements have been described in detail (7, 8). In brief, a voltage-clamp trial consists of a series of episodes. Generally a voltage step occurs during each episode and a light flash occurs during one or more of the episodes. Agonist-induced currents are isolated as follows. Electrically excitable Na$^+$ currents are suppressed with 10^{-6} M tetrodotoxin,

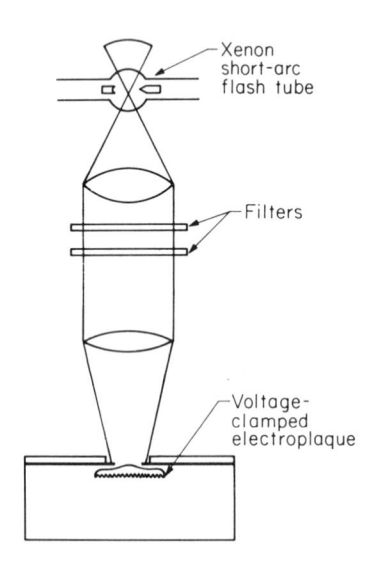

FIGURE 1 Schematic view of apparatus for studying electrophysiological action of photosensitive compounds on single electroplaques from *Electrophorus electricus*. Flash tube was 1 m above preparation. See text.

the anomalous rectifier currents are eliminated with 3 mM Ba^{++}, and passive currents are subtracted. The latter are measured in a range of membrane potentials where there is no agonist-induced current or in the absence of agonist.

A few modifications were made for the present experiments. Currents were passed across the electroplaque between chlorided silver wires or between platinum wires coated with platinum black. Compared with the previously used platinum plates, these electrodes had satisfactorily low resistance and only slightly obscured the light beam.

Some light flashes produced fractionally small changes in voltage-clamp currents (I_m). The following arrangement allowed us to amplify such signals without distortion by capacitive coupling. A voltage proportional to I_m was led to the inverting input of a differential amplifier (gain of 5) as well as to the input of a track-and-hold circuit; the output of the latter circuit was led to the inverting input of the amplifier. The tracked signal was held, starting a few milliseconds before the flash and for the remainder of the episode. The computer received both the original I_m signal and the amplified version with zeroed base line (see Fig. 4b).

Optical Arrangements

The flash tube, lamp housing, and trigger circuit were respectively the models 35S, 71, and 99 manufactured by Chadwick-Helmuth Co., Inc., Monrovia, Calif. The final element of the condenser (represented by the top lens in Fig. 1) served as the field aperture in a Koehler-type illumination system. For the "standard flash" this aperture was imaged onto a Mylar sheet containing the window (1 × 3 mm), which exposed the electroplaque's innervated face. The spot had a diameter of 7 mm and nearly filled the upper chamber (pool A). Usually the discharge was delivered from a bank of electrolytic capacitors (500 μF total) charged to 900 V.

In some experiments, we used an additional lens to increase the intensity at the electroplaque by a factor of 4.2 over that of the standard flash (photodiode measurement; see below). To insure that this smaller spot uniformly illuminated the exposed portion of the innervated face, we used a circular window with a diameter of only 1.2 mm.

Actinometric Calibrations

Cis photoequilibrium solutions of Bis-Q were produced by exposure to a 4-W, long-wave ultraviolet (UV) lamp (Pen-Ray 11SC-1L, Ultra-Violet Products, San Gabriel, Calif.), or to the flash tube filtered through a Chance OX1 filter (called "UV-only" light) (Ealing Corp., South Natick, Mass.). The *trans*-equilibrium state was produced with the unfiltered flash or with an in-

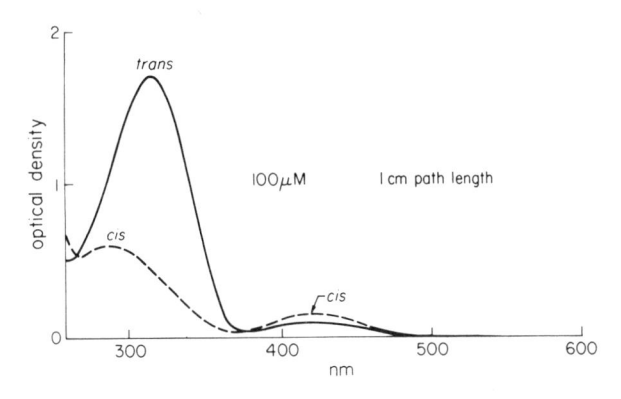

FIGURE 2 Absorption spectra of Bis-Q *cis*- and *trans*- photostationary states. See text for details.

candescent bulb. Fig. 2 presents absorption spectra of the photoequilibrium states. By measuring absorbance at 320 nm it was established that the photoequilibrium states are approached exponentially (with the number of flashes; see below) and with complete reversibility (J. Weissberg and H. A. Lester, unpublished). Both isomers appear to be thermally stable for at least 1 day at room temperature.

Let the concentrations of *cis* and *trans* isomers be C and T, respectively. We now define two "rate" constants, k_C and k_T, by the following statements: (*a*) the photoequilibrium state has the composition $C(\infty)/T(\infty) = k_T/k_C$ and (*b*) the approach to this state is described by $\exp[-(k_T + k_C)n]$, where n is the number of flashes. We measured k_C and k_T as follows.

Solutions of Bis-Q (20 μM) were exposed to flashes in the experimental chamber, optical densities were then measured at 320 nm in a spectrophotometer. For the calibrations the standard flash was attenuated by calibrated neutral density filters, if necessary, so that $k_C + k_T$ equaled 0.015–0.03 flash^{-1}, and 10 flashes were given; solutions were stirred between flashes. This procedure allowed us to minimize variations from one flash to the next and among different regions of the chamber. As the optical density of the solution in the chamber never exceeded 0.15 at any wavelength, no corrections were applied for absorption of the beam. The measurements showed that (*a*) with the unfiltered standard flash, the *trans*-equilibrium state was approached at a rate $k_T + k_C = 0.59$ flash^{-1}; and (*b*) with the standard flash filtered to give UV-only light, the *cis*-equilibrium state was approached at a rate $k_T + k_C = 0.03$ flash^{-1}.

Unfortunately, we do not know $C(\infty)/T(\infty)$ since we have not yet isolated a pure *cis* isomer or *trans* isomer of Bis-Q for determination of physical properties. Most work on azobenzene photochemistry has shown that the photoequilibrium states are about 85% pure (9–11; B. F. Erlanger and N. H. Wasserman, personal communication) and we have adopted these values for calculating k_C/k_T.

These measurements and assumptions yield a value for k_C of 0.50 flash^{-1} with the unfiltered standard flash. This value may be converted to photon flux. Consider the band of wavelengths 100 nm wide centered at 420 nm, the absorption maximum of *cis*-Bis-Q. Most of the *cis* → *trans* isomerization was apparently caused by light in this region since (*a*) k_C was only slightly smaller when the UV light was removed with a Chance OY10 ("No-UV") filter; (*b*) k_C was nearly zero with a yellow Schott glass filter that absorbed light of wavelengths less than 520 nm. The 420 nm peak has a width of 74 nm at half-maximal absorption. If the light has a uniform spectral density, it may be calculated that 75% of the *cis* → *trans* isomerizations are caused by photons with wavelengths in the 100 nm band of interest; as the optical system was inefficient in the UV, the true figure was probably about 85%. The average molar absorptivity $\bar{\epsilon}'$ (defined using natural logarithms and concentrations in moles per cubic centimeter) is 2.35×10^6 over this band. Zimmerman et al. (10) measured a quantum yield for *cis* → *trans* photoisomerization, Φ_T, of 0.55 in the band of interest. Let us now assume that only the intensity, not the spectral distribution, of the flashlamp changes with time (at least for the wavelengths where photons are absorbed by Bis-Q). We may then calculate the time integral of intensity I for a given flash:

$$\int_t I dt = 0.85 \, k_C / \Phi_C \bar{\epsilon}' = 2.0 \times 10^{17} \text{ photons/cm}^2, \tag{1}$$

in the 100-nm band centered at 420 nm. Since the spot area was 0.38 cm^2, the flash delivered 7.6×10^{16} photons to the experimental chamber.

The standard flash was also measured with a photodiode (PIN 10DB/541, United Detector Technology, Inc., Santa Monica, Calif.) placed at the chamber. The following calibrated filters were used: 420 nm interference, heat-reflecting interference ("No-IR"), and neutral density. The flash rose to its peak in about 40 μs and decayed with a time constant of 250 μs. Again assuming uniform photon flux in the band from 370 to 470 nm, the measurements gave a value of 8.4×10^{16} photons (40 mJ) in this band for the unfiltered flash, in satisfactory agreement with the actinometric results.

Temperature Changes

These were measured with a Yellow Springs model 520 thermistor probe (Yellow Springs Instrument Co., Yellow Springs, Ohio). The standard flash caused a temperature rise of less than 0.1°C and of roughly half this amount when the beam was passed through the no-IR filter.

RESULTS

Survey of Relaxations

The records in Fig. 3 provide a general survey of the signals we have observed. The *cis*-photoequilibrium solution of Bis-Q (400 nM) was added to the upper chamber about 1 min before the voltage-clamp trial. In each episode both a voltage jump and, about 37 ms later, a light flash occur. On the basis of the measurements and assumptions described in Methods, it is calculated that the *trans*-Bis-Q concentration is initially 60 nM and that the flashes produce "concentration jumps" to 186, 255, and 293 nM *trans*-Bis-Q.

The first voltage-jump relaxation, in the *cis*-equilibrium solution, resembles those seen with Bis-Q or other agonists (4, 7, 12–15). The conductance increases along an exponential time-course; the reciprocal time constant is 0.15 ms^{-1}. The first light flash

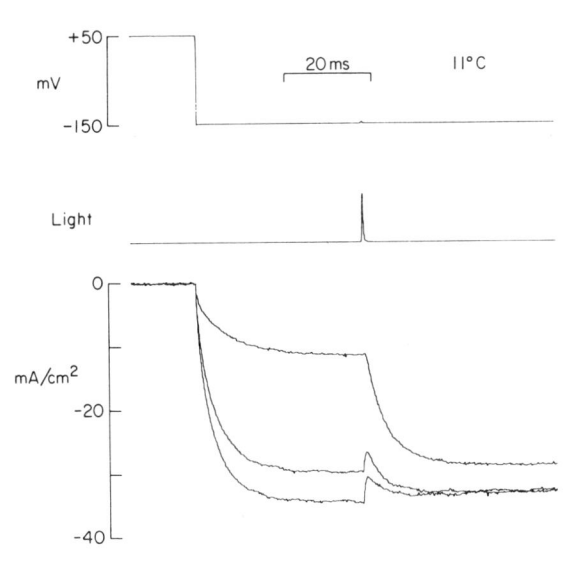

FIGURE 3 Voltage-jump relaxations and "agonist concentration-jump" relaxations in the presence of Bis-Q; see text. Three superimposed episodes at intervals of 20 s. Top trace shows voltage; during each episode, the voltage was jumped from +50 to −150 mV. An unfiltered standard light flash also occurred later in each episode and was monitored by a phototransistor (middle trace). Bottom traces show agonist-induced currents (see Methods for details, including procedures for subtracting leakage and capacitative currents). Cell 45-42 T22,24,26.

then produces roughly a 2.5-fold increase in the agonist-induced current. Such "concentration-jump" relaxations yield information on dose-response relations; these studies will be described more fully in a later paper. For the moment we note that the concentration-jump relaxation follows an exponential time-course and that its reciprocal time constant equals that of the second voltage-jump relaxation (0.20 ms^{-1}). These data extend to higher concentrations the observation that concentration-jump and voltage-jump relaxations have the same time constant under identical conditions (4), and provide further proof that the reciprocal relaxation times increase with agonist concentration (7, 14).

In Fig. 3 the second flash produces an initial rapid decrease in conductance (phase 1, to be described in detail below), followed by a "concentration-jump" increase like that of the first flash. However, the second concentration-jump relaxation is much smaller than the first. There are at least three reasons for this difference. (*a*) As the *trans*-photoequilibrium state is approached, successive flashes produce successively smaller increases of the *trans*-Bis-Q concentration. (*b*) At this voltage half the available receptor channels are open when the *trans*-Bis-Q concentration is 150 nM; thus the dose-conductance relation has begun to saturate (Nass, Lester, and Krouse. In preparation). (*c*) Flashes also induce a slower, temporary closing of receptor channels (phases 3 and 4, to be described below). Indeed, for the third flash, these effects become so important that at the end of the episode the agonist-induced current is less than before the flash, even though the flash produces a small increase in *trans*-Bis-Q concentration.

The Four Phases

This paper concerns relaxations such as those in the final two episodes of Fig. 3. To simplify the kinetic analysis, many of our experiments were performed with flashes that isomerized individual molecules but led to no macroscopic changes in the concentrations of *cis*- and *trans*-Bis-Q. This was usually accomplished by working with the *trans*-equilibrium solution and exposing the preparation to unfiltered flashes. These conditions have almost been attained by the end of the trial in Fig. 3. Under these circumstances a flash produces a sequence of four alternating conductance changes (Fig. 4). We have named these phases 1 through 4 in order of their appearance. Eventually the agonist-induced conductance returns to its value before the flash.

When the *cis*-equilibrium solution is exposed to UV-only flashes, there is again no macroscopic change in *cis* and *trans* concentrations. With electroplaques tested under these conditions, we also observed four phases with very similar kinetics and relative amplitudes (Fig. 5). However, since the UV-only flash produced fairly small conductance changes, we have only limited quantitative measurements on such data.

These phases are not artifacts. None of the phases are seen with light outside the wavelengths (300–500 nm) where Bis-Q absorbs, or with photostable agonists such as carbachol or suberyldicholine. All four phases are seen as usual in electroplaques treated with methanesulfonyl fluoride to inactivate acetylcholinesterase (8). Flashes have no effects on electrically excitable Na$^+$ currents (studied in the absence of tetrodotoxin).

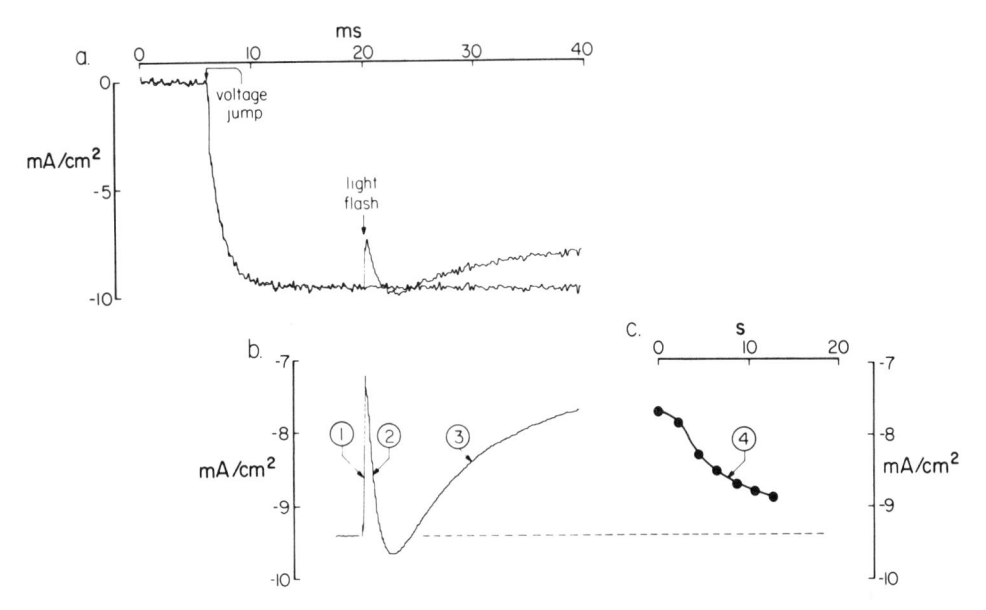

FIGURE 4 The four phases. *Trans*-photoequilibrium solution of Bis-Q (400 nM); temperature 24°C. (*a*) Agonist-induced currents from two episodes 2 s apart. A voltage jump (from +51 to −150 mV) occurred during both episodes; light flash (unfiltered) occurred only during the second episode. (*b*) Amplified trace, 2.4 ms before the flash to 20 ms after flash (see Methods); same time axis as in (*a*). Note phases 1, 2, and 3. (*c*) Time-course for most of phase 4. Additional voltage-clamp episodes, without flashes, occurred at intervals of 2 s after those of panels (*a*) and (*b*). Time axis starts at flash. Points give agonist-induced current at end of each episode. Agonist-induced currents eventually recovered (not shown) to level of dashed line. See text. Cell 47-42 T30, 31.

Phase 1 Results from trans → cis *Photoisomerization of Bound Agonist*

If receptor channels are already opened by *trans*-Bis-Q molecules just before a flash, the flash always produces the rapid decrease in conductance that we call phase 1. We emphasize that phase 1 occurs regardless of any macroscopic change in the *cis/trans* concentration ratio produced by the flash. Fig. 5 shows that phase 1 even occurs when a flash is applied to the *cis*-equilibrium solution (this solution contains about 15% *trans*-Bis-Q molecules). (For the experiment of Fig. 3, there was an electrical artifact which mostly obscured the small phase 1 in the *cis*-equilibrium solution.)

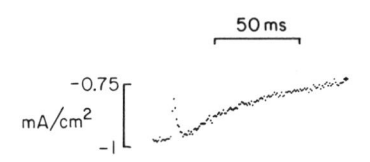

FIGURE 5 Record like that of Fig. 4*b*. *Cis*-photoequilibrium solution of Bis-Q (200 nM). UV-only flash at 4.2 times standard intensity. Trace is the average of four episodes. Temperature 19°C. Cell 51-12 T34.

We have tested the hypothesis that phase 1 results from *trans* → *cis* photoisomerizations of bound agonist molecules. The following assumptions have been made:

(a) More than one *trans*-Bis-Q molecule may need to bind to the receptor for channel opening. However we assume that, if one such bound molecule is isomerized to the *cis* isomer, the receptor channel closes.

(b) We assume that *cis*-Bis-Q molecules induce a channel closing rate so great that phase 1 reaches completion on the time scale of the flash (50–500 μs). Our clamping circuit has a settling time of 100 μs; therefore this picture would be accurate if *cis*-Bis-Q is such a poor agonist that *cis*-Bis-Q channels have a duration of less than 100 μs and very little probability of reopening.

(c) We assume that bound *trans*-Bis-Q molecules have the same "rate" constant for *trans* → *cis* photoisomerization, k_T, as do *trans*-Bis-Q molecules in solution. In photochemical terms the absorption cross-section, ϵ, and the quantum yield for photoisomerization, Φ_T, are not changed by binding (see Methods).

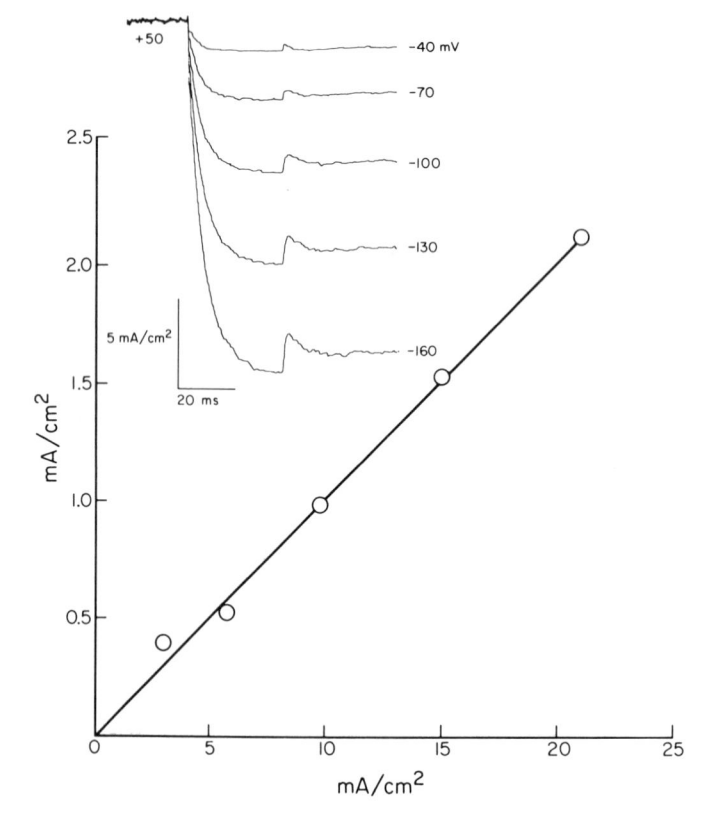

FIGURE 6 Agonist-induced current just before the flash (horizontal axis) compared with amplitude of phase 1 (vertical axis). Equimolar study using *trans*-photoequilibrium solution of Bis-Q (600 nM). For each episode the flash occurred while the voltage was clamped to a different level; see inset. Unfiltered standard light flash; temperature 11°C. Cell 46-12 T 48.

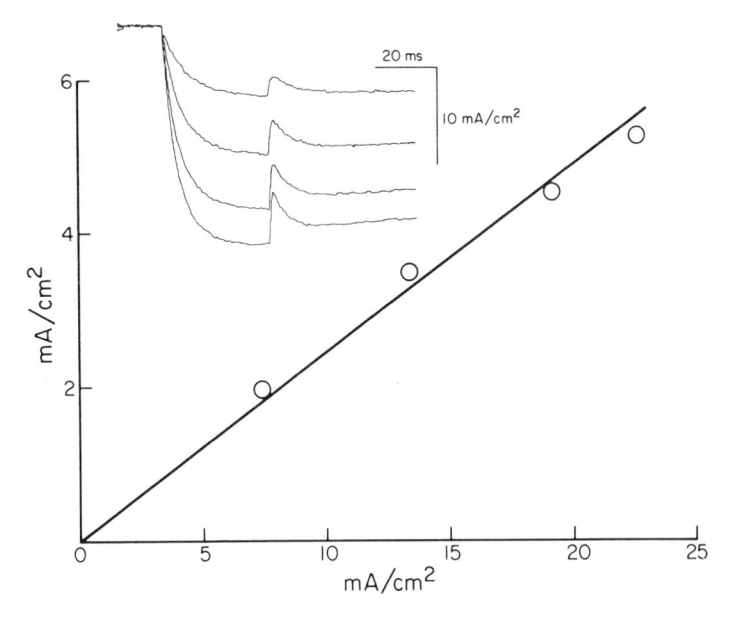

FIGURE 7 Equipotential study complementing Fig. 6. *Trans*-equilibrium solution of Bis-Q (600 nM) was slowly washed into the experimental chamber during the series. In each episode, voltage was jumped from +50 to −150 mV; an unfiltered flash occurred 37 ms later. Temperature 11°C. Cell 46-51.

These assumptions lead one to predict that the amplitude of phase 1 is proportional to the agonist-induced conductance that existed just before the flash. Furthermore, the constant of proportionality should equal the fraction: (molecules in the *trans* state before the flash that undergo at least one photoisomerization)/(molecules in the *trans* state before the flash). This fraction is $1-\exp(-k_T)$.

The proportionality has been verified with two types of experiments. In the experiment of Fig. 6 we varied the agonist-induced conductance among episodes by varying the membrane voltage (7, 16, 17). In another experiment (Fig. 7) the agonist-induced conductance slowly increased as the *trans*-equilibrium solution was washed into the experimental chamber. Finally, the hypothesis of *trans* → *cis* photoisomerization also predicts the constant of proportionality, both for UV-only and for unfiltered flashes (Fig. 8).

TIME-COURSE OF PHASE 1 Our electrophysiological methods are too slow to resolve the time-course of phase 1. However, using low-resistance electrodes and recording differentially the voltage across the electroplaque, we have determined that phase 1 begins—conductance starts to decrease—less than 80 μs after the flash begins. Voltage-clamp measurements with brief flashes (20 μs decay time constant) have also allowed us to conclude that phase 1 ends and phase 2 starts to dominate—conductance stops decreasing and begins to increase—less than 100 μs after the flash ends. Under these conditions, the amplitude of phase 1 still agrees with the *trans* → *cis* isomeriza-

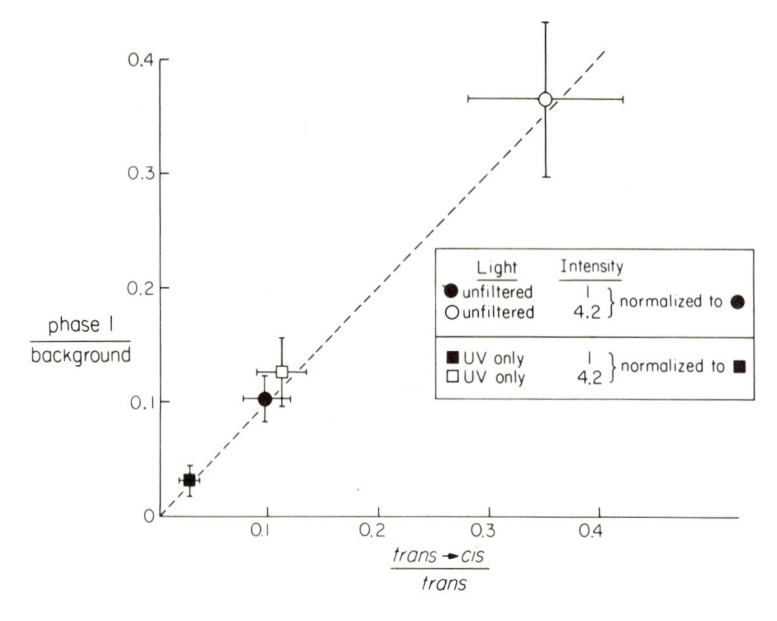

FIGURE 8 Phase 1 versus *trans* → *cis* photoisomerization flux. Horizontal axis is $1 - \exp(-k_T)$; see text. Error bars reflect uncertainties in actinometric calibrations; see Methods. Vertical axis: phase 1 as fraction of agonist-induced conductance just before flash; see Figs. 6 and 7 and text. Error bars give SD (three to six cells per point). *Trans*-photoequilibrium solution of Bis-Q (usually 200 nM); voltage -150 mV; temperatures 10°–25°C. Dashed line at 45° gives prediction of *trans* → *cis* photoisomerization hypothesis (see text).

tion mechanism given above. These observations imply that during phase 1, channels close at a rate in excess of 10 ms^{-1}, even at the lowest temperatures used (10°C).

Phase 2 Is the Relaxation Resulting from the Deficit of Open Channels after the Flash

The unfiltered flash does not affect the concentrations in the *trans*-photostationary solution. If this is also true for the concentration of *trans*-Bis-Q near the receptors (see Discussion), then at the end of phase 1 there are fewer open channels than appropriate for the concentration of agonist. We suggest that during phase 2, the channel population relaxes towards equilibrium again. If this is the case, phase 2 should have an exponential time-course whose time constant equals that of a concentration-jump or voltage-jump relaxation under identical conditions. There are some uncertainties in measuring phase 2, since it is partially obscured by phase 3. We have assumed that phase 3 can be fit as an exponential decrease in membrane conductance superimposed on a sloping base line; and this fit has been subtracted from the voltage-clamp currents (Fig. 9). Phase 2 is usually revealed as an exponential increase in conductance (Fig. 9b); at various temperatures its reciprocal time constant agrees well with that for the voltage-jump relaxation (Fig. 10). In Fig. 10 most of the data points lie close to, but not on, the dashed line corresponding to equal time constants. Apparently the discrepancies arise chiefly because of errors in the subtraction of phase 3: where phase 3 has been eliminated by curare, the data points lie on the theoretical line (squares in Fig. 10; see section on phase 3, below). In the *trans*-photoequilibrium solution, voltage jumps immediately after the flash had slightly smaller rates (5–10%) than those immediately preceding it.

ELIMINATION OF PHASE 2 WITH UV-ONLY FLASHES If a flash reduces the concentration of agonist near the receptors, the new equilibrium should consist of fewer open channels than before the flash. In such a case phase 2 might be expected to become smaller than phase 1, to disappear, or even to reverse its sign; the exact results would depend on the size of the concentration jump and on the local shape of the dose-conductance relation. Phase 2 does indeed become less evident with UV-only flashes that reduce the concentration of *trans*-Bis-Q (Fig. 11); but as expected, phase 2 reappears when the *cis*-photoequilibrium solution has been reached (Fig. 5). At present quantitative measurements are vitiated by the relative weakness of the UV-only flash; by the large phase 3; and by the effect described in the next section. However, the results with unfiltered, UV-only, and yellow light agree with the simple hypotheses presented: (*a*) The size of phase 1 depends only on the number of *trans* → *cis* isomerizations. (*b*)

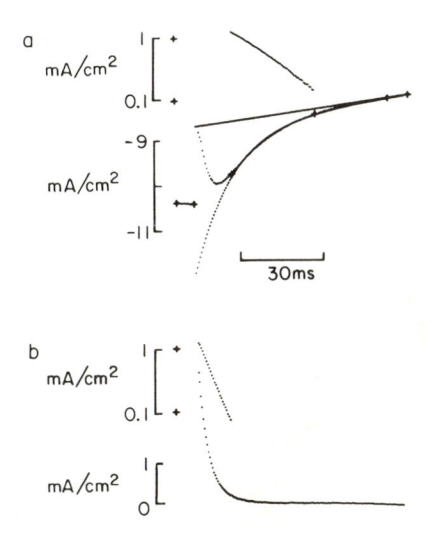

FIGURE 9 Analysis of phases 2 and 3. *Trans*-equilibrium solution of Bis-Q (200 nM), unfiltered flash. Same time axis in all panels. Voltage −150 mV; temperature 13°C. (*a*) Phase 3. Analysis of amplified trace like that of Fig. 4*b*. At start of trace, agonist-induced current was −10.4 mA/cm^2; flash occurred 9 ms later. As a first step in the analysis, a least-squares straight line is fit to final 9 ms of record between two cursors at right. This serves as sloping base line for next step. Beginning at inflection point of the trace, a semilog plot (top panel) is nearly linear. A single exponential is fit (8) to this part of the trace; it superimposes on the data. The exponential plus linear base line are extrapolated to the time of the flash at lower left of this panel. (*b*) Phase 2. Exponential fit to phase 3 is subtracted from the data. Result is shown in bottom panel; semilog plot (top panel) is linear. Cell 47-32 T30.

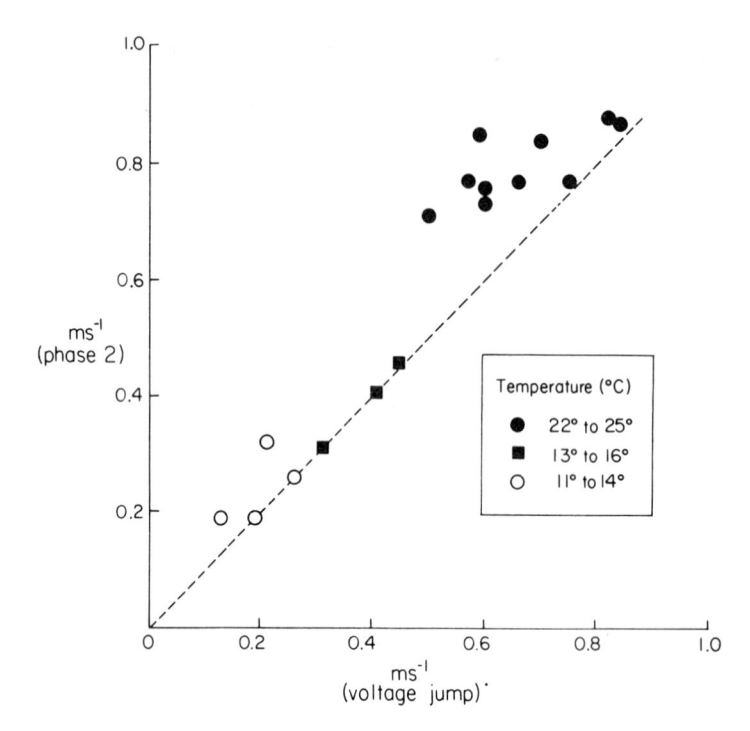

FIGURE 10 Reciprocal time constants for phase 2 (vertical axis, see Fig. 9b) and for the voltage-jump relaxation immediately preceding the flash in the same episode (horizontal axis; see 8, 4). Squares refer to data in presence of dTC (2–4 μM). *Trans*-photoequilibrium solution of Bis-Q (usually 200 nM, but 2 μM in presence of dTC); voltage − 150 mV; unfiltered flashes.

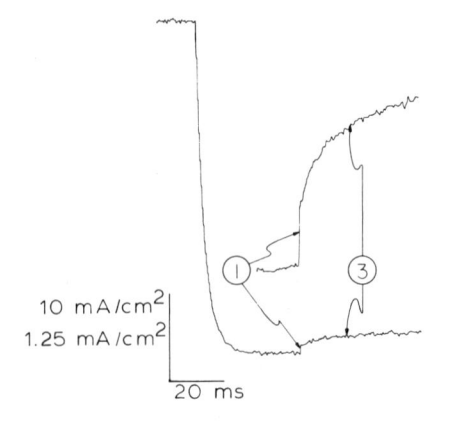

FIGURE 11 Agonist-induced current for a voltage jump from +50 to −150 mV, followed by a UV-only standard flash; average of seven episodes. Trial was started with *trans*-photoequilibrium solution of Bis-Q (600 nM). Same time axis for complete and amplified (inset; see Fig. 4b) traces. Note absence of discernible phase 2. Temperature 11°C. Cell 46-21 T35.

Phase 2 depends on the magnitude and sense of the light-induced disequilibrium between the number of open channels and the concentration of *trans*-Bis-Q.

The Transition between Phases 2 and 3

There is a peak of agonist-induced conductance between phases 2 and 3 (see Fig. 3*b*). The temporary maximum of inward (negative) current had a variable amplitude relative to the current just before the flash. Perhaps the solution near the receptors had a different *cis/trans* ratio from that of the photoequilibrium state in the bulk solution, because of buffering by fixed structures in or near the membrane or local filtering of light. More importantly, however, phase 3 itself appears to be reset by each light flash (Fig. 12). We have not been able to investigate this phenomenon in detail because the brightest flashes could be delivered at intervals of no less than 5 s. What seems clear at this point is that the "overshoot" (Fig. 12) never returns the agonist-induced conductance to a level significantly greater than that just preceding the first flash applied to the preparation.

Phases 3 and 4

Phases 3 and 4 disappear almost entirely when *d*-turbocurarine (dTC) is present at concentrations that partially block agonist-induced currents (Fig. 13; see ref. 7). The smaller currents per se do not underlie this action: in the absence of dTC similarly small agonist-induced currents are measured at low *trans*-Bis-Q concentrations, but phases 3 and 4 are still clearly present (Fig. 13*b*, *c*). In preliminary experiments we find that all four phases are still present, although uniformly reduced in size, after treatment with cobra neurotoxin (*Naja naja siamensis* T3), an irreversible postsynaptic blocker (18).

WAVEFORM OF PHASES 3 AND 4   None of our voltage-clamp episodes clearly reveal the transition between phases 3 and 4. In records like those of Figs. 4*b* and 12, the conductance continues to decrease slowly for as much as 150 ms after the flash. At this point phase 3 is already larger than phase 1 (typically by 20–40%). Similar slow

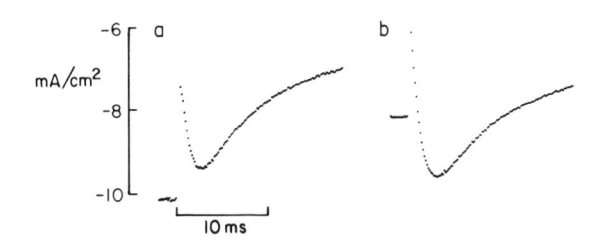

FIGURE 12   A second flash appears to reset phase 3. Amplified records as in Fig. 4*b*. Unfiltered flash occurred during both first (*a*) and second (*b*) episodes, 5 s apart. Note incomplete recovery from phase 3 just before the second flash: the agonist-induced current is smaller than before the first flash (see also Fig. 4*c*). On the other hand, beginning about 3 ms *after* the flash, the conductances are similar in the two episodes. *Trans*-equilibrium solution of Bis-Q (200 nM); voltage −150 mV; unfiltered flashes at about twice standard intensity. Temperature 22°C. Cell 47-51 T51.

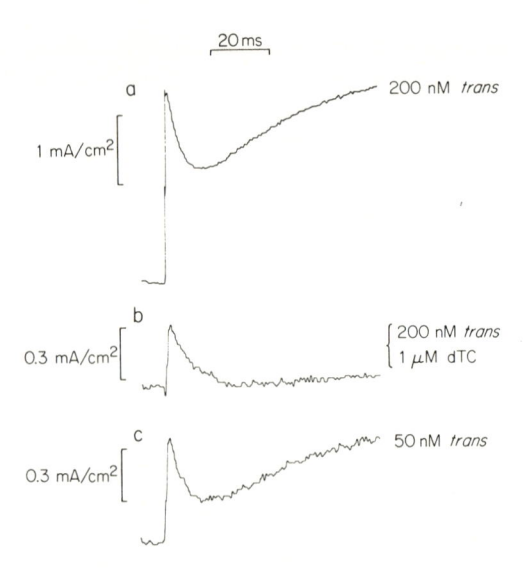

FIGURE 13 Curare abolishes phase 3. Amplified records as in Fig. 4b. Unfiltered flashes, 4.2 times standard intensity. Voltage −150 mV. Temperature 11°C. Cell 51-21 T23, 32, 50.

drifts do not usually appear in control episodes where no flash occurs. Phase 3 cannot be characterized as a simple exponential; its time-course sometimes resembles that of a diffusion process.

Despite these quantitative uncertainties, we can offer some preliminary observations based on analyses like that of Fig. 9a. In most cases phase 3 has a reciprocal time constant two to four times smaller than that of phase 2 or of the voltage-jump relaxation. The values for phase 3 were 0.090 and 0.1 ms^{-1} (2 cells) at 13°, and 0.24 ± 0.05 ms^{-1} (mean ± SD, 10 cells) at 22–25°. Thus phase 3 clearly proceeds faster at higher temperatures.

PHASE 3 DOES NOT DEPEND ON THE HISTORY OF THE MEMBRANE VOLTAGE. Whether a voltage jump precedes or follows the light flash, phase 3 has the same time-course, relative to the flash, at the final voltage (Fig. 14).

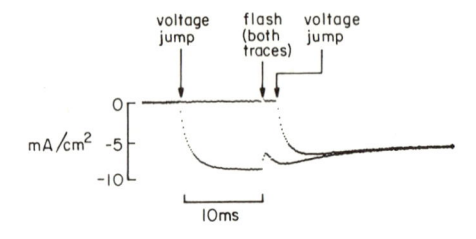

FIGURE 14 Phase 3 depends on the time since the flash and on instantaneous value of the membrane voltage. Agonist-induced currents from two episodes, 36 s apart. Flash occurred at same time in both episodes. Voltage-jump (+50 to −150 mV) preceded the flash in one episode and followed it in the other. Trans-equilibrium solution of Bis-Q (200 nM); temperature 23°C. Cell 47-61 T52, 53.

We have incomplete data on the concentration dependence of the rate of phase 3; however, it depends much less strongly on agonist concentration than does either the voltage jump or phase 2. In one cell phase 3 was studied as a function of flash intensity with the *trans*-equilibrium solution of Bis-Q (200 nM). Over a 30-fold intensity range, phase 3 varied in amplitude from 0.15 to 0.43 times the agonist-induced conductance just before the flash; but the reciprocal time constant changed by less than 10% over the series.

PHASE 4 We have relatively little information on phase 4. It has a time-course of seconds (Fig. 4d) and it too disappears in the presence of dTC.

DISCUSSION

Which Phases Go Together?

In our view the phases occur in pairs. During phase 2 the receptor population recovers from the process that caused phase 1; and likewise for phases 3 and 4. Phase 4 could also be the recovery from phase 1; phases 2 and 3 would constitute a temporary perturbation, perhaps caused by brief pulse of extra agonist near the receptors. But this seems unlikely for three reasons. (*a*) We describe a condition—curare treatment—that eliminates phases 3 and 4 while leaving 1 and 2 unaffected (Fig. 13). (*b*) Phases 3 and 4 are observed even when the voltage has been set to eliminate or reduce phases 1 and 2 (Fig. 14). (*c*) We have studied certain other photoisomerizable azobenzene derivatives which are not agonists. These compounds do not produce phases 1 and 2; but phenomena similar to phases 3 and 4 still occur (M. M. Nass, H. A. Lester, B. F. Erlanger, and N. H. Wassermann, unpublished observations).

Phases 1 and 2

The cartoon in Fig. 15 summarizes our conception of the molecular basis for phases 1 and 2. We have proposed that phase 1 is caused by a specific event, the *trans* → *cis* photoisomerization of reversibly bound Bis-Q molecules (Fig. 15*a*, *b*). Could channels close simply because an absorbed photon raises a bound *trans*-Bis-Q molecule, or the entire agonist-receptor complex, to an excited state? In the absence of specific information about the agonist-receptor complex, we have assumed that the photochemistry of this complex resembles that of *trans*-Bis-Q in solution. If so, the excited state hypothesis predicts a much larger phase 1 than actually observed. With UV light, only 10% of photons absorbed by *trans*-Bis-Q lead to an isomerization; with visible light the figure is 25% ($\Phi_T = 0.10$ and 0.25, respectively; ref. 10). Therefore with the weaker flashes in Fig. 8 the discrepancies would amount to a factor of 10 for UV-only flashes and a factor of 4 for unfiltered flashes. Of course we cannot rule out the possibility that phase 1 occurs via some other photochemical event whose quantum yield fortuitously matches that of *trans* → *cis* isomerization at the wavelengths tested.

THE BINDING SITE AND CHANNEL ARE CLOSELY COUPLED It is often asked whether channel opening and closing are rate-limited by binding and dissociation of the agonist, or by separate conformational changes that might occur. The results

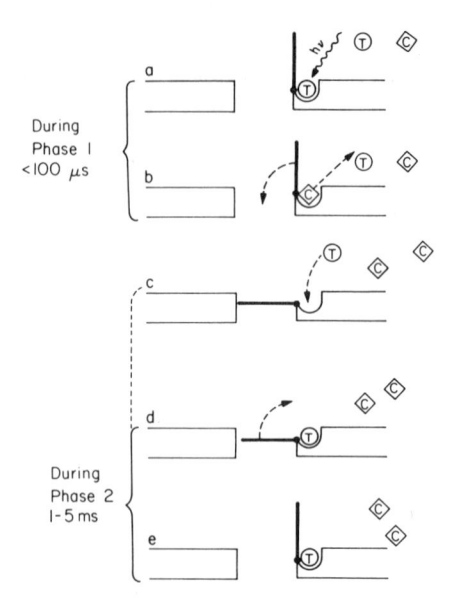

FIGURE 15 Simplified view of events during phases 1 and 2. (a) A *trans*-Bis-Q molecule is bound to an acetylcholine receptor and the channel is open. The *trans* molecule will shortly absorb a photon. (b) The absorbed photon has led to *trans* → *cis* photoisomerization. Within 100 µs the channel closes as the newly created *cis*-Bis-Q molecule leaves the binding site. (c) A *trans*-Bis-Q molecule from the solution binds to the site vacated by the *cis*-Bis-Q molecule. We do not know whether the binding step limits the rate of channel reopening; hence the dashed bracket. (d) The channel reopens. (e) Channel remains open for the usual lifetime characteristic of *trans*-Bis-Q. See text.

do not resolve this question. Nonetheless the present studies bear more directly on this point than do studies of relaxations resulting from jumps or agonist concentration (4) or of membrane voltage (8, 12–15, 19). A concentration-jump relaxation is caused by perturbations in the nearby solution rather than at the receptors themselves; voltage-jump experiments almost certainly affect the receptor-channel complex, but at an unknown site. By contrast, the present data argue strongly that phase 1 results from an event at the agonist-receptor binding site.

If this is the case, we note that phase 1 has the amplitude predicted by the *trans* → *cis* photoisomerization hypothesis, and that this decrease of conductance attains its full value with the same time-course as the flash intensity (the resolution is defined by the settling time of our clamp, roughly 100 µs). As pointed out in Results, this implies that channels close at a rate of at least 10 ms^{-1} during phase 1, even at 10°C. By contrast, if Bis-Q molecules are not perturbed by a light flash, channels close at a rate of about 0.09 ms^{-1} under the same conditions. Thus a change in agonist structure has led to a 100-fold increase in closing rate. Furthermore, the new closing rate takes effect at the channel with a time delay of less than 80 µs. Other studies have shown that agonist structure influences the opening and closing rates of ACh receptor channels (8, 14, 20–23); of interest here is the large size of this effect and the speed with which information moves between the binding site and the channel, if indeed these are separate structures. This information fits well with estimates that receptor channels begin to open less than 150–200 μs after the agonist appears in the synaptic cleft (24, 25).

The details of the photochemical events lead to further conclusions. The studies of Fig. 8 were performed with the *trans*-equilibrium solutions. For the UV-only flashes, most of the photoisomerizations had the direction *trans* → *cis,* and there was a net increase of *cis*-Bis-Q concentration. For the unfiltered flashes there was an even larger rate of *trans* → *cis* isomerizations. However, during these latter flashes, newly created *cis* molecules underwent rapid photoisomerizations back to the *trans* state. During the first 250 μs of the brightest flash, this *cis* → *trans* reisomerization proceeded at a rate approaching 10 ms^{-1} (calculated from k_C and from the time-course of the flash). Suppose that the reisomerization occurred when the *cis*-Bis-Q molecule was still bound to the receptor, with the channel either open or closed. If open, the channel would presumably never close and phase 1 would be smaller, for unfiltered flashes, than predicted by our model. If closed, channels would rapidly reopen, and phase 2 would be more rapid than the voltage jump (26). These arguments suggest that Bis-Q leaves the receptors less than 100 μs after being isomerized to the *cis* state.

There is further evidence for this statement. According to the cartoon, *cis*-Bis-Q molecules must leave the binding site (Fig. 15*b*) before *trans*-Bis-Q molecules can rebind to receptors (Fig. 15*c*). Channels may then reopen (Fig. 15*d*). But with brief (50 μs) flashes, phase 2 dominates (channels reopen) less than 150 μs after the flash begins. Thus at most 150 μs intervenes between channel closure and departure of a *cis*-Bis-Q molecule, though one cannot say which occurs first, even at 10°C. The present conclusion derives from a study of *cis*-Bis-Q, which is a poor agonist; but it may apply to all interactions between ligands and the nicotinic receptor.

The data on phases 1 and 2 all seem consistent with the suggestion that the ACh receptor channel opens as the agonist molecule binds to the receptor (or as the second of two molecules does so) and that the channel closes again as the agonist leaves the receptor (or as either of the two molecules does so) (8, 14, 27).

Phases 3 and 4

We cannot offer a completely satisfactory explanation for phases 3 and 4. One might postulate two general classes of phenomenon: (*a*) after the flash, the receptor or the agonist-receptor complex undergoes intramolecular changes of state; or (*b*) the flash produces pharmacologically active molecules near receptors.

INTRAMOLECULAR CHANGES   Certain observations agree with such a mechanism. The time-course of phase 3 depends little on the concentration of Bis-Q or on the number of molecules isomerized (strength of the flash). In many cases phases 1 and 3 have similar amplitudes, suggesting that a receptor participating in phase 1 later participates in phase 3.

One possibility is that the flash converts some receptors to a "desensitized" state, similar to the one obtained after prolonged exposure to high agonist concentrations.

Perhaps Bis-Q molecules, already bound to receptors but not necessarily at the site associated with channel activation, move to a blocking position after absorbing a photon. A subsequent flash would photoisomerize and dislodge some of the blocking molecules, so that the two slow phases would appear to be reset as in Fig. 12. Noncompetitive blockade, presumably at the channel, is seen with many molecules containing both lipid-soluble and amino groups (7, 19, 28–31), and with Bis-Q itself at concentrations greater than 2 μM (unpublished). The original binding would not depend on voltage (Fig. 14), but either the binding or movement would be blocked by curare (Fig. 13). For such a model one would like to know whether the putative blocking molecule is in the *cis* or *trans* conformation before the flash, its conformation after the flash, and where and how tightly it binds both before and during blockade. At present no simple model of this sort explains all the data.

LIGHT-INDUCED CHANGES IN THE CONCENTRATION OF DRUGS NEAR RECEPTORS
In evaluating this possibility we note that high-affinity drugs are buffered by binding to receptors in the synaptic cleft (32–35). Dose-response studies show that *trans*-Bis-Q has an apparent dissociation constant of about 150 nM at −150 mV (M. M. Nass, H. A. Lester, and M. E. Krouse. In preparation). If half the receptors in the synaptic cleft have *trans*-Bis-Q molecules bound when the agonist activity is 150 nM, then the total number of *trans*-Bis-Q molecules per unit volume of the cleft will exceed this activity by at least 100-fold. If our hypothesis for phase 1 is correct, from 10% to 40% of these molecules would be released into the cleft by flashes of the strength used in these studies. By analogy with the equally affine molecule, dTC, buffering by receptors would retain *trans*-Bis-Q molecules in the synaptic cleft for up to 1 s (36). Thus these molecules would have ample opportunity to act as agonists or channel blockers. If dTC binds to the same site, it would reduce the buffering.

There are difficulties with such a mechanism. (*a*) If the receptors are buffering Bis-Q, the extent of buffering should depend on voltage (37); thus phase 3 should depend on the history of the voltage. This is not the case (Fig. 14). (*b*) If acetylcholinesterase is buffering Bis-Q (38), phase 3 should disappear after inactivation of the esterase. This is also not the case.

Phases 3 and 4 deserve further study. In particular one would like to know the pharmacology of the pure *cis* isomer of Bis-Q; and one would like to have comparable experiments with other photoisomerizable drugs.

We thank D. Williams for help with the animals, M. Walsh for help with the optics and electronics, H. W. Chang for furnishing the Bis-Q, and D. Armstrong, R. K. Clayton, B. F. Erlanger, R. E. Sheridan, and N. H. Wassermann for helpful discussion.

This research was supported by a postdoctoral fellowship from the Muscular Dystrophy Association of America to M.M.N., by an Alfred P. Sloan Research Fellowship and National Institutes of Health Career Development Award (NS-272) to H.A.L., by an N.I.H. predoctoral training grant to M.E.K., and by research grants from the Muscular Dystrophy Association and the N.I.H. (NS-11756).

Received for publication 1 December 1977.

REFERENCES

1. GAGE, P. 1976. Generation of endplate potentials. *Physiol. Rev.* **56**:177–247.
2. COLQUHOUN, D., and A. G. HAWKES. 1977. Relaxations and fluctuations of membrane currents that flow through-drug operated channels. *Proc. R. Soc. Lond. B. Biol. Sci.* **199**:231–262.
3. BARTELS, E., N. H. WASSERMANN, and B. F. ERLANGER. 1971. Photochromic activators of the acetylcholine receptor. *Proc. Natl. Acad. Sci. U.S.A.* **68**:1820–1823.
4. LESTER, H. A., and H. W. CHANG. 1977. Response of acetylcholine receptors to rapid, photochemically produced increases in agonist concentration. *Nature (Lond.).* **266**:373–374.
5. ROSS, D. L., and J. BLANC. 1971. Photochromism by *cis-trans* isomerization. *In* Techniques of Chemistry. Vol. III. Photochromism. G. H. BROWN, editor. John Wiley & Sons, Inc., New York. 471–556.
6. NASS, M. M., and H. A. LESTER. 1977. A photochemical probe for the nicotinic drug-receptor complex. *J. Gen. Physiol.* **70**:13a. (Abstr.).
7. LESTER, H. A., J.-P. CHANGEUX, and R. E. SHERIDAN. 1975. Conductance increases produced by bath application of cholinergic agonists to *Electrophorus* electroplaques. *J. Gen. Physiol.* **65**:797–816.
8. SHERIDAN, R. E., and H. A. LESTER. 1977. Rates and equilibria at the acetylcholine receptor of *Electrophorus* electroplaques. A study of neurally evoked postsynaptic currents and of voltage-jump relaxations. *J. Gen. Physiol.* **70**:187–219.
9. FISCHER, E., and Y. FREI. 1957. Photoisomerization equilibria in azodyes. *J. Chem. Physiol.* **27**:328–330.
10. ZIMMERMAN, G., L-Y. CHOW, and U-J. PAIK. 1958. The photochemical isomerization of azobenzene. *J. Am. Chem. Soc.* **80**:3528–3531.
11. FISHER, E. 1959. Temperature dependence of photoisomerization equilibria. Part I. Azobenzene and the azonaphthalenes. *J. Am. Chem. Soc.* **82**:3249–3252.
12. ADAMS, P. R. 1975. Kinetics of agonist conductance changes during hyperpolarization at frog endplates. *Br. J. Pharmacol.* **53**:308–310.
13. NEHER, E., and B. SAKMANN. 1975. Voltage-dependence of drug-induced conductance in frog neuromuscular junction. *Proc. Natl. Acad. Sci. U.S.A.* **72**:2140–2144.
14. SHERIDAN, R. E., and H. A. LESTER. 1975. Relaxation measurements on the acetylcholine receptor. *Proc. Natl. Acad. Sci. U.S.A.* **72**:3496–3500.
15. ADAMS, P. R. 1977. Relaxation experiments using bath-applied suberyldicholine. *J. Physiol. (Lond.).* **268**:271–289.
16. RUIZ-MANRESA, F., and H. GRUNDFEST. 1971. Synaptic electrogenesis in eel electroplaques. *J. Gen. Physiol.* **57**:71–92.
17. DIONNE, V. E., and C. F. STEVENS. 1975. Voltage dependence of agonist effectiveness at the frog neuromuscular junction: resolution of a paradox. *J. Physiol. (Lond.).* **251**:245–270.
18. LESTER, H. A. 1972. Blockade of acetylcholine receptors by cobra toxin: electrophysiological studies. *Mol. Pharmacol.* **8**:623–631.
19. ADAMS, P. R. 1977. Voltage jump analysis of procaine action at frog endplate. *J. Physiol. (Lond.).* **268**:291–318.
20. KATZ, B., and R. MILEDI. 1973. The characteristics of "end-plate noise" produced by different depolarizing drugs. *J. Physiol. (Lond.).* **230**:707–717.
21. COLQUHOUN, D., V. E. DIONNE, J. H. STEINBACH, and C. F. STEVENS. 1975. Conductance of channels opened by acetylcholine-like drugs in muscle end-plate. *Nature (Lond.).* **253**:204–206.
22. NEHER, E., and B. SAKMANN. 1976. Single-channel currents recorded from membrane of denervated frog muscle fibers. *Nature (Lond.).* **260**:799–802.
23. DREYER, F., C. WALTHER, and K. PEPER. 1976. Junctional and extrajunctional acetylcholine receptors in normal and denervated frog muscle fibers: noise analysis with different agonists. *Pflügers Arch. Eur. J. Physiol.* **366**:1–9.
24. KATZ, B., and R. MILEDI. 1965. The measurement of synaptic delay and the time course of acetylcholine release at the neuromuscular junction. *Proc. R. Soc. Lond. B Biol. Sci.* **161**:483–495.
25. LLINAS, R., I. Z. STEINBERG, and K. WALTON. 1976. Presynaptic calcium currents and their relation to synaptic transmission: voltage clamp study in squid giant synapse and theoretical model for the calcium gate. *Proc. Natl. Acad. Sci. U.S.A.* **73**:2918–2922.

26. MAGLEBY, K. L., and C. F. STEVENS. 1972. A quantitative description of end-plate currents. *J. Physiol. (Lond.).* **223**:173–197.
27. KORDAS, M. 1972. An attempt at an analysis of the factors determining the time course of the end-plate potential. II. Temperature. *J. Physiol. (Lond.).* **224**:333–348.
28. STEINBACH, A. B. 1968. Alteration by Xylocaine (Lidocaine) and its derivatives of the time course of the end plate potential. *J. Gen. Physiol.* **52**:144–161.
29. STEINBACH, A. B. 1968. A kinetic model for the action of Xylocaine on receptors for acetylcholine. *J. Gen. Physiol.* **52**:162–180.
30. ADAMS, P. R. 1976. Drug blockage of open endplate channels. *J. Physiol. (Lond.).* **260**:531–552.
31. STEINBACH, J. H. 1977. Local anesthetic molecules transiently block currents through individual open acetylcholine receptor channels. *Biophys. J.* **18**:357–358.
32. KATZ, B., and R. MILEDI. 1973. The binding of acetylcholine to receptors and its removal from the synaptic cleft. *J. Physiol. (Lond.).* **231**:549–574.
33. COLQUHOUN, D. 1975. Mechanisms of drug action at the voluntary muscle end plate. *Annu. Rev. Pharmacol.* **15**:307–320.
34. HARTZELL, H. C., S. W. KUFFLER, and D. YOSHIKAMI. 1975. Post-synaptic potentiation: interaction between quanta of acetylcholine at the skeletal neuromuscular synapse. *J. Physiol. (Lond.).* **251**:427–463.
35. MAGLEBY, K. L., and D. A. TERRAR. 1975. Factors affecting the time-course of decay of end-plate currents: a possible cooperative action of acetylcholine on receptors at the frog neuromuscular junction. *J. Physiol. (Lond.).* **244**:467–495.
36. ARMSTRONG, D., and H. A. LESTER. 1977. Kinetics of curare action at the frog nerve-muscle synapse. *Neuroscience Abstracts.* **3**:369.
37. LESTER, H. A., D. D. KOBLIN, and R. E. SHERIDAN. 1978. Role of voltage-sensitive receptors in nicotinic transmission. *Biophys. J.* **21**:181–194.
38. ROSENBERRY, T. L., H. W. CHANG, and Y. T. CHEN. 1972. Purification of acetylcholinesterase by affinity chromatography and determination of active site stoichiometry. *J. Biol. Chem.* **247**:1555–1565.

NOTE ADDED IN PROOF

We have recently conducted experiments with a pure *cis* isomer of Bis-Q, synthesized by N. H. Wasserman and B. F. Erlanger. At concentrations less than 1 μM, this compound has no effect on receptors.

DISCUSSION

SCHECHTER: We begin with several questions submitted by the referees. Question 1: Does a light flash cause any change in either open or closed receptors other than that mediated by isomerization of Bis-Q?

NASS: Well, none of the phases I've described occurs with photo-stable agonists; that's an important control. Bis-Q itself does not absorb enough energy from the flash to cause a significant temperature jump in the bulk solution. In fact we calculate and measure a change of less than 50 millidegrees per incident light flash. But it is an interesting hypothesis that because of the high receptor concentration in this synaptic cleft, enough Bis-Q accumulates to cause significant local heating when photons are absorbed. This temperature rise would of course be conducted away, but we do not know how rapidly.

SCHECHTER: Question 2: Does *cis*-Bis-Q bind to receptors?

NASS: Well, just recently, working with Norbert Wasserman and Bernard Erlanger at Columbia, we found that pure *cis*-Bis-Q is neither an agonist, nor an antagonist, nor a local anesthetic at concentrations less than 1 μM. This suggests to us that *cis*-Bis-Q does not bind at all to receptors, but we plan direct binding studies to clarify this point.

SCHECHTER: Question 3: What about possible local anesthetic effects?

NASS: The different actions of cobra toxin and curare, especially on phases 3 and 4, might indicate that after activation of receptors, a second site of agonist binding is exposed. Interaction of agonists and this site would lead to channel closure. Curare might have a high affinity for this site, and prevent channel closure, while cobra toxin might have no access to it, possibly through steric hindrance. This mechanism has been put forward by several investigators to explain how local anesthetics and related drugs block acetylcholine receptor channels.

Here we refer to the work of Neher and Steinbach (ref. 31). Traces B through D were taken in the presence of a drug, QS-222. Open channels seem to chatter; they look as though rats have eaten away at the single-channel event. And as we go from traces B through C to D, the concentration of QS-222 has increased, and we see in fact that the channel chatters all the more with increasing concentrations of local anesthetics. With regard to local anesthetics as they apply to the experiments we've described here, and again with the help of Erlanger and Wasserman at Columbia, we have done a few studies on a light-activated local anesthetic of this sort, *N-p*-phenyloxophenyl carbamylcholine (we call it EW-1). Upon activation by our light flash, this drug becomes a local anesthetic. Its action is really quite different from phases 1 and 2, and we don't think that a local anesthetic effect will explain either phases one or two; but we do note that the relaxations we see with EW-1 show some similarities to phases 3 and 4.

SCHECHTER: There is just one other submitted question, by Charles Stevens: *Cis* and *trans*-Bis-Q will, in general, have different binding constants for any sites to which they might bind, both hydrophobic (through the azobenzene moiety) and polar (via the trimethylammonium). The environment near receptors is a complicated one, with membrane lipids, membrane proteins, and ground substance. When isomerizations occur, then, a complicated concentration transient will result, representing binding to a large variety of sites, and diffusion processes. A flash—even an infinitely rapid one—cannot then actually cause a step change in concentration, and rather complex concentration transients must occur even when the isomer equilibrium of free Bis-Q is unperturbed by the flash. The photoisomerization technique is useful, then, in producing step perturbations in concentration only insofar as the concentration transients occur with time scale outside the range of interest, and/or are of insignificant magnitude relative to the other effects under investigation.

What is the time-course and magnitude of concentration changes that follow a light flash? This is really a very important question. What is the most significant source of the transient?

NASS: We think that no significant activity transients occur because of binding to sites other than the receptors themselves. Because *trans*-Bis-Q seems to bind so tightly to receptors, we are able to use very low concentrations of Bis-Q that would not be expected to saturate low-affinity, nonspecific binding sites. Furthermore, as we pointed out, voltage-jump and concentration-jump relaxations have identical kinetics under many conditions. One would not expect such agreement if a flash released active molecules from nonspecific binding sites.

On the other hand, we think that the receptors themselves cause the most significant activity transients by buffering the local concentration of *trans*-Bis-Q. With pure *cis*-Bis-Q we have been studying the response to flashes that increase the *trans*-Bis-Q concentration from 10 nM to more than 500 nM. There is an initial fast increase of conductance like that of the first trace (Fig. 3 in our paper); however, there are further, slower increases lasting several seconds. We believe that the slow increases occur because (*a*) the receptor concentration in the synaptic cleft is several hundred micromolar (this is discussed in our paper) and (*b*) that the binding is really quite tight, with a K_d for *trans*-Bis-Q of approximately 200 nM. Thus, receptors bind newly created *trans*-Bis-Q molecules, and the *trans*-Bis-Q activity in the synaptic cleft itself lags behind the instantaneous jump that has occurred in the external solution. That is, receptors buffer the concentration of Bis-Q: we believe this is the most important sense in which the concentration jump may not in fact be a simple jump.

BARRANTES: My first question concerns the feasibility of single-channel recording using the patch technique of Neher and Steinbach in the electroplaques. Have you attempted its implementation? My second question is related to the pharmacological activity of Bis-Q. Bis-quaternary ammonium compounds are in general very potent agonists, but have additional effects that mimic those of local anesthetics. In addition, Bis-Q in its *cis* form could easily be a local anesthetic-like substance. Would you care to comment on this point?

NASS: I will answer the last question first. Bis-Q, we can assure you, has no effect on receptors and we hope that that also means that it doesn't bind to receptors. As for the first question, we would like to do single-channel recording on electroplaques, but many technical difficulties accompany that kind of recording.

JARDETZKY: What do you think is the mechanism of the chatter in the local anesthetic effect?

NASS: Neher and Steinbach's view can be summarized by picturing a bathtub with a plug which moves rapidly in and out of a drain; presumably there is a binding involved to sites in that drain. The chatter occurs as the channel is simply plugged and unplugged rapidly by these local anesthetics.

LESTER: Neher and Steinbach have done a very careful analysis of the local anesthetic effect with respect to the concentration dependencies and the voltage dependencies of the postulated mechanism, i.e., direct blockade of the channel. The remaining question from such analyses is whether channels blocked by local anesthetics can nonetheless close as normal or whether the local anesthetic molecule must remove itself from the drain before the channel closes. The situation may differ among various blocking molecules.

JARDETZKY: It doesn't really matter whether you have a binding phenomenon or a conformational phenomenon, you will get exactly the same results. Your bathtub plug could be the binding of the molecule or it could be the conformational change in the channel where the channel gets stuck in a half-open position in the presence of the local anesthetic; you will have exactly the same kind of phenomenological result.

LESTER: In macroscopic relaxation data and even in fluctuation data it is theoretically impossible to distinguish between these two kinds of theories. However, with single-channel recording, assume that a very small member of local anesthetic molecules are involved per channel, one must take a mass action viewpoint and explain increasing chatter frequencies in terms of more collisions by the blocking molecule rather than a higher probability of transitions. Again let me emphasize that one is not dealing with a population measure, but rather with single molecular events.

JAIN: Have you considered the possibility of adjacent channels in which the anesthetic affects the coupling of the adjacent channels? For example, a situation where there exists a combination of four channels and the anesthetic blocks only one channel or affects the coupling between adjacent channels?

LESTER: One could make a great many models along the lines that you have suggested. Sheridan and I performed a series of voltage-jump relaxation experiments in which the major puzzle was that the voltage-jump relaxations remained exponential with time even when very large relaxations were involved: that is, when the system could not be treated as an infinitesimally small perturbation. This behavior severely constrained any "concerted" model of the Monod-Wyman-Changeux type or any type of dimerization mode of the Bamberg-Läuger type. One doesn't expect exponential relaxations from models like that and so we think it fairly unlikely that the rate-limiting step in the opening and closing of acetylcholine receptor channel is a dimerization or a concerted transition of the entire protein.

JAIN: Yes and no. The stacking kind of aggregation, as in gramicidin, is taken care of by what you suggested. However, one can think of other types of aggregation. Suppose there are four channels in parallel and they are aggregated together and give fourfold conductance changes. These small channels are not seen under normal conditions, when you see tetrameric channels giving fourfold conductance. In the presence of anesthetics you are cutting off the coupling between the four channels and you are looking at only the uncoupled single channels. This coupling may be reflected in the time (frequency) or amplitude domains.

LESTER: The point raised by Dr. Jain is quite an important one. The answers are not yet definite, but the fact that single-channel recordings do reveal only one open state of the channel, as contrasted, let's say, with alamethicin channels, makes it fairly unlikely that different states of the proteins can make channels with different conductances. One must emphasize that all the experiments and all of the facts are not in. Within present levels of resolution, let's say 5-10%, there seems to be only one open state of the channel.

NASS: If I understood your question correctly, it would imply that the single-channel conductance in the presence of local anesthetics would be different, when in fact it is the lifetime that is different.

JAIN: I think that is exactly what I said.

WEINSTEIN: There are several points that I would like you to help me clarify. First, does the *cis* isomer have a measurable affinity for the receptor in your model?

NASS: Could you refer to a particular figure?

WEINSTEIN: Fig. 15 is where you have *cis* isomer bound to the receptor in your model.

NASS: The idea is that the *cis* isomer has been created instantaneously on the time scale of membrane events. It has been produced from the *trans* isomer by our light flashes but has not yet had time to dissociate from the receptor complex.

WEINSTEIN: Yes, that is another question: can this isomerization take place while the molecule is bound? Could you show that such isomerization is obtained when Bis-Q is bound to serum albumin?

NASS: Well, studies don't directly address themselves to the question but they do lead to an important implication about the receptor and channel. That is, we know the channel closes less than 100 μs after the agonists of the binding site have been removed, or at least been isomerized to an inactive form. This implies that there is at most a 100-μs lag in the propagation of information from one site to another.

LESTER: We don't know at this point what complex of subunits do in fact make up what we call the acetylcholine receptor. We have been assuming throughout this discussion that the receptor consists of something that binds agonist and something that allows ions to go through—that is, the receptor has both a binding site and a channel.

We have been tooling up to do experiments with Bis-Q and its analogues bound to soluble proteins; probably the soluble protein we would use is the acetylcholine esterase. But all that we will measure, again, are spectral changes in Bis-Q: unless one has an actual measure of binding—perhaps as a precipitate on a Millipore filter—on the millisecond time scale, one is still up in the air with regards to these questions.

WEINSTEIN: That is why the enzyme may not be the best bet. Another question about your experimental procedure: is acetylcholine present in your preparation during the experiment?

LESTER: No, we hope there isn't any. In some experiments we have eliminated the acetylcholine esterases and that produced absolutely no difference in the result. But acetylcholine is not present.

WEINSTEIN: That new local anesthetic you mentioned—is it a derivative of carbamylcholine? Could carbamylcholine be obtained by photodissociation and what you observe simply result from the action of this drug to release acetylcholine or from an action directly on the cholinergic receptor?

LESTER: I think Dr. Weinstein has addressed two very important points. The first is the existence of the putative presynaptic acetylcholine receptor. Dr. Weinstein was wondering whether the acetylcholine left around the nerve terminals is going to enter into the experiments. We performed several controls to make sure it is not around and these are listed in the paper.

We performed preliminary flash photolysis experiments with *N-p*-phenyl-azo-phenyl-carbamylcholine and it does seem to be capable of being isomerized back and forth indefinitely. Therefore we think there is no hydrolysis of the sort that you suggest.

WEINSTEIN: Finally, is your system fully reversible after Bis-Q is washed out?

LESTER: With regard to the data presented here, yes. However, prolonged exposure to agonist produces a prolonged permeability increase in the electroplaque. Eventually, this depletes the gradients for sodium and potassium across the member and leads to the sustained depolarizations generally measured by using bath application of drugs in many laboratories. So this artifact has constrained and vitiated the interpretation of a great deal of data over the past 20 years but we don't think that it enters into any of the effects we have described here.

SCHECHTER: With reference to one of these last questions: What is your estimate for the binding constant of the *trans*-Bis-Q and what is your estimate of the difference in energy between the *cis* and *trans* forms? Do you have any data to form an estimate of these values?

LESTER: Questions like that were one of the reasons we came here, because we didn't even pose them before. The binding constants of all agonists are voltage dependent. We have described this in a recent paper in the *Biophysical Journal* (ref. 37). At minus 150 mV, where these experiments are conducted, the apparent dissociation constant for *trans* Bis-Q is roughly 150 mM. For *cis*-Bis-Q, if there are any effects on receptors, it is safe to say that these effects occur with apparent dissociation constant at least 100 times greater.

SCHECHTER: What is the ratio of the *cis* and *trans* forms at equilibrium?

LESTER: After a flash of light which produces the change in the *cis- trans* conformation ratios, there will be no further change with time, on a time scale of a week.

JARDETZKY: Changeux has some very reasonable evidence that the receptor itself can be separated from the channel; that is, the bungarotoxin binding entity is in a different polypeptide chain from the sodium channel. Now I don't think that will change your model very much. Specifically, the plug model would require a conformational lock in this case. This could still depend on the rate of binding and unbinding of the agonist.

LESTER: It must first of all be emphasized that our contribution to the discussion mentioned local anesthetics only peripherally. We do not think that local anesthetic effects play a role in phases 1 and 2. Perhaps these contribute to phases 3 and 4. The probe used by Changeux is histrionicotoxin. Albuquerque and his collaborators have demonstrated that histrionicotoxin acts like a typical local anesthetic; that is, it blocks open channels and therefore it seems quite reasonable to expect that this drug is indeed a probe of the acetylcholine receptor channel and doesn't bind to the binding site.

NASS: One would have difficulty, I think, imagining local anesthetic effects that occur on a time scale of 100 μs after liberation of local anesthetics by a light flash. Furthermore, even if it could be this fast, why would the recovery from the local anesthetic effect proceed with kinetics like that of standard voltage jump? So in short, quantitatively we have been able to account for both phases 1 and 2 without invoking local anesthetics, although local anesthetic effects may play some role in phases 3 and 4.

SIMIC: Which particular features of Bis-Q do you think are pertinent to each action? The quarternary ammonium? The electronic structure?

LESTER: It would be best to tell an anecdote in this regard. When Bis-Q was first synthesized by Erlanger and Wasserman, they looked at structure-activity relations and they looked at dose-response curves and they said that the *cis* isomer will be an agonist and the *trans* isomer won't. Of course the reverse happened. I really would not like to speculate on the structural features responsible for making a molecule an agonist or a local anesthetic.

SIMIC: But do you think perhaps the size or the charges do play a role?

LESTER: Yes.

JAKOBSSON: I am still a little puzzled by one feature of your earlier discussion of the chatter effect induced by the local anesthetic. It seems to me that to suppose an increased rapidity of conformational change is to suppose an anesthetic-reduced activation energy for the transformation. This would seem to me to be reasonable, if the anesthetic gets into the lipid phase of the membrane and modifies the local environment of the channel.

LESTER: The rate-limiting steps in activation and in blockade by local anesthetics are indeed not clearly sorted out and the mechanism you propose remains possible. We have recently done an experiment that tends to rule it out to some extent with EW-1 and *N-p*-phenyl-azo-phenyl carbamylcholine. The question is, does a local anesthetic have to be present in its active form in the membrane before blocking channels or can it appear suddenly from outside and block channels as it binds? The experiment starts with the channels open in the presence of *trans*-EW-1, which has no pharmacological effect on receptors. Channels are open and conductance is activated as usual. A light is then flashed producing some *cis* EW-1, which we believe is the active blocking form of this local anesthetic. Channels immediately start to close; the conductance relaxes to a lower value with a time-course completely consistent with the local anesthetic hypothesis. Furthermore, one can do the equivalent experiment on postsynaptic currents presumably caused by a nearly delta function of azo-phenyl carbamylcholine in the synaptic cleft. If one flashes a light just before the postsynaptic current, the local anesthetic effect proceeds as though the local anesthetic had been in the solution for many minutes beforehand. So this indicates to us that a local anesthetic need not be present in its active form before acting.

BARRANTES: In relation to the previous question of Dr. Simic, it is relevent to refer to the work of the Russian molecular pharmacologists (e.g., Michelson and Zeimal). Indeed, when the distance between the two quaternary ammonium ions extended, the potency of bis-choline compounds increases; a maximum is reached at about $n = 8-10$ for $(CH_2)n$ interposed between the two choline ends. In the case of Bis-Q it is of course difficult to predict what influence the additional modifications on "the alkyl chain region" will have on the pharmacological activity of this ligand. The results presented by Nass et al. today suggest, however, that Bis-Q behaves in the electroplaque preparation as a very potent agonist.

BERNHARD: With regard to the last question, it should be pointed out that at least for our small and medium-size organic molecules, agonist antagonists, and blockers of the acetycholine receptor, all are effectors of acetylcholine esterase. This enzyme appears to function just as well in every respect whether it is bound to a membrane fragment or not, and so it would seem that the affinity with which effectors are bound to the enzyme is related to the affinity with which these effectors act on the receptors. It seems unlikely that membrane effectors are acting on the membrane by the mechanism by which they affect the receptor.

LESTER: Dr. Bernhardt has been done a series of very elegant experiments on acetylcholine esterase and been very fortunate in being able to have it function similarly in the membrane as in solution. Moreover, there is an effect known as desensitization for acetylcholine receptors. What desensitization means is that after being exposed to agonists for a while, new receptor channels stop opening. No one knows why or how this occurs. There seem to be desensitizing effects merely by removing the receptor from its membrane environment. This is why we ordinarily study the receptor in a membrane.

HAPTEN-LINKED CONFORMATIONAL EQUILIBRIA IN IMMUNOGLOBULINS XRPC-24 AND J-539 OBSERVED BY CHEMICAL RELAXATION

S. VUK-PAVLOVIĆ, Y. BLATT, C. P. J. GLAUDEMANS, D. LANCET, AND
I. PECHT, *Department of Chemical Immunology, The Weizmann Institute of Science, Rehovot, Israel, and Laboratory of Chemistry, National Institute of Arthritis, Metabolism and Digestive Diseases, National Institutes of Health, Bethesda, Maryland 20014 U.S.A.*

ABSTRACT The interaction of oligogalactan haptens with the murine myeloma proteins XRPC-24 and J-539 has been investigated by the fluorescence temperature-jump method. The relaxation spectrum is composed of two processes, the faster representing hapten association and the slower a protein isomerization. In both cases the concentration dependence of relaxation times and amplitudes was consistent with the general mechanism formulated by Lancet and Pecht (1976, *Proc. Natl. Acad. Sci. U.S.A.* **73**:3549), in which the equilibrium between two conformations of the protein is shifted by hapten binding. The intact proteins and their Fab fragment had identical kinetic behavior, indicating that the conformational changes are located in the Fab region. Temperature dependence analysis for protein J-539 permitted the calculation of activation parameters and led to a consistent energy profile for all the elementary steps. The conformational states are separated by large activation barriers, but have similar free energies. The results suggest that hapten-induced conformational changes in immunoglobulins are more general phenomena than was previously thought.

INTRODUCTION

Immunoglobulins recognize antigens and bind them to the variable portions in their Fab parts by multiple noncovalent interactions. The antigen-antibody complexes thus formed are capable of activating various macromolecular and cellular functions related to the immune response. The triggering of these events involves sites on the Fc region of the immunoglobulin away from the antigen-binding site. The molecular mechanism of this signal transmission has not yet been clearly resolved, and several models, not mutually exclusive, have been proposed for it (1). In one of these, the allosteric model, it is assumed that antigen (or hapten) binding will impose structural changes in the hapten binding domain (Fv) or possibly in the whole Fab, which may then propagate to the Fc (2).

Dr. Vuk-Pavlović's permanent address is: Institute of Immunology, 41000 Zagreb, Yugoslavia. Address correspondence to Dr. Pecht, Department of Structural Biology, Sherman Fairchild Building, Stanford University School of Medicine, Stanford, Calif. 94305.

Recently we have reported kinetic evidence for an isomerization step in the homogeneous immunoglobulin MOPC 460 which is linked to the binding of hapten (3). In the present paper we similarly analyze the elementary steps in the interaction of the homogeneous immunoglobulins secreted by murine plasmacytomas J-539 and XRPC-24 with their specific (1 → 6)-β-D-oligogalactan haptens (4). The observed chemical relaxation spectra are shown to be as expected for an allosteric monomer, where two interconvertible conformations of the hapten-binding domains exist, and hapten shifts the equilibrium between them towards the better binding one. The present findings are therefore consistent with an allosteric model for antibody action. They provide a detailed dynamic and energetic picture of the processes involved in such a possible mechanism.

MATERIALS AND METHODS

Mildly reduced and alkylated homogeneous murine plasmacytoma proteins J-539 and XRPC-24 (X-24), their Fab fragments, and the haptens (1-6)-β-D-galactotriose (Gal_3) and (1-6)-β-D-galactotetraose (Gal_4) were prepared as previously described (4). In all experiments phosphate-buffered saline (PBS; 0.01 M phosphate, 0.15 M NaCl, pH 7.0) was used.

Kinetic measurements used a temperature jump apparatus, as described by Rigler et al. (5), operating in the fluorescence mode. Capacitor discharge of 20 kV raised the temperature by 5.2° from 20.0 ± 0.1° or 30.0 ± 1°C. The analog signal was digitized by a Biomation 802 transient recorder (Biomation, Cupertino, Calif.). The sum of at least five relaxation curves was analyzed by a computer program using a modified Marquardt routine (6). The program yielded the relaxation times and the corresponding amplitudes. These amplitudes were normalized by a generalized version of the formula used by Lancet and Pecht (3): $A = (\Delta F_{obs}/F)(1 + \Delta f_{max}\theta)$, where A is the normalized amplitude, ΔF_{obs} and F are the observed changes in fluorescence and the total fluorescence after the temperature jump both in the same arbitrary units, Δf_{max} is the maximal fractional fluorescence change at full saturation (negative for quenching and positive for enhancement), and θ is the fractional saturation of the protein at the particular hapten concentration. The experimental concentration dependence of the relaxation times and amplitudes was fitted to expressions according to Lancet and Pecht (3), using the algorithm of Powell (7).

RESULTS AND INTERPRETATION

The relaxation spectrum for both proteins XRPC-24 and J-539 with Gal_3 consists of a biexponential decay. The two relaxation times are observed throughout the hapten concentration range and are well separated. No relaxation process was observed for any of the reactants alone. In the case of XRPC-24, the analysis was problematic due to the interference of the cooling of the sample with the slower relaxation component. The amplitudes were more sensitive to such deviations, and therefore only a limited analysis of relaxation times is presented here. In the case of protein J-539 the slower relaxation was completed within 1 s, and the data allowed full analysis, which resulted in a complete and self-consistent picture of the binding and conformational equilibria.

The concentration dependences of the fast ($1/\tau_f$) and slow ($1/\tau_s$) relaxation rates for proteins XRPC-24 and J-539 binding Gal_3 at 25°C are shown in Figs. 1 and 2, re-

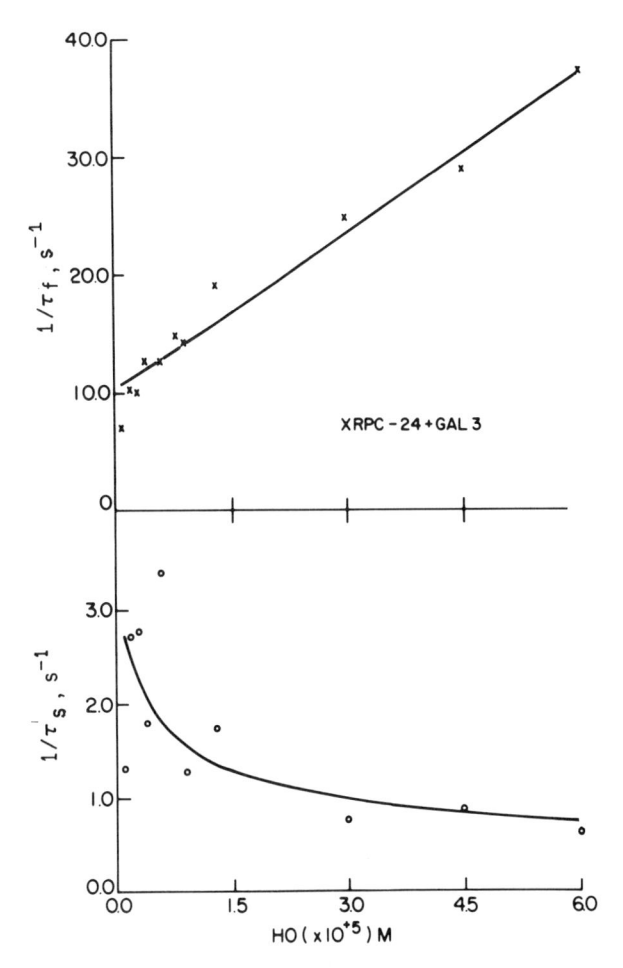

FIGURE 1 The dependence of fast (×) and slow (○) reciprocal relaxation times of XRPC-24 on total Gal$_3$ concentration at 25°C. Intact protein concentration was 9.52×10^{-7} M sites. The lines were calculated by using the best fit parameters for mechanism 3, as listed in Table I. Individual values (in all figures) are ± 5%.

spectively. It can be seen that for both proteins $1/\tau_f$ increases linearly, while $1/\tau_s$ levels off with increasing hapten concentration. Such behavior indicates the presence of a bimolecular association together with a slower monomolecular step, which we attribute to protein isomerization.

Lancet and Pecht (3) have amply discussed the mechanisms which may be considered when evaluating antibody-hapten interactions, for which also a monomolecular slow step is observed. These mechanisms are (in the standard nomenclature of Monod et al. [8]):

$$H + T_0 \underset{k_{-T}}{\overset{k_T}{\rightleftarrows}} \quad \begin{matrix} T_1 \\ k_1 \| k_{-1}, \\ R_1 \end{matrix} \tag{1}$$

$$\begin{array}{c} & T_0 & \\ k_0 \| & k_{-0} & k_R \\ H + R_0 & \underset{k_{-R}}{\overset{}{\rightleftarrows}} & R_1, \end{array} \qquad (2)$$

$$\begin{array}{ccc} H + T_0 & \overset{k_T}{\underset{k_{-T}}{\rightleftarrows}} & T_1 \\ k_0 \| \ k_{-0} & & k_1 \| \ k_{-1}, \\ H + R_0 & \overset{k_R}{\underset{k_{-R}}{\rightleftarrows}} & R_1 \end{array} \qquad (3)$$

where the isomerizations occur in the hapten-antibody complex (mechanism 1), in the free antibody (mechanism 2), or in both (mechanism 3). For a comprehensive table listing the detailed kinetic and thermodynamic features of these mechanisms, the reader is referred to ref. 3.

Kinetics of Gal_3 Binding to XRPC-24

The concentration dependence plots in Fig. 1 represent experiments with intact protein XRPC-24, but the Fab fragment gave practically identical results. A qualitative in-

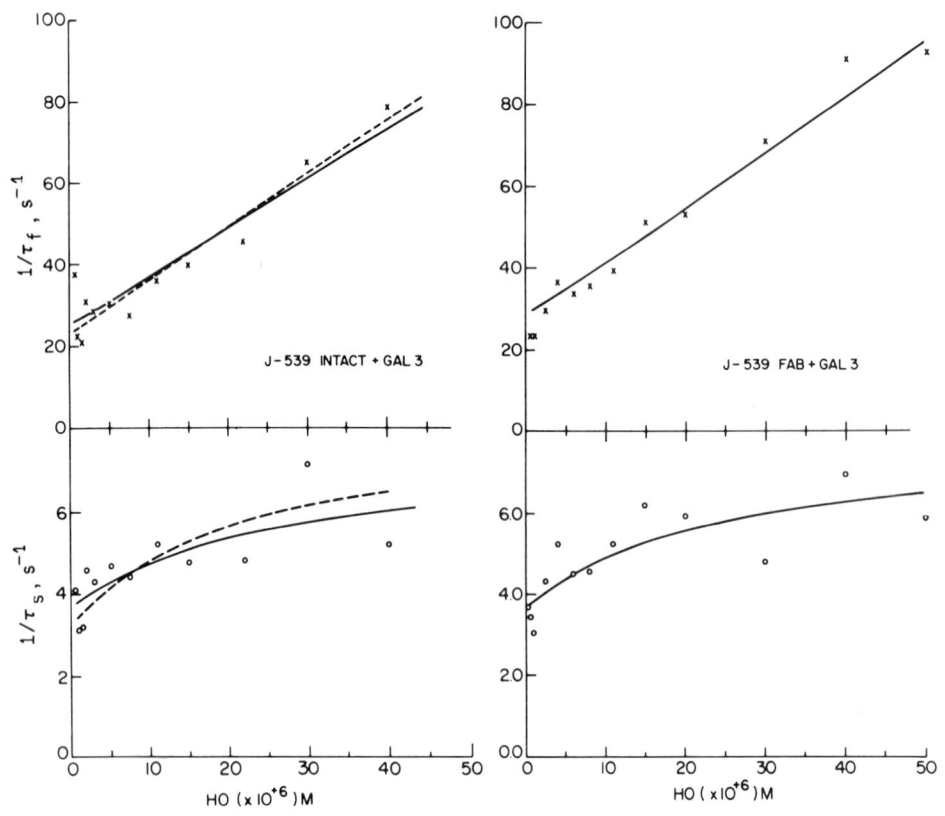

FIGURE 2 The dependence of fast (x) and slow (o) reciprocal relaxation times of J-539 on total Gal_3 concentration at 25°C. Left: intact protein, 9.52×10^{-7} M sites. Right: Fab fragment, 1.33×10^{-6} M sites. The broken and solid lines were calculated by using the best fit parameters for mechanism 1 and 3, respectively.

TABLE I
KINETIC PARAMETERS FOR THE REACTION BETWEEN INTACT XRPC-24 AND GAL$_3$ AT 25°C

i	K_i	k_i	k_{-i}	ΔG_i
		s^{-1}	s^{-1}	$kcal/mol$
T	4.5×10^4	4.5×10^5	10.0	−6.32
R	2.7×10^5	—	—	−7.38
0	0.5	1.05	2.09	0.41
1	3.0	0.3	0.1	−0.65
Overall	1.2×10^5			−6.90

spection of these plots leads to the rejection of mechanisms 1 and 2. This may be seen as follows: $1/\tau_s$ decreases to a plateau, behavior consistent with mechanism 2, but not with 1. On the other hand, the slope over intercept of the $1/\tau_f$ line yields a rough estimate of the binding constant for the fast association. This is found to be $k_{ass} = 5 \times 10^4$ M^{-1}, smaller than the known overall association constant $K = 1.75 \times 10^5$ M^{-1} for XRPC-24 Fab (4). Such relation may be obtained for mechanism 1, but not for 2. Only mechanism 3 is compatible with both these observations simultaneously.

The lines in Fig. 1 represent the best fit of the data to Eqs. 4 and 5 below. The corresponding best fit parameters are listed in Table I. We have also performed experiments with Gal$_4$ hapten. The data obtained were identical to those with Gal$_3$ within the experimental error and, therefore, are not presented here. However, this finding supports the notion of Jolley et al., who found that the fourth galactose in the oligomer does not contribute much to the total binding energy (9).

Kinetics of Gal$_3$ Binding to J-539

For the protein J-539 the relaxation times (τ_f, τ_s) as well as the fast (A_f) and slow (A_s) relaxation amplitudes were analyzed in detail. The data for the intact protein and Fab at 25°C are depicted in Figs. 2 and 3. Here, the relation between K_{ass} and overall equilibrium constant, K, is as found for XRPC-24 ($K_{ass} = 5.8 \times 10^4$ M^{-1}, $K = 1.5 \times 10^5$ M^{-1} [4]), while the behavior of the slow time is different: $1/\tau_s$ increases to a plateau. This means that mechanism 1 may be used to account for the relaxation phenomena. Analysis in terms of this mechanism revealed fairly good agreement with the data (dashed lines in Figs. 2 and 3) for all four relaxation quantities (τ_f, τ_s, A_f, A_s) yielding the following parameter values: $k_T = 1.31 \times 10^6$ M^{-1} s^{-1}, $k_{-T} = 23.4$ s^{-1}, $K_T = 5.77 \times 10^4$ M^{-1}, $k_1 = 4.82$ s^{-1}, $k_{-1} = 3.01$ s^{-1}, $K_1 = 1.68$ with the overall equilibrium constant fixed at the value $K = 1.5 \times 10^5$ M^{-1} (4).

Mechanism 1 was shown to be a special limiting case of mechanism 3 (3). A system that fits mechanism 1 may in principle also be described in terms of the thermodynamically general mechanism 3. Since two other immunoglobulin-hapten equilibria (MOPC-460 and XRPC-24) were shown to fit only the latter mechanism, it forms the only basis for comparison of all three systems. Furthermore, only mechanism 3 provides a full insight into the energy and activation attributes of hapten-induced con-

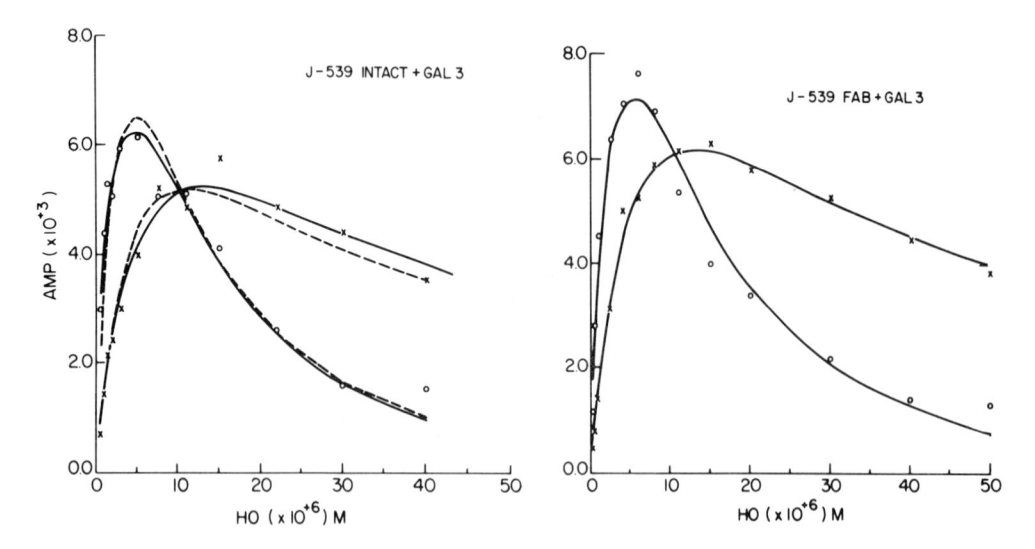

FIGURE 3 The dependence of fast (x) and slow (o) amplitude of J-539 intact (left) and Fab (right) on total Gal$_3$ concentration. The points are $\Delta F_{obs}/F$ (see Materials and Methods) and the normalization is done on the fitted lines. Other details are the same as in Fig. 2.

formational changes (see below). We therefore chose to analyze the relaxation data presented above also in terms of mechanism 3.

For this mechanism one expects three relaxation times. When the associations are much faster than the isomerizations, two of these represent the T and R association steps and one the isomerization steps (10). For both XRPC-24 and J-539 (as also found in the case of MOPC-460, ref. 3) one observes only two relaxation times. An attempt to fit the relaxation spectra to a sum of three exponents gave a very large scatter in the concentration dependence, implying that such procedure was chemically meaningless and that only two relaxation times really occur. We explain this discrepancy assuming that the spectral change of one of the steps is zero. The concentration dependences of τ_f and A_f are found by the fitting procedure to be compatible only with those of a T association, and a fit with $\Delta f_R = 0$ indeed yields consistent results as seen below.

The concentration dependences were fitted to the following equations derived according to Castellan (10) and Jovin (11):

$$1/\tau_f = k_T \cdot g_{11} \cdot T_0 \cdot H, \tag{4}$$

$$1/\tau_s = |g_3| \cdot (k_1 \cdot T_1 + k_0 \cdot T_0)/|g_2|, \tag{5}$$

$$A_f = \Delta H_T \cdot \Delta f_T \cdot \Delta T/(g_{11} \cdot A_0 \cdot R \cdot T^2), \tag{6}$$

$$A_s = |g_2| \cdot (Q_{13} \cdot \Delta f_T + Q_{23} \cdot \Delta f_R + \Delta f_0) \cdot (Q_{13} \cdot \Delta H_T + Q_{23} \cdot \Delta H_R + \Delta H_0).$$
$$\cdot \Delta T/(|g_3| \cdot R \cdot T^2 \cdot A_0), \tag{7}$$

where Δf_i and ΔH_i are the normalized fluorescence change and standard enthalpy change of step i, A_0 is the total protein concentration, and $Q_{13} = (g_{12} \cdot g_{23} - g_{13} \cdot g_{22})/$

$|g_2|$, $Q_{23} = (g_{12} \cdot g_{13} - g_{11} \cdot g_{23})/|g_2|$. g_{ij} and $|g_i|$ are an element and a principal partial determinant of the Castellan g-matrix (10). These equations are identical to those used by Lancet and Pecht (3): both there and here the Castellan form was used in the computer fit, being concise and easy to formulate. The equations may then be transformed by somewhat tedious algebraic manipulations into the explicit form given by Lancet and Pecht.

The best-fit lines for J-539 (intact and Fab at 25°C) are shown in Fig. 2 (full line curves), and the corresponding parameters are given in Table II. It is clearly seen that the intact protein and the Fab fragment give very similar relaxation patterns and parameters. The following points should be stressed in the context of these fits: (a) As mentioned above, only the T association gives rise to an observable relaxation process. Eq. 4 is derived by assuming that this process is the fastest in the system. We also attempted a fit assuming that the R association is faster, and that the T association is coupled to it, but a worse fit was obtained. (b) The information on the parameters of the unobservable R association is obtained through the fit of the slow time and amplitude (Eqs. 6 and 7). (c) The overall Δf was fixed in the fitting procedure to the value measured previously (4). K and ΔH on the other hand, were left as free parameters and their values in Table II represent the best fit. No independent information is available for ΔH, while K agrees fairly well with $K = 1.5 \times 10^5$ M^{-1} previously reported (4).

Experiments were also performed at 15° and 35°C. At 15°C the relaxation was so slow that numerical evaluation was unreliable. A good fit was obtained for the data at 35°C with parameters as listed in Table II. From the change of k_i with temperature,

TABLE II

KINETIC AND THERMODYNAMIC PARAMETERS FOR THE REACTION BETWEEN INTACT OR Fab FRAGMENT OF J-539 AND GAL$_3$ AT 25 AND 35°C ACCORDING TO MECHANISM 3

i		T	K_i	k_i	k_{-i}	ΔF_i	ΔH_i	ΔG_i	ΔS_i
		°C		s^{-1}	s^{-1}		kcal/mol		cal/mol·°
T	Intact	25	4.96×10^4	1.23×10^6	24.8	0.2	−12.8	−6.38	−21.5
	Fab	25	4.96×10^4	1.36×10^6	27.4	0.25	−12.5	−6.38	−20.5
	Fab	35	2.95×10^4	2.17×10^6	73.3	0.27	−12.7	−6.07	−22.2
R	Intact	25	1.96×10^5	—	—	0.0*	−10.6	−7.19	−11.4
	Fab	25	1.96×10^5	—	—	0.0*	−10.7	−7.19	−11.8
	Fab	35	1.15×10^5	—	—	0.0*	−10.0	−6.88	−10.5
0	Intact	25	0.93	1.77	1.91	0.3	−0.63	0.043	−3.56
	Fab	25	0.93	1.77	1.91	0.45	−0.26	0.043	−1.02
	Fab	35	0.9	5.16	5.74	0.37	−0.27	0.062	−1.02
1	Intact	25	3.67	5.72	1.56	0.009	0.53	−0.77	4.36
	Fab	25	3.67	6.09	1.66	−0.09	−0.48	−0.77	0.97
	Fab	35	3.5	20.6	5.88	0.64	−0.31	−0.74	1.44
Overall	Intact	25	1.2×10^5	—	—	0.196	−11.3	−6.9	−14.8
	Fab	25	1.2×10^5	—	—	0.255	−10.7	−6.9	−12.7
	Fab	35	7×10^4	—	—	0.255	−10.7	−6.58	−13.8

Error ± 20%.
*Assumption.

TABLE III

ACTIVATION PARAMETERS FOR THE REACTION BETWEEN PROTEIN J-539 Fab AND GAL$_3$ ACCORDING TO MECHANISM 3, CALCULATED FROM THE KINETIC PARAMETERS AT 25 AND 35°C.

i	T	$-T$	0	-0	1	-1
$\Delta G_i^\ddagger$, kcal/mol	9.05	15.4	17.0	17.0	16.3	17.1
$\Delta H_i^\ddagger$, kcal/mol	7.88	17.3	18.8	19.4	21.5	22.4
$\Delta S_i^\ddagger$, cal/mol·°	-3.92	6.28	5.89	8.06	17.4	17.8
$\Delta G_i^\ddagger - \Delta G_{-i}^\ddagger$, kcal/mol		-6.38		0.04		-0.78
$\Delta H_i^\ddagger - \Delta H_{-i}^\ddagger$, kcal/mol		-9.42		-0.59		-0.87
$\Delta S_i^\ddagger - \Delta S_{-i}^\ddagger$, cal/mol·°		-10.2		-2.17		-0.46

the activation parameters were calculated by the Eyring equation (12). The resultant values are given in Table III. The difference $\Delta H_i^\ddagger - \Delta H_{-i}^\ddagger$ is the overall enthalpy change in the step, and it is obtained here in an independent way. Comparison to the values obtained from the amplitude analysis (Table II) reveals satisfactory agreement, and this is true also for $\Delta G^\ddagger$ and $\Delta S^\ddagger$. Thus, the internal consistency of the analysis performed here is clearly brought out.

DISCUSSION

The present kinetic study provides a detailed insight into the dynamics and energetics of hapten-induced conformational transitions in two sugar-binding immunoglobulins. For both XRPC-24 and J-539 the results are consistent with the "allosteric monomer" model (mechanism 3) where two conformational states of the protein, T and R, bind the hapten differently ($K_R > K_T$). $\Delta(\Delta G) = \Delta G_R - \Delta G_T$ serves as the driving force for the shift in the conformational equilibrium, as it is also equal to $\Delta G_1 - \Delta G_0$, and therefore

$$\Delta G_1 = \Delta G_0 + \Delta(\Delta G). \tag{8}$$

For both XRPC-24 and J-539 ΔG_0 is found to be small and positive, (0.41 and 0.06 kcal/mol, respectively), implying T_0 is more stable and that the two states differ only slightly in their free energy. The value of $\Delta(\Delta G)$, although not very large (-1.1 and -0.8 kcal/mol, respectively), is sufficient to make ΔG_1 negative in both cases, so that R_1 becomes the predominant conformation. The extent of the hapten-induced shift in the conformational equilibrium is given by K_1/K_0 ($= K_R/K_T = 1/c$ in the Monod et al. model, ref. 8), and is 4.0 and 6.0 for J-539 and XRPC-24. This is somewhat smaller than 11.4 found for MOPC 460 (3).

The rate constants for the conformational processes are found to be in the range of 1–10 s^{-1} at 25°, so that $\Delta G^\ddagger = 16$–17 kcal/mol. The k values are 3–4-fold higher at 35°, due to the high positive $\Delta H^\ddagger$ for the isomerizations (18–22 kcal/mol). The low rates and high activation parameters most probably reflect the participation of a significant portion of the Fab in the conformational transition, i.e., many interactions are disrupted to attain the transition state. The positive $\Delta S^{\ddagger}$, on the other hand, implies that entropically favorable processes are involved in its formation.

It thus appears that, although the two conformations have almost identical G and H values (Table II), they are separated by a significant energetic barrier. This barrier is somewhat higher here than that found for MOPC 460, where isomerization rate constants are 10-100 s^{-1}, and $\Delta H^{\ddagger}$ = 10 kcal/mol (3, 13).

The values of k_{on} for the protein-oligosaccharide associations are found here to be two orders of magnitude lower than those for nitroaromatic-binding immunoglobulins (13), and comparable to those found for other sugar-binding proteins (13, 14). This may stem from the necessity to overcome the flexibility of the ligand (reflected in $\Delta S^{\ddagger}_T < 0$) and/or from the need to disrupt and form many hydrogen bonds upon association, expressed in the large positive $\Delta H^{\ddagger}_T$ (see ref. 13 for a discussion of this point).

For another sugar-binding immunoglobulin, Maeda et al. (14) suggest that such a slow association step includes a monomolecular (possibly conformational) rearrangement of the hapten-immunoglobulin complex. If this mechanism, originally proposed by Haselkorn et al. (15) to describe any hapten-antibody association, holds also in our case, then the full binding scheme (written for simplicity in terms of our mechanism 1) should be:

$$H + T_0 \overset{1}{\rightleftharpoons} (HT_0) \overset{2}{\rightleftharpoons} T_1 \overset{3}{\rightleftharpoons} R_1. \qquad (9)$$

In this scheme, step 1 represents the formation of the encounter complex (13) or a process of labile association (14). Step 2 may involve hydrogen-bond exchange and/or fast conformational rearrangements of both the ligand and protein contact residues, which do not give rise to a resolvable reaction step. Step 1 and 2 together result in a single fast relaxation step (cf. ref. 13). Step 3 represents those conformational changes in the protein which involve higher energetic barriers, and therefore give rise to a distinct slow relaxation time.

The present study extends the kinetic evidence for hapten-induced conformational changes in immunoglobulins. Thus, such phenomena are not restricted only to rare cases or to a particular type of hapten. Our data suggest that a large portion of the Fab is involved in the kinetically observed conformational transition, but that it does not depend on the Fc. It is however possible that high molecular weight antigens, having a larger $\Delta(\Delta G)$, are able to shift the conformational equilibrium more significantly and by a mechanism such as that proposed by Huber et al. (2) will lead to changes also in the Fc. In addition, the observed conformational changes may be related to the attainment of optimal antibody-antigen complementarity, and may thus be relevant to the question of antibody diversity.

S.V.-P. gratefully acknowledges a long-term fellowship from the European Molecular Biology Organization.

Received for publication 10 December 1977.

REFERENCES

1. METZGER, H. 1974. Effect of antigen binding on the properties of antibody. *Adv. Immunol.* **18**:169.
2. HUBER, R., J. DEISENHOFER, P. M. COLMAN, M. MATSUSHIMA, and W. PALM. 1976. Crystallographic structure studies of an IgG molecule and an Fc fragment. *Nature (Lond.).* **264**:415.
   PECHT, I. 1976. Recognition and allostery in the mechanism of antibody action. *Collog. Ges. Biol. Chem. Mosbach.* **27**:41.
3. LANCET, D., and I. PECHT. 1976. Kinetic evidence for hapten induced conformational transition in immunoglobulin MOPC 460. *Proc. Natl. Acad. Sci. U.S.A.* **73**:3549.
4. JOLLY, M. E., S. RUDIKOFF, M. POTTER, and C. P. J. GLAUDEMANS. 1973. Spectral changes on binding of oligosaccharides to murine immunoglobulin A myeloma proteins. *Biochemistry.* **12**:3039.
5. RIGLER, R., C. R. RABL, and T. M. JOVIN. 1974. A temperature-jump apparatus for fluorescence measurements. *Rev. Sci. Instrum.* **45**:580.
6. FLETCHER, R. 1971. *In* Harwell Subroutine Library, Atomic Energy Research Establishment, Harwell, U.K. (Subroutine VB01A).
7. POWELL, M. J. D. 1971. *In* Harwell Subroutine Library, Atomic Energy Research Establishment, Harwell, U.K. (Subroutine VA04A).
8. MONOD, J., J. WYMAN, and J. P. CHANGEUX. 1965. On the nature of allosteric transitions: a plausible model. *J. Mol. Biol.* **12**:88.
9. JOLLEY, M. E., C. P. J. GLAUDEMANS, S. RUDIKOFF, and M. POTTER. 1974. Structural requirements for the binding of derivatives of D-galactose to two homogeneous murine immunoglobulins. *Biochemistry.* **13**:3179.
10. CASTELLAN, G. W. 1963. Calculation of the spectrum of chemical relaxation times for a general reaction mechanism. *Ber. Bunsenges. Phys. Chem.* **67**:898.
11. JOVIN, T. 1975. Fluorimetric kinetic techniques: chemical relaxation and stopped flow. *In* Biochemical Fluorescence Concepts. R. F. CHEN, and H. EDELHOCH, editors. Marcel Dekker, Inc., New York. Vol. 1, 305.
12. GLASSTONE, S., K. J. LAIDLER, and H. EYRING. 1941. The theory of rate processes. McGraw-Hill Book Company, New York. 14.
13. PECHT, I., and D. LANCET. 1977. Kinetics of antibody-hapten interaction. *In* Chemical Relaxation in Molecular Biology. I. Pecht and R. Rigler, Editors. Springer-Verlag, Berlin. 306.
14. MAEDA, H., A. SCHMIDT-KESSEN, J. ENGEL, and J. C. JATON. 1977. Kinetics of binding of oligosaccharides to a homogeneous pneumococcal antibody: dependence on antigen chain length suggests a labile intermediate complex. *Biochemistry.* **16**:4086.
15. HASELKORN, D., S. FRIEDMAN, D. GIVOL, and I. PECHT. 1974. Kinetic mapping of the antibody combining site by chemical relaxation spectrometry. *Biochemistry.* **13**:2210.

DISCUSSION

SCHECHTER: The first two questions were submitted by a referee: What are the best numerical procedures now available for the analysis of chemical relaxation data of the type presented here? Within what confidence limit can the relaxation decay law be determined and what are some of the main artifacts to be avoided in studies of this type?

PECHT: The analysis includes two major consecutive steps: (*a*) Analysis of the relaxation curves and evaluation of relaxation times and amplitudes (and base lines): Routinely, a sum of six or more relaxation curves has been analyzed. In principle any least-squares fit procedure may be used. We found that the modified algorithm of Marquardt is particularly effective (cf. ref. 6 for the subroutine). This subroutine requires the derivative of the exponential function with respect to the parameters, but the first guesses supplied to the program need not be very accurate. (*b*) Analysis of the concentration dependence of the relaxation times and amplitude: Here the functions are often more complex, and analytical derivations are not always convenient. Therefore we use a "simplex" nonlinear least-squares algorithm (Powell, ref. 7), which does not require the function derivatives and handles many parameters quite efficiently.

As for the second question, regarding confidence limits for relaxation times and amplitude: τ's separated more than a factor of three can be resolved with good confidence limits. Also if their respective amplitudes are similar in size, better resolution can be achieved. The absolute magnitude of the amplitude should not be lower than 0.01%. Although the base line (final value of the signal) can be found as a parameter by the computer program, having an experimentally reliable value improves the confidence of the values. To examine the possibility of more than two exponents, a separation of a factor of 5-10 between the τ's is required. As stated in the text, the error limits were in the range of ± 5-10%.

Artifacts in the measurements mainly arise due to the relatively long relaxation times which often had interference from cooling and convection. To overcome this difficulty, the base lines have been examined by the fitting procedure.

CZERLINSKI: Your data on Gal_3 binding to J-539 protein led you to the cyclic mechanism 3. You find only two relaxation times where three are expected, and you assume that the spectral change of one of the steps is zero. Have you considered the possibility that both bimolecular steps proceed at about the same speed? As error bars are not shown in your figures, I cannot judge what difference between two relaxation times would still permit you to distinguish them as two and determine them with any precision (the amplitude of the R-association may be as much as a factor of 4 below that of the T-association).

PECHT: Yes, we have considered this possibility. We have tried to fit the observed fast process to a sum of two processes having similar τ's. An important check is the "total amplitude analysis" of the two fast association steps. With the small ligand coupling (at the high hapten excess employed), a sum of two individual amplitude expressions for the T and R association is obtained. Analysis of such terms showed that in this case the R contribution is less than 5%. We cannot exclude a smaller contribution of association with the K state.

CHOCK: Based on your mechanism 3, the results of your analysis show an essentially equal population for T_0 and R_0 and the rates for the isomerization between these two species are significantly slower than the rate of hapten binding. Do you care to comment on the physiological role of this slow isomerization step? In fact your data probably will fit better with your mechanism 1 with an additional step; i.e. $H + T_0 \rightleftharpoons HT \rightleftharpoons HR \rightleftharpoons HR_1$ (similar to your Eq. 9). This is particularly true for your J-539 system.

PECHT: Well, the question is really very relevant and I am grateful that it was raised. I mentioned before that the link between the changes that we see and the actual biological activity of the antibodies has to be established. And that's why I said that it can be interpreted at least in two ways. First, that the conformational changes are confined to the Fab. In this case we just have a case of an induced fit, a way by which the antibody can interact better with the given hapten and prefer it over another one. That would be the minimalistic way to look at it.

A more interesting way to consider our findings is that the structural transition is a trigger for further biological activities. I stressed a moment ago that we can't establish this link at present; furthermore, I should remind the audience that the hapten per se does not trigger the immunological reaction. The haptens should be attached to a macromolecular carrier or, preferably, be polyvalent. That is a very interesting feature of immunochemistry—that small molecules usually do not induce the immune response. So if the hapten induces the actual changes related to triggering of immune phenomena, there are ways of explaining why the conformational equilibrium lies where it does, equally distributed between two states. One way would be to try and fit it into a mechanism that will amplify the effect by binding a third component. We have not examined that as yet.

The answer to the question whether we tried to add a further step is that we didn't try to fit it to a more complex scheme. The data fit very well to the minimal scheme and we did not feel it necessary to go beyond this. In summary, the really interesting question is to what extent do we expect to see further changes depending on interaction with other components of the immune system, for example, interactions with complement components. In other words, the binding of the hapten is not the end of the process. There are further components that interact with the immunoglobin.

R. P. TAYLOR: My questions relate to the potential generality of your observation. The first is, have you been able to confirm the existence of the slow transition by observing it by stopped-flow methods? And second, has anyone detected the transition in these systems, either by T-jump or stopped flow?

PECHT: Well, to answer your first question, in the case of the galactan-specific antibodies (XRP-24 and J-539), we didn't check it by stopped-flow. We did check it for the 2,4-dinitrophenol (DNP)-specific MOPC-460, and there you definitely see the slow transition by stopped-flow. I should perhaps mention at this point that although we monitored only the intrinsic fluorescence of the antibody for the saccharide-binding antibodies, in the case of MOPC-460, which binds DNP derivatives, we have been monitoring the reaction by following the quenching, or enhancement, of fluorescence of the antibody. We have also studied the changes in transmission due to the formation of the complex of the DNP with the protein; by using an analogue of the DNP, namely a nitrobenzoxadiazol derivative, we could follow it also by quenching of the hapten fluorescence. All three modes of monitoring the reaction gave us essentially the same kinetic behavior. We have also shown the stopped-flow trace for the reaction of protein 460 with nitrobenzoxadiazol alanine and the results are very much in agreement with what the chemical relaxation T-jump analysis gave.

I think I commented already to a certain extent about the generality of the results. More specifically, the question is would we expect multi-step reactions to occur also with normally induced antibodies in contrast to myeloma proteins, which are somewhat suspect, being of tumor origin. The reservation about the immunoglobulins produced by the tumor line has been quite amply proved to be unjustified. These are definitely legitimate representatives of the immunoglobulin antibodies that every animal produces. Still, work has been done also on normally induced antibodies; difficulties emerge with a heterogeneous population of antibodies. In other words an animal challenged with a certain antigen will produce a multitude of different antibodies. They will all recognize a DNP attached to bovine serum albumin, but they will recognize it in rather different ways, and one expression of this will be the rather broad range of binding constants that those antibodies will express against DNP. To expect a clear-cut answer from a broad range of different combining sites is perhaps asking too much. That is the main reason that we have confined ourselves to the homogeneous proteins. There is now a way of circumventing this problem, namely producing homogeneous antibodies from hybrid cells produced from tumor lines and normal antigen-induced cells.

ROMINE: Could you elaborate on what particular immunoglobulins were used in your experiments?

PECHT: Those that I spoke about were all IgAs; however quite similar results were also obtained recently with an IgM (MOPC IONE, specific for nitrosyl oligosaccharides). Unfortunately no measurements on IgG have been done.

ROMINE: IgM is known to bind complement and will itself activate complement in the CH_2 region. Is that not true? Would it not be possible to design experiments using galactan polymers larger than three units to mimic a large macromolecular antigen, as opposed to using the time? That way you can add complement to the system and see if this induces a further relaxation process to check the biological activity?

PECHT: The answer is yes, it has to be done.

ROMINE: The other question is what other experiments along these lines might reveal biological activity in relation to the conformational changes involved? This is a rather general question.

PECHT: Well, we can spend the next hour on this, obviously. I think one very interesting line of work, which Dr. Barisas has pursued in the last few years, is examining the reactions of antibodies with divalent haptens and trying to resolve further changes induced by the fact that the hapten is divalent. Porter and his associates have shown that oligomers formed by the interaction of antibodies with divalent haptens can induce complement activation. I refer now to IgG antibodies.

CZERLINSKI: I have two parts to my comment. Part I: Your mechanistic scheme 3 actually has four limiting cases, while you presented only two. I shall employ the terminology of your mechanism: A bimolecular step in mechanism 3 may be coupled to a molecular interconversion either at T_0 or at R_0. One would in both cases observe a decrease of $1/\tau_s$ vs. concentrations and an approach to a plateau level. Furthermore, your bimolecular step could couple to either T_1 or R_1 of your biomolecular interconversion of complexes. These two cases would lead to the same observation: $1/\tau_s$ would increase with increasing concentrations and approach a "saturation"-level. Why did you not discuss the additional two cases?

Part II: You concentrate on cyclic mechanism 3. However, to interpret your data, you assume that one of the steps is not connected with any fluorescence change. This is certainly an acceptable assumption. However, I feel you should also consider the possibility that you do have three relaxation times, with the two fast ones directly connected with the bimolecular steps and closely spaced. I have to ask in this connection how closely could your relaxation times be spaced with the signal-to-noise ratio which you have in your experiments. Could they be spaced within a factor of 2, or less, or a factor of 4? I do not expect that the amplitudes of the two relaxation processes would be the same; they may be different by up to a factor of 4, on the basis of the difference in the equilibria of the two interconversions alone.

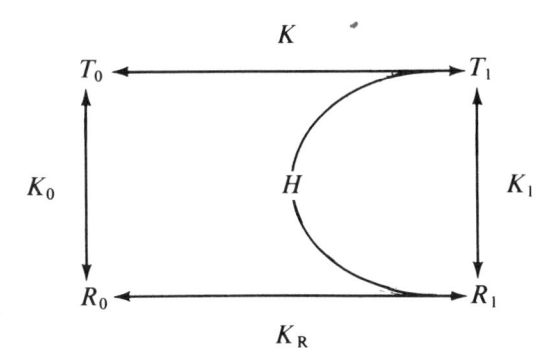

LANCET: This general scheme has the following implication: We assume that there are two different conformations for any kind of protein, including our antibody, that bind the hapten (or the ligand in general) differently, and by that we mean that one has a higher association constant than the other. The idea behind this scheme is that it is symmetric; when we say that T is the low association constant one, and R is the higher association constant (mechanism 1) this is just part of the naming process. When we say, for example, that we have the isomerization on the bound state only (that is mechanism 1 in our paper), we can show that we have implied in it that the second conformation in the sequence is of higher affinity. Actually it has infinite affinity since it does not dissociate at all. We therefore couple the K_T association with the K_1 isomerization. The mechanism with K_R association coupled to K_1 isomerization is just the same, only with affinity R. It's not a different mechanism; it is simply that we use different names. Chemically they are not different. Actually, only two cases out of the four are physically interesting, although in terms of notation there are four. I think this is the answer to the first question.

Now, as to the second question, I would like to refer you to a figure in our ref. 3. These curves are taken again from the very well characterized case of protein 460, but the analysis very much resembles the case of the galactan-binding proteins presented in this conference. The idea is to make a total amplitude analysis for the fast relaxation time. In the general scheme drawn here by Dr. Czerlinski, we expect, because of the cycle, only one slow relaxation time that represents both isomerization steps, and two fast relaxation times. According to what Dr. Czerlinski has proposed, which is very reasonable, we may have one fast relaxation time that represents the two association processes, due to the closeness of the values of the two times. But the amplitudes should show the behavior of a mixture of two different associations characterized by two different association constants. Thus we take the amplitude of the observed fast relaxation and analyze it in terms of concentration dependence, as shown here. Now, for the two association constants, which may be derived by means other than the amplitude analysis, we may expect different amplitude behaviors and we try to fit the observed fast relaxation amplitude to each of those expected behaviors. For protein 460 this is what we expect for what we call T binding, and this is for the R binding, and we see that it fits to the R binding. When we did the same thing for the two proteins presented here we found that the amplitude behavior could be fitted to almost pure T binding behavior; by that I mean that less than 5%, within experimental error, could be ascribed to R binding. Now I wouldn't like to say that there is no residual contribution from the R association. There very probably is, but it could probably not be more than a 5% contribution. Thus we can say that this is essentially pure T binding, and that the R association is not represented in the observed fast association. Therefore, we ascribe a small or zero fluorescence change to this step.

CZERLINKSI: There is one little problem when you assume only one fast relaxation time and thus one amplitude. Two relaxation times are expected to fit your data better, if you trade the assumption of "no signal change" for the assumption of "one more relaxation time." You ought to compare equivalent assumptions. You may use (as I do) the sum of squares of residues in your nonlinear least squares analysis as a criterion.

LANCET: Just a very short answer. One has to remember that here we perform *total* amplitude analysis. Such analysis does not depend at all on kinetics. Total amplitudes are pure thermodynamic magnitudes and we assume that the fast steps are completely uncoupled from the slow ones. We could even not look at all at the kinetics, but just at the magnitude of the effect versus concentration, and get this representation that I have shown, and derive from it the pure T nature of the association.

THE RATE OF ENTRY OF DIOXYGEN AND CARBON MONOXIDE INTO MYOGLOBIN

ROBERT H. AUSTIN AND SHIRLEY SUILING CHAN, *Max Planck Institute for Biophysical Chemistry, D-3400 Göttingen, West Germany*

ABSTRACT The model for carbon monoxide or dioxygen recombination with heme proteins developed by the group at the University of Illinois is reexamined. We propose that the carbon monoxide or dioxygen molecule enters the protein at essentially a diffusion-limited rate determined by the solvent viscosity and that the protein offers no important barriers to this entry. The viscosity dependence of the entry rate k_{ED}, its magnitude (1×10^{10} $M^{-1}s^{-1}$), and the rate of quenching of triplet states of protoprophyrin IX in apomyoglobin by dioxygen are used as supporting evidence. Comparison is made to the model of a fluctuating protein developed by G. Weber (1).

INTRODUCTION

The classic picture of a protein molecule as presented by X-ray crystallography is a static one, yet most protein molecules spend their working lives in fluid environments. How rigid is a globular protein in solution? How readily can small molecules penetrate the interior of a globular protein? Basically, there are two opposing models of protein permeability in solution, with supporting experiments. (*a*) The fluorescence quenching experiments of Lakowicz and Weber (1) showed that tryptophan groups buried within protein molecules were quenched by dioxygen at nearly the rate they were quenched free in solution. The model that comes from these carefully performed experiments is one of a floppy protein molecule, with large fluctuations in conformation on the nanosecond time scale. The recent calculations of Karplus (2) also seem to support this view. (*b*) An opposite view comes from the phosphorescence work of Saviotti and Galley (3). In their experiments, the phosphorescence of tryptophan groups in the interiors of some proteins can have lifetimes in the tenths of seconds range at room temperature in air-saturated water. This implies an access rate of dioxygen to the interior of the protein very much less than the rates seen by Lakowicz and Weber.

The work of the group at Illinois (4–6) using the recombination kinetics of carbon-monoxide or dioxygen with heme proteins also provides information on the access of a small ligand with the interior of a protein. In this communication we would like to examine the Illinois model and see whether it can be interpreted in terms of either of the two opposing models just stated.

The particular question we would like answered by the carbon monoxide recombination work is: What is the rate at which the carbon monoxide molecules enter the heme pocket of the protein? In the Illinois experiment, it was possible to separate internal recombination from external, concentration-dependent recombination. This

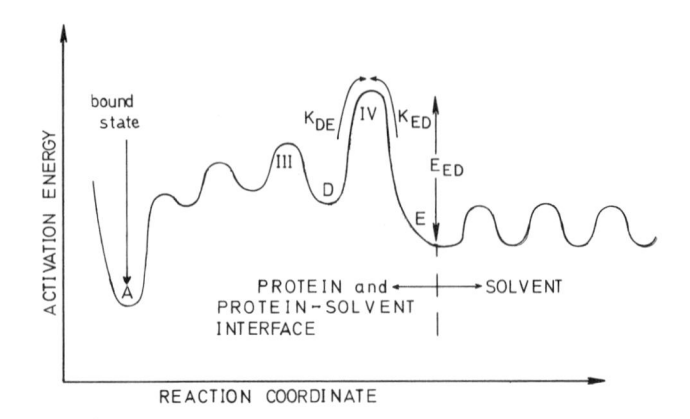

FIGURE 1 The postulated energy barrier scheme for CO recombination with Mb, taken from Austin et al. (4). Barrier IV represents the solvent-protein interface that the CO must surmount to enter the heme crevice. The solvent is represented by the smaller activation energy barriers to the right. To be consistent with Alberding et al. (5,6), where barrier III is lacking, the reader should rename barrier III barrier II and well D well C for those cases where no evidence for barrier III exists.

allowed the construction of a set of differential equations with one of the rate parameters interpreted as the actual rate of entry of the carbon monoxide molecule into the binding area near the heme. Fig. 1 shows the scheme of activation energy barriers and their associated rates as postulated by the Illinois group. The barrier of most interest to us here is barrier IV, and the rate k_{ED} for passage over it. Ideally, k_{ED} should represent the rate of entry of the carbon monoxide into the protein (as distinguished from binding at the iron site) and k_{DE} the rate of exit. This means that the rate k_{ED} should be roughly the same as the dioxygen quenching rate measured in the fluorescence quenching experiments. We have used the techniques of transient optical anisotropy decay and triplet lifetime quenching in an attempt to test this proposal.

METHODS AND MATERIALS

The myoglobin (Mb) was purchased from Sigma Chemical Co. (St. Louis, Mo.). The protein was freshly dissolved in distilled water and centrifuged to remove any precipitates.

Glycerol, reagent grade, was purchased from the J. T. Baker Chemical Co. (Phillipsburg, N.J.). Glycerol-water mixture ratios were determined by measurement of the viscosity of the sample with a Schott Capillary Viscometer (Schott, Inc., New York) and compared to viscosities of known mixtures (Chemical Rubber Company Handbook of Chemistry and Physics, 1962). The protein sample was dialyzed for 2 days at 4°C against the desired glycerol-water mixture.

The carbon monoxide (CO) and argon gases used are 99.995% pure. Equilibration of the gas was via slow bubbling (CO) or by stirring under an atmosphere (Argon) for $\frac{1}{2}$ h. Dioxygen (O_2) was similarly introduced by slowly stirring the sample with exposure to air. Thus, the external O_2 pressure in these experiments was 0.21 bar, while the CO pressure was 1.0 bar. The solubilities of CO and O_2 in our solvents were taken from the Landolt-Boernstein tables (7).

The disodium salt of protoporphyrin IX was purchased from ICN Nutritional Biochemicals Div., International Chemical & Nuclear Corp. (Cleveland, Ohio). It was freshly dissolved in

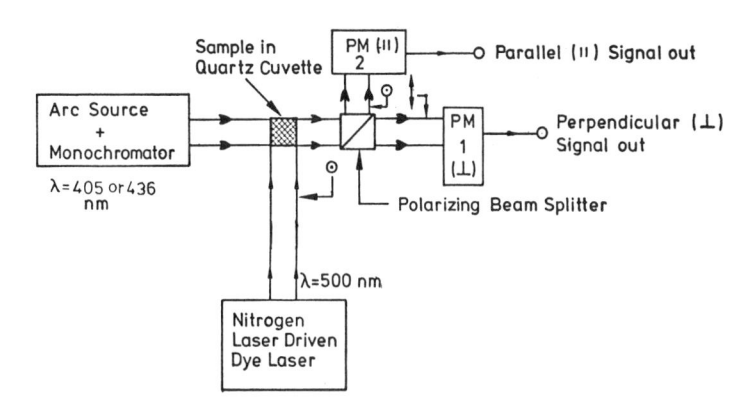

FIGURE 2 Block diagram of the apparatus. The polarization of the exciting laser pulse was out of the plane of the paper (o). The monochromator was set to 405 nm for PP9-Mb triplet state measurements (singlet depletion signal measured) or 436 nm for Mb-CO photolysis (creation of Fe[+2] state with no sixth ligand). The repetition rate of the laser was typically 10 Hz, and 1,000 shots were averaged at each temperature.

reagent grade dimethylformamide (DMF) purchased from Baker Chemical Co. to which 4 ml/ 1,000 ml of 1 N HC had been added. The apomyoglobin was prepared via the procedure of Rossi Fanelli et al. (8) and stored at $-20°C$ in 50% glycerol. A Bio-Gel P-4 column (Bio-Rad Laboratories, Richmond, Calif.) equilibrated with a 50% glycerol-10 mM phosphate buffer system, pH 7.5, was used to separate unbound protoporphyrin IX (PP9) from the bound PP9-Mb complex.

The photolysis system is shown in Fig. 2. A nitrogen laser (Lambda Physik GmbH, Göttingen, W. Germany) pumped a Coumarin 307 dye laser, emission at 500 nm and pulse duration 5 ns. The polarization of the laser output was approximately 90%, vertical. The unpolarized emission of a mercury-xenon arc lamp was passed through a monochromator and through the sample. The transmitted beam passed through a polarizing beam splitter into two 1P28A/V1 photomultipliers. The anode had a 1-kohm load resistor followed by a LH0033 high-speed buffer amplifier (National Semiconductor Corp., Santa Clara, Calif.) and a LeCroy VV100 amplifier (LeCroy Research Systems Corp., Spring Valley, N.Y.). The measured rise time of the entire system was 20 ns for a 5-ns light pulse. The outputs of the two channels were fed to a Tektronix 7A13 differential amplifier (Tektronix, Inc., Beaverton, Oreg.) and then to a Data Lab model DL920 transient recorder (Data Laboratories Ltd., Mitcham, Surrey, U.K.), maximum sample rate 20 MHz. The final rise time measured on the transient recorder was 100 ns, determined by the transient recorder amplifiers. The digital output of the transient recorder was averaged in a Fabritek averager (Fabri-Tek Instruments Inc., Madison, Wis.) and fed to a Univac computer for analysis.

The anisotropy was determined from the expression,

$$\alpha(t) = (\Delta A_\parallel - \Delta A_\perp)/(\Delta A_\parallel + 2\Delta A_\perp), \qquad (1)$$

where $\Delta A_\parallel$ and $\Delta A_\perp$ are the transient absorbance changes parallel and perpendicular, respectively, to the exciting laser polarization. The absorbance changes were measured at 436 nm for CO-Mb and 405 nm for protoporphyrin IX. The anisotropy corrected absorbance change was determined from the expression,

$$\Delta A = \Delta A_\parallel + 2\Delta A_\perp. \qquad (2)$$

RESULTS AND DISCUSSION

In the anisotropy decay experiments, the excitation laser pulse photolyzed the CO-Fe bond, creating an absorbance change. The photolyzing beam was polarized in the vertical direction and thus created a population of photolyzed Mb molecules whose heme planes were preferentially aligned along the exciting laser beam polarization vector. The unpolarized monitoring beam was at a right angle to the exciting beam and becomes polarized after passage through the photolyzed sample because of the induced optical anisotropy. A complete calculation for a chromophore of planar symmetry shows that the maximum anisotropy expected is 0.14 (9). In reality, our anisotropy 100 ns after the flash varied from 0.045 to 0.07, the higher initial anisotropy occurred in the most viscous solvents. The subsequent anisotropy decay was always a good fit to a single exponential even when the absorbance changes with time were complex, as is true for Mb-CO at low temperatures. This also implies that our corrected absorbance changes had no polarization effects in them. Corrections for polarization effects in optical photolysis or other techniques (such as T-jump) in viscous solvents or for large linear molecules like DNA are not generally done, although it is possible to get very large, apparent "absorbance changes" which are nothing more than induced dichroism changes due to alignment of an asymmetric chromophore.

The anisotropy decay rate is the rotational rate of the Mb molecule, k_{rot}. From the Stokes-Einstein relation, we can determine either the hydrodynamic radius of the protein molecule (r) or the viscosity of the medium η:

$$k_{rot} = 3k_B T/4\pi r^3 \eta. \quad (3)$$

Thus, if the viscosity of the medium is measured at some temperature and the hydrodynamic radius of the protein calculated, it is then possible to determine the viscosity of the medium at any other temperature if it is assumed that the radius of the protein does not change. In Fig. 3 we have plotted log k_{rot} vs. $1/T$ over the range of temperatures of interest. In this manner we obtain viscosities difficult to measure by normal means. We calculate the hydrodynamic radius of Mb to be 2.4 nm at 293 K, in good agreement with the approximate value of 2.0 nm as determined from X-ray crystallography.

The slope of log(k_{rot}/T) vs. $1/T$ is a measure of the activation energy for viscosity changes, E_η. Although we have plotted log k_{rot} vs. $1/T$, over our temperature range and with our error bars the curvature induced by the correct formulation is slight and the slope given remains valid. Because we determined the glycerol-water content of our sample as 84% by volume, we denote this activation energy as E_η (84%). In our preparation of the sample, we tried to duplicate the Illinois procedure as closely as possible. The percentage of glycerol is slightly different from the value quoted in refs. 4–6, although no absolute determination of the glycerol-water content was made at Illinois. This small difference should not affect the main conclusions.

Let us now look at barrier IV in Fig. 1 and its associated rates, k_{ED} and k_{DE}. We have plotted in column 1 in Table I the energy barrier height of IV for various heme

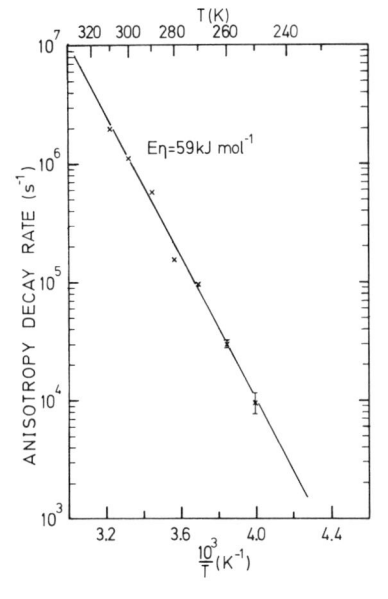

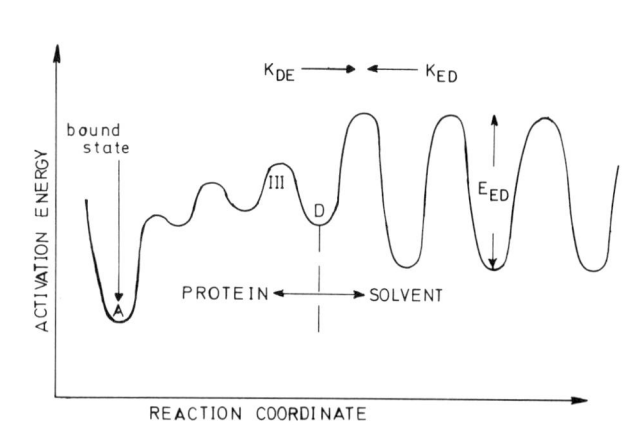

FIGURE 3

FIGURE 4

FIGURE 3 The log of the anisotropy decay rate of Mb, k_{rot}, vs. $1/T$. The solvent was 84% glycerol-water by volume, as determined by a Schott Capillary Viscometer. Although tables of glycerol-water viscosity do not extend down into the temperatures measured, extrapolations of existing tables gave reasonable agreement with our measured values using the radius of Mb (2.4 nm) determined at 293 K.

FIGURE 4 Revised model of the entry of CO or O_2 into the Mb molecule. Barrier IV no longer exists as a single barrier but instead is the activation energy for the viscosity of glycerol-water. The small gas molecule is postulated to enter the region near the binding site (barrier III) at a solvent diffusion-limited rate. The rate k_{DE} is the diffusion-limited dissociation rate. This picture may be modified at lower ratios of glycerol-water, where the small protection barrier provided by the protein may become larger than the activation energy for the viscosity of water (12.6 kJ/mol).

TABLE I
ROTATIONAL RATE AND LIGAND ENTRY RATES FOR VARIOUS PROTEINS

Substance	E_η or E_{ED}	k_{rot} or k_{DE} at 280K	k_{ED} at 280K	$\dfrac{k_{ED}}{[C]} \times \dfrac{\eta}{\eta_{H_2O}}$	Reference
	kJ/mol	(s^{-1})	(s^{-1})	$(M^{-1}s^{-1})$	
Viscosity of 84% glycerol-water	59	1.56×10^5 (Mb)	—	—	This work
Mb-O_2	53	2×10^5	2×10^3	1×10^{10}	4
Mb-CO	79	9×10^5	7×10^3	7×10^{10}	4
β_{SH}-CO	44	5×10^5	5×10^3	5×10^9	6
α_{SH}-CO	37	2×10^5	5×10^3	5×10^9	6
Heme-CO	74	7×10^8	2×10^4	2×10^{10}	5

proteins as determined by Austin et al. (4) and Alberding et al. (5,6), and in columns 2 and 3 the rate parameters k_{ED} and k_{DE} at 280 K. With the results of Lackowicz and Weber (1) in mind, one is immediately struck by the high value (34–75 kJ/mol) that barrier IV presents to the diffusion of a small ligand into the protein. However, it must be kept in mind that the measurements at Illinois were done in 75% glycerol. The well-known Smoluchowski equation (10) gives a reliable prediction of the maximum value that a bimolecular rate parameter can have (the diffusion-limited rate),

$$k_{diff} = (4\pi N_{AV}/1{,}000) D \cdot a, \tag{4}$$

where a is the sum of the molecular radii and D is the sum of the diffusion coefficients. Because the diffusion constant is inversely proportional to viscosity,

$$D = k_B T / 6\pi a \eta, \tag{5}$$

where η is the viscosity of the solvent in poise (cgs unit), to compare rates in different viscosities we will multiply the observed rates by the ratio of the measured viscosity to the viscosity of water at 280 K, and divide by the concentration of the gas in solution. The fourth column in Table I contains the raw and corrected k_{ED} for various proteins that have been measured. The final entry in Table I is our measured rate at which oxygen quenches the triplet state of PP9 in DMF, corrected for the measured viscosity and the known concentration of atmospheric O_2 in DMF. One can also calculate roughly the expected rate at which O_2 should quench the triplet PP9 from Eqs. 4 and 5. This value is given in Table II.

Several relationships can be seen in Table I. First, the activation energies of the k_{ED}s (34–75 kJ/mol) are similar to $E\eta$ (84%), 59 kJ/mol. Some of the activation energies, it must be confessed, are closer to $E\eta$ (84%) than others. The activation energy for Mb-CO is clearly too large (75 kJ/mol), whereas the value for Mb-O_2 is quite close

TABLE II
TRIPLET QUENCHING RATES

Parameter at 280K	Raw value	$\dfrac{\text{Rate}}{[O_2]} \times \dfrac{\eta}{\eta_{H_2O}}$	Reference
	(s^{-1})	$(M^{-1}s^{-1})$	
Quenching rate of PP9 triplet by O_2 in DMF	1.8×10^6	1.4×10^9	This work
Quenching rate of PP9-Mb triplet in air-saturated water	3.1×10^4	2.0×10^8	This work
Quenching rate of PP9-Mb triplet in air-saturated 75% glycerol-water	2.7×10^2	2.8×10^8	This work
Calculated rate of O_2 quenching of PP9 triplet in medium of 0.01 poise	—	9.2×10^9	
k_{ED} for Mb-O_2	2×10^3	1×10^{10}	4

(54 kJ/mol). It should be stressed that k_{ED} is a computer-derived fit, not directly measured. Second, the corrected bimolecular rates at 280 K are actually *greater* than the expected diffusion-limited rate, although it is not totally impossible for the diffusion-limited rate for arrival at the area near the heme group to be different from the diffusion-limited quenching rate of O_2 with PP9. Granting the imperfections, we feel that the evidence is sufficiently encouraging to propose the following: Barrier IV has little to do with the protein-solvent interface; it is simply the activation energy E_η for the diffusion of CO or O_2 in 75% glycerol-water. The O_2 or CO arrives at barrier III at the diffusion-limited rate.

There is a further test we can make, quite similar to a fluorescence quenching experiment. Because the lifetime of the triplet state of PP9 in water at 280 K is 1.3 ms, it is easy to measure the quenching of the PP9 triplet in a protein by O_2 at atmospheric pressure if the quench rate is in the diffusion-limited range. The unprotected quench rate of PP9 by O_2, corrected again for viscosity, is given in Table II. We have also measured the quenching rate of O_2 for PP9 bound to apomyoglobin in water and in 75% glycerol. From Table II we see that the reduction of the quenching rate by the protein (0.15) is in the range of values found in Lackowicz and Weber (1), 0.17-0.60. Because the O_2 must penetrate the protein to some extent to quench, it would not even be unreasonable to expect that $k_{ED} = k_{quench}$. They are not, but $k_{ED} > k_{quench}$, once again supporting the picture that k_{ED} is a diffusion-limited rate. There certainly is a barrier presented by the protein—the PP9 is protected to some extent—but not the 50- to 84-kJ/mol barrier originally associated with the protein-solvent interface. Our modified view of the potential barriers is given in Fig. 4.

The above statement is our main conclusion. There are further puzzles which need much more careful work. Why does k_{DE} have the same activation energy as k_{ED}? We would like to argue that k_{DE} represents the diffusion-limited dissociation of CO or O_2 from the protein, with its correspondingly greater "effective radius" and thus higher rate, but this remains speculative. We originally were struck by the equivalence of k_{DE} and k_{rot} and had thought that the solvent-cage barrier originally postulated for barrier IV had a lifetime given by the rotational rate of the Mb, but our further experiments have convinced us that barrier IV in effect is the solvent alone. Only further experiments, in particular determination of the barriers at other ratios of glycerol-water, can judge the correctness of our proposal. For now we would say that the fluctuating protein model of G. Weber seems to give the best explanation of our data, at least for access of small molecules to groups near the surface of a protein.

We would like to acknowledge helpful comments and discussions from Prof. H. Frauenfelder. We are indebted to G. Striker for his expert computer data analysis.

Received for publication 3 December 1977.

NOTE ADDED IN PROOF

It has been brought to our attention that the quenching of triplet protoporphyrin IX in proteins by oxygen has previously been done by B. Alpert and L. Lindqvist [1975. Porphyrin triplet state probing the diffusion of oxygen in hemoglobin. *Science (Wash. D.C.).* **187**:836–837; and 1976. Laser study of triplet porphyrin quenching by oxygen in porphyrins globins. *In* Excited States of Biological Molecules. J. B. Birks, editor. John Wiley & Sons, Inc., New York]. Their results for the triplet quenching are similar to ours and we apologize to them for our oversight.

REFERENCES

1. LAKOWICZ, J., and G. WEBER. 1973. Quenching of protein fluorescence by oxygen. *Biochemistry.* **12**:4171.
2. MACAMMON, J., B. GELIN, and M. KARPLUS. 1977. Dynamics of folded proteins. *Nature (Lond.).* **267**:585.
3. SAVIOTTI, M., and W. GALLEY. 1974. Room temperature phosphorescence and the dynamic aspects of protein structure. *Proc. Natl., Acad. Sci. U.S.A.* **71**:4154.
4. AUSTIN, R. H., K. W. BEESON, L. EISENSTEIN, H. FRAUENFELDER, and I. C. GUNSALUS. 1975. Dynamics of ligand binding to myoglobin. *Biochemistry.* **14**:5355.
5. ALBERDING, N., R. H. AUSTIN, S. S. CHAN, L. EISENSTEIN, H. FRAUENFELDER, I. C. GUNSALUS, and T. M. NORDLUND. 1976. Dynamics of carbon monoxide binding to protoheme. *J. Chem. Phys.* **65**:4701.
6. ALBERDING, N., S. S. CHAN, L. EISENSTEIN, H. FRAUENFELDER, D. GOOD, I. C. GUNSALUS, T. M. NORDLUND, M. F. PERUTZ, A. H. REYNOLDS, and L. B. SORENSEN. 1978. Binding of carbon monoxide to isolated hemoglobin chains. *Biochemistry.* **17**:43.
7. LANDOLT-BOERNSTEIN Tables. 1962. Lösungsgleichgewichte I, 2 Teil, Bandteil b. Springer-Verlag, Berlin.
8. ROSSI FANELLI, A., E. ANTONINI, and A. CAPUTO. 1958. Physicochemical properties of human globin. *Biochim. Biophys. Acta.* **30**:608.
9. STEPANOV, B., and V. GRIBKOVSKII. 1968. Theory of Luminescence. Iliffe, London. 60.
10. VON SMOLUCHOWSKI, M. 1917. Versuch einer mathematischen Theorie der Koagulationskinetik Kolloider Lösungen. *Z. Physik. Chem.* **92**:129.

DISCUSSION

VANDERKOOI: Oxygen is a small lipophilic molecule. Its partition coefficient for organic solvents is about 10 times greater than that for aqueous solution. Would you not expect oxygen to partition into the hydrophobic interior of proteins so that the effective concentration of oxygen in the interior is higher than in the aqueous phase? How would this affect your calculations?

AUSTIN: That is an excellent question. (*a*) Let us state the obvious and note that since our protein concentration was only about 10 μM, a net increase in solubility of the CO in the protein would not have affected our bulk overall solubility. (*b*) Let us try to calculate the probability of finding a CO molecule "dissolved" in a myoglobin molecule. The volume of a myoglobin molecule is about 6×10^{-20} cm^3. The volume for each O$_2$ molecule at 0.2 bar pressure in 75% glycerol-water is about 2.5×10^{-17} cm^3. If we assume that the actual local concentration of the O$_2$ in the protein is 10 times that in water (or 30 times that in 75% glycerol-water), then the probability of finding an O$_2$ molecule in the protein molecule is [(6 × 10^{-20})/(2.5 × 10^{-17})] × 30 = 0.08. This implies that the triplet quenching of the heme should show two phases: an oxygen concentration-independent quenching of 0.08 of the molecules and an oxygen concentration-dependent quenching of the remaining 0.92 molecules. What we see is the following: the single exponential triplet lifetime of the argon-saturated sample of several milliseconds changes to a single exponential with a lifetime of several microseconds in oxygen-saturated solution without a change in the zero time amplitude (50-ns time resolution). This seems in contradiction to the model and would suggest that oxygen does not partition itself into the interior of small proteins at higher ratios than the bulk solubility. Of course, our arguments here are hand-waving. We do think the problem needs a stronger theoretical analysis.

PERSOONS: First I want to make a comment on the symbols. In Eq. 4 a is the sum of the molecular radii, or the reaction distance, while in Eq. 5, a is the hydrodynamic radius of a single molecule: $D_i = k_B T/6\pi a_i \eta$, but $D_i + D_j \neq k_B T/6\pi(a_i + a_j)\eta'$. Also, the ordinates of Figs. 1 and 4 should be potential energy, not activation energy.

AUSTIN: Your comment about a is correct. We calculate the rate by using $a = a_1 + a_2$, and letting $D = D_1 + D_2 = (kT/6\pi\eta)(1/a_1 + 1/a_2)$. We did not assume $D = (kT/6\pi\eta)(1/a_1 + a_2)]$. You are also correct about Figs. 1 and 4. We should have labeled that axis "potential energy" or "enthalpy."

PERSOONS: You speak also of the rate constant for diffusion-limited dissociation. How is this calculated? What is the reason for a bimolecular rate constant greater than the theoretical value? What values of D_i, D_j, and a_{ij} are chosen?

AUSTIN: As we said in the paper, logical consistency would require k_{DE} also to be a diffusion-limited process, but this is speculative and we do not know how to calculate that process right now. There is no real hard limit to the diffusional rate. If we assumed a very large radius for the outer region in the protein—say the entire protein radius—then the predicted rate for k_{ED} would be considerably faster than our calculated one.

PERSOONS: I don't know if you can apply the eigen equations for it. Can the eigen equation for diffusion be used?

AUSTIN: Yes, the problem here is that the O_2 or CO are neutral molecules so that formalism is not directly applicable. Another problem is that the ratio k_{DE}/k_{ED} is 10 for myoglobin but 100 for heme. This can be rationalized by assuming there is more steric hindrance for outward diffusion in the case of the protein, but we don't know how to calculate that.

PERSOONS: How is the rate constant for a diffusion-limited dissociation calculated (see Eq. 4 of the paper)?

AUSTIN: It hasn't been, and we don't know how to do it. We just wanted to suggest that the diffusional rate *away* from the heme may be greater than the rate into it.

PERSOONS: If k_{ED} and k_{DE} have the same activation energy, the energy levels of D and E are equal, so the difference in magnitude between k_{ED} and k_{DE} should formally be explained from a difference in entropy of activation. Where is this difference coming from?

AUSTIN: It would come, if it exists, from the greater phase space available for outward diffusion as compared to inward diffusion. These numbers come out of fits where the amount of CO that escapes from the interior region into the solvent is determined by the ratio of the innermost barrier to k_{DE}. The innermost barrier rate is known by extrapolation from low temperatures. k_{ED} is determined from the bimolecular rebinding rate and again the ratio of internal rebinding to escape. In the case of the protein these calculations are quite involved.

VUK-PAVLOVIĆ: My questions are connected to some structural features of the barriers. (*a*) Assuming that the structure of the heme pockets in the ferrous and the ferric states are comparable, would you expect that barrier II in hemoglobin chains or either of the barriers II and III in myoglobin is related to barrier(s) modulating the solvent proton magnetic relaxation rates? (*b*) If the answer to the first question is positive, that there is a correlation between two types of experiment, would you agree that the distal histidine may be the structural basis of these barriers? If the answer is negative, what would you propose?

AUSTIN: (*a*) If the distance at which the iron spin relaxes the water spin is 1 Å or less, we would guess that the nuclear relaxation measurements should also sense the full barriers. However, if the spin relaxation can occur anywhere in the vicinity of the heme periphery, then the relaxation rate should be similar to the oxygen quenching rates and only sense the outer barriers, if they exist.

(*b*) We don't think we can say. If you look at the results of Dr. Traylor, where he sees large changes merely due to steric hindrance on the proximal side, then one might guess that the inner barrier is not predominantly affected by the distal histidine. As to the other barriers (II and III), which only seem to occur in proteins, I don't think we have any systematic knowledge right now. Dr. Mike Sherrock of Gustavus Adolphus College (St. Peter, Minn.) has done some very nice experiments where he has studied various hemes with different side groups that seem to lead to perturbation of the inner barrier but little influence on barrier III. So perhaps you're right; we don't know.

SIMIC: Hemin-*c*, a Cys-His derivative of heme and a simplest model of Mb, i.e.

shows the following: hemin-c(II) + O_2 → hemin-c(III) + $\cdot O_2^-$; $k = (4 \pm 1) \times 10^8 \text{M}^{-1}\text{s}^{-1}$, as determined by pulse radiolysis. The k value is less than diffusion-controlled, yet 20 times faster than $k(\text{Mb(II)} + O_2)$. This means also an energy barrier for hemin-*c* reaction, though much less than myoglobin.

AUSTIN: We think your results agree with what we are trying to say. A particularly interesting experiment would be to repeat your pulse radiolysis experiment with a heme protein as a test for the model of enriched local oxygen concentration. Also, since the technique is an effect rather like a very fast stopped-flow measurement of oxygen recombination, it would be interesting to see what happens at lower temperatures in mixed solvents.

LANCET: $\Delta H^{\ddagger}$ are in general not additive. If entering the site is rate limiting, then observed $\Delta H^{\ddagger}$ represents almost pure $\Delta H^{\ddagger}$ of binding.

AUSTIN: That is correct.

MCCRAY: We would like to point out that the protein does have an effect on the entrance of the ligand, either CO or O_2, into the pocket. We cite, for example, our work (Sono, Smith, McCray, and Asakura. 1976. *J. Biol Chem.* **251**:1418–1426.), where in phosphate buffer at room temperature different hemes modified at the 2,4-vinyl positions give rise to changes in rates of binding of up to a factor of 10. On the other hand, Moffet and Gibson have found that the R-state on rates for meso-, deutero-, and proto-hemoglobin are very similar. We have also found recently that a whole series of modified hemes, when substituted into α and β subunits, gives the same rate constants for CO binding as when a particular subunit apoglobin is considered. This is also true for O_2. Thus in one case, myoglobin, one sees the effects of changing the heme, while on others, α and β chains, modifying the heme does not lead to changes in the association rate of binding. This would seem to be an effect of the protein near the surface. Do you have any comment on this point?

AUSTIN: We completely agree with you that the protein can modify the inner barriers. Our argument was that the large outer barrier is much smaller than it appears, because of our failure to take the temperature dependence of the viscosity of glycerol-water into account. k_{ED} is not directly measurable but comes from model-dependent data fitting; you are measuring the actual rebinding of the CO to the heme (rate λ_{IV} in our terminology), which has a complicated interplay of barriers determining its value.

TRAYLOR: Comparisons of the kinetics of carbon monoxide binding to myoglobin with the kinetics of binding to R and T state hemoglobins suggest some differences in diffusion of carbon monoxide into myoglobin. The on-rate, l'_{Mb}, for myoglobin is like that of T-state

TABLE

Heme	l'	l
	$\mu M^{-1} s^{-1}$	s^{-1}
Hb^R	6	0.01
Hb^T	0.1	0.1
Mb	0.3	0.016
T-model*	0.5	0.16
R-model	6	0.01

*In 2% cetyltrimethylammonium bromide, micelles in water.

hemoglobin, whereas the off-rate resembles that of R-state hemoglobin (see Table). We have studied model systems for R and T hemoglobin states consisting of five-coordinated mesoheme dimethyl ester-base complexes using unhindered 1-methyl-imidzole (R-state model) and hindered 2-methyl imidazole (T-state model). These two models duplicate the kinetic behavior of R and T-state hemoglobin, respectively, with surprising accuracy. We conclude from this that the reduced ligand affinity of T-state hemoglobins can be accounted for without any distal side effects. But this leaves the slow reaction of myoglobin with carbon monoxide to be explained. Does myoglobin react slowly because carbon monoxide diffuses to the iron slowly? Is it possible that quenching by oxygen occurs at a larger distance than the CO-heme reaction and thus that the CO diffusion barriers within the pocket are not excluded by this finding of diffusion-controlled quenching?

AUSTIN: We feel that Dr. Traylor's model compounds could provide an excellent test of the inner-well model. We hope it is fair to say that so far we have not done a systematic study of chemically well-defined modifications of the heme and the effect on the inner barrier. Now, in Dr. Traylor's model compounds we would predict that the low temperature inner barrier of the

R model should have an average value 10 times faster than the T model's inner barrier. If this relation can be found, it would support the concept that pulling the iron out of the plane (as in the T model) increases the height of the inner barrier.

Myoglobin is difficult to compare since it also has intermediate wells, which seem to be not directly sensitive to heme modifications but instead are related to other parts of the protein. We would thus tentatively state that while drawing the iron out of the plane may well be a large part of differences between R and T hemoglobin, the intermediate wells can also affect binding parameters and these are determined by the protein structure. Both effects probably play a role.

Within our model the relatively slow reaction rate of myoglobin with carbon monoxide is due to the high value of the peak activation energy for the inner barrier. This leads to a net rejection of the CO molecule since it is more likely to go back out rather than in. We don't think diffusion is the right word to use for movement within the protein, since one is dealing presumably with a microscopic system here.

It is clear that triplet quenching by oxygen can occur at the heme periphery and thus one should not expect the triplet quenching rates to compare with the oxygen binding rates to the iron. What the triplet quench rates should be comparable to is the rate k_{ED} for entry into the heme pocket. Thus, the triplet quench rates do not sense the inner barriers.

The rebinding of CO to hemoglobin is not diffusion-limited at room temperature. We believe that the entry of the CO into the protein interior *is* diffusion limited. These are two different things.

SINGLE CELL OBSERVATIONS OF GAS REACTIONS AND SHAPE CHANGES IN NORMAL AND SICKLING ERYTHROCYTES

E. ANTONINI, M. BRUNORI, B. GIARDINA, *C.N.R. Centre of Molecular Biology, Institutes of Biochemistry and Chemistry, Faculty of Medicine, University of Rome, Italy,* and
P. A. BENEDETTI, G. BIANCHINI, AND S. GRASSI, *Laboratorio per lo Studio delle Proprietà Fisiche di Biomolecole e Cellule del C.N.R., Pisa, Italy*

ABSTRACT Microspectrophotometry has been applied to single red blood cells to reinvestigate the linked processes of diffusion of gases inside the erythrocyte and their combination with hemoglobin. The experiments took advantage of the photosensitivity of the carbon monoxide derivative of hemoglobin, which allows ligand release from the CO-saturated red cells under strong illumination and recombination when the light is switched off. The photochemical method was also used to study the kinetics of sickling on ligand removal in single erythrocytes of Hb S carriers. The results give new information on the mechanism of the sickling process.

INTRODUCTION

Single cell spectroscopy represents a powerful tool to obtain information on the time and space distribution of a great number of intracellular events.

In the case of red blood cells, this technique has now been applied to reinvestigate the linked processes of diffusion of gases inside the cell and their combination with hemoglobin (1–5). Taking advantage of the photosensitivity of the carbon monoxide derivative of hemoglobin, our experimental approach is based on the direct spectrophotometric observation of single cells, within which the reversible dissociation of the HbCO complex may be induced by light: $HbCO \xrightarrow{h\nu} Hb + CO$; $Hb + CO \xrightarrow{dark} HbCO$.

Thus any single red cell may be subjected to ligand dissociation and binding, with simultaneous monitoring and recording of optical density changes within small areas and of overall morphological changes. This photochemical approach is particularly useful in studying the time relationships between ligand removal and sickling in erythrocytes of Hb S carriers.

Thus, the kinetics of the polymer formation may be studied within any single "*SS*" cell under a variety of environmental and intracellular conditions.

INSTRUMENTATION AND METHODS

Measurements were carried out with an instrument based on a fast condenser-scanning technique described elsewhere (6, 7).

Dual-beam measurements of the absorption spectra, as well as optical density maps, can be taken in the range from 380 to 700 nm, while continuous infrared illumination allows the display of the cells on a TV monitor by means of a silicon-vidicon telecamera.

Photodissociation of intraerythrocytic hemoglobin was obtained with a Mercury lamp (546-nm line) as excitation source and a suitable filter combination to eliminate interferences from the analyzing beam.

Red cell suspensions were prepared in isotonic solutions of sodium chloride plus phosphate buffer at pH = 7.2 equilibrated with 1 atm of CO at 20°C. The CO concentration was changed by dilution of the red cell suspension with the same isotonic gas-free solution. Removal of oxygen was assured by addition of a slight excess of sodium dithionite. The total heme content of the samples used was always $\leq 10^{-5}$ M.

RESULTS

Kinetics of Ligand Photodissociation and Ligand Binding in Normal Erythrocytes

The photodissociation levels achieved were in all cases $\geq 60\%$, as shown from the analysis of the absorption spectra of the intraerythrocytic hemoglobin (Fig. 1). Fig. 2 reports typical oscillograph traces of experiments performed at two different CO concentrations, showing the light-activated transition and the dark recombination process, followed at $\lambda = 430$ nm.

DARK TO LIGHT RELAXATION The approach to the steady state in the light cannot be accounted for by a single exponential process and the half-time is independent, at least within a factor of about two, of CO concentration (10^{-4}M $\leq$ CO $\leq$ 10^{-3}M).

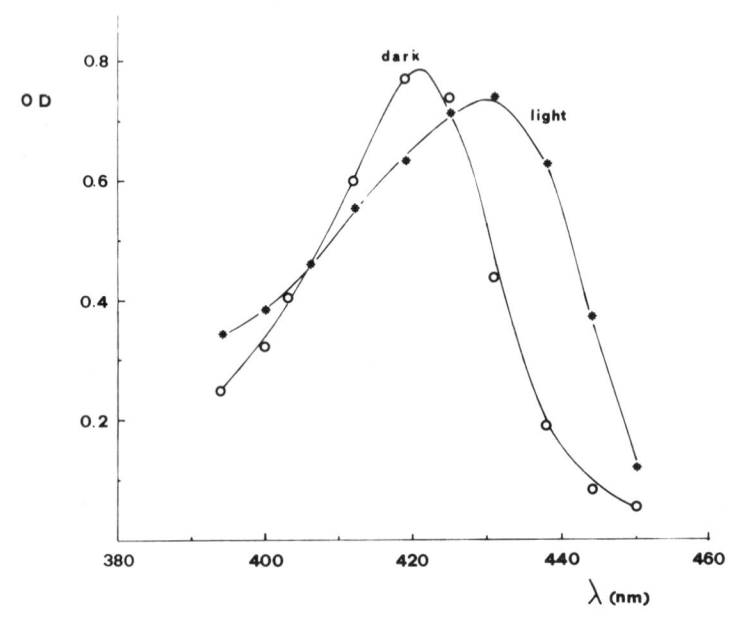

FIGURE 1 Soret absorption spectra of CO and deoxy Hb in a single human erythrocyte, as recorded in the dark and in the light. Conditions: $T \sim 25°$C; CO concentration = 10^{-3}M.

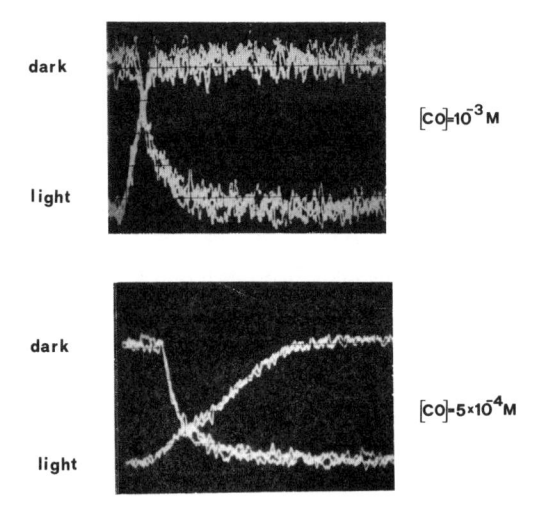

FIGURE 2 Typical oscillograph traces of experiments performed at pH = 7.2, λ = 430 nm, $T \sim$ 25°C, sweep = 100 ms per grid division; carbon monoxide concentration: 10^{-3} and 5×10^{-4}M.

LIGHT TO DARK RELAXATION The relaxation from the steady state in the light to the equilibrium in the dark corresponds under all the conditions explored to a zero-order process (Fig. 2). In the same range of CO concentration the reciprocal of the half-time is ligand concentration-dependent: i.e. increases linearly with carbon monoxide concentration (8). This body of results shows that combination of CO with hemoglobin within erythrocytes is about 30-fold slower than that measured with hemoglobin in solution, and an order of magnitude smaller than that obtained by stopped-flow experiments with red cells.

It should be pointed out that the present experiments are performed on erythrocytes under the limiting condition of complete stagnancy of the surrounding extracellular fluid. Under these conditions, it appears that a layer of unstirred solvent around the cells may become an important factor in determining the rate of equilibration of intraerythrocytic hemoglobin with respiratory gases. A more detailed analysis of data concerned with this particular problem is given elsewhere (8).

Hemoglobin S Aggregation

The photochemical approach described above may allow one, in principle, to obtain information on the time relationships between hemoglobin deoxygenation, polymer formation, and cell deformation within red blood cells of individuals carrying Hb S. It is, in fact, possible to investigate the sickling process by controlling with light the extent of carbon monoxide dissociation from hemoglobin within the SS cell, since sickling follows carbon monoxide dissociation, which is photochemically controlled. In general, the photochemical behavior of SS cells was similar to that reported above for normal cells. It was found that the time necessary for the cell to sickle completely, i.e., to undergo the full final deformation, ranged in all the cells examined from 3 to 5 s. By reducing the intensity of the photodissociating beam, once the sickling was obtained, we observed that the deformation could be maintained by a light intensity several times lower than that necessary to induce the sickling process. In addition, it should be pointed out that the phenomenon is apparently fully reversible, since full recombination of hemoglobin with CO in the dark is immediately followed by acquisition of the original erythrocyte shape.

Single SS cells in which sickling was induced several times by repeatedly turning the light on and off were observed to sickle always along the same axis and with the same apparent rate. This finding, indicating a preferential direction of sickling, may be interpreted as due (*a*) to stable alterations in the membrane, (*b*) to the presence of residual polymers which, triggering the stacking, could induce a preferential direction of sickling.

The results reported here should be taken as examples of the significance of this type of approach to the problem of sickling, for instance in respect to the effect of drugs that may interfere with the sickling process. Finally, since human erythrocytes in heterozygotes contain both Hb A and Hb S, it may be feasible to study the distribution of the two hemoglobins among erythrocytes.

Received for publication 19 December 1977.

REFERENCES

1. ROUGHTON, F. J. W. 1959. Diffusion and simultaneous chemical reaction velocity in haemoglobin solutions and red cell suspensions. *Progr. Biophys. Biophys. Chem.* **9**:55–104.
2. SIRS, J. A. 1963. Uptake of O_2 and CO by hemoglobin in sheep erythrocytes at various temperatures. *J. Appl. Physiol.* **18**:166–170.
3. SIRS, J. A. 1963. Effect of shape-volume changes on uptake of CO and O_2 by hemoglobin in erythrocytes. *J. Appl. Physiol.* **18**:171–174.
4. SIRS, J. A. 1963. Influence of metabolism on uptake of CO and O_2 by hemoglobin in erythrocytes. *J. Appl. Physiol.* **18**:175–178.
5. SIRS, J. A., and F. J. W. ROUGHTON. 1963. Stopped-flow measurements of CO and O_2 uptake by hemoglobin in sheep erythrocytes. *J. Appl. Physiol.* **18**:158–165.
6. BENEDETTI, P. A., G. BIANCHINI, and S. GRASSI. 1975. 5th International Biophysics Congress, Copenhagen.
7. BENEDETTI, P. A., G. BIANCHINI, and G. CHITI. 1976. Fast scanning microspectroscopy: an electrodynamic moving-condenser method. *Appl. Opt.* **15**:2554–2558.
8. ANTONINI, E., M. BRUNORI, B. GIARDINA, P. A. BENEDETTI, G. BIANCHINI, S. GRASSI. 1978. Kinetics of the reaction with CO of human erythrocytes: observations by single cell spectroscopy. *FEBS (Fed. Eur. Biochem. Soc.) Lett.* In press.

DISCUSSION

SCHECHTER: A comment submitted by a referee: With regard to the ligand photolysis experiments on normal erythrocytes, the authors report only that there was greater than 60% photodissociation in the steady state, which, from Fig. 2, appears to require between 300 and 400 ms to attain. Is this the best guess possible for the fractional saturation of normal erythrocytes under the photolysis conditions used for these experiments? Furthermore, is this also the best estimate of the amount of photolysis in the experiments on SS erythrocytes?

GIARDINA: I wish to make one point with reference to the question on the photodissociation levels achieved in normal as well as *SS* erythrocytes. A number of control experiments, under the same conditions reported in the paper, have been performed by cutting the intensity of the photodissociating light by known amounts and measuring the corresponding optical density changes at $\lambda = 435$ nm. Hence the deflection was calibrated in terms of the ratio of Hb CO to Hb within the cell. Therefore, from the experimentally determined relation between the optical density changes and the relative light intensity, it appears that the intensity of the light used is such as to produce full photodissociation of the ligand. In addition I wish to point out that the results reported in the paper have to be considered as the beginning of a considerably more detailed study, which in the case of SS erythrocytes will comprehensively study the effects of different variables, such as Hb concentration and specific drugs.

SCHECHTER: We have an extended written comment by Hiroshi Mizukami and Betsy Adams.

MIZUKAMI: Once again I am very impressed with the ingenious approach taken by Dr. Antonini's group to the problem of ligand interaction of sickle cell hemoglobin and the cellular structural change caused by de-ligation. Their approach is elegant, since it allows reversible interaction of the ligand. Our recent experiments resemble theirs. I would like to present some of our findings here.

We have constructed an instrument, the main body of which is shown in Fig. 1.

A single layer of erythrocytes is suspended within a thin liquid film enforced by EMI micro-mesh (Fig. 2) (EMI Gencom Inc., Plainview, N.Y.) and placed at the light beam.

The purging vaporized gases approaching from the left are quickly changed with a solenoid valve from a gas containing oxygen (25% O_2, 5% CO_2, and 70% N_2) to another without, and vice versa. The erythrocytes, being suspended in a single layer, interact quickly with these gases. A Bausch & Lomb high-intensity grating monochromator is used as the light source (Bausch & Lomb, Scientific Optical Products Div., Rochester, N.Y.) and the spectral changes are recorded on a storage oscilloscope.

The erythrocytes were suspended in Krebs-Henseleit buffer and all experiments were performed at $25 \pm 0.5°C$. The observations were made at 440, 542, and 522 nm. 522 nm is the isobestic point for the ligand reaction (Kiesow et al., 1972). The flow rates of both purging gases were kept constant at 435 ml/min. The estimated lag period for the gas to reach the erythrocyte suspension at this flow rate was 0.25 s.

The oscilloscope traces at 440 nm are shown in Fig. 3. The trace going downward is deoxygenation and that going upward is reoxygenation. The horizontal sweep is 0.2 s/div. Fig. 4 is for 542 nm. Fig. 4 A is deoxygenation at a sweep of 0.5 s/div, and Fig. 4 B is reoxygenation

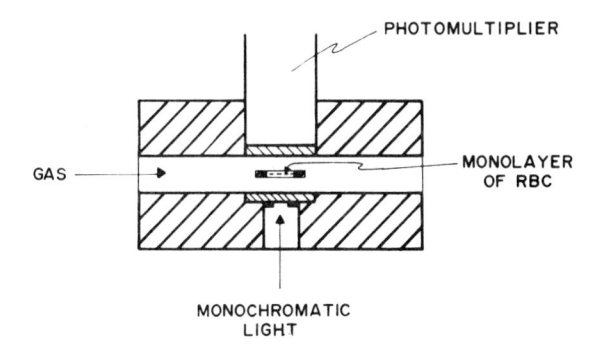

FIGURE 1 Schematic representation of the main components of the instrument. RBC, erythrocytes.

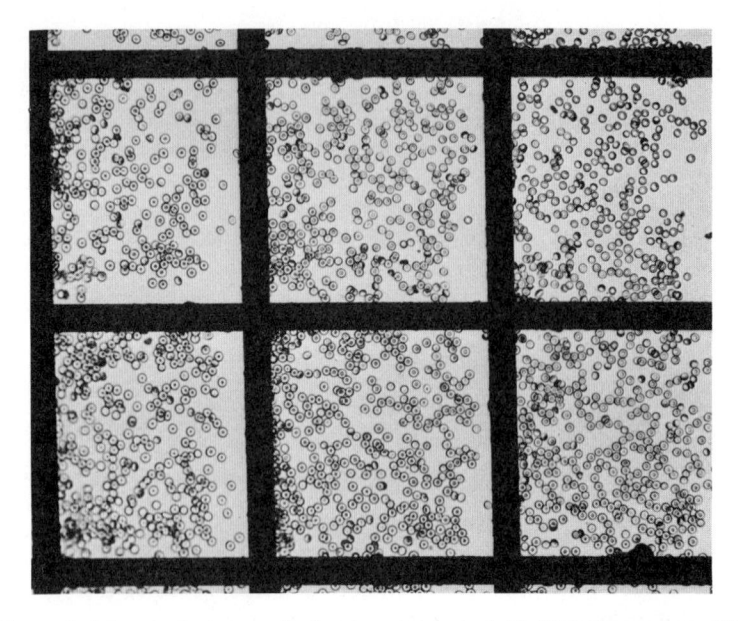

FIGURE 2 Microscopic photograph of erythrocytes suspended in EMI micromesh no. 100.

at 0.2 s/div. Fig. 5 shows similar traces at 522 nm. Fig. 5 A is deoxygenation at a sweep of 0.5 s/div, and Fig. 5 B is reoxygenation at 0.2 s/div.

Since the absorbance change at 440 nm (Soret band) is large, the electronic gain was set relatively low compared with other wavelengths used and most of the absorbance change did arise from the ligand interaction. However, due to the slow flow rates of the purging gases, the results are not a true representation of ligand kinetic interaction, but rather an indication of the degree of ligation of hemoglobin in the erythrocytes.

The spectral changes at 542 nm (Fig. 4) reveal at least two types of reactions. Upon deoxygenation, there is a rapid spectral change that corresponds to that observed at 440 nm, followed by a slow reverse spectral change. When the sample is reoxygenated, there is again a rapid

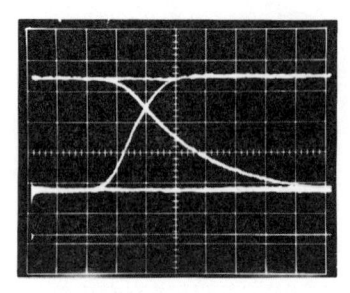

FIGURE 3 Oscilloscope traces at 440 nm of deoxygenation (downward curve) and reoxygenation (upward curve) of S-erythrocytes diluted 1/6 with Krebs-Henseleit buffer, pH 7.4, and temperature 25°C. The gas over the thin film of erythrocytes was exchanged from one composed of 20% O_2, 5% CO_2, and 70% N_2 to one containing 5% CO_2 and 95% N_2 for deoxygenation, and vice versa for reoxygenation. The flow rate of the gases for all reactions was 435 ml/min. Vertical input represents the change in absorbance. The straight horizontal trace is the dark current. Each horizontal division corresponds to 0.2 s.

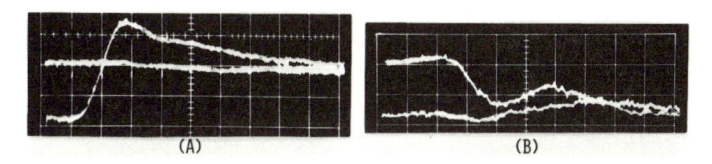

FIGURE 4   A. Oscilloscope trace of deoxygenation of sickle cell erythrocytes at 542 nm, with horizontal sweep of 0.5 s/div.   B. Oscilloscope trace of reoxygenation of sickle cell erythrocytes at 542 nm, with horizontal sweep of 0.2 s/div.

ligand interaction, almost immediately interrupted by a reverse spectral change, followed finally by a slow spectral change in the same direction as that of the ligand reaction. These observations suggest that the ligand interaction is followed by a light-scattering interaction. The light loss by scattering after deoxygenation is almost entirely recovered by reoxygenating.

The change of light-scattering alone can be measured if the wavelength is set at 522 nm, the isobestic point of ligand interaction. Fig. 5 clearly demonstrates this change of absorbance arising from only the light-scattering. The apparent lag periods are longer during both deoxygenation and reoxygenation reactions, indicating that no absorbance change takes place during the ligand interaction. Subsequent absorbance changes correspond closely to the reverse absorbance changes observed in Fig. 4 at 542 nm for each reaction.

The results shown in Figs. 3, 4, and 5 are normalized and summarized in Fig. 6. The direction of the change of absorbance at 440 nm has been reversed to that it can be compared with the results at the other two wavelengths. As expected, the rates of deoxygenation observed at 440 and 542 nm are nearly identical, but the scattering light loss at 542 nm takes place before complete deoxygenation of the sample. This scattering change coincides with that observed at 542 nm. Considering the additional lag period due to the time required for the new gas to reach the erythrocytes, the t_{50S} (time required to complete 50% scattering change) is 1.6 ± 0.2 s for the four samples observed.

Comparison of the rate of reoxygenation observed at 440 and 542 nm is difficult, since the scattering change also takes place shortly after the initiation of reoxygenation—the results at 542 nm being a complex of these two phenomena. For reoxygenation, the t_{50RS} (time taken to recover 50% loss of light scattering) determined at 522 nm) is 0.5 ± 0.04 s for the four samples examined.

The rate of scattering light intensity change presented here is similar to that of Zarkowsky and Hochmuth (1975) for sickling—their t_{50S} being about 1.5 s—and that of Messer et al., (1976) for desickling—their t_{50RS} being 0.5 s. However, the exact cause of the scattering change of light could be the optical change due to aggregation of hemoglobin molecules, the structural change of cells, or both. No such extent of light-scattering change has been observed for normal red blood cells, and light microscopic analysis with videotape has shown that after complete deoxygenation of the sample on micromesh, a large percentage of sickle cells (90%) become sickled. Reoxygenation of the same sample results in a large percentage

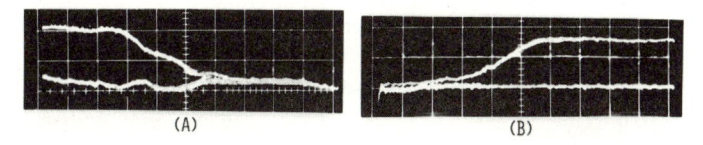

FIGURE 5   A. Oscilloscope trace of deoxygenation of sickle cell erythrocytes at 522 nm (the isosbestic point), with horizontal sweep of 0.5 s/div.   B. Oscilloscope trace of reoxygenation of sickle cell erythrocytes at 522 nm, with horizontal sweep of 0.2 s/div.

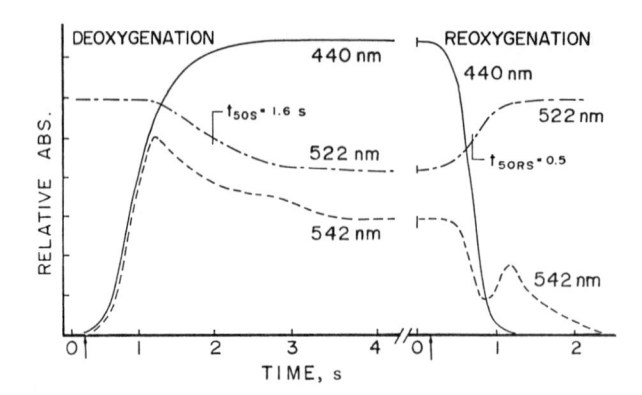

FIGURE 6 Normalized summary of Figs. 3, 4, and 5. The direction of change of absorbance at 440 nm has been reversed so that it can be compared with the results at 542 and 522 nm. The abscissa is time in seconds, and the ordinate is relative absorbance. The lag time (time required for the complete exchange of gases) is indicated by arrows along the abscissa. The t_{50S} for deoxygenation is 1.6 ± 0.2 s, and the t_{50RS} for reoxygenation is 0.5 ± 0.04 s at 522 nm. Each is indicated on its respective curve.

(90%) of the sickled cells recovering their normal shape, inducing little, if any, irreversibly sickled cells.

It can also be seen that the sickling process proceeds before complete deoxygenation and that desickling begins before complete oxygenation. The changes observed here are statistical averages of whole blood, and any effect of cell age remains to be studied. However, preliminary microscopic observation suggests that the cell age has a considerable effect on kinetic results. As the change in intracellular constituents is age-dependent, this is to be expected.

REFERENCES

KIESOW, L. A., J. W. BLESS, D. P. NELSON, and J. B. SHELTON. 1972. A new method of the rapid determination of oxygen dissociation curves in small blood samples by spectrophtometric titration. *Clin Chim Acta.* **41**:123.

MESSER, M. J., J. A. HAHN, and T. B. BRADLEY. 1976. The kinetics of sickling and unsickling of red cells under physiologic conditions: rheologic and ultrastructural correlations. *In* Proceedings of the Symposium on Molecular and Cellular Aspects of Sickle Cell Disease. U.S. Department of Health, Education, and Welfare, Bethesda, Md. 225.

ZARKOWSKY, H. S., and R. M. HOCHMUTH. 1975. Sickling times of individual erythrocytes at zero PO_2. *J. Clin. Invest.* **59**:1023.

GIARDINA: The results just reported by Dr. Mizukami do not disagree with our results. I wish to point out that our measurements were limited to the Soret region because we used as photodissociating light the 546-nm line of a mercury lamp. In addition, with reference to whether the scattering change of light is due to aggregation of hemoglobin molecules or to the cell's structural deformation, I think that microspectrophotometry could discriminate between these two hypotheses. Again, if the scattering derives from aggregation of hemoglobin molecules, it could be used to follow the kinetics of polymer formation within a single cell before overall morphological changes take place.

FERRONE: The polymerization of sickle cell hemoglobin exhibits some remarkable features. One of the most remarkable is that when a perturbation is applied to start polymerization, nothing seems to happen at first. Then, after a delay, the polymers assemble with great rapidity. This delay is more than just a curiosity: it has been hypothesized to be the primary determinant of the severity of sickle-cell disease. To see why, it's necessary to recognize that red cells must deform to pass through the capillaries, precisely the point at which oxygen release takes place. If the delay period is longer than the second or so that the cell takes to pass through the capillary, then sickling will have less severe consequences than if the cell sickles and sticks in the capillary. Clearly then, it is crucial to know if polymerization can occur faster than 1 s. Although the data presented this morning have not shown rapid sickling, experiments performed recently by James Hofrichter, William Eaton, and myself have shown that this reaction can occur at least 3 orders of magnitude faster.

Like Dr. Giardina and his colleagues, we felt that photolytic removal of CO would be a good way to initiate this reaction. We, however, chose as a photolysis source a continuous-wave argon ion laser to assure rapid and complete conversion of CO-hemoglobin to deoxy hemoglobin. In purified, highly concentrated samples we obtain complete photolysis in less than 1 ms.

It is also imperative to have a signal that probes the polymerization. A conventional approach is to use the birefringence of the aligned polymers; at short times, though, these signals are too weak. Fortunately, scattered photolysis light can give the same information at much higher signal-to-noise ratios. Using this technique, we induced and observed polymerization of sickle cell hemoglobin at times as short as 4 ms.

The final step was to show that in fact cells can sickle rapidly. The problems that attend imaging a single cell are not small; hence our data on cells leave room for improvement. Nonetheless, we have measured cell delay times of a few hundred milliseconds.

Why have others not seen this rapid a reaction? One reason is the difficulties in providing complete ligand removal rapidly. Another, though, is that of rigidification of intracellular hemoglobin may not cause gross morphological changes when polymerization occurs rapidly. In other words, if alignment does not accompany polymerization, the cells may not distort.

GIARDINA: In general I agree with Dr. Ferrone's comments. I wish to point out that future investigations must take into account the effect of different variables such as pH temperature and hemoglobin concentration.

NANOSECOND RELAXATION PROCESSES IN LIPOSOMES

MUGUREL G. BADEA, ROBERT P. DETOMA, AND LUDWIG BRAND,
*The Biology Department and McCollum-Pratt Institute,
The Johns Hopkins University, Baltimore, Maryland 21218 U.S.A.*

INTRODUCTION

The fluorescence characteristics of aromatic molecules are often strongly influenced by their immediate environment. For this reason, fluorescence probes have found wide application for monitoring changes occurring in biological or model membrane systems (Azzi, 1975; Radda, 1975; Badley, 1976). The majority of the reported studies have made use of steady-state fluorescence measurements. These reflect a time average of any excited-state interactions that occur on the nanosecond time scale. Fluorescence decay measurements, on the other hand, provide direct kinetic information regarding these interactions. Direct measurements are possible in a time interval limited on the short side by the time resolution of the instrumental technique and on the long side by the decay time of the fluorophore. As a consequence, the time-dependent characteristics of the fluorescent probe environment are better defined.

The parameters characterizing the excited state behavior of 2-anilinonaphthalene (2-AN) in fluid solutions such as cyclohexane, ethanol, and mixtures thereof, and in a viscous solvent (glycerol) have previously been reported (DeToma and Brand, 1977). They were found to be strongly influenced by the presence of relatively small amounts of polar molecules in a nonpolar solvent. This particular property makes 2-AN a valuable fluorescent probe for a heterogeneous polar-apolar liposomic environment. In addition, the information obtained in the model solvent systems mentioned above constitutes a firm basis for interpreting the results obtained with liposomes.

In the present communication we describe the results obtained by measuring the fluorescence decay, time-resolved emission spectra (TRES), and the decay of the emission anisotropy (DEA) of 2-AN adsorbed to dimyristoyllecithin (DML) single bilayer liposomes. It will be shown that the rotational motion of 2-AN as well as its excited-state interactions occur on the nanosecond time scale both above and below the gel-liquid transition temperature of these vesicles. The nanosecond relaxation processes found to occur with the liposomes will be compared to those observed with 2-AN in homogeneous solvents.

Dr. DeToma's present address is: University of Richmond, Department of Chemistry, Richmond, Va. 23173.

METHODS

Experimental

2-AN was obtained from Aldrich Chemical Co., Inc. (Milwaukee, Wis.) and purified as previously described (DeToma and Brand, 1977). The DML single bilayer liposomes were prepared as described (Chen et al., 1977), and the fluorescence measurements were carried out in 0.01 Tris-HCl buffer, pH 8.5, containing 0.1 M NaCl. 4 μl of a 1-mM solution of 2-AN in ethanol was added to 4 ml of liposomes in the Tris buffer followed by rapid vortexing for about 1 min. The mixture was left to equilibrate for 2-3 h before the optical measurements. The lipid to dye ratio was about 700. The glycerol was spectral grade for fluorescence microscopy and was obtained from AG Merck (Darmstadt, W. Germany).

The instrumental and computational techniques for obtaining deconvolved nanosecond time-resolved emission spectra and the decay of the emission anisotropy have been described by Easter et al. (1976) and Chen et al. (1977), respectively. The monophoton counting technique was used to obtain the fluorescence decay data. Each curve and its corresponding excitation profile were collected semi-simultaneously to minimize the errors associated with long-term timing variations in the excitation source. The excitation was performed with an air, nanosecond flash lamp through a Baird Atomic 3,150-Å filter (Baird Atomic, Inc., Bedford, Mass.).

Steady-state fluorescence spectra were obtained both with the monophoton counting instrument and with a Perkin-Elmer MPF4 spectrofluorometer Perker-Elmer Corp., Instrument Div., Norwalk, Conn.). They were measured at the start and end of each series of decay measurements for every sample and found to be identical. This indicated the absence of intervening photodecomposition. To avoid errors due to time-dependent changes in the emission anisotropy, all the decay measurements were carried out under "magic angle" conditions (Spencer and Weber, 1970). Excitation was with vertically polarized light, and the emission was observed through a polarizer whose transmission axis was at 54.7° to the vertical direction.

RESULTS

The interaction of 2-AN with the DML vesicles results in enhanced fluorescence intensity and a blue shift in the fluorescence emission spectrum. The normalized, corrected emission spectra shown in Fig. 1 indicate that 2-AN has a structured spectrum in cyclohexane with a maximum at 372 nm and that the emission maximum when adsorbed to DML vesicles is intermediate between that in cyclohexane and the red-shifted emission observed in water.

The nanosecond time-dependent behavior of 2-AN in DML vesicles was examined both below (1°C) and above (37°C) the crystalline-liquid crystalline phase transition temperature (22°C). The procedure of Easter et al. (1976) was used to obtain the nanosecond time-resolved fluorescence profile at each temperature. Fluorescence decay curves, together with the corresponding lamp profiles, were collected at selected wavelengths spanning the emission band. Impulse response functions for each decay curve were obtained by the method of nonlinear least squares. A sum of exponential terms was used as the fitting function. The results of a typical fit to experimental data are shown in Fig. 2. The good visual agreement between the experimental and theoretical decay, as well as the residuals and autocorrelation of the residuals, all indicate that a sum of three exponentials represents an adequate function to represent the impulse

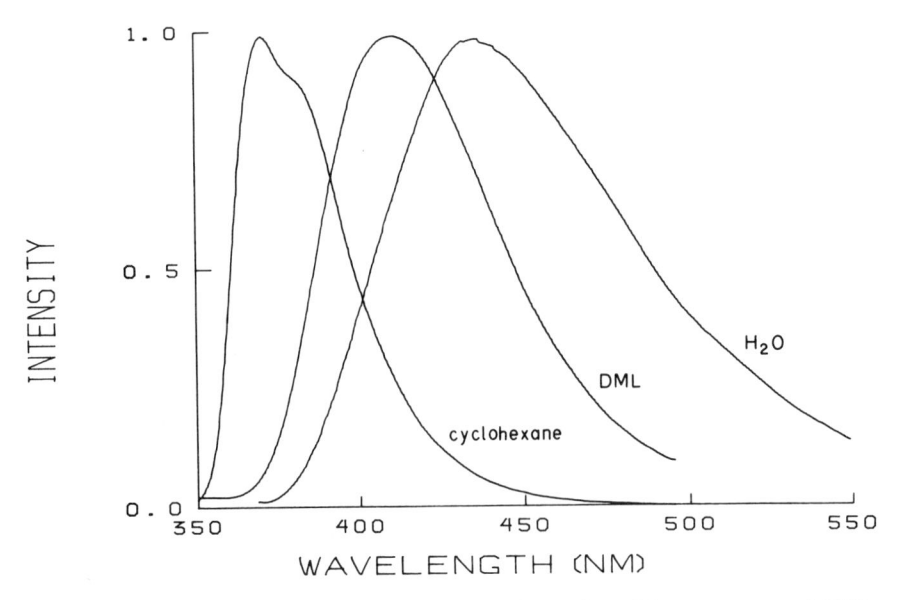

FIGURE 1 Steady-state fluorescence emission spectra of 2-AN in cyclohexane, water, and DML liposomes. The spectra have been normalized at the peak emission and are corrected for nonlinear transmission of the monochromator and sensitivity of the detector. The fluorescence spectra were obtained with the Perkin-Elmer MPF-4 spectrophotofluorometer with an excitation and emission bandwidth of 10 nm. The single bilayer DML vesicles (~0.5 mM in lipid) were suspended in 0.01 M Tris-HCl, 0.1 M NaCl at pH 8.5 at 20°C.

response. It can be seen that a fitting function of only two exponential terms is not sufficient to describe the decay in this wavelength region.

The impulse response (deconvolved decay) obtained at four representative wavelengths is shown in Fig. 3. Two salient features are immediately apparent. First, the decay at the red portion of the emission (500 nm) is characterized by an initial rise. This shows up as a negative preexponential term in the decay analysis (see Table I) and indicates that at least a portion of the emission at this wavelength region has its origin in species created during the lifetime of the excited state. The pattern of decay as a function of emission wavelength is similar to that observed with 2-AN dissolved in glycerol. A second feature of the impulse response curves is that the mean decay time $\langle \tau \rangle = \sum_j \alpha_j \tau_j^2 / \sum_j \alpha_j \tau_j$ increases with increasing wavelength. This is presented more explicitly in Fig. 4, which shows the variation in mean lifetime with emission wavelength at temperatures above and below the phase transition of the liposomes. It is of interest that in the long wavelength region the mean decay time decreases with increased temperature whereas a small enhancement of the mean decay time is found at short wavelengths. The impulse response curves (deconvolved fluorescence decay) at 380 and 460 nm at 1° and 37°C are shown in Fig. 5. The increase in mean decay time at 37°C is evident at 380 nm.

The results of an analysis of the decay data at five representative wavelengths are presented in Table I. More than two exponential terms are required to obtain a good

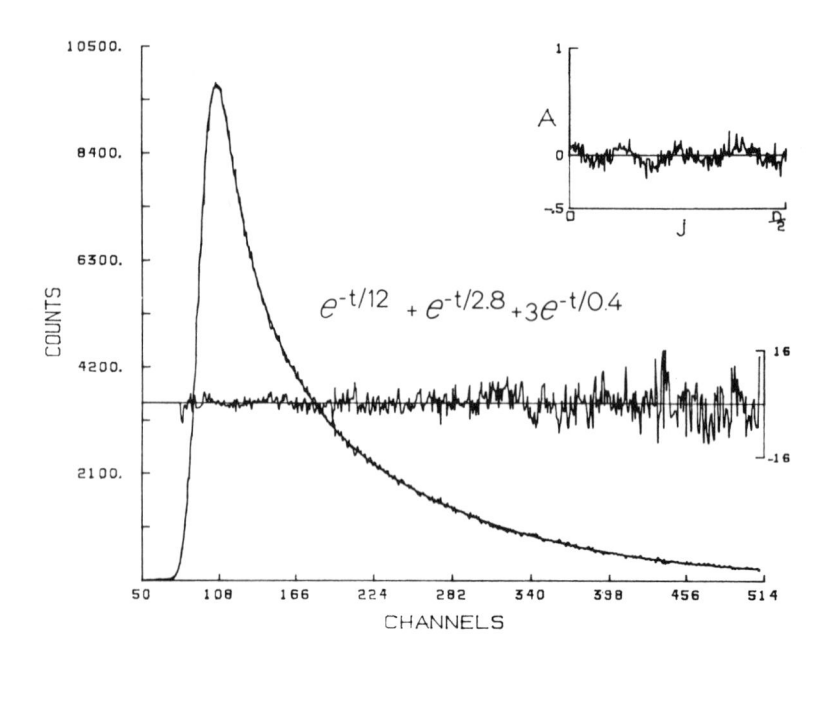

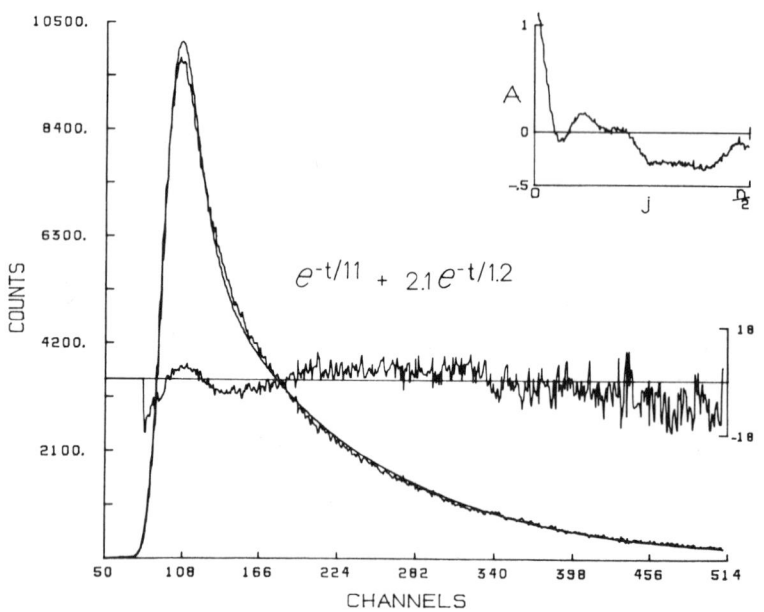

FIGURE 2 Double and triple exponential analysis of a typical decay curve (375 nm) of 2-AN in DML single bilayer liposomes. The smooth curve representing the impulse response convoluted with the lamp flash is superposed on the raw decay data. The weighted residuals of the fit are plotted along the horizontal line. The autocorrelation function of the residuals is shown in the upper right inset. (Above) three exponential fit. (Below) two exponential fit. The timing calibration was 0.101 ns/channel and the bandpass was 13 nm (4-mm slits). Conditions of 2-AN/DML mixture are as in Fig. 1. Temperature was 37°C.

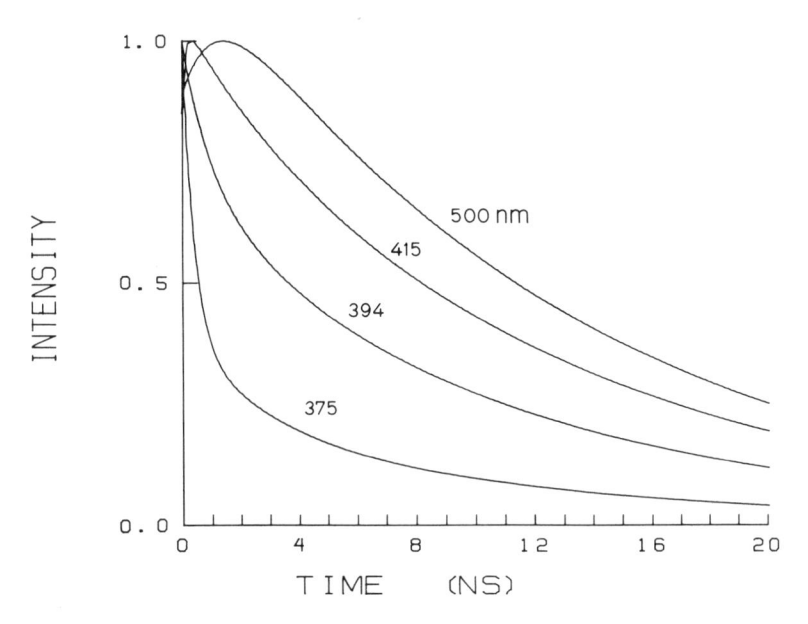

FIGURE 3 Normalized impulse response functions of 2-AN in DML single bilayer liposomes at 375, 394, 415, and 500 nm. Conditions are as in Fig. 2.

fit in the blue region of the emission band. Another representation of the decay surface is shown in Fig. 6 which shows the nanosecond time-resolved emission spectra obtained at three representative times, 0.4, 1.4, and 12 ns. There is a clear shift of the emission maxima to lower energy with time. The time-course of these spectral shifts is represented in Fig. 7, which shows the wave number of maximum emission intensity as a function of time. It is seen that after photoexcitation, 2-AN undergoes relaxation

TABLE I
EMPIRICAL DECAY PARAMETERS OF THE 2-AN/DML COMPLEX AT 1° AND 37°C

Temperature	Wavelength	α_1	τ_1	α_2	τ_2	α_3	τ_3	$\langle\tau\rangle$
			ns		ns		ns	
1°C	375	0.74	0.23	0.16	3.34	0.10	11.87	8.35
	394	0.30	0.68	0.33	4.23	0.37	13.11	10.78
	418	0.32	3.96	0.68	13.71	—	—	12.51
	460	−0.12	1.12	0.23	7.44	0.65	15.37	14.37
	500	−0.23	3.39	0.28	4.78	0.49	15.50	14.95
37°C	375	0.60	0.38	0.20	2.84	0.20	12.27	9.81
	394	0.26	0.90	0.19	4.31	0.55	12.77	11.56
	418	0.13	5.70	0.87	12.86	—	—	12.42
	460	−0.23	1.04	0.77	12.63	—	—	12.92
	500	−0.22	1.14	0.78	12.45	—	—	12.75

Empirical decay parameters (nanoseconds) obtained by a nonlinear least-square fit of a multiexponential model function to the fluorescence decay of the 2-AN/DML complex at 1° and 37°C. At each wavelength the sum of pre-exponential terms is normalized to 1.

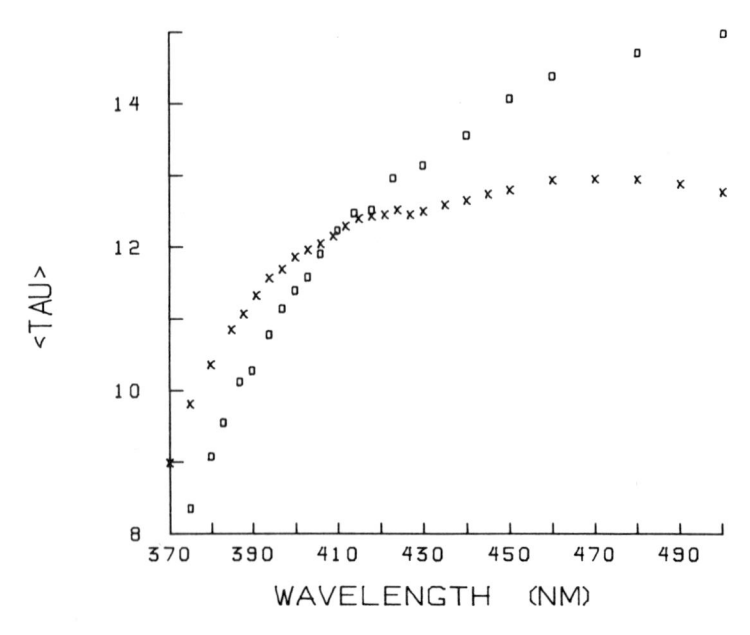

FIGURE 4 Mean lifetime of fluorescence emission decay of 2-AN in DML single bilayer liposomes vs. emission wavelength. □, 1°C; x, 37°C. Conditions are as in Fig. 2.

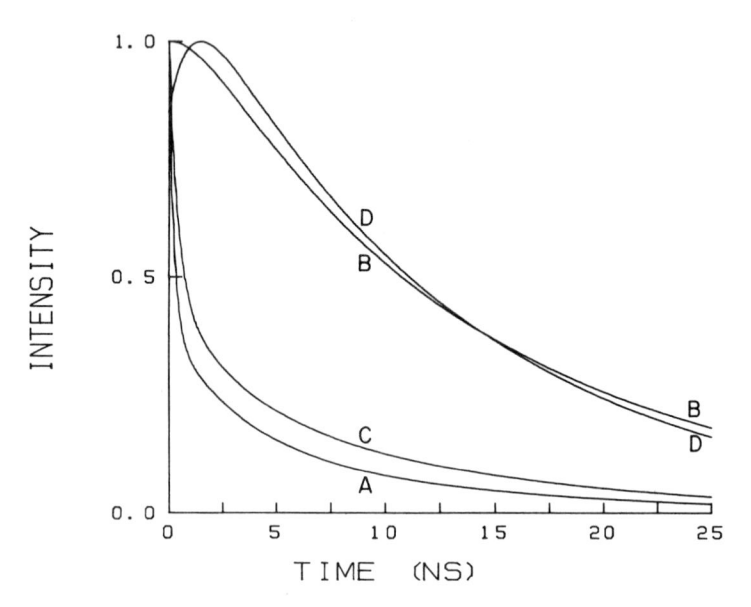

FIGURE 5 Normalized impulse response functions of 2-AN in DML single-bilayer liposomes at 1°C: (A) 380 nm, (B) 460 nm; and at 37°C: (C) 380 nm, (D) 460 nm. Conditions are as in Fig. 2.

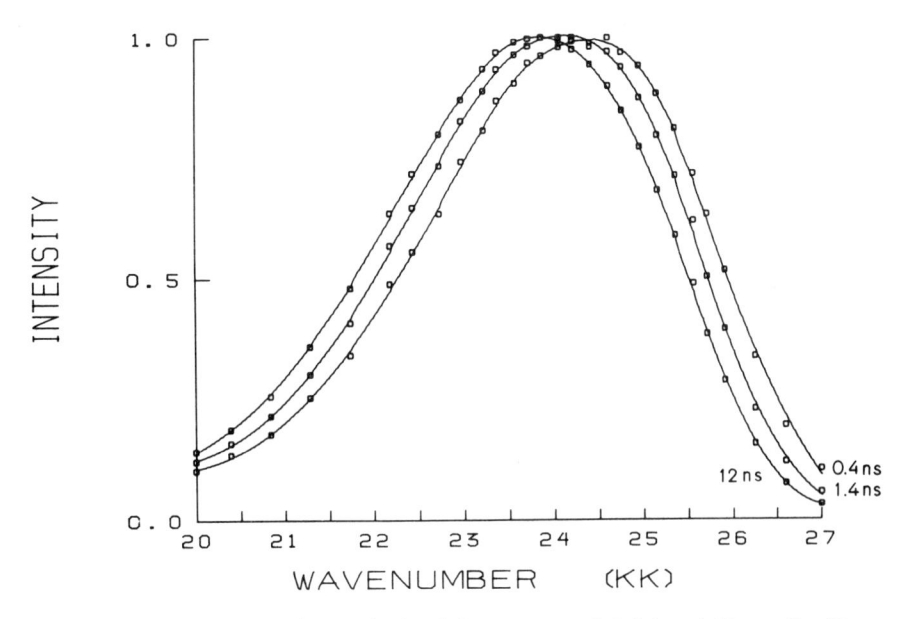

FIGURE 6 Peak normalized time-resolved emission spectra at 0.4, 1.4, and 12 ns. Conditions are as in Fig. 2.

processes both on the subnanosecond and the nanosecond time scales. The subnanosecond processes occur before the nanosecond time window of our experimental technique and are inferred from the apparent zero-time emission maxima observed on the nanosecond time scale. In the liposomes, for instance, 2-AN starts its nanosecond spectral relaxation from an apparent zero-time position close to that in glycerol and

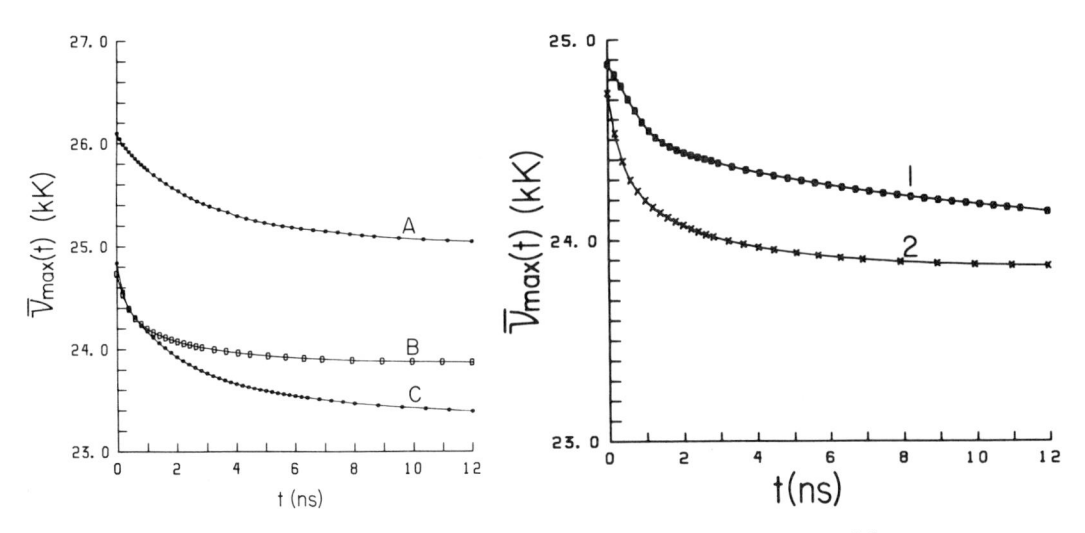

FIGURE 7 Right: wave number of maximum intensity vs. time of 2-AN in: (A) cyclohexane + 0.1 M ethanol mixture at 20°C; (B) DML single bilayer liposomes at 37°C; and (C) glycerol at 10°C. Left: variation of curve B with temperature: (*1*) at 1°C, (*2*) at 37°C.

more than 1 kK lower than the corresponding situation in cyclohexane containing 0.1 M EtOH. On the other hand, as shown in Fig. 7 (right), the zero-time spectral position of 2-AN in liposomes is practically independent of their physical state although the nanosecond relaxation shows a different behavior above and below the phase transition. To quantitate this difference, the curves representing the kinetics of the nanosecond spectral shift were analyzed in terms of an empirical multiexponential model, and the results are shown in Table II.

The best fit was obtained in each case with two exponentials and a constant. The constant represents the fully relaxed spectral position, which, as shown in Table II, is independent of temperature. In contrast, this position is reached approximately three times faster above the phase transition than below it. The additional, fast subnanosecond decay constant could be interpreted as the tail end of the inferred subnanosecond interaction mentioned above.

The data presented thus far indicate that 2-AN adsorbed to bilayer vesicles undergoes interactions with polar residues both on the subnanosecond and the nanosecond time scale. These interactions might involve mutual reorientational motion of the interactive species. It is, thus, of interest to follow the relaxation characteristics of the rotational motion of the fluorophore by measuring the decay of the emission anistropy both above and below the phase transition. These measurements can in principle be used to ascertain whether 2-AN undergoes rotational motion on the nanosecond time scale and whether this motion is related to the excited-state solvation reactions described above.

The fluorescence emission anisotropy is defined as follows:

$$r(t) = \frac{D(t)}{S(t)} = \frac{I_{VV}(t)G - I_{HV}(t)}{I_{VV}(t)G + 2I_{HV}(t)},$$

where $I_{VV}(t)$ is the observed decay with a vertical polarizer in both the excitation and emission beam. $I_{HV}(t)$ is the decay obtained with horizontally polarized excitation and

TABLE II
EMPIRICAL DECAY PARAMETERS ABOVE AND BELOW T_m

	Below T_m	Above T_m
Spectral relaxation	$\bar{\nu}_{max}(t) =$ $0.44e^{-t/0.51} + 0.61e^{-t/8.74} + 23.84$	$\bar{\nu}_{max}(t) =$ $0.44e^{-t/0.40} + 0.42e^{-t/2.81} + 23.86$
Anisotropy relaxation	$r(t) = 0.09\,e^{-t/6.5} + 0.07$	$r(t) = 0.13e^{-t/1.5} + 0.03$
Anisotropy in glycerol	$r(t) = 0.19\,e^{-t/397.0}$	$r(t) = 0.20e^{-t/28.8}$

Empirical decay parameters (nanoseconds) for spectral and anisotropy relaxation of 2-AN in DML bilayer liposomes above and below the crystalline-liquid crystalline phase transition ($T_m = 22°C$). The spectral relaxation experiments were carried out at 1° and 37°C and the anisotropy relaxation experiments at 2° and 40°C. Also shown is the decay of the emission anisotropy of 2-AN in glycerol at 0° and 28°C.

vertically polarized emission. $G = I_{HH}/I_{VH}$ and represents a small correction factor (Chen and Bowman, 1965).

The instrumental methods used to obtain the emission anisotropy data and the computational procedures utilized to extract the parameters characterizing $r(t)$ are similar to those described by Chen et al. (1977). Briefly the procedure was as follows. $I_{VV}(t)$ and $I_{HV}(t)$, together with a lamp profile, were collected during the same time period using the instrument in the sample alternation mode. The sum curve, $S(t)$, was analyzed in terms of a multiexponential decay law using the method of nonlinear least squares. A model function consisting of a sum of exponentials plus a constant was chosen for $r(t)$: $r(t) = \sum_i \beta_i e^{-t/\phi_i} + c$. The values of the variable parameters β_i, ϕ_i, and c were obtained by convolving the product $r(t) \cdot s(t)$ with the lamp profile and searching for the best fit to the experimental difference curve, $D(t)$, by the nonlinear least-squares method. The best parameters obtained for the fluorescence emission anisotropy of the 2-AN vesicle system below (2°C) and above (40°C) the melting transition are given in Table II. The impulse response curves (deconvolved anisotropy) are shown in Fig. 8. Also shown in Table II are the parameters for the decay of the emission anisotropy for 2-AN in glycerol at 0° and 28°C. At both temperatures the decay of the anisotropy gave excellent fits to a single exponential. In contrast, 2-AN in lipo-

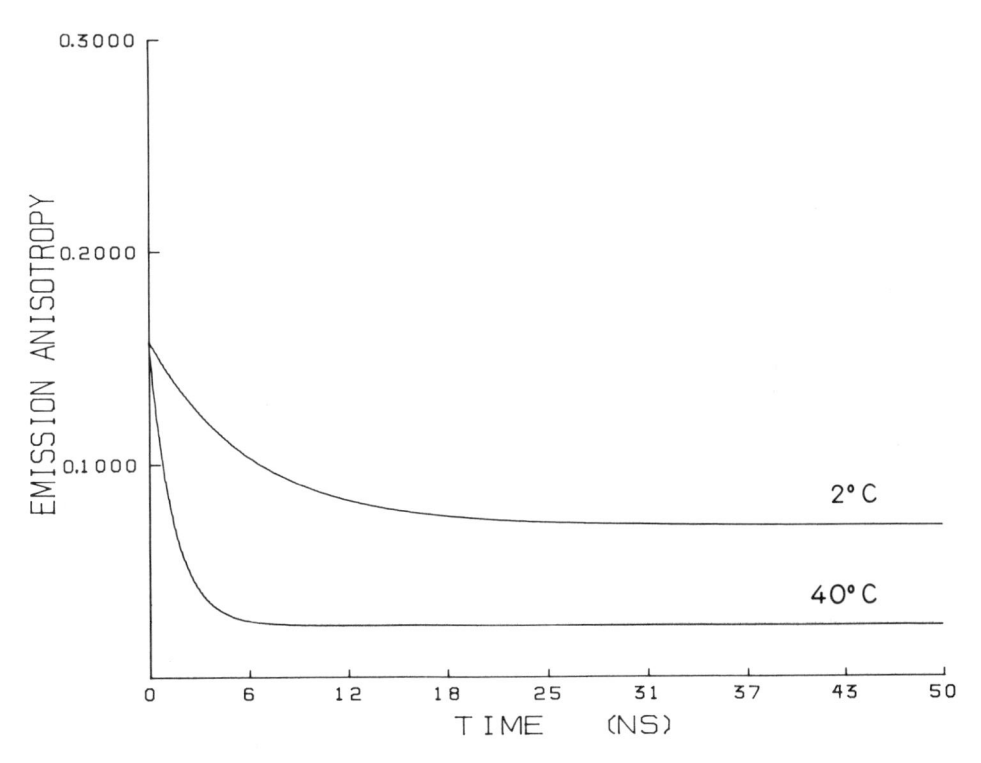

FIGURE 8 The impulse response curves for the decay of the emission anisotropy at 2° and 40°C. The parameters of these decays are given in Table II.

somes gave a decay of the emission anisotropy made up of a single exponential term plus a constant. This is consistent with the notion that the rotation of the dye in the liposomes is restricted. The restrictions are more stringent below the phase transition of the liposomes than above it. This is indicated by the higher value of the constant term at 2°C relative to that at 40°C. The rate at which these constant anisotropy levels are attained is also temperature dependent. Thus, anisotropy relaxation occurs about four times faster above the phase transition temperature than below it. As mentioned above, spectral relaxation had a similar temperature variation pattern. The potential implications of this similarity will be discussed below.

DISCUSSION

The results presented above show that 2-AN adsorbed to DML single-bilayer liposomes undergoes excited-state reactions on the nanosecond time scale. The decay kinetics across the emission band suggest a continuous type of relaxation rather than a simple two-state reaction. Thus the decay times are not wavelength independent, and more than two exponentials are required to fit the fluorescence decay data. Concomitant with this spectral relaxation, the molecule also rotates on the nanosecond time scale. Its emission anisotropy relaxation, initially exponential, was found to level off at a nonzero value at longer times. Thus, the rotation is restricted both above and below the phase transition of the liposomes. In each of the two physical states of the liposome, the initial rate of anisotropy relaxation is comparable to the nanosecond rate of spectral relaxation (Table II).

The interpretation of these data should be considered in relation to the location of 2-AN in liposomes. It is generally accepted that hydrophobic molecules not carrying a net charge such as 2-AN will be located in the apolar hydrocarbon region of the bilayer (Azzi, 1975; Badley, 1976). Thus the nuclear magnetic resonance study of Colley and Metcalfe (1972) located the related molecule 1 anilinonaphthalene (1-AN) deep inside the egg lecithin lipid bilayer, whereas the charged probe 1-anilinonaphthalene-8-sulfonate (1, 8 ANS) was found to be at the surface. Radda and Vanderkooi (1972) used the D_2O fluorescence enhancement effect exhibited by several fluorescence probes to estimate their location in different lipid bilayers. Again 1-AN was found to be better protected than 1, 8 ANS, suggesting its deeper location. Overath and Trauble (1973) have also concluded that 1-AN is a probe of the interior of the bilayer in contrast with charged molecules, such as ANS, which probe mainly the surface of the bilayer.

It seems reasonable to assume that 2-AN is also located in the interior of the bilayer. The characteristics of the relaxation of 2-AN emission anisotropy in glycerol and the liposomes are in agreement with this location. Whereas in glycerol the molecule rotates isotropically (monoexponential decay of the emission anisotropy), in the DML liposomes its rotation was found to be restricted both above and below the phase transition temperature. Similar results have been found with 1, 6-diphenyl-1,3,5-hexatriene (DPH) in DML liposomes (Chen et al., 1977; Kawato et al., 1977). This molecule has also been assumed in a variety of studies (see Chen et al., 1977, for a review) to probe the interior of the bilayer. The initial relaxation times of DPH emission anisotropy are 5.2 ns below the phase transition (14.8°C) and 1.5 ns above it (37.2°C). As shown in Table II, the corresponding parameters of 2-AN anisotropy relaxation are similar (1.5 ns at 40°C, 6.5 at 2°C). Thus the temperature-induced change in the local microenvironment of the probe is reported in a parallel manner by these two molecules. In contrast the results of 2,6 toluidinonaphthalene sulfonate (2,6 TNS) emission anisotropy decay (Chen et al., 1977) adsorbed to the same liposomes present a different picture. The initial anisotropy relaxation time varies from 1.5 ns at 37°C above the phase transition to only 2.9 ns below it (practically independent of temperature in the range 1°–14°C). The molecule 2,6 TNS is similar in structure to 2-AN except for the negatively charged sulfonate group at the 6 position of the naphthalene ring. Like 1,8 ANS its location in the liposomes is considered to be in the polar head region.

The anisotropy relaxation results can be summarized as follows. A tighter packing of the lipids in the crystalline phase affects more markedly the initial rate of the anisotropy decay for probes of the interior than for those of the liposome surface. This temperature differential effect could be used as an initial indication of a potential location. However, the fluorescence anisotropy measurements by themselves could not assign a definite location to a molecular probe in the liposome.

If, based on all the evidence presented above, it is assumed that 2-AN is located in the interior of the bilayer, the origin of the nanosecond time-dependent spectral shifts remains to be explained. DeToma and Brand (1977) found that 2-AN dissolved in cyclohexane solution containing 0.1 M ethanol undergoes a classic two-state excited-state reaction. Its spectral relaxation time profile is represented as curve A in Fig. 7. In contrast tht decay kinetics of 2-AN in glycerol are not consistent with a two-state exiplex reaction and were interpreted in terms of a continuous multistate, solvation scheme. As shown in Fig. 7 (curve C), the initial spectral position of 2-AN in glycerol is more than 1 kK lower than the corresponding position in the cyclohexane—0.1 M ethanol system. It is clearly seen that the spectral relaxation profile of 2-AN in liposomes (curve B, Fig. 7) follows more closely the glycerol model system than the two-state exiplex system. As mentioned above, the empirical decay parameters presented in Table I also support this conclusion. This behavior indicates that 2-AN in liposomes undergoes continuous relaxation during which many polar residues are simultaneously interacting with the probe. Thus, either the probe is located close to the polar head region or small polar molecules (water, for instance) have access to it. This latter possibility is in contradiction with the D_2O experiments of Radda and Vanderkooi (1972). It could, however, be argued that the considerable permeability to water of phospholipid bilayer liposomes (5–100 × 10^{-4} cm/s; Papahadjopoulos and Kimelberg, 1973) is sufficient to induce a positive time-resolved emission spectroscopy effect and insufficient for an isotope fluorescence enhancement.

Our data support the possibility that the probe is interacting with polar moieties more rigidly held in place than with highly mobile, small polar molecules. This tentative conclusion is based on the comparable relaxation rates of spectral position and anisotropy both above and below the phase transition (see Table II). Thus, the nanosecond spectral relaxation is characterized by an empirical time constant of 2.81 ns above the phase transition and 8.74 ns below it. The corresponding values for the initial anisotropy relaxation are 1.5 and 6.5 ns (Table II). In glycerol such a correlation is not found. The nanosecond part of the interaction between the probe and the polar moieties proceeds mainly through mutual reorientational motion of the interactive species (Bakhshiev et al., 1966). If this mutual reorientation motion is performed mainly by the probe, then both spectral and anisotropy relaxation would monitor it. It becomes the source of the common time scale for both effects. On the other hand, in a more mobile environment, such as glycerol, the dipolar reorientational motion is more likely to be performed mainly by the solvent molecules than by the excited chromophore. This would explain why in this case, when the relaxation processes are not monitoring a common motion, uncorrelated temperature variation patterns would be found.

Thus, the combination of time-resolved emission spectroscopy with the decay of the emission anisotropy provides a detailed kinetic picture of the interactions that small molecules undergo in biological systems. The parameters obtainable by steady-state measurements lump together many hidden variables each with its own physical state and temperature dependence. As shown by DeToma et al. (1976), the Bakhshiev theoretical approach (Bakhshiev et al., 1966) to solvent relaxation can be used to interpret in a self-consistent way the time-resolved emission spectroscopy data. In this analysis the time and energy dependence of the fluorescence intensity $F(\bar{\nu}, t)$ is described as a product of two terms, namely: $F(\bar{\nu}, t) = i(t)\rho(\bar{\nu}, t) = i(t)\rho(\bar{\nu} - \bar{\nu}_{max}(t))$, where $F(\bar{\nu}, t)$ is the observed fluorescence emission behavior, $i(t)$ is an electronic damping term that would characterize the decay law in the absence of spectral shifts, and $\rho(\bar{\nu} - \bar{\nu}_{max}(t))$ represents the normalized elementary fluorescence spectral contour shifted in energy by the amount $\bar{\nu}_{max}(t)$ at time t. This latter function, $\bar{\nu}_{max}(t)$ is obtained independently of the Bakhshiev theoretical model. In the present case its course in time has been empirically characterized in terms of two exponentials and a constant (Table II). The nanosecond decay constants are strongly temperature dependent. In contrast, the two decay constants characterizing the electronic damping term $i(t)$ are both temperature and wavelength independent (around 12.5 and 2.4 ns, respectively). Similar experiments carried out with 2,6 TNS absorbed to egg lecithin vesicles (Easter et al., 1977) showed that with this more polar probe the corresponding decay constants for $i(t)$ were temperature dependent. Their values decrease from 1.9 to 1.0 ns and from 9.7 to 4.8 ns as the temperature is increased from −1° to 32°C. It is proposed that this difference between 2,6 TNS and 2-AN also reflects the different location of the two probes in the liposome. The latter molecule being more deeply buried is better protected against temperature-dependent quenching processes.

Comparative studies at the level of these hidden variables are potentially much more meaningful. For instance, as shown in Fig. 4, 2-AN in DML liposomes exhibits a negative temperature coefficient of the mean decay time in the blue region, whereas a positive one is found in the red region. The tempting hypothesis of different locations of the probe in the liquid-crystalline and crystalline phases is, however, not warranted in this case. The decay constants for $i(t)$ being phase independent, the likelihood of such a situation is greatly diminished. Additional studies of this nature are needed to determine whether the fast relaxation processes discussed in this work are related to the ability of membranes to carry out their function of selective transport.

We thank Susan M. Thomas for assistance with some of the data reduction.

This work was supported by National Institutes of Health Grant GM 11632.

Received for publication 2 December 1977.

REFERENCES

AZZI, A. 1975. The application of fluorescent probes in membrane studies. *Qt. Rev. Biophys.* **8**:237.

BADLEY, R. A. 1976. Fluorescent probing of dynamic and molecular organization of biological membranes. Modern Fluorescence Spectroscopy. Vol. II. E. L. WEHRY, editor. Plenum Press, N.Y. 91–168.

BAKHSHIEV, N. G., Y. T. MAZURENKO, and I. Y. PITERSKAYA. 1966. Luminescence decay in different portions of the luminescence spectrum of molecules in viscous solutions. *Opt. Spectrosc. (U.S.S.R.).* **21**:307.

CHEN, L. A., R. E. DALE, S. ROTH, and L. BRAND. 1977. Nanosecond time-dependent fluorescence depolarization of diphenylhexatriene in dimyristoyllecithin vesicles and the determination of "microviscosity." *J. Biol. Chem.* **252**:2163.

CHEN, L. A. 1977. Ph.D. Thesis. Johns Hopkins University, Baltimore, Md.

CHEN, R. F., and R. L. BOWMAN. 1965. Fluorescence polarization: measurement with ultraviolet-polarizing filters in a spectrophotofluorometer. *Science (Wash. D.C.).* **147**:729.

COLLEY, C. M., and J. C. METCALFE. 1972. The localization of small molecules in lipid bilayers. *FEBS (Fed. Eur. Biochem. Soc.) Lett.* **24**:241.

DETOMA, R. P., J. H. EASTER, and L. BRAND. 1976. Dynamic interactions of fluorescence probes with the solvent environment. *J. Am. Chem. Soc.* **98**:5001.

DETOMA, R. P., and L. BRAND. 1977. Excited-state solvation dynamics of 2-anilinonaphthalene. *Chem. Phys. Lett.* **47**:231.

EASTER, J. H., R. P. DETOMA, and L. BRAND. 1976. Nanosecond time-resolved emission spectroscopy of a fluorescent probe adsorbed to L-α-egg lecithin vesicles. *Biophys. J.* **16**:571.

EASTER, J. H., R. P. DETOMA, and L. BRAND. 1977. Fluorescence measurements of environmental relaxation at the lipid-water interface region of bilayer membranes. B.B.A. *(Biochim. Biophys. Acta) Libr.* In press.

KAWATO, S., K. KINOSITA, JR., and A. IKEGAMI. 1977. Dynamic structure of lipid bilayers studied by nanosecond fluorescence techniques. *Biochemistry.* **16**:2319.

OVARATH, P., and H. TRAUBLE. 1973. Phase transitions in cells, membranes and lipids of *Escherichia coli:* detection by fluorescent probes, light scattering and dilatometry. *Biochemistry.* **12**:2625.

PAPAHADJOPOULOS, D., and K. K. KIMELBERG. 1973. Phospholipid vesicles (liposomes) as models for biological membranes: their properties and interactions with cholesterol and proteins. *Prog. Surf. Membr. Sci.* **4**:171.

RADDA, G. K., and J. VANDERKOOI. 1972. Can fluorescent probes tell us anything about membranes? *Biochem. Biophys. Acta.* **265**:509.

RADDA, G. K. 1975. Fluorescent probes in membrane studies. *Meth. Membrane Biol.* **4**:97–188.

SPENCER, R. D., and G. WEBER. 1970. Influence of Brownian rotations and energy transfer upon the measurement of fluorescence lifetime. *J. Chem. Phys.* **52**:1654.

DISCUSSION

SWENBERG: Why is a decrease in the energy of emission associated with an increase in lifetime? Are nonradiative decay routes being suppressed? If so, which ones?

BADEA: If we accept the continuum model for the solvation process, supported by our data, then we can account for the increase of the lifetime with the decrease in the energy of emission without assuming an opening or closing of any radiative or nonradiative decay channel during the solvation time interval. A simple parent-daughter model, involving a multitude of intermediate excited states partially solvated, explains the phenomena observed. As the solvation progresses, the lower energies are preponderantly emitted, so that at longer wavelengths the overall decay time measured reflects both the rate of formation and the rate of decay of that preponderant intermediate. In our particular case, we are fortunate to have a sufficient proof to indicate that this is what's happening. The negative amplitude (i.e. the rise) in the decay curve at 500 nm, for instance, shows clearly that the intermediates are formed in the excited state. We would like to emphasize, however, that this negative amplitude is not always observable experimentally. Therefore its absence does not necessarily indicate the absence of an excited state reaction.

SWENBERG: If I understand you, you are telling me that the reciprocal relationship between the lifetime and the energy of emission is purely due to a solvation energy effect. This leaves the mechanism unspecified.

BADEA: Our data are operationally interpreted in terms of an electronic relaxation function and a spectral relaxation function. The empirical relaxation times characterizing these functions are used to monitor different physical microenvironments around the probe by correlating the properties of the environment with the particular values of these relaxation times. This was our purpose in this study. The data analysis indicates the occurrence of time-dependent changes in the molecular manifold; the electronic relaxation function has in fact not been recovered as a monoexponential decaying function. The characterization of the mechanism in terms of intra- or inter-manifold conversion rates is simply an alternate way to describe the complex phenomena. It has not yet been attempted.

BECKER: Even though parameters obtained by steady-state polarization measurements lump together many hidden variables, aren't these measurements by themselves still valid for determining the existence and the temperature of a phase transition in lipid bilayers?

BADEA: The existence of, certainly, yes. This has been convincingly proved by the great number of steady-state fluorescence measurements on liposomes. Regarding its exact value, however, I do not exclude the possibility that some explicit time-resolved parameter can provide a more sensitive measure of the midpoint value. No such experiment has been performed up to now. It would involve the measurement of the time-resolved emission spectra and of the decay of the emission anisotrophy at closely spaced temperature points in the phase transition interval.

MANTULIN: The question of anilinonaphthalene diffusion from the interior of the phospholipid bilayer to the polar head group both above and below the lipid phase transition is most interesting. Would it not be possible to explore further this phenomenon or competitive relaxation processes by covalently attaching your fluorescence probe to the fatty acid chain of the phospholipid at various positions along the chain and then comparing the relaxation behavior of the free and bound fluorescence probe?

BADEA: This is a very valid point. Certainly, if you want to prove a particular liposome-like environment, this is the way to proceed, i.e. to attach the probe covalently at various positions along the phospholipid chain and see what the relaxation processes are at that particular point. In such a way you have not only an improved time resolution but also an improved space resolution. But our aim in this paper was to visualize in a time-resolved fashion the complexities involved in a standard fluorescence probe experiment; not to probe this particular DML liposome, but to make you aware that a lot of dynamic processes occur simultaneously on the nanosecond time scale.

VANDERKOOI: Have you looked for a D_2O effect on the spectral shift of 2-AN in liposomes?

BADEA: No, we have not as yet. This will be a very interesting experiment in the event that such an effect is first found in model solvent systems. It will help ascertain the relative sensitivity of time-resolved parameters versus steady-state parameters for picking up small changes in the immediate environment of the probe. However, I would like to add that neither D_2O nor other "modifiers" of the fluorescence characteristics of the probe can help solve the basic difficulty of water accessibility experiments performed with nonbound probes: is it the excited probe moving towards the lipid-water interface or is it the water already present in sufficient amounts at its deeper location that produces the effect observed? In this paper we try to solve this dilemma using the correlation we observed between the spectral relaxation pattern and the anisotropy relaxation pattern.

GEACINTOV: Have you looked at decay profiles of simple polycyclic molecules such as naphthalene or anthracene, for example, and are their decay profiles also nonexponential in liposomes?

BADEA: Up to now we haven't found a single example of a clear monoexponential decay in the liposomes. At this point our own bias should be acknowledged. We judge interesting the molecules that, in model solvents, do not exhibit monoexponential decay across the emission band. Time-resolved monitoring of biological substrates is more accurate with these molecules. However, to answer your question specifically, preliminary experiments on anthracene showed that it does not have a single-exponential decay in liposomes.

GEACINTOV: Why? Do you think there is an intrinsic heterogeneity in the structure of liposomes that could account for multiple sites occupied by the fluorophores and thus for their nonexponential decay?

BADEA: The liposomes are without any doubt heterogeneous aggregates. If the fluorescence probe is randomly distributed in different microenvironments, then in general you are bound to have a nonmonoexponential decay. Also, as exemplified in the present work, a number of processes may also take place during the lifetime of the excited state. It is a difficult question. Some progress toward distinguishing between microheterogeneity and homogeneous excited state reaction processes has been made in our laboratory in the case of DPH in egg-lecithin liposomes. However, in the general case the problem is still wide open. I would like to add that a *sine qua non* condition for successfully hunting monoexponential decays in the liposomes is the use of magic angle excitation-emission conditions. This is all too often neglected in the literature.

PRENDERGAST: If I've interpreted your statement correctly, you ascribe the time-resolved spectral changes predominantly, if not solely, to relaxations of the fluorophore rather than of the environment. The difficulty in assigning relaxation to either fluorophore or "solvent" must clearly relate to the problems associated with choosing an appropriate reference solvent. We are more interested, however, in assessing the relaxation behavior of the lipid environment. For example, if the probe were localized to the glycol backbone region of the bilayer, would you not expect that either the carbonyl function or "bound water" in this region might relax with time constants similar to the ones you have given? Is there a way, using your technique, to detect relaxations of this type?

BADEA: Yes, the comment is already in the paper in a way. We have an angular correlation function describing each relaxation. One of them measures the spectral relaxation, the other one the rotational relaxation. In the general case they should not correlate whatsoever. In fact we did not find any correlation in glycerol, for instance. But in this particular liposomic environment, from the experimental point of view, independent of any particular theoretical model used to interpret the data, we found a correlation. Based on this correlation, we are trying to say that this reflects an interaction with a rather immobile partner, so that the orientational motion of the probe during this dipolar interaction is mainly performed by the fluorophore and not by the environment. The relaxations of the environment can be similarly studied with fixed, covalently bound probes instead of adsorbed ones. In such a way the relaxation of the fluorophore reflects exclusively the motion of the phospholipid to which it is attached.

PRENDERGAST: We have performed analogous experiments using oxygen quenching to decrease the probes' lifetimes artifically and thereby effect "time-resolved" spectral changes and we are caught in the same dilemma as you in not having an appropriate solvent model. But one of our approaches is to perform our studies using a lipid that lacks the ester dipole. We are using an ether lipid to see, for example, if the relaxation process are different for the smaller dipole than you would expect for the analogous ester function. I can't give you the results yet, but such experiments may be useful for you to perform also. Other questions obviously arise out of this. For example, what would the effect of cholesterol be?

BADEA: We have not performed an independent measurement on cholesterol yet. We have this now planned and we will let you know about the results. Ideally the cholesterol should inhibit differentially different modes of rotational relaxation of the particular fluorophore in the liposome, but we don't have any data as to this effect yet.

PRENDERGAST: You see, the reason why I keep on with this question is that it directly relates to the function of the membrane. The rotational behaviors of the phospholipid head group and glycerol backbone are probably going to be critical in the determination of the membrane permeabilities of many small molecules.

BADEA: This is also my feeling at the present time.

DETECTION OF HINDERED ROTATIONS OF 1,6-DIPHENYL-1,3,5-HEXATRIENE IN LIPID BILAYERS BY DIFFERENTIAL POLARIZED PHASE FLUOROMETRY

J. R. LAKOWICZ, *Freshwater Biological Institute and Department of Biochemistry, University of Minnesota, Navarre, Minnesota 55392, and*
F. G. PRENDERGAST, *Department of Pharmacology, Mayo Medical School, Rochester, Minnesota 55901 U.S.A.*

ABSTRACT Differential polarized phase fluorometry has been used to investigate the depolarizing motions of 1,6-diphenyl-1,3,5-hexatriene (DPH) in the isotropic solvent propylene glycol and in lipid bilayers of dimyristoyl-L-α-phosphatidylcholine (DMPC), dipalmitoyl-L-α-phosphatidylcholine (DPPC), and other phosphatidylcholines. Differential phase fluorometry is the measurement of differences in the phase angles between the parallel and perpendicular components of the fluorescence emission of a sample excited with sinusoidally modulated light. The maximum value of the tangent of the phase angle (tan Δ_{max}) is known to be a function of the isotropy of the depolarizing motions. For DPH in propylene glycol the maximum tangent is observed at 18°C, and this tangent value corresponds precisely with the value expected for an isotropic rotator. Additionally, the rotational rates determined by steady-state polarization measurements are in precise agreement with the differential phase measurements. These results indicate that differential phase fluorometry provides a reliable measure of the probe's rotational rate under conditions where these rotations are isotropic and unhindered.

Rotational rates of DPH obtained from steady-state polarizaton and differential phase measurements do not agree when this probe is placed in lipid bilayers. The temperature profile of the tan Δ measurements of DPH in DMPC and DPPC bilayers is characterized by a rapid increase of tan Δ at the transition temperature (T_c), followed by a gradual decline in tan Δ at temperatures above T_c. The observed tanΔ_{max} values are only 62 and 43% of the theoretical maximum. This defect in tanΔ_{max} is too large to be explained by any degree of rotational anisotropy. However, these defects are explicable by a new theory that describes the tan Δ values under conditions where the probe's rotational motions are restricted to a limiting anisotropy value, r_∞. Theoretical calculations using this new theory indicate that the temperature dependence of the depolarizing motions of DPH in these saturated bilayers could be explained by a rapid increase in its rotational rate (R) at the transition temperature, coupled with a simultaneous decrease in r_∞ at this same temperature. The sensitivity of the tan Δ values to both R and r_∞ indicates that differential phase fluorometry will provide a method to describe more completely the depolarizing motion of probes in lipid bilayers.

INTRODUCTION

Measurements of the steady-state fluorescence polarization of probes embedded in lipid bilayers have been widely used to estimate the microviscosity of the hydrophobic regions of both natural and model cell membranes (Bashford et al. 1976; Cogan et al. 1973; Jacobson and Wobschall, 1974; Moore et al. 1976; Shinitzky et al. 1971; Shinitzky and Inbar, 1974). All these estimates of microviscosity assume that the depolarizing rotational motion of the fluorophores in lipid bilayers are equivalent to those in isotropic solvents. This assumption is likely to be in error since lipid bilayers are inherently anisotropic and are therefore likely to provide an environment that hinders diffusion motions selectively.

Diphenylhexatriene (1,6-diphenyl-1,3,5-hexatriene, DPH) has become widely used in microviscosity studies as a result of its favorable fluorescence polarization and spectral properties (Shinitzky and Barenholz, 1974), but perhaps more importantly as a result of the dramatic changes in fluorescence polarization that occurs at the solid-to-liquid phase transition of lipid bilayers (Lentz et al. 1976a,b). These highly temperature-dependent polarization changes appear to be unique to DPH. Other fluorescent probes, such as 1-anilino-8-naphthalene sulfonic acid, perylene, 9-vinylanthracene, 2-methylanthracene, 12-anthroyl stearate, and N-phenyl-1-naphthylamine, undergo less dramatic changes in the fluorescence polarization at the transition temperature when embedded in lipid bilayers.

The high fluorescence anisotropy value observed for DPH in the absence of depolarizing rotations indicates that the axes of the absorption and emission transition moments are parallel (Shinitzky and Barenholz, 1974) and probably lie along the long axis of the fluorophore. As a result, only rotations which displace this axis will be depolarizing, and it appears that depolarization is governed by a single rotational rate. For these reasons, steady-state polarization measurements using DPH are considered to reflect accurately the microviscosity of lipid bilayers.

Recently, Chen et al. (1977) have used time-resolved decays of fluorescence anisotropy to demonstrate that DPH undergoes only hindered torsional motions below the phase transition temperature of DMPC vesicles, and a highly nonexponential decay of anisotropy above this temperature. These results indicate an environment in the lipid bilayer which does not permit free probe rotation, even though these limited rotations may be isotropic. Hence, the determinations of membrane microviscosities are likely to be in error since such determinations compare polarization values of a probe undergoing unhindered isotropic rotations with the polarization values for a probe undergoing hindered rotations.

We clearly require a more detailed understanding of the types of probe motion responsible for fluorescence depolarization in order to interpret these data in terms of membrane microviscosity. Additionally, the molecular details of depolarizing rotations in bilayers should reflect to some extent the molecular organization and the segmental motions of the fatty acyl chains.

THEORY

The theory of differential phase fluorometry for isotropic and anisotropic depolarizing rotations has been recently described by Weber (1977). DPH is thought to be an isotropic rotator. Under these conditions the tangent of the differential phase angle (tan Δ) is given by

$$\tan \Delta = \frac{(2R\tau)\omega\tau r_0}{(1/9)(1 + 2r_0)(1 - r_0)(1 + \omega^2\tau^2) + (2R\tau)/3)\,(2 + r_0) + (2R\tau)^2}. \quad (1)$$

The rotational rate of the fluorophore may be obtained from the quadratic form of the equation.

$$(2R\tau)^2 + (2R\tau)[(2 + r_0)/3 - |r_0/\tan \Delta|\,\omega\tau] + m(1 + \omega^2\tau^2) = 0$$
$$m = (1/9)(1 + 2r_0)(1 - r_0). \quad (2)$$

R is the rotational rate of the probe in radians per second, r_0 the anisotropy value in the absence of rotational diffusion, ω the circular modulation frequency, and τ the fluorescence lifetime. For isotropic rotations the maximum value for tan Δ is a function of r_0, ω, and τ only, such that

$$\tan \Delta_{\max} = \frac{3\omega\tau r_0}{(2 + r_0) + 2[(1 + 2r_0)(1 - r_0)(1 + \omega^2\tau^2)]^{1/2}}. \quad (3)$$

The above equations apply to spherical molecules, or to molecules whose depolarizing rotations are isotropic. Under these conditions the tan $\Delta_{\max}$ values will agree with that predicted by Eq. 3. Anisotropic rotations, or a population of isotropic rotators with more than a single rotational rate, results in tan $\Delta_{\max}$ values less than those predicted by Eq. 3. The studies of Mantulin and Weber (1977) demonstrate the usefulness of differential phase fluorometry in the detection of anisotropic rotations.

Theoretical calculations have shown that one should not expect defects in the differential tangent resulting from anisotropic rotations to exceed 25%. However, our studies of DPH in vesicles of dimyristoyl-L-α-phosphatidylcholine (DMPC) or dipalmitoyl-L-α-phosphatidylcholine (DPPC) showed tangent defects of about 40 and 60%, respectively. We deduced that such defects could result from hindered torsional motions of DPH. Weber (1977)[1] has recently obtained solutions for the tan Δ under such conditions. If the rotations are limited to a nonzero anisotropy value (r_∞) at times long compared to the fluorescence lifetime, then the parallel ($I_\parallel(t)$) and perpendicular ($I_\perp(t)$) components of the fluorescence emission are given by[1]

$$I_\parallel(t) = (1 + 2r_\infty)e^{-t/\tau} + 2(r_0 - r_\infty)e^{-(1/\tau + 6R)t}, \quad (4)$$

$$I_\perp(t) = 1 - 2r_\infty)e^{-t/\tau} - (r_0 - r_\infty)e^{-(1/\tau + 6R)t}. \quad (5)$$

---

[1] Weber, G. 1977. Personal communication.

From these equations Weber has obtained

$$\tan \Delta = \frac{\omega\tau(r_0 - r_\infty)(2R\tau)}{\begin{bmatrix} (1/9)(1 + 2r_0)(1 - r_0)(1 + \omega^2\tau^2) \\ + (2R\tau/3)[2 + r_0 - r_\infty(4r_0 - 1)] + (2R\tau)^2(1 + 2r_\infty)(1 - r_\infty) \end{bmatrix}}, \quad (6)$$

and

$$\tan \Delta_{max} = \frac{\omega\tau(r_0 - r_\infty)}{\begin{bmatrix} (1/3)[2 + r_0 - r_\infty(4r_0 - 1)] \\ + (2/3)[(1 + 2r_0)(1 - r_0)(1 + 2r_\infty)(1 - r_\infty)(1 + \omega^2\tau^2)]^{1/2} \end{bmatrix}}, \quad (7)$$

For $r_\infty = 0$ these equations reduce to those applicable to isotropic rotations.

Rotational rates of probes may also be determined by steady state polarization measurements. In this case the Perrin-Weber equation was utilized

$$r_0/r = 1 + 6R\tau, \quad (8)$$

where r is the fluorescence anisotropy in the presence of rotational diffusion. Fluorescence polarization (P) and anisotropy values are interchangeable with

$$P = 3r/(2 + r). \quad (9)$$

METHODS

Instrumentation

Fluorescence lifetimes and differential tangents were measured by the phase shift method (Spencer and Weber, 1969, 1970), using a light modulation frequency of either 10 or 30 MHz. A schematic drawing of the T-format differential phase fluorometer is shown in Fig. 1. Fluorescence lifetimes were determined from the phase angle of the fluorescence emission compared with the phase angle of the exciting light when scattered at right angles by a glycogen suspension. For these lifetime measurements only a single photomultiplier tube was used. The effects of depolarizing Brownian rotations on the observed lifetimes was eliminated by setting the excitation and emission polarizers at 0° and 55° from the vertical, respectively (Spencer and Weber, 1970). All measurements were performed on equipment obtained from SLM Instruments, Inc., Urbana, Ill. The instrumental conditions were: excitation wavelength, 360 nm; excitation filter, Corning 7-54; emission filter, Corning 3-73 (Corning Glass Works, Corning, N.Y.) and 2 mm of 1 M $NaNO_2$. Steady-state anisotropy values were obtained directly in the subnanosecond spectrofluorometer with all the radio frequency electronics and light modulation turned off. At $-57°C$ in propylene glycol we found $r_0 = 0.392$ for DPH. This r_0 value was used in all our calculations.

Differential tangent values were obtained by measuring the phase difference when one emission polarizer was rotated 90° from the parallel to the perpendicular orientation. This phase difference was always measured relative to the second photomultiplier, which constantly observed the phase of the perpendicular emission. These measurements were continued until a satisfactory average was obtained.

In measurements of the differential tangent, systematic errors which could result from the detection system being sensitive to the polarization of the fluorescent signal must be avoided. We

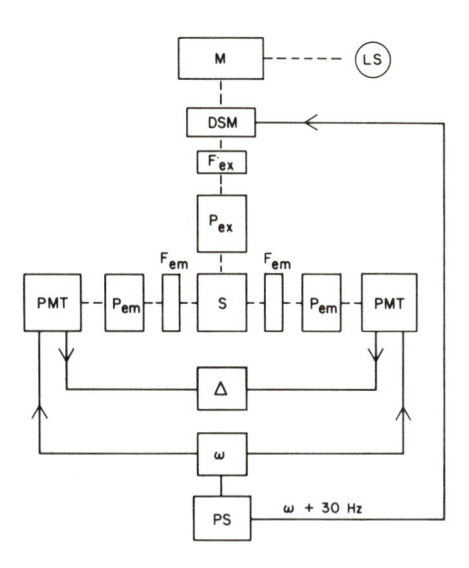

FIGURE 1 Schematic of a differential phase fluorometer. *LS*, light source; *M*, monochromator; *DSM*, Debye-Sears ultrasonic light modulator; F_{ex}, excitation filter; P_{ex}, excitation polarizer; *S*, sample; F_{em}, emission filter; P_{em}, emission polarizer; *PMT*, phototubes; *PS*, phase shifter for adding 10 or 30 Hz to the high-frequency input of the light modulator; Δ, low-frequency electronics to measure the phase of the cross-correlated fluorescence emission.

determined that such systematic errors were not significant in our measurements. Upon excitation of our samples with horizontally polarized light, the phase difference between the vertically and horizontally polarized components of the emission within experimental limits was zero. These control measurements were performed with each sample at a temperature which yielded the maximum differential tangent.

Materials

DMPC, DPPC, and dioleoyl-L-α-phosphatidylcholine (DOPC), synthetic lipids obtained from Sigma Chemical Company (St. Louis, Mo.), were used without further purification. Chromatography of these phospholipids on silica in chloroform/methanol/water (65:25:4) and ethyl ether/benzene/ethanol/acetic acid/H_2O (40:40:20:8:4) showed a single spot by both phosphate and dichromate char staining.

The phospholipid vesicles were prepared by addition of benzene solutions of the probe and lipid to a stainless steel beaker. The probe-to-lipid molar ratio is 1:500. The benzene was evaporated by gently warming the solution while maintaining a constant flow of argon over the materials. Buffer (0.01 M Tris, 0.05 M KCl, pH = 7.5) was added to the dried lipid to establish a concentration of 0.17 mg lipid/ml buffer. Sonication was effected with a Heat Systems model 350 sonicator (Heat Systems-Ultrasonics, Inc., Plainview, N.Y.) at 200 W using a 0.5-inch diameter tip. The temperature of the solution was maintained near 40°C during the 15-min sonication period. This preparation was annealed for 1 hr at 40°C and then centrifuged at 48,000 g for 90 min. Unsonicated bilayers were prepared in an identical fashion except that the sonication and centrifugation steps were eliminated, and the sample was agitated on a vortex mixer for 10 min after addition of buffer. Phosphate assays (Kates, 1972) performed on these preparations indicate a phospholipid concentration of at least 90% of the expected concentration.

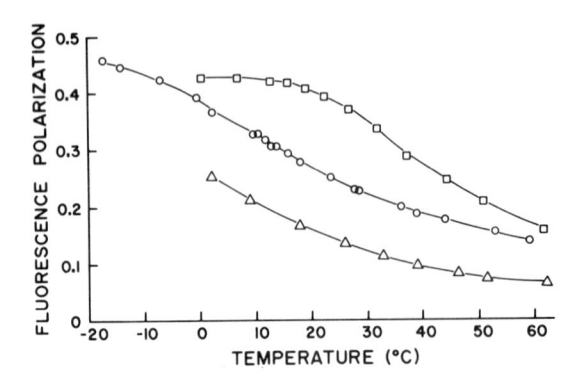

FIGURE 2 Steady-state polarization values of DPH-labeled lipid bilayers and solutions. These measurements were made on the same samples used in the tan Δ measurements, directly in the subnanosecond spectrofluorometer, as is described in the Methods section. Data are shown for DPH in propylene glycol (o), DOPC (Δ), and DMPC/cholesterol, 3/1, (□).

No significant fluorescent impurities were observed in unlabeled vesicles prepared in an identical manner. Additionally, no scattered light at the excitation wavelength, or at the wavelength for Raman scatter, was observed through the filters used for the lifetime and differential phase measurements.

The optical density of DPH in propylene glycol was 0.4 at 360 nm. A blank solution of propylene glycol showed no significant fluorescence at equivalent instrumental conditions.

Figs. 2 and 3 show the temperature profiles of the steady-state polarization values obtained from our vesicle preparations. These results are comparable to those obtained by others.

RESULTS

Fluorescence lifetimes and differential tangent values for DPH in propylene glycol are shown in Fig. 4. The agreement of the maximum observed tan Δ value with that predicted from Eq. 3 (see Table I) indicates that the depolarizing rotations of DPH in propylene glycol are isotropic. Additionally, these rotations must be unhindered since

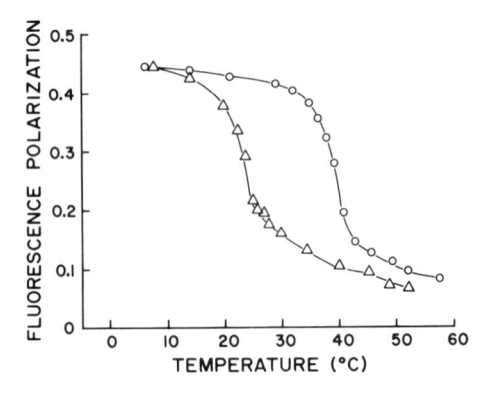

FIGURE 3 Steady-state polarization measurements of DPH-labeled lipid bilayers. Data are shown for DPH in DMPC (Δ) and DPPC (o) bilayers.

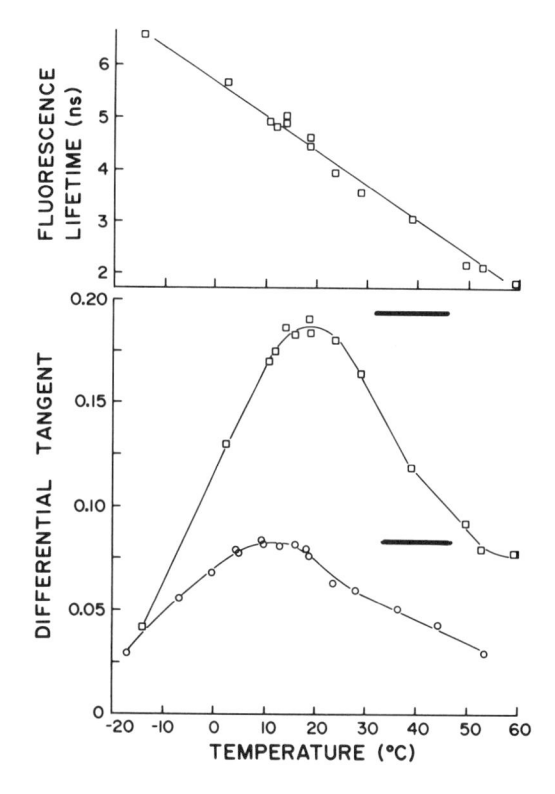

FIGURE 4 Fluorescence lifetimes and differential tangent values of DPH in propylene glycol. Data are shown for 10 (o) and 30 (□) MHz. The optical density of DPH at 360 nm was 0.40. The solid bars indicate tan Δ_{max} for an isotropic rotator with $r_0 = 0.392$ and a fluorescence lifetime equal to the observed lifetime at the temperature of the maximum observed differential tangent (10 MHz, 5 ns; 30 MHz, 4.3 ns).

a nonzero value of r_∞ also results in a decrease in tan Δ_{max} (Fig. 5). In Fig. 6 we compare the rotation rates obtained from steady-state polarization measurements (Eq. 8) with those obtained from the differential phase measurements (Eq. 2). These data show that there is precise agreement between steady-state polarization and differential phase fluorometric estimates of rotational rates, and validate the assertion that in solvent such as propylene glycol DPH does indeed behave as a free isotropic rotator. Additionally, it is clear that in such situations the extrapolation to a value for the microviscosity of the medium may be quite valid.

The differential phase measurements for DPH in lipid bilayers prepared from saturated phosphatidylcholine bilayers are distinctly different from the results obtained for DPH in isotropic solvents. Figs. 7-9 show the temperature dependence of tan Δ for single-lamellar DMPC vesicles, multilamellar DMPC bilayer, and single-lamellar DPPC vesicles, respectively. The dominant characteristics of all these data are (a) a rapid increase in tan Δ at the transition temperature, followed by a decrease in tan Δ above this temperature, and (b) failure of tan Δ to reach its theoretical maximum. In agreement with steady-state polarization measurements by ourselves and others (Lentz

TABLE I
DIFFERENTIAL PHASE MEASUREMENTS OF DPH IN PROPYLENE GLYCOL AND LIPID BILAYERS

Sample	Frequency	τ*	tan Δ_{max} Observed	tan Δ_{max} Calculated‡	Percent of theory
	MHz	ns			%
Propylene	10	5.0	0.082	0.081	101
glycol	30	4.3	0.185	0.195	95
DMPC	10	8.8	0.084	0.136	68
	30	8.7	0.188	0.302	62
DMPC§	10	8.5	0.073	0.132	62
	30	8.5	0.170	0.298	52
DPPC	10	8.4	0.057	0.131	43
DOPC	10	7.7	0.105	0.127	83
	30	6.9	0.200	0.274	73
DMPC/cholesterol	10	8.8	0.025	0.135	19
(3:1)	30	7.9	0.068	0.288	24

*Fluorescence lifetime observed at the temperature of maximum differential tangent.
‡For unhindered isotropic rotations (Eq. 3).
§Unsonicated, multilamellar bilayers. All other lipids are sonicated to form single-lamellar vesicles.

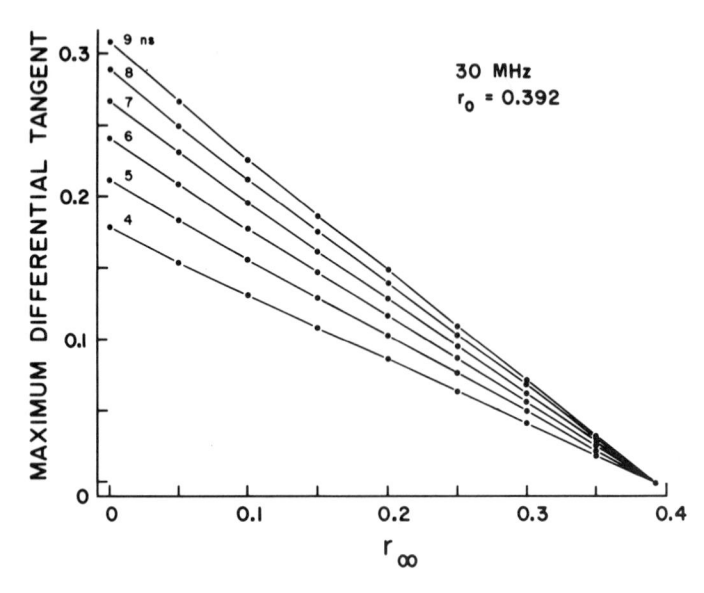

FIGURE 5 Maximum theoretical differential tangent for an isotropic rotator in a hindred environment. The plot is meant to simulate DPH at 30 MHz. Data for the plot were derived from Eq. 7 with r_0 set at 0.392 and the fluorescence lifetimes indicated on the figure.

FIGURE 6

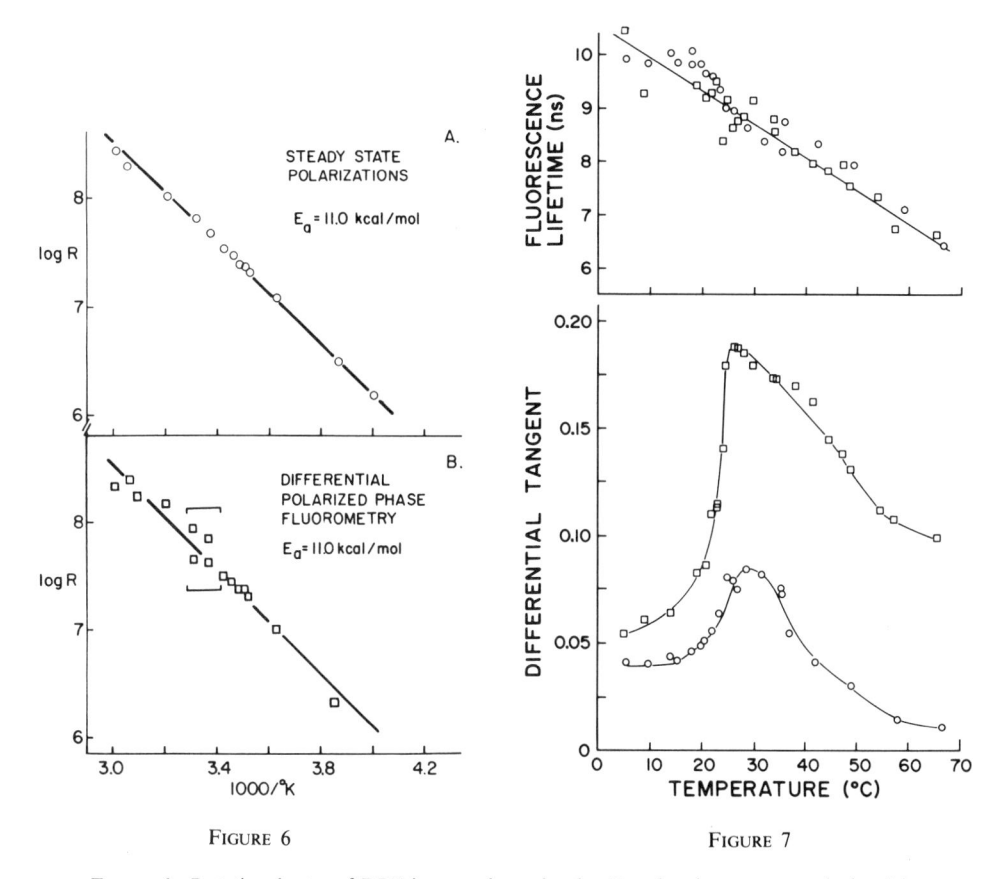

FIGURE 7

FIGURE 6 Rotational rates of DPH in propylene glycol. Rotational rates were calculated by both steady state polarization measurements (A, ○) and by differential polarized phase fluorometry (B, □). The duplicate points between the bars on part B are a result of the two possible solutions to Eq. 2. Near tan Δ_{max} the choice of the proper solution is unclear. However, the correct choice is often unimportant since the rotational rates are similar.

FIGURE 7 Fluorescence lifetimes and differential tangents for DPH in small, single-lamellar bilayers of DMPC. 10 (○) and 30 (□) MHz data are shown.

et al., 1976a, b), the transition, as observed by differential phase measurements, is sharper in the large, multilamellar bilayers than in the small single-lamellar vesicles.

A summary of the observed and theoretical tan Δ_{max} values is provided in Table I. The observed values of tan Δ max in DMPC and DPPC vesicles are only about 62 and 42%, respectively, of that expected for a free isotropic rotator. In contrast to the results obtained for DPH in propylene glycol, we could not obtain agreement between the rotational rates observed by steady-state polarization measurements with those observed by differential phase measurements when we assumed DPH was an unhindered isotropic rotator. Indeed, the observed tangent defects in the lipid bilayers are too large to be explained by any degree of rotational anisotropy (Weber, 1977). However, as can be seen from Fig. 5, the maximum differential tangent decreases to

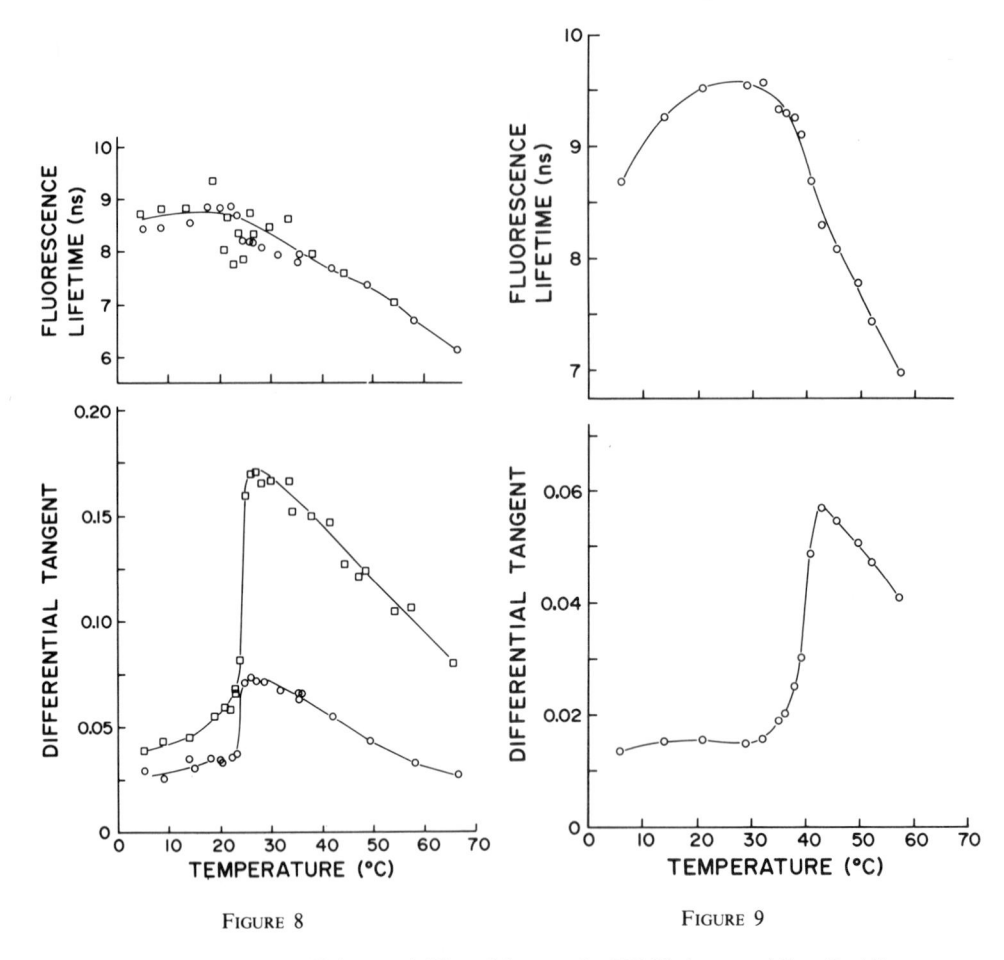

FIGURE 8 Fluorescence lifetimes and differential tangets for DPM in large, multilamellar bilayers of DMPC. 10 (○) and 30 (□) MHz data are shown.

FIGURE 9 Fluorescence lifetimes and differential tangets for DPH in small, single-lamellar bilayers of DPPC. The data shown were collected at 10 MHz.

zero in an approximately linear fashion as r_∞ increases from zero to r_0. Hence for $r_\infty - 0.5\ r_0$ one obtains an approximate 50% tangent defect. Thus, an anisotropic environment that limits the probe's rotations to a nonzero anisotropy could easily result in tangent defects of the magnitude we observed in DMPC and DPPC bilayers.

Figs. 10 and 11 show the tan Δ/temperature profiles for DPH in vesicles of DOPC and DMPC/cholesterol (3:1), respectively. The depolarizing rotations of DPH in DOPC vesicles appear less hindered and more rapid than in vesicles of the saturated phosphatidylcholines. These results are inferred from the small tangent defect observed in these bilayers (Table I), and the occurrence of the maximum tan Δ at low temperatures. The rotational motion of DPH in cholesterol-containing bilayers appears to be greatly hindered and less rapid than in bilayers without cholesterol. The

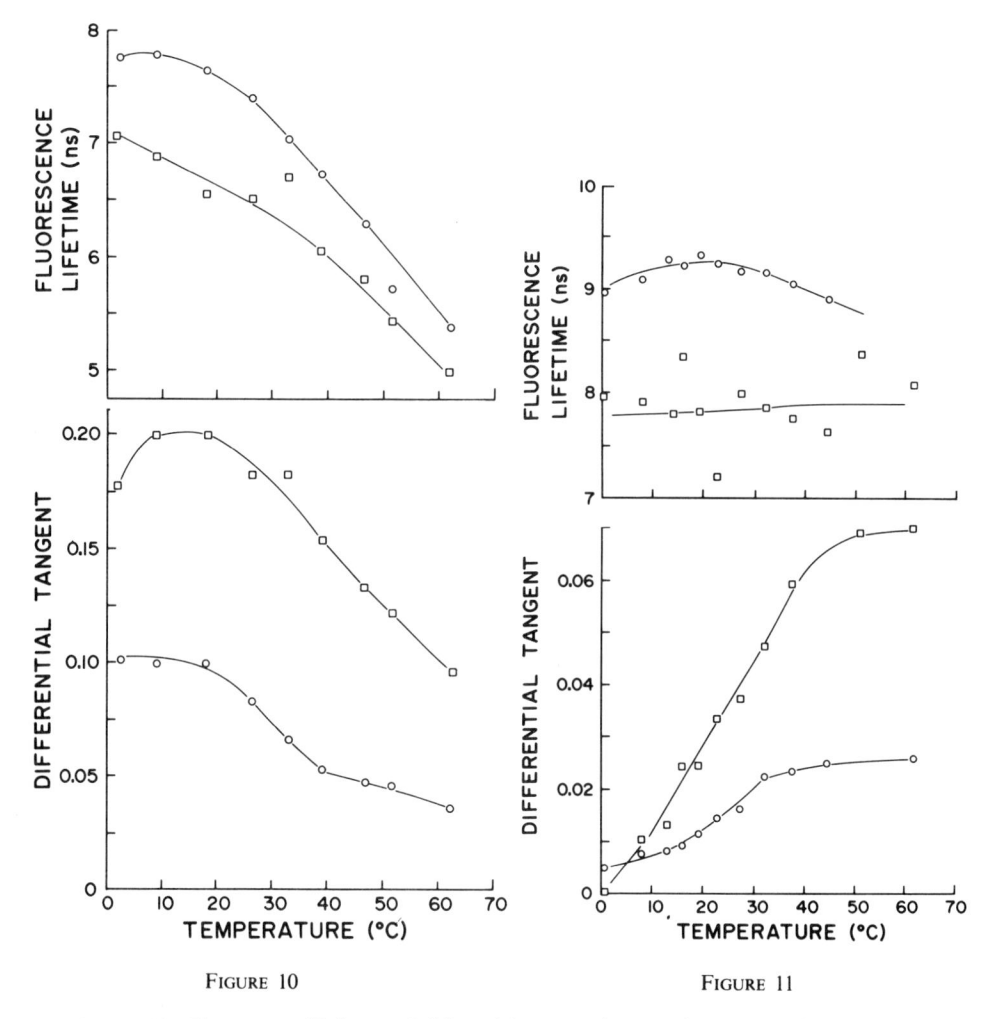

FIGURE 10 Fluorescence lifetimes and differential tangents for DPH in DOPC vesicles. 10 (o) and 30 (□) MHz data are shown.

FIGURE 11 Fluorescence lifetimes and differential tangents for DPH in DMPC/cholesterol vesicles. The molar ratio of DMPC to cholesterol is 3/1. 10 (o) and 30 (□) MHz data are shown.

maximum differential tangent is only about 20% of the theoretical maximum, and this value occurs only at high temperatures.

DISCUSSION

What characteristics of the depolarizing motions of DPH could result in the temperature profiles for tan Δ observed in lipid bilayers? The time-resolved polarization studies of Chen et al. (1977) indicate that DPH has a nonzero limiting anisotropy below the transition temperature of DMPC, and that this r_∞ value decreases above the transition temperature. Additionally, steady-state polarization measurements have indicated that the rotational rate of DPH increases rapidly through the phase transition, and that the temperature dependence of this rotational rate is similar at temperatures far above and below this temperature. However, in light of the data presented in this paper which indicates a hindered environment for DPH in bilayers, this apparent increase in the rotational rate may be primarily a result of a decrease in r_∞ coupled with a smoothly changing rotational rate. The observed temperature profiles could be the result of a simultaneous increase in R and decrease in r_∞.

In an attempt to understand these data more fully, we devised a mathematical model. We assumed a transition temperature of 22°C; we further assumed that this transition was Gaussian with a standard deviation (σ) of 5°C. At a temperature T the fraction of the transition which is complete is given by the area ($A(T)$) under the normal curve from low temperatures to T. Thus,

$$A(T) = (1/\sigma\sqrt{2\pi}) \int_{-\infty}^{T} \exp\{-1/2[(T - Tc)/\sigma]^2\} dT. \qquad (10)$$

Two further assumptions were made. First, r_∞ was assumed to decrease from 0.3 for temperatures far below T_c, to 0.1 for temperatures far above T_c, according to

$$r_\infty = 0.3 - 0.2 A(T). \qquad (11)$$

The activation energy for DPH rotations was assumed to be 10 kcal/mol at temperatures far from the transition temperature, and R was assumed to increase by one additional order of magnitude as a result of the phase transition. Hence,

$$\log R = -Ea/2.303 R_g T + C + A(T). \qquad (12)$$

The pre-exponential factor C (equal to 14.407) was chosen to obtain $\log R = 7.5$ at $T = T_c$. R_g is the gas constant and T the absolute temperature.

The temperature profile of the tan Δ values expected under these conditions was calculated with Eq. 6, and is shown in Fig. 12. No attempt was made to adjust the functional forms of R and r_∞ so as to minimize deviations from our experimental results. We feel the most important point of this calculation is that our simple model for the behavior of DPH in a restricted environment reproduces the general characteristics of the observed temperature profile of DPH in lipid bilayers prepared from saturated phosphatidylcholines. These characteristics are a rapid increase in tan Δ at $T = T_c$, followed by a decrease in tan Δ at higher temperatures, and maximum differential tangents of approximately 50% of the value expected for $r_\infty = 0$.

The shapes of these tan Δ temperature profiles are sensitive to the temperature dependence of both R and r_∞. Suppose, for example, the phase transition results in a rapid increase in R (Eq. 12) but no change in r_∞. Under these conditions the result of a constant nonzero r_∞ value is a decrease in tan Δ_{max} and a rather symmetrical temperature profile of decreased half-width (Fig. 13). (These profiles are symmetrical when tan Δ is plotted versus $1/T$.) It should be noted that an assumed constant r_∞ value results in much poorer simulations of the experimental data. Because tan Δ is sensitive

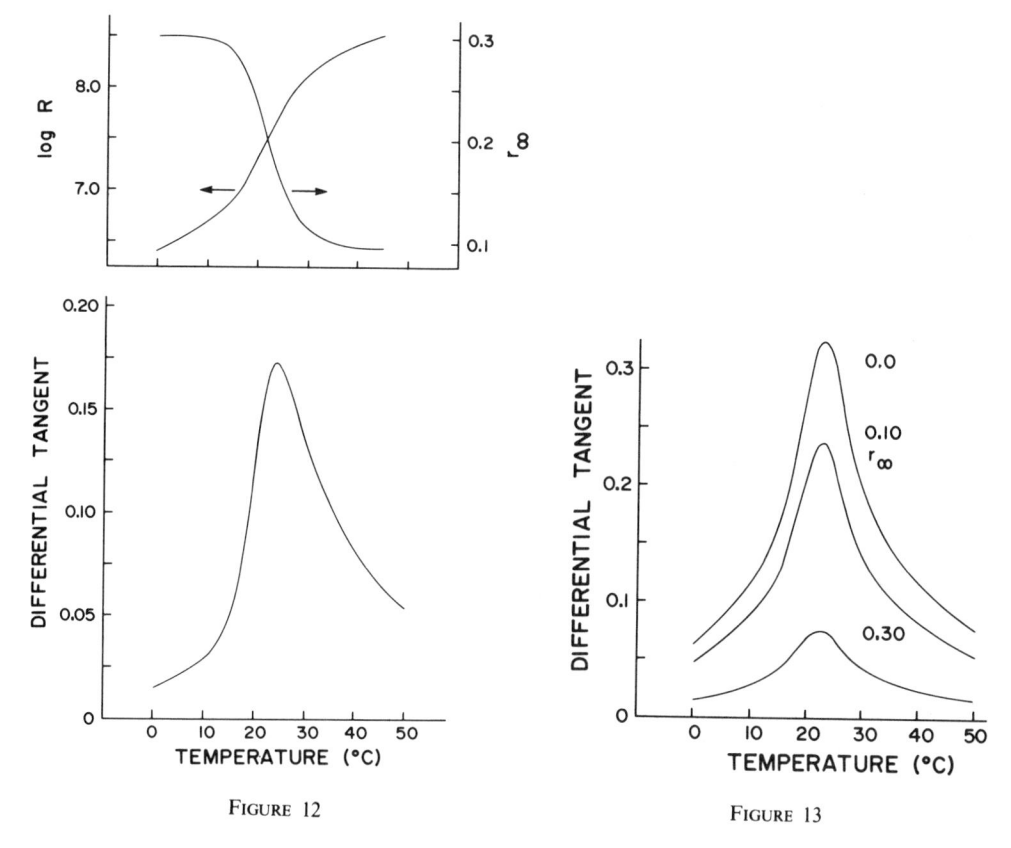

FIGURE 12 Model calculations of the differential tangent of a fluorophore in a restricted environment. The top panel shows the values of log R and r_∞ obtained under the assumptions described in Eq. 9–12. The bottom panel shows the expected differential tangent values (Eq. 6) under these conditions, and the following representative values $\omega = 2\pi \times 30$ MHz, $\tau = 10$ ns, and $r_0 = 0.392$.

FIGURE 13 Model calculations of the differential tangent of a fluorophore in a restricted environment. For these calculations we assumed that the temperature dependence of log R was identical to that given for Fig. 12. r_∞ was assumed to be independent of temperature. The r_∞ values used are shown on the figure.

to both R and r_∞ it appears likely that further experimentation and data analysis will allow us to determine more precisely the nature of the rotational motions responsible for fluorescence depolarization of DPH in lipid bilayers. Combined analysis of both steady-state and differential phase data promises to provide even greater insight.

The data we have provided illustrate the difficulties inherent in the extrapolation of DPH fluorescence polarization data to estimation of membrane microviscosity. These difficulties are determined by geometric constraints imposed on the molecule in the lipid bilayer and are unavoidable. To some considerable extent the further development and refining of a theory of hindered and/or anistropic rotations may allow us to describe better the molecular motions of this probe in membranes. However, experiments such as we have described should also be valuable in helping us to select better fluorescence probes whose depolarizing rotations in lipid bilayers are truly similar to those in homogenous solvents.

Our special thanks are due to Professor G. Weber for having supplied us with the theory of hindered rotations (as yet unpublished), and for his continuous generous support of this work.

We thank the Freshwater Biological Research Foundation, and especially its founder Mr. Richard Gray, Sr., without whose assistance this work would not have been possible. Additionally, we acknowledge the generous support of the American Heart Association, Mayo Foundation, and the National Institutes of Health grant ES GM 01238-01A1 to J.R.L. and CA 150 83-00 to F.G.P. J.R.L. is an Established Investigator of the American Heart Association.

Received for publication 8 December 1977.

REFERENCES

BASHFORD, C. L., C. G. MORGAN, and G. K. RADDA. 1976. Measurement and interpretation of fluorescence polarizations in phospholipid dispersions. *Biochim. Biophys. Acta.* **426**:157–172.

CHEN, L. A., R. E. DALE, S. ROTH, and L. BRAND. 1977. Nanosecond time-dependent fluorescence depolarization of diphenyl hexatriene in dimyristoyllecithin vesicles and the determination of "microviscosity." *J. Biol. Chem.* **252**:2163–2169.

COGAN, U., M. SHINITZKY, G. WEBER, and T. NISHIDA. 1973. Microviscosity and order in the hydrocarbon region of phospholipid and phospholipid-cholesterol dispersions determined with fluorescent probes. *Biochemistry.* **12**:521–527.

JACOBSON, K., and D. WOBSCHALL. 1974. Rotation of fluorescent probes localized within lipid bilayer membranes. *Chem. Phys. Lipids.* **12**:117–131.

KATES, M. 1972. Techniques in Lipidology. Elsevier North-Holland, Inc., New York. 355–356.

LENTZ, B. R., Y. BARENHOLZ, and T. E. THOMPSON. 1976a. Fluorescence depolarization studies of phase transitions and fluidity in phospholipid bilayers. 1. Single component phosphatidylcholine liposomes. *Biochemistry.* **15**:4521–4528.

LENTZ, B. R., Y. BARENHOLZ, and T. E. THOMPSON. 1976b. Fluorescence depolarization studies of phase transitions and fluidity in phospholipid bilayers. 2. Two component phosphatidylcholine liposomes. *Biochemistry.* **15**:4529–4537.

MANTULIN, W. W., and G. WEBER. 1977. Rotational anisotropy and solvent-fluorophore bonds: an investigation by differential polarized phase fluorometry. *J. Chem. Phys.* **66**:4091–4099.

MOORE, N. F., Y. BARENHOLZ, and R. R. WAGNER. 1976. Microviscosity of togavirus membranes studied by fluorescence depolarization: influence of envelope proteins and the host cell. *J. Virol.* **19**:126–135.

SHINITZKY, M., A. C. DIANOUX, C. GITLER, and G. WEBER. 1971. Microviscosity and order in the hydrocarbon region of micelles and membranes determined with fluorescent probes. *Biochemistry.* **10**:2106–2113.

SHINITZKY, M., and Y. BARENHOLZ. 1974. Dynamics of the hydrocarbon layer in liposomes of lecithin and sphingomyelin containing dicetylphosphate. *J. Biol. Chem.* **249**:2652–2657.

SHINITZKY, M., and M. INBAR. 1974. Difference in microviscosity induced by different cholesterol levels in the surface membrane lipid layer of normal lymphocytes and malignant lymphoma cells. *J. Mol. Biol.* **85**:603–615.

SPENCER, R. D., and G. WEBER. 1969. Measurements of subnanosecond fluorescence lifetimes with a cross-correlation phase fluorometer. *Ann. N.Y. Acad. Sci.* **158**:361–376.

SPENCER, R. D., and G. WEBER. 1970. Influence of Brownian rotations and energy transfer upon the measurements of fluorescence lifetime. *J. Chem. Phys.* **52**:1654–1663.

WEBER, G. 1977. Theory of differential phase fluorometry: detection of anisotropic molecular rotations. *J. Chem. Phys.* **66**:4081–4091.

NOTE ADDED IN PROOF

We have recently determined that one can combine steady-state anisotropy measurements with the differential phase measurements to obtain unique solutions for both R and r_∞. (Lakowicz and Prendergast, *Science*, in press). These calculations indicate that DPH behaves as a highly hindered rotator below the lipid's transition temperature ($r_\infty \simeq 0.33$), and that these rotations become less hindered above this temperature ($r_\infty \simeq 0.05$). The rotational rate R does not change dramatically at the transition temperature. However, because of the high value of r_∞ below the transition temperature, the errors in the calculated values of R are large below T_c. Overall these observations indicate that the observed steady-state anisotropy values are determined primarily by the degree to which the rotations are hindered, and not by the fluorophore's rotational rate. We conclude that the assumption that the depolarizing rotations of DPH in isotropic solvents are the same as in lipid bilayers is not correct, and hence advise caution in the extrapolation of anisotropy measurements to the membrane's microviscosity.

DISCUSSION

GEACINTOV: The theory of differential phase fluorometry used in your paper applies to the case of a single exponential decay. We have seen in the previous paper, however, the decay of fluorescent molecules in liposomes described by several exponentials (Badea et al., page 197). How does your analysis apply to such cases of nonexponential decay?

LAKOWICZ: We have addressed ourselves to that question and have several answers. Firstly, our decays are not as nonexponential as those in the previous paper. The decays of DPH and lipid bilayers are generally quite a bit closer to a single exponential. Secondly, we observe data at two different frequencies, both 10 and 30 MHz, and we have found precise agreement in the calculated values of the limiting anisotropies (r_∞) at these two frequencies. We feel, but are not certain at this time, that this agreement indicates the appropriateness of a harmonic method in this case. Thirdly, only small differences are observed in the lifetimes measured at these two frequencies by both the phase shift and the demodulation method. The maximum difference we have observed is 1.7 ns between two different frequencies. The differences do not, when propagated into the final results, significantly affect the results compared to the errors in the differential tangent measurements.

GEACINTOV: I am satisfied that in your case the analysis is probably right, but what about other situations in which there is nonexponential decay? Would it be possible to modify the theory to take such cases into account?

LAKOWICZ: Yes. We are presently working on a method that gives an independent determination of the time-resolved decays of anisotropy without actually time-resolving the decays. I think this method will be very powerful for complex decays. We measure steady-state anisotropies in the lifetime domain, rather than in the time domain. We vary the lifetime by oxygen quenching. This allows us to investigate primarily the short side of the decay, and to collect very precise data. The data we collect are the Laplace transforms of the time-resolved decay. This independent method yields results consistent with our harmonic method. Additionally, data collected at both 10 and 30 MHz agree. This result indicates our harmonic method is appropriate.

PRENDERGAST: There is an alternative possibility also, in that a theoretical treatment is now being developed to permit analysis of heterogeneous decays by the phase-modulation technique. With the availability of three frequencies, we might be able to determine individual lifetimes directly in a three-component system.

HALVORSON: For your model calculations you assumed (in Eq. 11) that the equilibrium anisotropy could be expressed as a simple weight-average of anisotropies pertaining to the two phases. In fact, the weight averaging should be done over the product of anisotropy and intensity Since fluorescence lifetimes change significantly over the temperature range of these experiments, implying similar changes in intensity, would it not have been worthwhile to incorporate this in the simulations?

LAKOWICZ: Yes, I think we should have taken that into account in our calculations. When the present manuscript was written, we were interested in obtaining a simple model to explain the general temperature profiles of the differential tangents. In the light of the present possibility to solve uniquely for the limiting anisotropies (r_∞), we feel that the curve fitting, which as you point out would be more appropriately done with the proper weighted averages, will not be necessary.

MANTULIN: As the lipids pass through their transition temperature, residual pockets of crystallinity exist. How well do R and r_∞, as measured by differential polarized phase fluorometry, reflect such heterogeneity in the lipid bilayer?

LAKOWICZ: That is a difficult question, because we are all faced with the same experimental limitations, that we determine the average angular distribution of the molecule only at times long compared to the fluorophore's lifetime.[2] At this time I do not know whether the average angular distribution results from one or two populations. With this angular distribution being the experimental observable, I do not know if such a heterogeneity can be reached. We do have one indication that resolution of a heterogeneous population may be possible. We have seen the resulting partial crystallization differences in the 10 or 30 MHz data in a system that become heterogeneous as a result of temperature. This system is mineral oil at −6°C. I hasten to point out until we have made calculations or studied the effect of frequency from a heterogeneous population, I am not convinced that it will yield any information whatsoever.

PRENDERGAST: Just to comment a little bit further on that. You'll realize the problem if you recall the profiles that we obtained for cholesterol at a ratio 3:1, meaning three phospholipid molecules to one cholesterol molecule. You can see immediately from that broad and essentially structureless profile that we are going to have problems. If you assume a two-to-one relationship for the interaction of phospholipid and cholesterol, then there should be some islands in the bilayer rich in, and some islands free of, cholesterol. We might have expected at least a biphasic curve; but we didn't get such a result. It is clearly going to be difficult to study heterogeneous systems.

MANTULIN: Could you try some kind of analysis looking at the band width, for instance, of your differential tangent graphs?

LAKOWICZ: I am not sure at this time that we will be able to do that. There are some indications

---

[2]See note 1.

that we may. In particular, all of the phospholipids at the same temperature, when they are all above their transition temperatures, behave precisely in the same manner, i.e. the limiting anisotropies are the same and the rotational rates of the fluorophore are the same. Below the phase transition temperature they also behave alike, in that the rotations are highly hindered.

VANDERKOOI: The refractive index of lipids is different from that of water. How will this affect the measurement of polarization? I am not asking about the total refractive index of the solution, but rather the refractive index of the lipid bilayer itself. What about volume changes in the vesicle?

LAKOWICZ: I don't know how much the refractive index changes at the transition temperature. However, I feel that the effect of refractive index on polarization measurements will be the same for a photon leaving a bilayer as for a photon passing through another bilayer in the same solution. Therefore effects of refractive index should be revealed in a phospholipid concentration dependence on the observed anisotropy. At high concentrations a photon leaving one bilayer will still have to pass through other bilayers and, in general, at the low phospholipid concentrations that we use there is no concentration dependance. If you feel that there is something special about the photon leaving the initial bilayer, I am not prepared to answer your question in that regard.

PRENDERGAST: With respect to the second part of your question, about the volume change in the lipid bilayer of liposomes or of vesicles, I don't think it is likely to be very large. If I remember correctly, Trauble and Haynes (*Chem. Phys. Lipids.* 1971. 7. 324.) calculated the volume change ($\Delta V/V$) at the phase transition temperature to be only 1.4% for lipid bilayer lamellae prepared from dipalmitoyl phosphatidylcholine.

VANDERKOOI: I think it would be considerably larger than that, and the question is not the property of the whole solution. If the light hits a curved surface of the bilayer, the light will be refracted within the bilayer and therefore the polarization will be changed. And how will this affect the result?

PRENDERGAST: Do you have any figures to tell me what the change in angular curvature of the surface would be?

VANDERKOOI: No, I don't.

PRENDERGAST: I have thought about this question and believe that given the small volume change in the bilayer, there will be no substantial difference in curvature of the vesicle below and above the phase transition temperature.

KUCHAI: You made a very good argument that because of the hindered rotation of DPH and because of its anisotropy, it is not a very good indicator of bulk viscosity. I would like to know if you have any candidates for a fluorescent probe that will give us information more analogous to bulk viscosity? Which of the available "membrane probes" with favorable spectral properties is also likely to show isotropic motion in lipid bilayer membranes?

LAKOWICZ: We have investigated the diffusional rate of molecular oxygen in the bilayers, above and below their phase transition. As observed by DPH anisotropy, there is a large change in microviscosity. Our data (Lakowicz and Weber, unpublished observations) indicate a change in microviscosity, (i.e. the microviscosity as calculated by the oxygen diffusivity) of no more than a factor of two, and perhaps less. The point of mentioning this is that I feel every microviscosity measurement is very strongly tied into the exact process being measured. I'll give you another example. We (Lakowicz, Hogan, and Omann. 1977. *Biochim. Biophys. Acta.*) have been looking at the diffusion of the small molecule hexachlorocyclohexane, a common pesticide, in phospholipid bilayers. Again, this measurement does not indicate a microviscosity change at the transition temperature as great as observed by diphenylhexatriene. The second part of your question is whether I can suggest a more appropriate probe. No, I cannot, because so many properties must be taken into account, such as lifetime, spectral properties, and polarization spectra. No single probe may ever be appropriate.

MENDELSON: I respond to an earlier question about nonexponential fluorescence depolarization. We have numerically simulated the (time-resolved) problem of a rotor moving freely within a cone-shaped barrier (Mendelson and Cheung. 1978. *Biochemistry*. In press). When the dipoles are oriented off the rotor's major axis, multiexponential behavior is observed in both the constrained and unconstrained cases. In the constrained case this behavior arises from anisotropic motion as well as the presence of the barrier. These calculations might be useful for interpreting the kinds of experiments presented here.

LAKOWICZ: The question of the simulation beings to my mind a very important point. I hoped someone would ask this question, which relates to the meaning of the cone angle. The meaning of this cone angle confused me for close to a week, and I presume it would confuse some other people. There are two common definitions for the cone angle, that is the angle through which a rotator can diffuse. One is that given by Kawato and Kinosita. These workers define the cone angle to be the angle at which the fluorophore strikes an infinite potential barrier. For a free rotator, for which there is no energy barrier, this angle is 90°. The definition of the cone angle that we are using is subtly different from the former definition. We define the cone angle to be the average angular distribution of the probe at times long compared to the fluorescence lifetime. Now, this is experimentally observable and therein lies the reason we think this is preferable to the angle used by Kawato. In our case we measure a limiting anisotropy (r_∞). It is experimentally observable, related to the average angular distribution at times long compared to the fluorescence lifetime, and requires no assumptions concerning the form of the potential well. The other angle defined by Kawato et al. takes an experimental observable and interprets this angle in terms of an infinite potential well mode model. This is, of course, valid only if the assumed model is correct. Our interpretation of r_∞ makes no assumptions concerning the shape of the potential well: $r_\infty/r_0 = \frac{1}{2}\cos\theta_c(1 + \cos\theta_c)^2$ (Kawato and Kinosita); $r_\infty/r_0 = (3\cos^2\theta - 1)/2$ (Weber et al.).

BADEA: The 25% upper limit for the tangent deflect reflecting anisotropic motion is obtained only by considering a hydrodynamic diffusional model. This model implies small and frequent angular jumps. In the liposome, where the "solvent" molecules are larger than the fluorophore, the angular jumps are likely to be neither small nor frequent. Finite angular jump rotational models have been described in the literature. Would the authors comment on the upper limit values for the tangent deflect in these cases? In the event that some of those models could yield tangent deflect values up to 60%, what would be the basis for preferring the hindered rotation model?

LAKOWICZ: First of all, we do not know whether the other models will indeed yield the tangent deflect greater than 25% resulting from rotational anisotropy. My own impression is that they will yield similar tangent deflects. But if they yield greater deflects, I think there are independent observations of the anisotropies measured as a function of fluorescence lifetime. We have measured steady-state anisotropies as the fluorescence lifetime is decreased by oxygen quenching. I point out that these results are in precise agreement with the results presented here. The differential phase techniques concentrate the measurements at the larger rotational angles, whereas the oxygen quenching measurements concentrate the information content at the smaller rotational angles. This is circumstantial evidence for the propriety of the model we employed. But it is an independent observation.

BADEA: So in effect you need an independent measurement to ascertain the validity of this particular model of interpreting your data, which is oxygen quenching in your case?

LAKOWICZ: No, we do not need an independent model; I added that as supporting evidence. I think in both cases we have the same experimental observable—the average distribution at long times—and this is all we can ever observe. I don't know if it is experimentally detectable whether that results from short rotations at small angles or bigger ones at larger angles. Additionally, we do not know that jump models could yield larger tangent deflects.

PRENDERGAST: One of the questions implicit in your question is the issue of the best model for the membrane motion, i.e. for the motion of the phospholipid tails. I would like to take a different viewpoint from yours and say that we can't be sure at the present time which of the several models proposed to date is most accurate; indeed, is any single model valid? But it may be that an "angle-jump" model and a migrating "kink" model would give different patterns of probe movement. One thing is certain, that the rate of kink migration across the lipid chain could easily be very rapid. Such rapid movement would make your suggestion less of a problem for our model.

KING: We have been working on similar problems with NMR data and have found that mathematically, up to the first order, the angular jump model and the hindered rotation model with the angle representing the standard deviation of the fluctuations of that rotation give identical results. There is no way to distinguish the two models.

JAKOBSSON: How slow can hindered rotation of molecules in bilayers be? In particular, can it possibly be as slow as the millisecond time scales in which we are interested?

LAKOWICZ: I don't know, because all of our measurements are limited to times shorter than three fluorescence lifetimes. One would have to go to alternative, longer-lived, probes and also attempt to fix the liposomes themselves from rotation, which should be possible. A different choice of probes could extend the observable time range.

PRENDERGAST: To answer the question further it is required that we understand exactly what is relaxing in the membrane and where, because the relaxation times for moieties located at different positions across the bilayer will all differ quite markedly. Most relevant to an electrophysiologist, however, is the relaxation of proteins in the membrane, for example, the relaxation of ion channels. If you look at relaxation data for gramicidin channels, you may get a pretty good idea what relaxation times are for small proteins in membranes. I don't think, however, that we can realistically extrapolate from the behavior of a simple probe, a small molecule lodged in the middle of the membrane, to the relaxation times expected of very large molecules such as proteins. Indeed, extrapolation is difficult even to relatively small molecules, such as the commonly used ionophores rotating in lipid bilayers.

EXTENDED ABSTRACTS

THE APPLICATION OF SELECTIVE EXCITATION DOUBLE MÖSSBAUER TO TIME-DEPENDENT EFFECTS IN BIOLOGICAL MATERIALS

BOHDAN BALKO, *National Heart, Lung, and Blood Institute,*
Laboratory of Technical Development, Bethesda, Maryland 20014 and
EUGENIE V. MIELCZAREK, *George Mason University,*
Fairfax, Virginia 22030 U. S. A.

This paper is concerned with the development and application of selective excitation double Mössbauer (SEDM) techniques to the study of time-dependent hyperfine interactions in biological systems. The Mössbauer effect (ME) is the recoilless emission and absorption of nuclear resonant radiation. Usually these are γ-rays of energy from 10 to 100 keV. Under special and quite well understood conditions the γ-ray is emitted and absorbed without loss of energy and is highly monochromatic (line width for ^{57}Fe ~10^{-9} eV). All Mössbauer experiments are based on the utilization of this extremely well-defined energy to measure small energy differences. It is a convenient coincidence that iron, important biologically, also has the most suitable isotope, ^{57}Fe, for Mössbauer spectroscopy. This coincidence allows us to study heme proteins, iron sulfur proteins, iron transport and storage proteins, and other important systems.

The spectra of biological materials very often are quite complex and difficult to interpret by the standard Mössbauer techniques. SEDM will help us resolve some of the complex spectra and allow us to obtain more specific information not available from the usual ME investigations. The SEDM apparatus requires two Doppler modulators. The first one drives a single-line Mössbauer source at a constant velocity, thereby inducing a transition of the nuclei in the scatterer to a predetermined excited substate. The other drive moves a single-line absorber and is used in the usual transmission geometry to analyze the scattered radiation. In this way we measure the nuclear resonant differential scattering cross-section of a material. The other Mössbauer techniques usually measure the absorption cross-section.

We have used SEDM to obtain new information on several inorganic materials

---

Dr. Mielczarek is presently assigned to Laboratory of Technical Development, National Heart, Lung, and Blood Institute, under IGPA.

(1–4). We have also developed theoretical expressions to calculate the effect of electronic relaxation on the SEDM spectra. In this paper we will present the results of a SEDM investigation of such processes in ferrichrome A (5).

REFERENCES

1. BALKO, B., and G. R. HOY. 1974. *Phys. Rev. B.* **10**:36.
2. BALKO, B., and G. R. HOY. 1976. *Phys. Rev. B.* **13**:2729.
3. BALKO, B., and G. R. HOY. 1977. *Physica.* **86–88B**:953.
4. BALKO, B., and G. R. HOY. 1976. *J. Phys. Paris Colloq. C6. Suppl. 12.* **37**:89.
5. WICKMAN, H. H. 1968. *In* Mössbauer Effect Methodology. I. J. Gruverman, editor. Plenum Publishing Corporation, New York. Second edition. 316.

ON MAGNETICALLY INDUCED TEMPERATURE JUMPS

GEORGE H. CZERLINSKI, *Department of Biochemistry,*
Northwestern University, Chicago, Illinois 60611 U. S. A.

Any generator of an alternating magnetic field also produces an alternating electric field. These fields may interact directly with ionic solutions or via (ferro-electric and/or -magnetic) mediators. The (local) energy density then has four contributing terms:

$$_xQ'_a = {_xQ'_{H,L}} + {_xQ'_{E,L}} + {_xQ'_{H,M}} + {_xQ'_{E,M}}. \tag{1}$$

The various symbols in this equation—as well as in all later equations—are defined in the Glossary of Symbols (Table I). The subscript before Q refers to the x-axis, the direction of the propagation of the field. Fig. 1 below describes the simplified model used and defines the directions of the electric and magnetic field vectors.

At the boundary plane the magnetic field strength is defined by:

$$H_Z(0, t) = H_0 \sin \omega t. \tag{2}$$

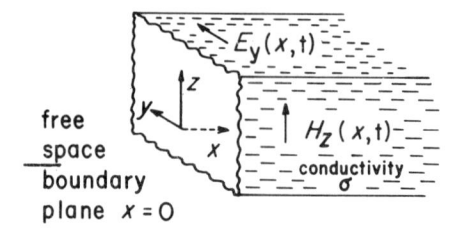

FIGURE 1 Set of definitions of an electromagnetic wave, propagating in the x-direction, and entering a "slab" at $x = 0$.

TABLE I
GLOSSARY OF SYMBOLS

Symbol	Units	Description	Equation
$_xQ'$	Jm^{-3}	Energy density in x direction	1
H_Z	Am^{-1}	Magnetic field strength in z direction	2
H_0	Am^{-1}	Amplitude of magnetic field strength	2
ω	s^{-1}	Angular frequency	2
t	s	Time of observation	2
δ_e	m	Electric skin depth	3
μ	$Vsm^{-1}A^{-1}$	Magnetic permeability	3
E_0	Vm^{-1}	Amplitude of electric field strength	4
σ	$AV^{-1}m^{-1}$	Electric conductivity	4
D_e	m^2s^{-1}	Diffusion constant of electric field	5
l	m	Effective dimension of inductor-coil	7
λ	m	Wavelength of alternating field	7
ϵ	$Asm^{-1}V^{-1}$	Electric permittivity	8
D_h	m^2s^{-1}	Diffusion constant of heat	11
δ_h	m	Thermal skin depth	11
δ_a	m	Absorption depth	12
κ'	—	Compound loss coefficient	14
κ_3	—	Loss coefficient of magnetization	16
$c_{p,L}$	$Jdeg^{-1}m^{-3}$	Specific heat of liquid component	17
$c_{p,M}$	$Jdeg^{-1}m^{-3}$	Specific heat of particulate material	17
κ_1	—	Loss coefficient of dilution	17
V_L	m^3	Volume of liquid component	17
V_M	m^3	Volume of particle material	17
κ_2	—	Loss coefficient of frequency match	18
τ	s	Relaxation time of heat diffusion	18
r_M	m	Radius of magnetic particle	19
d_{LNK}	m	Half-width of cell with suspension	20
Q''	Jm^{-3}	Loss energy density	21
a	m	Radius of cell with suspension	22
r	m	Radius direction in cylinder	22
$_\phi Q'$	Jm^{-3}	Energy density in cylinder	23

Symbols are aligned in sequence of appearance with reference equation given in the last column.

The stationary state expressions for the first two terms are given by (1–4):

$$_xQ'_{H,L} = \tfrac{1}{2}\mu H_0^2 \exp(-2x/\delta_e)\omega t, \quad (3)$$

$$_xQ'_{E,L} = \tfrac{1}{2}\sigma E_0^2 \exp(-2x/\delta_e)t. \quad (4)$$

The electrical skin depth in these equations is defined by:

$$\delta_e = \sqrt{2D_e\omega^{-1}}, \quad (5)$$

with the electrical diffusion constant:

$$D_e = (\sigma\mu)^{-1}. \quad (6)$$

Eqs. 3 and 4 imply the following conditions:

$$\lambda \gg l, \tag{7}$$

$$\omega^{-1} \gg \epsilon/\sigma, \tag{8}$$

$$t \gg (2\omega)^{-1}, \tag{9}$$

$$\delta_e \gg \delta_h, \tag{10}$$

where

$$\delta_h = \sqrt{2D_h\omega^{-1}}. \tag{11}$$

The last two terms in Eq. 1 refer to energy absorption by a mediator. One obtains under the same conditions stated:

$$_xQ'_{H,M} = \tfrac{1}{2}\mu H_0^2 \exp(-x/\delta_a)\omega t, \tag{12}$$

$$_xQ'_{E,M} = \tfrac{1}{2}\epsilon E_0^2 \exp(-x/\delta_a)\omega t. \tag{13}$$

The absorption depth for condition 7 is given by:

$$\delta_a = (\partial \kappa'/\partial x). \tag{14}$$

If condition 7 were not fulfilled, the absorption depth is defined by (5):

$$\delta_a = \lambda(4\pi\kappa')^{-1}. \tag{15}$$

However, condition 7 is fulfilled and Eq. 14 thus applies.

Parameter κ' consists of a product of three terms:

$$\kappa' = \prod_{i=1}^{3} \kappa_i. \tag{16}$$

The first two terms are the same for electric and magnetic fields:

$$\kappa_1 = \frac{c_{p,M} V_M}{c_{p,M} V_M + c_{p,L} V_L}, \tag{17}$$

$$\kappa_2 = \frac{\omega\tau}{1 + \omega^2\tau^2}, \tag{18}$$

with the dissipative time constant given by (6):

$$\tau = \frac{2r_M^2}{\pi^2 D_h}. \tag{19}$$

The third term represents the ratio of two energy (density) terms. The numerator contains the energy reversibly absorbed by the mediator particles, the denominator the energy supplied by the field. Mediator absorption has thus far only been described for magnetic fields (6, 7).[1]

---

[1] Czerlinski, G. H. In preparation.

If d is the width of the test cell with the particulate suspension, the exponents in Eqs. 12 and 13 become 0.01 κ' for $x = d = 0.01$ m in zero approximation. The lost energy density may easily be derived from Eq. 12 with the condition:

$$2x \leq 2d < l \ll \lambda; \kappa' \ll 1, \tag{20}$$

One obtains for the lost energy density:

$$_xQ''_{H,M} = \tfrac{1}{2}\mu H_0^2 \kappa' 2\omega t. \tag{21}$$

For cylindrical geometry the conditions of Eq. 20 become:

$$2r \leq 2a < l \ll \lambda \text{ and } \kappa' \ll 1. \tag{22}$$

One obtains then for the *lost* energy density:

$$_\phi Q''_{H,M} = \tfrac{1}{2}\mu H_0^2 \kappa' \tfrac{1}{2}\omega t. \tag{23}$$

Quite similar relations are obtained when the lost energy density is computed for systems without particulate mediators. Eqs. 3 and 4 are utilized with conditions 20 and 22, respectively.

Table II presents the results of computer simulations based on parameter values derived from the literature. It is apparent that the magnitude of the field strength determines which one of the terms in Eq. 1 predominates. The absorption by ferroelectric mediators is judged unrealistic and therefore not considered in Table II.

TABLE II
FIELD DIFFUSION LOSSES IN MATTER

Medium	Field amplitude Electric	Field amplitude Magnetic	ΔT for one term only Electric field	ΔT for one term only Magnetic field	ΔT for one term only Magnetic particles
	V/m	A/m	°C		
0.1 M NaCl,	0.16	0.16	0.000028	0.00011	0.65
10% suspension	1.16	1.6	0.0028	0.011	3.77
	16	16	0.28	1.07	10.5
0.01 M NaCl,	0.16	0.16	2.8×10^{-7}	0.000011	0.65
10% suspension	1.6	1.6	0.000028	0.0011	3.77
	16	16	0.028	0.107	10.5
0.1 M NaCl,	0.16	0.16	0.000028	0.00011	0.065
1% suspension	1.6	1.6	0.0028	0.011	0.377
	16	16	0.28	1.07	1.02

The suspensions consist of particles with nickel-like properties (and Curie temperature at room temperature, 293°K) of optimum diameter for 0.3 MHz field oscillation. The pulse width is 10^{-4} s, the total cell width $d = 0.01$ m, the simple case denoted in Fig. 1. The temperature rises ΔT refer to the isolated cases ("as if only one loss-process is present") and are obtained from the appropriate loss energy densities (Q'') through dividing them by the specific heats (c_p) involved.

Particulate mediators are not useful at very high field strengths, at very low concentrations, or in a poor match between angular frequency and dissipative time constant (and thus particle size, Eq. 19).

The introductory Eq. 1 treats the individual contributions as additive terms. This is valid only as long as the input energy is insignificantly changed along x. If there is both strong absorption due to electrolyte conductivity and particle mediation, the (local) magnetic energy density is given by:

$$_xQ'_{H,M} = \tfrac{1}{2}\mu H_0^2 \exp(-2x/\delta_e)\exp(-x/\delta_a)\omega t. \tag{24}$$

This last equation is thus more general than either Eq. 3 or Eq. 12.

REFERENCES

1. KNOEPFEL, H. 1970. Pulsed High Magnetic Fields. North-Holland Publishing Co., Amsterdam.
2. GUY, H. W., J. F. LEHMANN, and J. B. STONEBRIDGE. 1974. *Proc. IEEE.* **62:**55-75.
3. SCHWAN, H. P. 1965. *In* Therapeutic Heat and Cold. S. Licht, editor. Licht Publishing Co., New Haven, Conn.
4. COLE, K., and R. COLE. 1941. *J. Chem. Phys.* **9:**34.
5. ROSSI, B. 1959. Optics. Addison-Wesley Publishing Co., Inc., Reading, Mass.
6. CZERLINSKI, G. H. 1977. *Bull. Am. Phys. Soc.* 43.
7. CZERLINSKI, G. H. 1977. *Biophys. J.* **17:**301a. (Abstr.).

KINETIC AND TRANSIENT ELECTRIC DICHROISM STUDIES OF THE IREHDIAMINE-DNA COMPLEX

N. DATTAGUPTA, M. HOGAN, AND D. CROTHERS, *Yale University, New Haven, Connecticut 06472 U. S. A.*

We have used transient electric dichroism and temperature-jump relaxation kinetics to characterize the complex formed between irehdiamine A (preg-5-ene-3β-20α-diamine, IDA) and DNA from a variety of sources. Transient electric dichroism was used to monitor the change in DNA length and base tilt angle due to complex formation. A 5% length increase at saturation with IDA was observed for DNAs from *M. luteus* (70% G·C), *E. coli* (50% G·C), and *C. perfringens* (30% G·C), and also for poly dA·poly dT. The base ultraviolet transition moments in the complex are inclined at an average angle of about 60° to the helix axis. These properties are consistent with the β-kinked B-DNA structure proposed for the complex by Sobell et al. Two DNAs from eukaryotic sources (human placenta and calf thymus) showed a 13% length increase at saturation, whereas poly dG·poly dC showed no length increase. The base tilt angle in the complex was found to be independent of the DNA source.

Temperature-jump relaxation times for the DNA-IDA complex were generally faster than 1 ms, and showed a concentration dependence consistent with a simple bimolecular reaction mechanism. The measured forward reaction rate constants ($10^7 - 10^8$ $M^{-1}s^{-1}$) are too large to permit a mechanism in which IDA reacts with pre-existing kinks that occur transiently at a small fraction (less than 10^{-2}) of the DNA base pairs. The equilibrium constant calculated from the ratio of the forward and reverse binding rate constants agrees well with results obtained from equilibrium dialysis. Although the extent of hyperchromicity induced in DNA due to IDA binding is similar for DNAs from calf thymus and *M. luteus,* the activation energies and reaction enthalpies are clearly different. For example, the binding enthalpy change is 14 kcal mol^{-1} for *M. luteus* DNA, and only about 4 kcal mol^{-1} for calf thymus DNA.

Our observations are generally consistent with the β-kinked structure for the complex as proposed by Sobell et al., but differences, presumably sequence-dependent, must exist in the detailed mode of binding to DNAs from different sources.

REFERENCE

SOBELL, H. M., S. C. JAIN, C. TSAI, and S. G. GILBERT. 1977. *J. Mol. Biol.* **114**:333–365.

OPTICAL DETECTION OF COMPRESSIBILITY DISPERSION

RELAXATION KINETICS OF GLUTAMATE DEHYDROGENASE SELF-ASSOCIATION

HERBERT R. HALVORSON, *Biochemistry and Microbiology, Edsel B. Ford Institute for Medical Research, Henry Ford Hospital, Detroit, Michigan 48202 U. S. A.*

Pressure perturbation techniques of chemical relaxation exploit the augmented compressibility arising from the finite ΔV for the chemical process being observed. Compared to temperature jump, pressure jump suffers from a greatly decreased sensitivity of the chemical system to the perturbation. A sinusoidal pressure perturbation (traveling sound wave) allows phase-sensitive detection and time-averaging of the response ("stationary method" of Eigen and deMaeyer), thereby enhancing the signal-to-noise ratio. A finite relaxation time manifests itself as a phase shift of the response relative to the perturbation, and the signal can then be decomposed into the in-phase and quadrature components. The quadrature or imaginary part of the signal is directly proportional to $\omega\tau/\{1 + (\omega\tau)^2\}$, where ω is the angular frequency and τ is the relaxation time. Measuring the frequency dependence (dispersion) of the imaginary compressibility reveals a maximum at $\omega = 1/\tau$. Measurements of the real component at frequencies remote from the relaxation region (both above and below) permit the determination of ΔV for the process in question.

An instrument has been built to measure the kinetics of subunit self-association reactions by monitoring the intensity of light scattered at 90° (proportional to weight-average molecular weight). Although relative volume changes are small (change in partial specific volume is typically 10^{-3} ml/g), the relevant thermodynamic parameter is the difference between the molar volumes of products and reactants. These differences tend to be large (50-100 ml/mol of subunit), making this experimental approach feasible. Sinusoidal pressure perturbation of 3 atm amplitude is achieved with a stack of piezoelectric ceramic elements. Lamp noise and static light scattering are removed by subtracting the output from a beam splitter-reference photomultiplier tube combination. The signal is then processed by a home-built phase-sensitive detector and gated integrator. The instrument displays the real and quadrature components of the light-scattering signal as the frequency is varied from 0.0005 ($\tau \simeq 300$ s) to 30 kHz ($\tau \simeq 5$ μs). Frequencies above 150 Hz ($\tau < 1$ ms) are not currently accessible.

Measurements on the kinetics of the indefinite linear self-association of bovine liver glutamate dehydrogenase will be compared with literature values

This work was supported by National Institutes of Health grant GM 23302.

CREATION OF A NONEQUILIBRIUM STATE IN SODIUM CHANNELS BY A STEP CHANGE IN ELECTRIC FIELD

Eric Jakobsson, *Department of Physiology and Biophysics and Program in Bioengineering, University of Illinois, Urbana, Illinois 61801 U. S. A.*

Voltage clamp of electrically excitable membranes can be considered an E jump, in which the macromolecules associated with the specific ion conductance changes are subjected to a sudden change in electric field of up to approximately 10^7 V/m. If the conductance and activatability of the channels are taken as measures of the populations of different states of the gating molecules, then at any constant voltage the potassium conductance gates are distributed between two states (conducting and nonconducting), while the sodium conductance gates are distributed among three states (conducting, activatable, and inactivated). Kinetic analysis of the sodium conductance changes, taking into account the phenomena of inactivation shift and τ_c-τ_h separation (anomalous with respect to the Hodgkin-Huxley model), suggests that a fourth state, which has no population at rest, may be the direct precursor of the conducting state (1). In this paper a physical model of the fourth state is suggested. The fourth state may consist of a non-Boltzmann distribution of gating molecule conformations and orientations relative to the energy minima that characterize the steady state at any particular value of average electric field. This state might be created when the electric field changes too rapidly for gating molecule orientation and conformation to

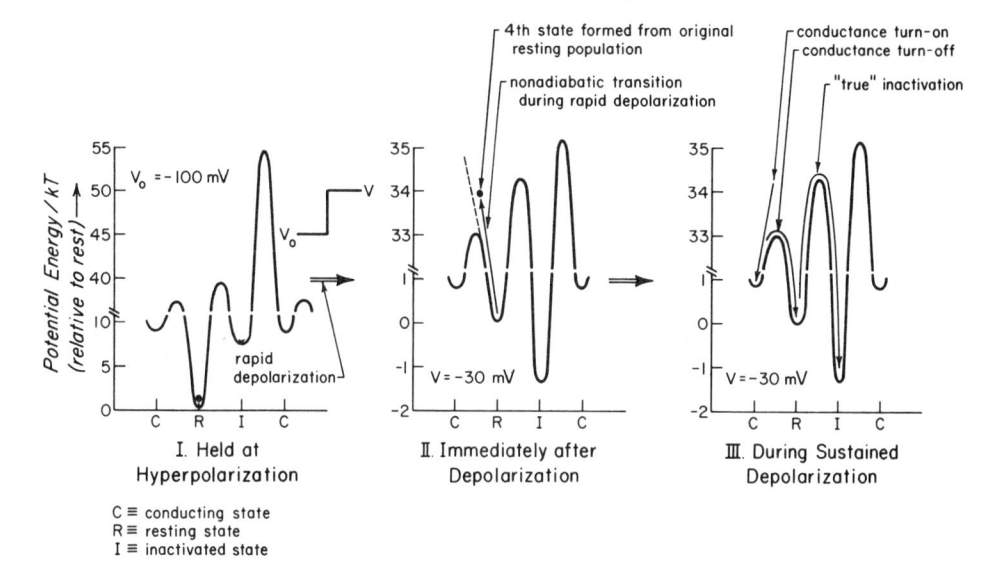

FIGURE 1 Energy-level diagram showing heights of potential barriers, depths of potential wells, and dominant transitions (nonadiabatic during rapid voltage change and adiabatic at constant voltage), hypothesized to be associated with the response of an axonal membrane to a step voltage depolarization.

follow, and would be similar to the state of "nonequilibrium dielectric polarization" postulated to be involved in electron-transfer reactions (2). Carrying the analogy with the existing electron-transfer formalism a step further, the creation of the fourth state is hypothesized to be a "nonadiabatic transition," which occurs during the brief time that the electric field is changing very rapidly.

Fig. 1 illustrates by energy level diagrams the sequence of events hypothesized in the gating mechanism of an ensemble of sodium channels in response to a depolarizing voltage step from -100 to -30 mV. The heights of the energy barriers and depths of the energy wells are consistent with rate constants fit to the kinetics of giant axon sodium channels (1), together with the observation that the Q_{10} of sodium channel kinetics is about three (3).

The arrows in sections II and III of Fig. 1 show the dominant transitions during rapid voltage change, and during a constant voltage after a sudden depolarization, respectively. The kinetic properties of the fourth state (i.e., the rate constants for transitions from it) will be a function of both V_0 and V for the voltage step shown above. The following differential equations include terms for other transitions between states in addition to the dominant ones represented in Fig. 1, but neglect transitions within the R-state:

$$dC/dt = -(k_{CI} + k_{CR})C + k_{IC}I + k_{R(V_0)C}R(V_0) + k_{R(V)C}R(V), \quad (1)$$

$$dI/dt = -(k_{IC} + k_{IR})I + k_{CI}C + k_{RI}R(V), \quad (2)$$

$$dR(V_0)/dt = -k_{R(V_0)C} R(V_0), \tag{3}$$

$$dR(V)/dt = k_{CR}C + k_{IR}I - (k_{RI} + k_{R(V)C})R(V). \tag{4}$$

Where all the k's are voltage-dependent rate constants, and the $R(V_0)$ and $R(V)$ denote those populations of the resting state that have appropriate resting conformations for V_0 and V, respectively. (The excited state at V is presumed to have the same conformation as the resting state at V_0.) For a single voltage step, in which the voltage change may be assumed to be instantaneous (i.e., fast compared to any other processes), the kinetic description of the system by Eqs. 1–4 is (relatively) simple. On the other hand, if the voltage is changing continuously (as it does in the normal functioning of the animal) then we have a continuous distribution of conformations of the R-state (i.e., a continuous distribution of "fourth states"), with each element of the distribution having its own rate constant for emptying. To write the equations for the case of continuously varying voltage so as to be amenable to numerical computation, it is necessary to divide the continuous distribution of R-state conformations into a finite number of sections. If we label each element of the R-state according to the voltage for which it has the steady-state minimum energy conformation, then we can create the discrete spectrum of R-states by:

$$R'(V') = \int_{V=V'-\Delta/2}^{V=V'+\Delta/2} dV\, R(V), \tag{5}$$

where Δ is some small increment.

The number of $R'(V')$'s depends on the membrane history and the size of Δ. For example, if $\Delta = 1$ mV and the membrane potential has varied from -70 to $+40$ mV, then there will be 110 values of $R'(V')$, one for each integral value of V' from -70 to $+40$.

The differential equations describing time courses in the continuously varying voltage case are:

$$dR'(V')/dt = -(k_{R(V')C} + k_{RR})R'(V') + k_{RR}R'(V' + \Delta|V - V'|/(V - V'))$$
$$+ \int_{V'=V-\Delta/2}^{V'=V+\Delta/2} dV'\,\delta(V - V')(k_{CR}C + k_{IR}I + (k_{RR} - k_{RI})R'(V')), \tag{6}$$

$$dC/dt = k_{R(V')C}R(V') + k_{IC}I - (k_{CR} + k_{CI})C, \tag{7}$$

$$dI/dt = -(k_{IC} + k_{RC})I + k_{CI}C + \int_{V'=V-\Delta/2}^{V=-V'+\Delta/2} k_{RI}R'(V'), \tag{8}$$

where $\delta(V - V')$ is the Dirac delta function and k_{RR} is the rate constant for relaxation *within* the R-state, such that after a long time at one voltage the R-state population becomes concentrated at $V = V'$.

It should be noted that Eq. 6 in practice is not one differential equation, but rather an algorithm for writing $(V_{max} - V_{min})/\Delta$ simultaneous differential equations, where V_{max} and V_{min} are the range of voltages the membrane has experienced. Thus description of the sodium gating by this formalism involves the numerical integration of many more simultaneous equations than do other models used to describe this process. The need for the extra equations comes directly from the hypothesis of nonequilibrium dielectric polarization, the need for which hypothesis in turn comes directly from the observation of the τ_c-τ_h separation.

Partial support for this work was received from the Bioengineering Program of the University of Illinois and from Grant HEW PHS HL 21342 from the National Institutes of Health.

REFERENCES

1. JAKOBSSON, E. 1978. A fully coupled transient excited state model for the sodium channel. I. Conductance in the voltage clamped case. *J. Math. Biol.* In press.
2. MARCUS, R. A. 1964. Chemical and electrochemical electron-transfer theory. *Annu. Rev. Phys. Chem.* **15**:155.
3. HODGKIN, A. L., and A. F. HUXLEY. 1952. A quantitative description of membrane current and its application to conduction and excitation in nerve. *J. Physiol. (Lond.).* **117**:500.

HIGH-FREQUENCY DIELECTRIC SPECTROSCOPY OF CONCENTRATED MEMBRANE SUSPENSIONS

DONALD S. KIRKPATRICK, JOHN E. MCGINNESS, WILLIAM D. MOORHEAD, PETER M. CORRY, AND PETER H. PROCTOR, *Department of Ophthalmology, Baylor College of Medicine, and the Department of Physics, The University of Texas System Cancer Center, M. D. Anderson Hospital and Tumor Institute, Houston, Texas 77030, and the Department of Physics, Youngstown State University, Youngstown, Ohio 44555*

The interfaces between biological membranes and their aqueous environments are of special interest to biology. This is because the many substances that bind to membranes, react with membranes, or pass through membranes must first enter and interact with these interfaces. We investigated the physiological chemistry and physics of the membrane-aqueous interface by high-frequency dielectric spectroscopy of concentrated suspensions of membranes. We adapted the recently developed "time-domain" technique for dielectric measurements (1). We observed a previously unreported dielectric absorption in concentrated suspensions of membranes obtained from either the outer segments of rod photoreceptor cells of the retina or blood erythrocytes.

We observed an anomalous dielectric absorption in these membranes having a dielectric constant of about 1,000 and a characteristic frequency, f_c, of about 170 MHz. It is a discreet absorption, well separated from the closest characteristic frequencies, 1 and 20,000 MHz, previously reported in tissue or cell suspensions (2). The discreet nature of the absorption ruled out the possibility that this absorption is the tail end of one of the neighboring absorptions. It also ruled out an artifact of the conductivity contribution to dielectric loss (3).

The anomalous dielectric absorption was not observed in suspensions of erythrocyte ghosts as dilute as 1:10. It became progressively stronger as the concentration of membranes increased, and grew to a dielectric constant of about 1,000 with centrifugally packed suspensions of erythrocyte ghosts or rod outer segments. As more aqueous solution was removed from the membrane suspension, the dielectric constant dropped. Thus, with rod outer segments the dielectric constant demonstrated a maximum value at about 60% water (wt/wt), corresponding to the reported values of water content of rod outer segments in vivo. The anomalous dielectric constant of membrane suspensions varied directly with conductivity as well as with the sodium chloride content of the aqueous solution.

We investigated the temperature dependence of the anomalous dielectric constant in erythrocyte ghost suspensions. Plotted against $1/T$, the logarithm of the dielectric constant produced a straight line. The dielectric constant increased by a factor of 1.6 as the temperature increased from 4° to 36°C. This dependence was opposite to the temperature dependence of dielectric constants exhibited by dipole orientation polarizations, and it was much larger than that exhibited by Maxwell-Wagner polarizations (4). The characteristic frequency of 170 MHz, together with the temperature dependence of the dielectric constant, ruled out the Maxwell-Wagner mechanism for this polarization. The temperature dependence of the anomalous dielectric constant had the same sign and approximate magnitude as the temperature dependence of ionic conductivity. Although we showed strong correlation of dielectric constant with conductivity, previous conductivity based dielectric mechanisms have been relevant only to much lower frequencies, i.e., about 10^4 Hz (5).

The value of 170 MHz for f_c corresponds to the characteristic frequency attributed to the tightly bound first "monolayer" of water of hydration on protein (6). This correspondence, along with its dependence upon water concentration, suggests that the anomalous dielectric absorption is due to water of hydration. Its additional dependence upon salt concentration and conductivity indicates that the anomalous dielectric absorption depends upon mutual or collective interactions of the membranes, water, and ions.

The characteristic frequency of aged erythrocyte ghosts or those prepared from outdated donor blood was higher by a factor of 1.5 than the f_c of fresh membranes. Consequently, we can correlate changes in the dielectric properties with changes in structure (7) and function known to occur with age in erythrocyte membranes. Furthermore, the age dependence of the characteristic frequency suggests that dielectric measurements are relevant to membranes that constantly regenerate, e.g., photoreceptor membranes and erythrocytes.

We observed this new dielectric absorption with a variety of different types of membrane preparations, viz., erythrocytes, erythrocyte ghosts, rod outer segments, isolated photoreceptor membranes, and photoreceptor membrane vesicles. Apparently, this dielectric absorption is a general phenomenon occurring with membranes. It may have been overlooked previously because it becomes noticeable only at high cell or membrane concentrations, because of interference by the enormous Maxwell-Wagner polarization, and because of the difficulty that the high conductivity of biological samples causes with dielectric measurements.

A dielectric polarization process with a large dielectric constant associated with the aqueous environment of membranes may exert a dominant effect upon the mobilities of ions and charged functional groups near the membrane surfaces. Consequently, the dielectric process that we observed may exert a major effect upon both the structure and function of biological membranes. We can now interpret these physical measurements in terms of biochemical properties of membranes. We have shown that high-frequency dielectric measurements are physiologically relevant to active membrane systems that constantly regenerate, especially if they are closely spaced in vivo, e.g., photoreceptor membranes in rod outer segments of the retina.

This work was supported by the Retina Research Foundation, Houston, Texas.

REFERENCES

1. CLARKSON, T. S., L. GLASSER, R. W. TUXWORTH, and G. WILLIAMS. 1977. *Adv. Mol. Relax. Interact. Proc.* **10**:173–202.
2. SCHWAN, H. P. 1957. *Adv. Biol. Med. Phys.* **5**:147–209.
3. DAVIES, M. 1969. *In* Dielectric Properties and Molecular Behavior. T. M. SUGDEN, editor. Van Nostrand Reinhold Co. Ltd., London. 280.
4. SCHWAN, H. P. 1948. *Z. Naturforsch.* **3b**:361.
5. SCHWAN, H. P., G. SCHWARZ, J. MACZUK, and H. PAULY. 1962. *J. Phys. Chem.* **66**:2626–2635.
6. HARVEY, S., and P. HOEKSTRA. 1972. *J. Phys. Chem.* **76**:2987–2994.
7. ALBERTSSON, P.-A. 1971. Partition of Cell Particles and Macromolecules. John Wiley & Sons, Inc., New York. Second edition. 192.

SUBNANOSECOND FLUORESCENCE LIFETIMES BY TIME-CORRELATED SINGLE PHOTON COUNTING USING SYNCHRONOUSLY PUMPED DYE LASER EXCITATION

VAUGHN J. KOESTER AND ROBERT M. DOWBEN, *University of Texas Health Center at Dallas Texas 75235 U. S. A.*

The measurement of fluorescence parameters, particularly fluorescence lifetimes, represents one of the principal techniques for studying the photophysical properties of organic molecules and for elucidating the dynamic chemical and physical processes seminally important in molecular biology. Particularly informative are changes in lifetimes with different environmental factors that mimic physiological states. Time-correlated single-photon counting has been applied to nanosecond fluorescence measurements since the mid-1960s. Because of the inherent characteristics of the traditional excitation source, the air gap-discharge arc, there are major problems associated with lifetime measurements when fluorescence intensities are low. Recently a synchronously pumped tunable dye laser has been developed (1) that is ideally suited as an excitation source for measurements of relaxation phenomena on the nanosecond and subnanosecond time scales (2). We have used a modified version of such a mode-locked laser with a time-correlated single-photon counting system to extend markedly the possibilities of determining nanosecond and subnanosecond lifetimes under experimental conditions where such measurements have thus far not been achieved (3). The particular advantages of this excitation source are: (*a*) very short (<35 ps full width at half maximum [FWHM]) excitation pulses are produced, (*b*) the peak pulse power is high (140 W at 580 nm), (*c*) high repetition rates of 10 kHz to 10 MHz are possible, (*d*) the pulse profiles are uniform over short and long times, (*e*) the light is monochromatic, wavelength-adjustable, and polarized, (*f*) frequency doubling provides tunable UV wavelengths, and (*g*) there is no radio frequency noise. Very narrow excitation pulse widths are necessary to accurately determine subnanosecond lifetimes as well as multiple exponential decay parameters. The mode-locked pump laser output, a 76.802 MHz pulse train with 13.020-ns period, was used to pump rhodamine 6G. A Bragg cell in the dye laser cavity was used to "dump" a preselected fraction of the mode-locked pulses. Single-pass and double-pass beams arise from the same dye laser pump and provide two optical output channels. The cavity-dumped output consisted of a main pulse with preceding and trailing pulses of 98% smaller amplitude, spaced by 13.020 ns, and with spatial separation permitting spatial filtering. Optimum timing resolution of the photon counting system was achieved by using start channel pulses of -2.5 V obtained from a fast photodiode that monitors the single-pass cavity dumped beam and by connecting the photomultiplier (PM) tube (RCA 8850, RCA Solid State, Somerville, N.J.) directly to the stop channel input with no preamplifier. The overall timing resolution of the photon counting system was 25.1 ± 2.5 ps when fast photodiodes were used to monitor the single- and double-pass beams from the Bragg cell, with one output going to the start channel and the other output going to the stop channel. Thus the start channel timing resolution was <25 ps. The laser pulse profile obtained by scattering from a Ludox solution (E. I. DuPont, Wilmington, Del.) had a pulse width of 0.34 ± 0.01 ns FWHM. The principal reason for the measured light pulse FWHM appearing greater than the actual <35 ps value is the inherent transit time dispersion in the PM tube, 0.32 ns FWHM minimum (4). With excitation at 570 nm, the fluorescence emission of 10^{-6} M rhodamine B in the presence of various amounts of KI quencher was monitored through a sharp cut-off red filter. Data acquisition times never exceeded 5 min with 40,000 counts in the peak channel. Fluorescence lifetimes were calculated by deconvolution of the decay profile by the method of moments (5). Using fabricated data, we obtained a precision of $\pm 20\%$ for for lifetimes that correspond to $\frac{1}{6}$ of the excitation pulse FWHM. For this study we analyzed for a single exponential decay component and obtained the following lifetimes (ns): 1.48 (0.00 M KI), 0.921 (0.075 M KI), 0.671 (0.15 M KI), 0.352 (0.30 M KI), 0.162 (0.60 M KI), and 0.068 (1.20 M KI). These results are presented to illustrate fluorescence relaxation behavior on the subnanosecond time scale. It should be pointed out that the observed fluorescence relaxation actually displays multiple exponential behavior principally due to static and dynamic KI quenching effects and time-dependent depolarization effects arising from molecular rotation due to Brownian motion. Also, systematic errors were introduced because of the wavelength dependence of the PM tube temporal response (excitation and fluorescence profiles were determined at different wavelengths) and convolution by the Ludox scattering solution compared to the clear sample solution of the excitation profile. In conclusion, we obtained fluorescence excitation pulse widths of 225 ps FWHM using a static crossed-field PM tube.[1] Because the transit time dispersion of a static crossed-field PM tube is <30 ps[2] it should be possible to display excitation pulse widths limited by the timing resolution of the photon-counting electronics, and with refined experimental and deconvolution methods, to determine relaxation times of the order of 10 ps with better than 20% accuracy.

This work was supported by the National Institutes of Health grant HL-11678 and the Sid W. Richardson Foundation.

REFERENCES

1. HARRIS, J. M., R. W. CHRISMAN, and F. E. LYTLE. 1975. *Appl. Phys. Lett.* **26**:16.
2. HARRIS, J. M., and F. E. LYTLE. 1977. *Rev. Sci. Instrum.* **48**:1469.
3. KOESTER, V. J., and R. M. DOWBEN. 1978. *Rev. Sci. Instrum.* **49**:41.
4. LESKOVAR, B., and C. C. LO. 1975. *Nucl. Instrum. Methods.* **123**:145.
5. ISENBERG, I. 1973. *J. Chem. Phys.* **59**:5693.

---

[1] Koester, V. J. In preparation.
[2] Abshire, J. 1977. Personal communication.

ALLOSTERY IN AN IMMUNOGLOBULIN LIGHT-CHAIN DIMER

A Chemical Relaxation Study

D. LANCET, A. LICHT, AND I. PECHT, *The Weizmann Institute of Science, Rehovot, Israel*

The light chain dimer of the murine immunoglobulin MOPC 315 (L_2 315) has been previously shown to bind the same nitroaromatic haptens as the parent molecules (HL), with similar fine specificity (1, 2). 2 mol of hapten were found to bind per 1 mol of the dimer (1). This finding may be explained in terms of a local twofold rotation axis, with both "original" light chain residues and "new" heavy-chain-homologous residues forming two distinct symmetry-related binding sites for the hapten (2). Recently we reported (3) that hapten binding to L_2 315 involves positive cooperativity, most probably mediated by an allosteric transition of the protein. Two haptens, ϵ-N-2-4-dinitrophenyl-L-lysine (DNPL) and 4-(α-N-L-alanine)-7-nitro-benz-2-oxa-1,3-

diazole (NBDA) were found to reveal such behavior. The changes in the added haptens' light absorption yielded a sigmoidal saturation curve, the shape of the difference spectrum being constant throughout the titration. This meant that the two binding sites were identical, and that the observed sigmoidity should be accounted for by a symmetric allosteric model. Analysis in terms of such a model, that of Monod, Wyman, and Changeux (MWC), with exclusive binding to the R conformation, was performed, yielding $K_R = 1.6 \times 10^4$ M^{-1} with $L = 3$ for DNPL and $K_R = 5.6 \times 10^4$ M^{-1}, with $L = 110$ for NBDA. Analysis of the DNPL system using the mathematically equivalent nonexclusive model (with binding to both the T and R forms) enabled us to obtain a fit with the same value of $L = 110$ as for DNBA, with $K_t = 2.9 \times 10^3$ and $K_R = 6.7 \times 10^4$ M^{-1}. We concluded (3) that the first hapten binding to L$_2$ 315 possibly brings about a concerted, symmetry-conserving transition in the dimer, which involves changes in the relative chain position, and which results in a higher binding constant for the second hapten. Such a process in L$_2$ may resemble antigen-induced changes in the conformation of the structurally homologous intact immunoglobulin.

To establish our mechanistic interpretation further, we also carried out a chemical relaxation study of this system. For L$_2$ 315 with DNPL at 4°C and pH 7.4 only one relaxation time of 2–20 ms was observed. The concentration dependence of this relaxation time is shown in Fig. 1. $1/\tau$ is plotted vs. the sum of total L$_2$ (A_0) and total hapten (H_0) concentrations. The three almost parallel lines are obtained for a roughly identical range of H_0 (1×10^{-5}–2×10^{-4} M), each line representing a different (constant) A_0 value (0, 17; Δ, 38; +, 120 μM). For a simple association mechanism $1/\tau$ depends equally on the free concentration of both protein (A) and hapten (H), as $1/\tau = k_{on}(A + H) + k_{off}$. Since in our experiment A_0 is comparable to H_0, and the concentration of the complex is relatively low, $1/\tau$ value is expected to be similar for all points with the same $A_0 + H_0$, even when they have different A_0. This is clearly not the case here, and actually, when $1/\tau$ is plotted vs. H_0 alone, all three lines almost coincide, implying that $1/\tau$ is practically independent of A_0. The simplest mechanism compatible with such behavior is $T_0 \rightleftharpoons R_0$ ($K_0 = k_0/k_{-0} = R_0/T_0 \ll 1$); $H + R_0 \rightleftharpoons R_1$ ($K_R = k_R/k_{-R} = R_1/(R_0H)$). Here the pro-

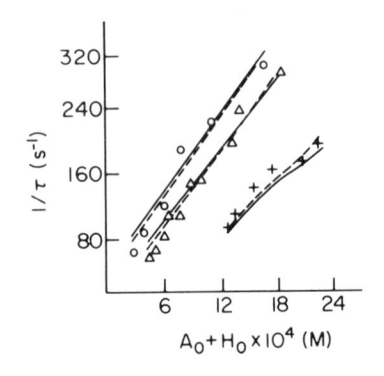

FIGURE 1

tein exists in two conformations T and R, of which only R, which comprises only a small fraction of the free protein, binds the hapten. For this system the relaxation time for the association is $1/\tau = K_R (R_0 + H) + k_{-R}$, and since $R_0 \ll H$ always, $1/\tau \simeq k_R H + k_{-R}$. The broken lines in the figure represent the best fit to this mechanism, with $K_R = 1.9 \times 10^6$ M^{-1} s^{-1}, $k_{-R} = 45$ s^{-1}, and $K_0 = 0.14$. An additional relaxation time representing the monomolecular $R_0 \rightleftharpoons T_0$ equilibrium is also expected. We suspect that this time is very slow and lies outside the time range of the temperature-jump method. This mechanism is not, however, in accord with the static measurements, since it predicts a hyperbolic saturation curve. We then proceeded to analyze the kinetic data using the MWC mechanism. For an allosteric dimer this model is redundant, i.e., many different K_T, K_R, and L sets may conform with a particular binding curve, as far as they all arise from the same pair of Adair constants. For the titrations of DNPL these were found to be $K_1 = 3.5 \times 10^3$ and $K_2 = 1.4 \times 10^4$ M^{-1}. A fit to the nonexclusive MWC model using the formulas of Kirschner et al. (4) and assuming that only an intermediate (R association) relaxation time is observed, gave a good agreement (full curve in the figure) the best fit parameters being $k_R = 1.9 \times 10^6$ M^{-1} s^{-1}, $k_{-R} = 52$ s^{-1}, $K_T = 1.9 \times 10^3$ M^{-1}, $K_R = 3.7 \times 10^4$, and $L = 36$. These represent $K_1 = 2.8 \times 10^3$, $K_2 = 1.5 \times 10^4$ M^{-1}, in good agreement with the titrations. The value of L obtained is intermediate between those found when exclusive MWC, or when equality of L for both haptens, was assumed. An analysis of relaxation amplitudes gave only a moderate agreement with the MWC model, implying that the actual mechanism may be somewhat more complicated. However, the very good fit of the relaxation time data suggests that the MWC model is a good first-approximation description. Relaxation measurements with NBDA are now underway. A complex spectrum, containing at least two relaxation times, is observed, which serves as a preliminary indication for the complexity of the binding mechanism.

REFERENCES

1. SCHECHTER, I., E. ZIV, and A. LICHT. 1976. *Biochemistry.* **15**:2785–2790.
2. LICHT, A., D. LANCET, I. SCHECHTER, and I. PECHT. 1977. *FEBS Lett.* **78**:211–215.
3. LANCET, D., A. LICHT, I. SCHECHTER, and I. PECHT. 1977. *Nature.* **269**:827–829.
4. KIRSCHNER, K., M. EIGEN, R. BITTMAN, and B. VOIGT. 1966. *Proc. Natl. Acad. Sci. U.S.A.* **56**:1661–1667.

THE STRUCTURE OF THE RETINYLIDENE CHROMOPHORE IN BATHORHODOPSIN

AARON LEWIS, *School of Applied and Engineering Physics, Cornell University, Ithaca, New York 14853 U. S. A.*

ABSTRACT Resonance Raman data on bathorhodopsin (bovine and squid) at 95, 77, and 4°K support a mechanism of excitation proposed by Lewis in which both a protein conformational transition and chromophore structural alteration to a "dicisoid" configuration are required to generate the bathorhodopsin species observed in steady-state photostationary mixtures. However, these results also suggest that the molecular structure with a red-shifted chromophore absorption detected at room temperatures in 1 ps using picosecond absorption spectroscopy may not necessarily have the same chromophore conformation as the steady-state bathorhodopsin species.

Recently, I proposed a mechanism of excitation in visual transduction which accounts for all the presently available photophysical and photochemical data on rhodopsin (1). In this mechanism, the primary action of light is to cause significant electron redistribution in the retinylidene (retinal) chromophore, which induces a protein conformational rearrangement (such as proton translocation in the opsin matrix) and subsequent structural alteration in the chromophore isoprenoid chain. The original description of this visual excitation mechanism (1) proposed a distorted retinal structure in bathorhodopsin that could be generated from either rhodopsin (containing the chromophore in an initial 11-*cis* isomeric configuration) or isorhodopsin (containing the chromophore in an initial 9-*cis* isomeric configuration) by a similar out-of-plane torsional motion. Furthermore, although our proposed bathorhodopsin chromophore structure is not consistent with mechanisms that predict a photochemical 11-*cis* to all-*trans* isomerization (2), the structure we propose can readily relax to an all-*trans* conformation by metarhodopsin I and is consistent with experiments that demonstrate the bathorhodopsin retinal structure is common to both rhodopsin and isorhodopsin (3, 4). In this note we demonstrate that our proposed bathorhodopsin chromophore structure is also consistent with the resonance Raman data on bovine and squid visual pigments.

Resonance Raman spectra of bovine and squid photostationary mixtures obtained with 514.5 nm illumination at 77 and 95°K, respectively, are displayed in Fig. 1. When our high-resolution bovine spectra (Fig. 1 B) are compared to the squid spectra (Fig. 1 A), detailed similarities are revealed that are not apparent from comparing our squid results to the earlier low-resolution experiments of Oseroff and Callender (5). The various peak assignments shown in Fig. 1 are made on the basis of dual-beam and variable temperature-experiments to be described in more detail elsewhere.[1] These assignments are in general agreement with our earlier results on squid visual pigments (6), except for the reassignment in this paper of the 1,147 cm^{-1} peak to isorhodopsin. Our earlier conclusion on the assignment of this band was made on the basis of observations by Oseroff and Callender (5) and Mathies et al. (7). that bovine isorhodopsin contained only one vibrational mode in this region. However, our higher resolution bovine data show two vibrational modes at 1,143 and 1,153 cm^{-1} even when the photostationary mixture is 96% isorhodopsin. Therefore, even though a pure squid isorhodopsin spectrum cannot be obtained at low temperatures, we are confident

---

[1]Sulkes, M., A. Lewis, and M. A. Marcus. 1978. Resonance Raman spectroscopy of squid and bovine visual pigments; the primary photochemistry in visual transduction. Submitted for publication.

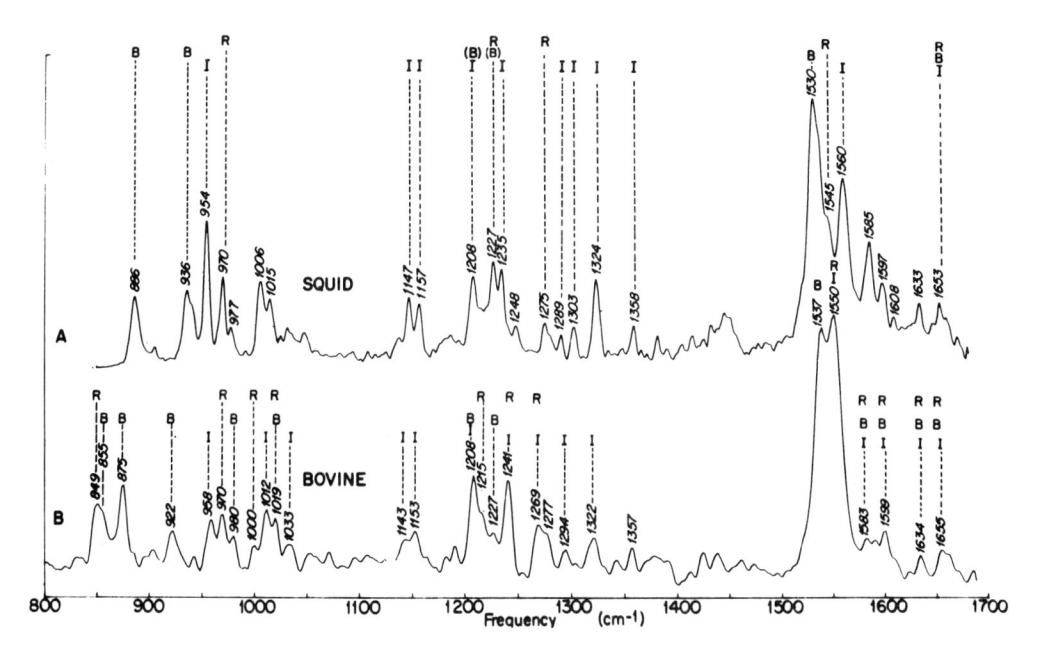

FIGURE 1 Photostationary state resonance Raman spectra of (A) Squid retinular membrane fragments prepared by the method of Hubbard and St. George (17) and (B) bovine rod outer segments prepared by the method of Appleburry et al. (18). These spectra were recorded at 77°K (A) and 95°K (B), with a 514.5-nm laser beam, and are virtually identical to spectra recorded at 4°K. The intensity of the C—H bond in squid isorhodopsin indicates that this species makes a large contribution to the photostationary state mixture at 514.5 nm illumination and this obscures certain rhodopsin bands (such as the 1,215 cm^{-1} mode) in squid rhodopsin, which can be detected in the bovine spectra. It is important to note that peaks are assigned to either rhodopsin (R), isorhodopsin (I), or bathorhodopsin (B), with tentative assignments in brackets, and that squid spectra of these species exhibit no bands between 800 cm^{-1} and the peak at 886 cm^{-1} seen in A. In addition, there is a shoulder at 1,277 cm^{-1} in bovine spectra and a peak at 1,248 cm^{-1} in squid spectra, which may be associated with bathorhodopsin.

that both the frequencies in this spectral region (see Fig. 1 A) arise mainly from squid isorhodopsin.

Based on these assignments, it is obvious that the isorhodopsin, rhodopsin, and bathorhodopsin bands exhibit little overlap. In addition, there are no bathorhodopsin vibrational modes in the 1,170–1,200 cm^{-1} region, where metarhodopsin exhibits strong scattering modeled well by all-*trans* protonated Schiff base resonance Raman spectra (6). These data indicate that the configuration of the chromophore in bathorhodopsin is not all-*trans,* and although it is common to rhodopsin and isorhodopsin (3, 4), it is structurally distinct from either the 11-*cis* or 9-*cis* isomeric configurations. I have proposed (1) a structure that meets all of the above criteria and can relax to an all *trans* conformation by metarhodopsin I; it is shown in Fig. 2. Notice that the C$_9$ and C$_{13}$ methyl groups in this structure are in very similar structural configurations and that the 11-*cis* 12-s-*trans*/ 11-*cis* 12-s-*cis* C$_{13}$ methyl-hydrogen interactions are absent in this "dicisoid" configuration. In these terms one can understand the absence

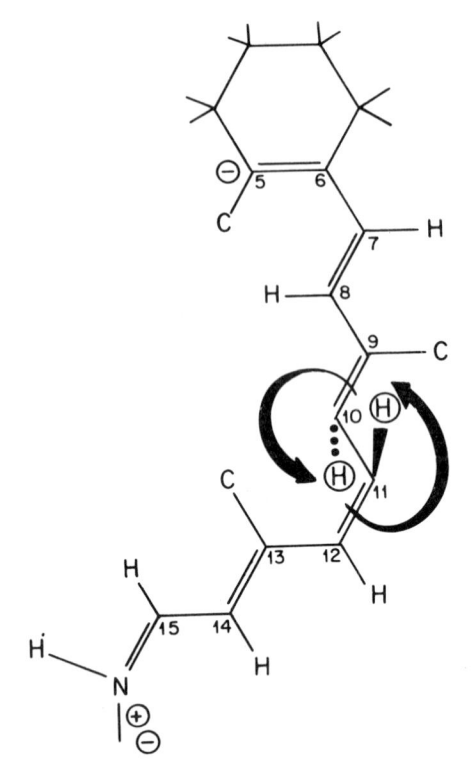

FIGURE 2 A proposed structure for the retinylidene chromophore in bathorhodopsin. The Hs at positions 10 and 11 are depicted out of the plane of the paper. This structure is generated by out-of-plane torsional motions from either rhodopsin or isorhodopsin by rotating as a unit the C_9—C_{10} and C_{11}—C_{12} ground-state double bonds, which are single bonds in the excited state. The exact out-of-plane rotation is not known.

in bathorhodopsin of the 1,000 and 1,019 cm^{-1} splitting in the C—CH$_3$ vibrational mode in rhodopsin (associated with the 11-*cis* 12-s-*trans* and 12-s-*cis* C$_{13}$ methyl-hydrogen interactions [8]) coalesces to one band at ~1,019 cm^{-1}. Furthermore, the distortions around double bonds in this structure may be the cause of the lowered C=C stretching frequency in bathorhodopsin in Fig. 1[2] and may also account for the appearance only in bathorhodopsin of an intense out-of-plane torsional mode between 860 and 900 cm^{-1}. It is interesting in this regard that, even though squid and bovine pigment spectra exhibit detailed similarities in other spectral regions (see Fig. 1 A), bovine pigments (rhodopsin and bathorhodopsin) have an additional mode at ~850 cm^{-1} not detected in squid photostationary state spectra. The calculations of Warshel

---

[2]This is suggested by the observation that bacteriorhodopsin with a 570 nm absorption and squid bathorhodopsin with a 545 nm absorption both have C=C stretching frequencies at 1,530 cm^{-1} (see ref. 9 and Fig. 1 A). This may indicate that in the highly strained bathorhodopsin chromophore structure the C=C stretching frequency is not linearly correlated to the chromophore absorption maximum, as some workers have suggested (10.)

and Karplus (11) indicate that this could occur as a result of ring constraints and without structural alterations in the isoprenoid chain configuration.

Finally, the $-\overset{\underset{\oplus}{|}}{\underset{|}{C}}=\overset{\underset{|}{H}}{N}-$ stretching frequency at 1,655 cm^{-1} (see Fig. 1) is identical in

rhodopsin, bathorhodopsin, and isorhodopsin. Since this is the only exchangeable proton on the chromophore, deuterium isotope effects on the time evolution of bathorhodopsin must result from a proton movement in the protein, as I recently proposed (1), and must be unrelated to changes in the state of protonation of the Schiff base (12). Thus, these results suggest that bathorhodopsin is produced by structural transitions in both the protein and chromophore, in support of the excitation mechanism and energy surface I proposed (1).

In conclusion, it should be noted that there are now three possible explanations for the red shift in bathorhodopsin. This red shift could be the result of: (*a*) protein structural alterations; (*b*) chromophore structural alterations; or (*c*) chromophore and protein structural alterations. However, it is known that widely differing (9- and 11-*cis*) chromophore structures in similar protein conformations have nearly identical absorption maxima (3, 4). This can be explained by the suggestion (6, 13, 14) that the color of visual pigments is controlled by the charged environment of the protein, which causes chromophore excited-state stabilization. Thus, it is quite possible that proton transfer in the protein changes the protein charge environment, which results in the bathorhodopsin red-shifted absorption. If this is so, subsequent chromophore structural alteration may not detectably alter the bathorhodopsin visible absorption and in these terms picosecond absorption spectroscopy (4, 12, 15) may be monitoring only the protein structural alteration and not the chromophore structural changes detected, even at 4°K in steady-state resonance Raman spectra. Kinetic resonance Raman measurements (16) with mode-locked lasers should be capable of testing this hypothesis.

REFERENCES

1. LEWIS, A. 1978. The molecular mechanism of excitation in visual transduction and bacteriorhodopsin. *Proc. Natl. Acad. Sci. U.S.A.* **75**:549.
2. WALD, G. 1968. Molecular basis of visual excitation. *Science (Wash. D.C.).* **162**:230.
3. YOSHIZAWA, T., and G. WALD. 1963. Pre-lumirhodopsin and the bleaching of visual pigments. *Nature (Lond.).* **197**:1279.
4. GREEN, B. H., T. G. MONGER, R. R. ALFANO, B. ATON, and R. H. CALLENDER. 1977. Cis-trans isomerization in rhodopsin occurs in picoseconds. *Nature (Lond.).* **269**:179.
5. OSEROFF, A. R., and R. H. CALLENDER. 1974. Resonance Raman spectroscopy of rhodopsin in retinal disc membranes, *Biochemistry.* **13**:4243.
6. SULKES, M., A. LEWIS, A. LEMLEY, and R. COOKINGHAM. 1976. Modeling the resonance Raman spectrum of a metarhodopsin: implications for the color of visual pigments. *Proc. Natl. Acad. Sci. U.S.A.* **73**:4266.
7. MATHIES, R., A. R. OSEROFF, and L. STRYER. 1976. Rapid flow resonance Raman spectroscopy of photolabile molecules: rhodopsin and isorhodopsin. *Proc. Natl. Acad. Sci. U.S.A.* **73**:1.
8. COOKINGHAM, R., and A. LEWIS. 1978. Resonance Raman spectroscopy of chemically modified reti-

nals: assigning the carbon-methyl vibrations in the resonance Raman spectrum of rhodopsin. Raman spectra of 11-cis retinal analogs. *J. Mol. Biol.* **119**:569.

9. LEWIS, A., J. SPOONHOWER, R. A. BOGOMOLNI, R. LOZIER, and W. STOECKENIUS. 1974. Tunable laser resonance Raman spectroscopy of bacteriorhodopsin. *Proc. Natl. Acad. Sci. U.S.A.* **71**:4462.
10. ATON, B., A. G. DOUKAS, R. H. CALLENDER, B. BECHER, and T. G. EBREY. 1977. Resonance Raman studies of the purple membrane. *Biochemistry.* **16**:2295.
11. WARSHEL, A. 1977. Interpretation of resonance Raman spectra of biological molecules. *Annu. Rev. Biophys. Bioeng.* **6**:273.
12. PETERS, K., M. L. APPLEBURY, and P. M. RENTZEPIS. 1977. Primary photochemical event in vision: proton translocation. *Proc. Natl. Acad. Sci. U.S.A.* **74**:3119.
13. MATHIES, R., and L. STRYER. 1976. Retinal has a highly dipolar-vertical excited singlet state: implications for vision. *Proc. Natl. Acad. Sci. U.S.A.* **73**:2169.
14. HONIG, B., A. B. GREENBERG, D. DINUR, and T. G. EBREY. 1976. Visual pigment spectra: implications of the protonation of the retinal Schiff base. *Biochemistry.* **15**:4593.
15. BUSCH, G. E., M. L. APPLEBURY, A. LAMOLA, and P. M. RENTZEPIS. 1972. Formation and decay of prelumirhodopsin at room temperatures. *Proc. Natl. Acad. Sci. U.S.A.* **69**:2802.
16. MARCUS, M. A., and A. LEWIS. 1977. Kinetic resonance Raman spectroscopy: dynamics of deprotonation of the Schiff base of bacteriorhodopsin. *Science (Wash. D.C.).* **195**:1328.
17. HUBBARD, R., and R. ST. GEORGE. 1958. The rhodopsin system of the squid. *J. Gen. Physiol.* **41**:501.
18. APPLEBURY, M. L., O. M. ZUCKERMAN, A. LAMOLA, and T. M. JOVIN. 1974. Rhodopsin purification and recombination with phospholipids assayed by the meta I → meta II transition. *Biochemistry.* **13**:3448.

CONTRACTILE DEACTIVATION BY RAPID, MICROWAVE-INDUCED TEMPERATURE JUMPS

BARRY D. LINDLEY and BIROL KUYEL, *Department of Physiology, Case Western Reserve University, Cleveland, Ohio 44106 U.S.A.*

Activation of muscle is controlled by Ca^{++} (released by changes in membrane potential), myofilament overlap, and muscle load. Understanding has been hindered by inability in many cases to achieve rapid changes in potential or intracellular concentration of agents. Biochemical approaches are hindered by the loss of the physiological ordered lattice and by loss of the coupling to external load.

A microwave temperature-jump system has been developed for the study of rapid-heating relaxation of single frog striated muscle cells, using a modified magnetron conditioner (4J50, 9.4 GHz, 200 kW peak power, 0.002 duty cycle) (Fig. 1). The isolated muscle cell is bathed in saline contained in an acrylic chamber (150 μl volume) passing through a waveguide (RG51) one quarter wavelength from the shorted end. Slits are provided to allow continuous measurement of sarcomere spacing by means of the diffraction of a HeNe laser beam, and isometric force is recorded using an AME silicon transducer element. Temperature increase was linear with duration of the heating train 0.2°K/ms, and cooling of the ambient temperature had a half-time exceeding 10 s. No cellular damage is apparent even with scores of T-jumps over some hours, if excessive heating ($T < 28°C$) is avoided. Heating for 5–10 ms preceding a twitch altered the twitch in the same fashion as altered steady T; heating during the rising phase caused an increase in the rate of tension development, followed by the more rapid relaxation

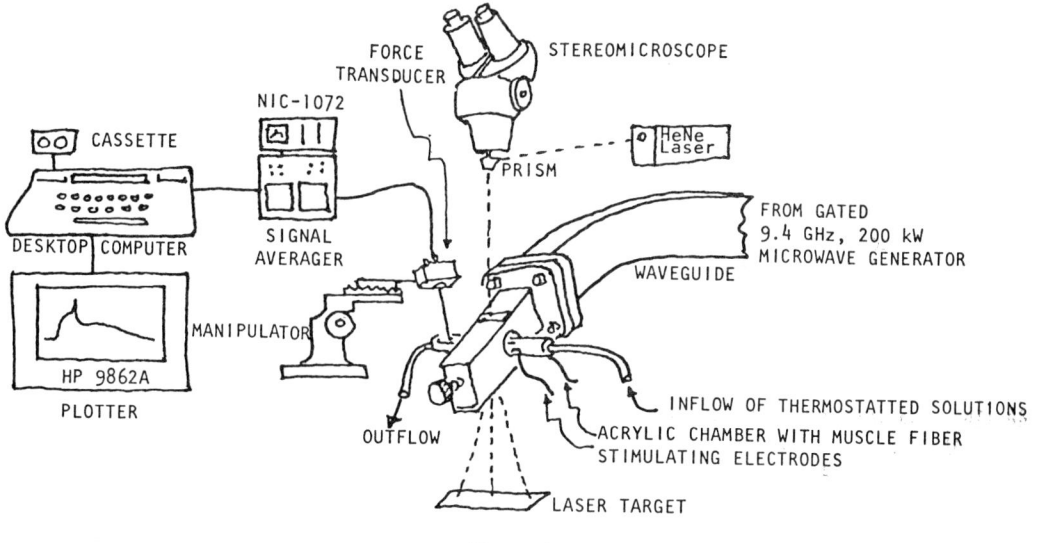

FIGURE 1

characteristic of twitches in warmer muscle; heating during relaxation resulted simply in more rapid relaxation. During the plateau of a tetanus, brief heating caused a two-phase rise to greater force. The initial phase of about 2% per degree, lagging slightly behind the heating (time constant ~5 ms at 10°C), varied with sarcomere spacing in the same way as tetanic force; the slower phase had a time constant of about 20 ms. Muscle fibers depolarized with elevated potassium concentration were briefly heated (10–20 ms) at the peak of force development. At maximal depolarization (100 mM K^+), there was a rapid rise to greater force, followed by the relatively slow spontaneous relaxation characteristic of the new T. At submaximal depolarization (25 mM K^+), there was a rapid activation beginning with the onset of heating and peaking in 50–80 ms, followed by a deactivation relaxation to the (lower) peak force characteristic of the new T at that depolarization. Both activation (order of 0.03/ms) and deactivation (order of 0.005/ms) rate constants decreased as the membrane potential (or preheating level of activation) became less negative. Speed of deactivation increased with increasing sarcomere spacing, in contrast to the rate of relaxation from tetanic contractions.

There are thus three major processes to be studied by this method, empirically identified by the time constants at 10°C—5 ms, 30 ms (activation), and 100–200 ms (deactivation). The practical limit of resolution with single-sweep operation is presently processes with time constants of 5 ms or greater. With signal averaging the present limit is 1 ms, a time scale at which the frequency response of the present transducer is itself limiting.

This work was supported by grants from U.S. Public Health Service (HL 19848) and the Northeast Ohio Affiliate of the American Heart Association.

CALCULATION OF DIELECTRIC PARAMETERS FROM TIME DOMAIN SPECTROSCOPY DATA

W. D. MOORHEAD, *Physics Department, Youngstown State University, Youngstown, Ohio 44555 U.S.A.*

To obtain accuracy in the determination of the dynamic dielectric constant with time domain spectroscopy (TDS) equipment, detector face reference voltages must be corrected for the effective source impedance and must include the phase correction due to the delay line between the detector and sample (1, 2). We report the corresponding algorithm for σ, ϵ_s, the conductivity and static dielectric constant. For sample termination, one may write:

$$\sqrt{\epsilon}\coth\lambda\,\frac{l}{c}\sqrt{\epsilon} = \epsilon\,\frac{V + R_s X + e^{2\gamma L}(X - V)}{V + R_s X - e^{2\gamma L}(X - V)},$$

$$V = A\,\frac{(1 - R_s e^{-2\gamma L_A})}{(1 + e^{-2\gamma L_A})}; \gamma = \lambda/c,$$

where ϵ, λ, l, L_A, A, L, X, and R_s, are respectively: dielectric constant, transform parameter, sample length, tube lengths and transformed voltages for sample out and in place, and effective source reflection coefficient. At low frequency, the effective source reflection coefficient may be expanded as:

$$R_s = R_0 + \lambda R_1.$$

The result of the calculation is:

$$\sigma/\epsilon_0 = [C(A_\infty - X_\infty)(1 - R_0)]/[lX_\infty(1 + R_0)],$$

$$\epsilon_s(1 + R_0)X_\infty + \frac{\sigma}{\epsilon_0}\left[A_\infty \frac{L}{C}(R_0 - 1) + \Delta X(R_0 + 1)\right.$$

$$+ X_\infty\left(R_1 + \frac{L_A - 2L - R_0 L_A}{C}\right)\bigg] = \frac{C}{l}\left[(\Delta A - \Delta X)(1 - R_0)\right.$$

$$+ A_\infty\left(\frac{2L_A - L}{C}R_0 + \frac{L}{C} - R_1\right) + X_\infty\left(R_1 + \frac{L_A - 2L - R_0 L_A}{C}\right)\bigg]$$

$$+ \frac{1}{3}\frac{l}{C}\frac{\sigma}{\epsilon_0}[A_\infty - X_\infty](1 - R_0).$$

Where $\Delta X = \int_0^\infty [X(t) - X_\infty]\,dt$, $\Delta A = \int_0^\infty [A(t) - A_\infty]\,dt$ and X_∞, A_∞ are the

---

The author of this paper did not attend the actual Discussion. The present text was submitted and circulated to participants before the meeting.

observed voltages at long times. The parameters R_0, R_1 are obtained from measurement of standard samples and then applied as corrections to unknowns.

REFERENCES

CLARKSON, et al. 1977. *Adv. Mol. Relax. Proc.* **10**:173.
KUZNETSOV. *The Propagation of Electromagnetic Waves in Multiconductor Transmission Lines,*

LIGHT-JUMP PERTURBATION OF CARBON MONOXIDE BINDING BY VARIOUS HEME PROTEINS

EMILIA R. PANDOLFELLI, CELIA BONAVENTURA, *and* JOSEPH BONAVENTURA,
  Duke University Marine Laboratory, Beaufort,
  North Carolina 28516 U.S.A. and
MAURIZIO BRUNORI, *Istituto di Chimica Biologica, Centro di Biologia*
  Molecolare, Citta Universitaria, Rome, Italy

Relaxation spectroscopy, a technique which measures the relaxation of a system back to equilibrium after a small perturbation, has found useful application in the kinetic analysis of hemoglobin function. One type of relaxation method, the light jump, has been employed in the present study as a means of investigating the extreme pH dependence of ligand binding by the hemoglobin of a marine teleost, *Leiostomus xanthurus*, commonly known as the spot. The light-jump method is in principle similar to the more familiar technique of temperature-jump spectroscopy, except that light of sufficient intensity to partially dissociate CO from hemoglobin is used to perturb the equilibrium between free and bound ligand. The measurements consequently involve analysis of the approach to a steady state in the light and the decay to the dark equilibrium when the photodissociating light is turned off. Variable parameters include the intensity and duration of the photodissociating beam. A comparison between the ligand affinity of the protein under photodissociating conditions and that measured in the absence of photodissociating illumination provides information on the net rate of light-induced ligand dissociation and thereby permits calculation of the quantum yield for photodissociation characteristic of the protein under investigation. Light-jump perturbations have also been used to investigate the contributions of the "on" and "off" kinetic rate constants to the homotropic and heterotropic interactions of the Root effect hemoglobin of the spot. The method is illustrated by presentation of results from simple heme proteins and various fish hemoglobins where more complex kinetics are observed. The pH sensitivity of Root effect fish hemoglobins makes these molecules particularly interesting cases to investigate by the light jump method.

MEASUREMENT OF INTERCONVERSION RATES OF BOUND SUBSTRATES OF PHOSPHORYL TRANSFER ENZYMES BY ^{31}P NUCLEAR MAGNETIC RESONANCE

B. D. NAGESWARA RAO and MILDRED COHN, *University of Pennsylvania, Philadelphia, Pennsylvania 19104 U.S.A.*

Kinases are phosphoryl transfer enzymes that catalyze the reaction ATP + X ⇌ ADP + XP, where X is the second substrate, using a divalent metal ion (usually Mg^{2+}) as an obligatory component. On the basis of known dissociation constants of the substrates and products from their binary (or ternary) enzyme-bound complexes, sufficiently high concentrations of enzyme can be chosen in excess of those of the substrates or products so that either E-X-ATP or E-XP-ADP account for the predominant fraction of the enzyme-bound complexes. Such a mixture can be generated by setting up an equilibrium mixture with a small amount of Mg^{2+} and later removing Mg^{2+} from the reaction by adding a complexing agent like EDTA. A ^{31}P-NMR spectrum of such a solution consists of resonances that may be readily assigned to the six phosphate groups in the above complexes with line widths and chemical shifts governed by the interaction of the substrates and the enzyme.

Addition of sufficient Mg^{2+} to the sample sets up an equilibrium mixture of the reaction E-X-MgATP ⇌ E-XP-MgADP. This interconversion of the reactants and products on the enzyme involves the following exchanges of the phosphate groups: $\alpha - P\,(ATP) \rightleftharpoons \alpha - P\,(ADP)$, $\beta - P\,(ATP) \rightleftharpoons \beta - P\,(ADP)$, and $\gamma - P\,(ATP) \rightleftharpoons P\,(XP)$, and the ^{31}P resonances in the spectrum display line shape changes. If τ_A and τ_B are the lifetimes at two sites for which ω_A and ω_B are, respectively, the resonance frequencies for $\tau_A^{-1}, \tau_B^{-1} \gg |\omega_A - \omega_B|$ (fast exchange), a single resonance of frequency and line width given, by the weighted averages of frequencies and line widths in the absence of the exchange, is obtained. If $\tau_A^{-1}, \tau_B^{-1} \ll |\omega_A - \omega_B|$ (slow exchange) the resonances are not shifted, but they become broadened due to the exchange by $\Delta\omega_A (= \tau_A^{-1}/\pi)$ and $\Delta\omega_B (= \tau_B^{-1}/\pi)$. In intermediate exchange condition both shifts and line width changes result. The $\alpha - P$ resonances of ATP and ADP are usually separated by <0.5 ppm whereas the corresponding $\beta - P$ resonances are separated by >12.0 ppm. For ^{31}P-NMR operating frequencies of ~40 MHz, exchange rates (τ_A^{-1}, τ_B^{-1}) of over 100 s^{-1} generally bring the $\alpha - P$ resonances of ATP and ADP into fast exchange, but the $\beta - P$ resonances easily satisfy slow exchange condition for rates up to ~1,000 s^{-1}. Therefore, from the difference in the line widths of the $\beta - P$ resonances of ATP in the presence and absence of exchange, the lifetime of the E-X-MgATP complex, τ_A (reciprocal of the rate of ADP formation), may be determined. The $\beta - P$ resonance of ATP is particularly suitable for line width measurements, since it is isolated at the high-field side of the ^{31}P-NMR spectrum. The lifetime of E-XP-MgADP complex, τ_B, may then be deduced on the basis of the equilibrium constant, K'_{eq} = [E-XP-MgADP]/[E-X-MgATP] = τ_B/τ_A. The concentrations of the complexes at equilibrium are determined directly from the spectrum.

The rates determined above are the rates of interconversion of reactants and products on the surface of the enzyme. The NMR experiment on the bound substrates isolates and monitors exclusively the interconversion step in the reaction. The rate is thus obtained in a straightforward manner compared to the relatively cumbersome isotope exchange methods. A comparison of the measured rates with the corresponding overall rates of the reaction determines whether the interconversion step is the rate-limiting step. It is also possible to determine, by this method, whether compounds that alter the overall rate of the reaction (e.g. inhibitors or modifiers) do so by modifying the rate of the interconversion step or some other step in the overall kinetic scheme.

Measurements of interconversion rates were made for five different kinases: arginine kinase, adenylate kinase, creatine kinase, pyruvate kinase, and 3-phosphoglycerate kinase. The rates of ATP and ADP formation were 154 and 200 ± 15 s^{-1}, respectively for arginine kinase (at pH 7.25 and 12°C) and for adenylate kinase (at pH 7.0 and 4°C) 420 ± 40 s^{-1} and 690 ± 50^{-1}, respectively, and are the fastest among the enzymes studied. The overall rates of the reactions in either direction were an order of magnitude smaller. A similar result is obtained for creatine kinase. The interconversion process is, therefore, not rate-limiting for these reactions.

Pyruvate kinase and 3-phosphoglycerate kinase are enzymes responsible for ATP formation in glycolysis with equilibrium constants at catalytic enzyme concentrations of the order of 10^3 in favor of ATP and negligible overall rates for ADP formation. The ^{31}P-NMR experiments on bound substrates reveal similar interconversion rates for forward and reverse directions, indicating that (*a*) the equilibrium constant on the enzyme is ~1, and (*b*) while the interconversion step may be a rate-determining step for ATP formation, it is certainly not so in the reverse direction. Thus, the predominance of equilibrium in favor of ATP formation at low enzyme concentrations seems to be due to the very slow rates of dissociation of the products in the reverse direction especially of phosphoenol-pyruvate from pyruvate kinase and 1,3-diphosphoglycerate from 3-phosphoglycerate kinase.

Whenever the sample conditions may be easily chosen, such that the enzyme-bound complexes are predominantly in the two forms E-X-MgATP or E-XP-MgADP, the interconversion rates for these reactions may be measured simply and straightforwardly. The ability to obtain this condition is the only serious limitation for this method. The scope and accuracy of the method may be significantly improved by using the temperature to alter the magnitudes of the rates and employing higher NMR operating frequencies that provide both larger chemical shifts and better sensitivity.

DISSOCIATION RATE OF SERUM ALBUMIN-FATTY ACID COMPLEX FROM STOP-FLOW DIELECTRIC STUDY OF LIGAND EXCHANGE

WALTER SCHEIDER, *Biophysics Research Division, Institute of Science and Technology, University of Michigan, Ann Arbor, Michigan 48109 U.S.A.*

The magnitude of the dipole vector of the serum albumin molecule is a sensitive indicator of the number of moles of fatty acid (FA) bound by the albumin (1). Other than chemical analysis, this is the most accurate indicator of such binding, and, in the case of human serum albumin (HSA), which lacks a fluorescent chromophore near a principal binding site, the dipole vector is probably the only measure suitable for rapid reaction techniques.

Moreover, it is known (1) that the change in dipole vector upon binding fatty acid is considerably different in bovine serum albumin (BSA) from that in HSA. The equilibrium constants of fatty acid binding are similar in the two proteins (2, 3), and it is therefore to be expected that ligand exchange will occur between the two species of protein if they are mixed. This, together with the differential dielectric effect, thus provides a means of measuring the dissociation rate constant, k_d, of the binding reaction,

$$\text{HSA} + \text{FA} \underset{k_d}{\overset{k_a}{\rightleftharpoons}} \text{HSA--FA}. \tag{1}$$

Svenson et al. (4) measured the k_d of this binding by ligand exchange with matrix-bound albumin, and obtained a rate constant of $4.2 \times 10^{-2} \text{s}^{-1}$ for palmitic acid at 25°C. Certain assumptions about the mechanisms of the exchange and the behavior of the matrix-bound protein are avoided in the single-phase system reported here.

In our experiment, a rapid-flow device with a conventional mixer is used. The solution from syringe I contains HSA defatted by the method of Chen (5), and subsequently re-lipidated with 1 M/M oleate added as the sodium salt. Solution II is a defatted BSA solution. Both solutions are deionized by passage over mixed-bed ion exchange resin (MB-1, Mallinckrodt Inc., St. Louis, Mo.).

The mixture is passed into an electrode chamber in which the dielectric constant is measured at intervals after mixing, by recording and analyzing the response to a step function in applied potential at the electrodes.

The stop-flow device used and the theory and technique of the time domain dielectric measurement have been recently reported (6). The dielectric measurement is made in real-time, rather than by the conventional scan of the frequency domain, but its relation to the molecular dipole vector is based on the classical analysis of Debye (7) and Perrin (8).

The use of this technique has already showed that the forward reaction of Eq. 1 takes place in at least two steps, the first too fast to measure by stop-flow methods, and the second, a first-order process of $k_a^{II} \simeq 3$ s^{-1} at 0°C, presumed to be a rearrangement of the protein molecule with the ligand.

Likewise, in studying the ligand exchange reaction, it is not a priori obvious whether the reaction proceeds by way of a (very low) free fatty acid concentration:

$$\text{HSA—FA} \underset{k_2}{\overset{k_1}{\rightleftharpoons}} \text{HSA} + \text{FA} \quad (2)$$

$$\text{FA} + \text{BSA} \underset{k_4}{\overset{k_3}{\rightleftharpoons}} \text{BSA—FA},$$

or whether it proceeds by ligand transfer during protein-protein encounter,

$$\text{HSA—FA} + \text{BSA} \underset{k_2'}{\overset{k_1'}{\rightleftharpoons}} \text{HSA} + \text{BSA—FA}. \quad (3)$$

In either case, what is monitored is the amount of apo-HSA and BSA-FA formed, as the increase in dipole vector of the HSA upon losing ligand is greater than the decrease in dipole vector of the BSA upon binding ligand.

The end products of Eqs. 2 and 3 are identical, but the kinetics are different. If, for simplicity, the assumptions are made that $k_1 = k_4$, $k_2 = k_3$, and $k_1' = k_2'$ (in fact very nearly valid assumptions), then

$$\text{Eq. 2} \rightarrow \tau^{-1} = k_1(1 + [\text{HSA}]_{\text{total}}/[\text{BSA}]_{\text{total}}),$$

$$\text{Eq. 3} \rightarrow \tau^{-1} = k_1'([\text{HSA}]_{\text{total}} + [\text{BSA}]_{\text{total}}). \quad (4)$$

In these equations, τ is a time constant measured from the relation, $\Delta\epsilon/\Delta\epsilon_\infty = 1 - e^{-t/\tau}$, in which $\Delta\epsilon$ is the change in dielectric constant of the mixture at time t after mixing, and $\Delta\epsilon_\infty$ is the change after the mixture has reached equilibrium.

Fig. 1 shows the result of mixing a 0.135 mM solution of defatted HSA to which 1 M/M oleic acid had been added, with a 0.223 mM solution of defatted BSA, at 1°C. Each point in this graph is calculated from the response curve to a square pulse in the electric field applied to the mixture.

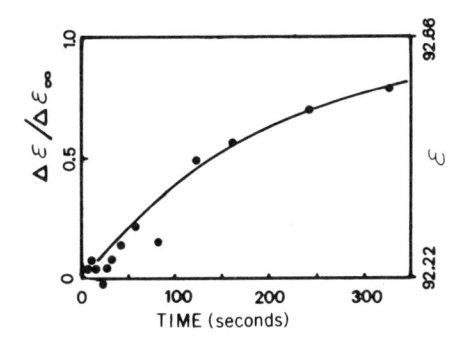

FIGURE 1 Change in dielectric constant as a function of time after mixing of a solution of HSA-FA with a solution of BSA.

The data are consistent with either Eq. 3, $k'_1 = k'_2 = 14$ s^{-1} mol^{-1}; or with Eq. 2, $k_1 = k_4 = 3 \times 10^{-3}$ s^{-1}. The two mechanisms have different implications for the physiology of the albumin-fatty acid complexation, but the experiments with a protein concentration series which can distinguish the two have not yet been done.

This work was supported by the Institute of General Medical Studies, National Institute of Health grant GM22309.

REFERENCES

1. SCHEIDER, W., H. M. DINTZIS, and J. L. ONCLEY. 1976. Changes in the electric dipole vector of human serum albumin due to complexing with fatty acids. *Biophys. J.* **16**:417.
2. ASHBROOK, J. D., A. A. SPECTOR, E. C. SANTOS, and J. E. FLETCHER. 1974. Long chain fatty acid binding to human plasma albumin. *J. Biol. Chem.* **250**:2333-2338.
3. SPECTOR, A. A., J. E. FLETCHER, and J. D. ASHBROOK. 1971. Analysis of long-chain free fatty acid binding to bovine serum albumin by determination of stepwise equilibrium constants. *Biochemistry.* **10**:3229-3232.
4. SVENSON, A., E. HOLMER, and L. O. ANDERSSON. 1974. A new method for the measurement of dissociation rates for complexes between small ligands and proteins as applied to the palmitate and bilirubin complexes with serum albumin. *Biochim. Biophys. Acta.* **342**:54-59.
5. CHEN, R. F. 1967. Removal of fatty acids from serum albumin by charcoal treatment. *J. Biol. Chem.* **242**:173.
6. SCHEIDER, W. 1977. Real-time measurement of dielectric relaxation of biomolecules: Kinetics of a protein-ligand binding reaction. *Ann. N.Y. Acad. Sci.* **303**:47-58.
7. DEBYE, P. 1929. Polar molecules. Dover Publications, Inc., New York.
8. PERRIN, F. 1936. Brownian movement of an ellipsoid. II. Free rotation and depolarization of fluorescence. Translation and diffusion of ellipsoidal molecules. *J. Phys. Radium.* **7**:1.

TIME-RESOLVED RESONANCE RAMAN CHARACTERIZATION OF THE INTERMEDIATES OF BACTERIORHODOPSIN

JAMES TERNER AND M. A. EL-SAYED, *Department of Chemistry, University of California, Los Angeles, California 90024 U.S.A.*

In an attempt to determine eventually the structure of the chromophore of the intermediates involved in the primary process of the photosynthetic proton-pumping cycle of bacteriorhodopsin, new simple techniques have been developed (1-4) to obtain time-resolved resonance Raman spectra of this system. The techniques used in this report involve chopping continuous wave (CW) laser light to produce pulses of variable width and separation to obtain temporal information on the microsecond and millisecond time scales. An optical multichannel analyzer (Princeton Applied Research Corp., Princeton, N.J.) with a Dry Ice-cooled silicon-intensified vidicon is used for detection. By using these techniques as well as different laser frequencies to take advantage of differing resonance enhancements of the intermediates, flow techniques (5-7), and computer subtraction methods, the resonance Raman spectra of the retinal chromophore of the individual intermediates are extracted.

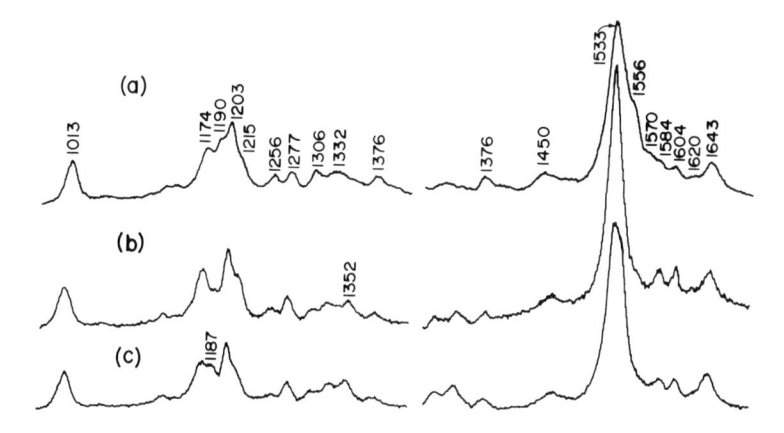

FIGURE 1 (a) 100 mW CW excitation from a Spectra-Physics model 165 argon ion laser at 5,145 Å of 100 μM bacteriorhodopsin in a melting point capillary (Spectra-Physics Inc., Laser Products Div., Mountain View, Calif.). (b) Light-adapted recirculated flow spectrum of bacteriorhodopsin with 4 mW of 5,145 Å excitation. (c) Same as in (b) but dark-adapted.

Resonance Raman spectra of bacteriorhodopsin are shown in Fig. 1. A CW spectrum is shown in Fig. 1 a. This spectrum is a superposition of several intermediates, bR_{570}, bL_{550}, bM_{412}, and at least one other intermediate, possibly bO_{640}. A flow spectrum, shown in Fig. 1 b, contains bR_{570} with minimal contributions from the other intermediates. A dark-adapted (DA) flow spectrum is shown in Fig. 1 c, giving the resonance Raman spectrum of bR_{560}^{DA}. Some conclusions about the various forms of bacteriorhodopsin are given below.

bR_{570}. The so called "fingerprint region" (1,100–1,400 cm^{-1}) has been shown to be very sensitive to the isomeric configuration of the retinal (7). The similarity between the fingerprint regions of bR_{570} and the protonated Schiff base of all-*trans* retinal is not as close as that between the fingerprint regions of the unphotolyzed rhodopsin and the protonated Schiff base of 11-*cis* retinal (5). The fingerprint region of bR_{570} bears a closer resemblance to that of the protonated Schiff base of 13-*cis* retinal (2).

bL_{550}. We have recently reported the observation of a band at 1,620 cm^{-1} (3) that grows at a similar rate as the C=C stretch at 1,556 cm^{-1} assigned to bL_{550} (2). If this band is the same as that assigned for the unprotonated C=N vibration (8), the results suggest that deprotonation might occur with the appearance of bL_{550} or earlier. The fingerprint region of bL_{550} is characterized by a strong band at 1,190 cm^{-1}. Other bands are too weak to report conclusively at this time.

bR_{560}^{DA}. This form is presently thought to be an equilibrium between 13-*cis* and all-*trans* retinal (9). The resonance Raman spectrum of bR_{560}^{DA} in Fig. 1 c confirms the existence of two isomers as evidenced by the broadening of the C=C stretch at 1,533 cm^{-1}. An examination of the fingerprint region reveals the presence of another isomer in addition to the isomer present in bR_{570} form (see Fig. 1 b).

Possible conclusions are: (a) the exact isomeric form of the chromophore in bacteriorhodopsin (as concluded from resonance Raman studies) is not firmly established.

The fingerprint regions of bR_{570}, bL_{550}, bR_{560}^{DA}, and bM_{412} (2, 6) are all different. Whether these changes are due to differences in isomeric configuration, state of protonation, or other changes in the electronic structure of the olefinic system of the retinal chromophore is not yet clear. (*b*) Deprotonation of the retinal Schiff base may occur earlier than previously thought from optical absorption data.

The authors wish to thank the U.S. Department of Energy for financial support. This is contribution 3996 from the Department of Chemistry, UCLA.

REFERENCES

1. CAMPION, A., J. TERNER, and M. A. EL-SAYED. 1977. Time-resolved resonance Raman spectroscopy of bacteriorhodopsin. *Nature (Lond.).* **265**:659.
2. TERNER, J., A. CAMPION, and M. A. EL-SAYED. 1977. Time-resolved resonance Raman spectroscopy of bacteriorhodopsin on the millisecond time-scale. *Proc. Natl. Acad. Sci. U.S.A.* **74**:5512.
3. CAMPION, A., M. A. EL-SAYED, and J. TERNER. 1977. Resonance Raman kinetic spectroscopy of bacteriorhodopsin on the microsecond time-scale. *Biophys. J.* **20**:369.
4. MARCUS, M. A., and A. LEWIS. 1977. Kinetic resonance Raman spectroscopy: dynamics of deprotonation of the Schiff base of bacteriorhodopsin. *Science (Wash. D.C.).* **195**:1328.
5. MATHIES, R., R. B. FREEDMAN, and L. STRYER, 1977. Resonance Raman studies of the conformation of retinal in rhodopsin and isorhodopsin, *J. Mol. Biol.* **109**:367.
6. ATON, B., A. G. DOUKAS, R. H. CALLENDER, B. BECHER, and T. G. EBREY. 1977. Resonance Raman studies of the purple membrane. *Biochemistry.* **16**:2995.
7. RIMAI, L., D. GILL, and J. L. PARSONS. 1971. Raman spectra of dilute solutions of some stereoisomers of vitamin A type molecules. *J. Am. Chem. Soc.* **93**:1353.
8. LEWIS, A., J. SPOONHOWER, R. A. BOGOMOLNI, R. H. LOZIER, and W. STOECKENIUS. 1974. Tunable laser resonance Raman spectroscopy of bacteriorhodopsin. *Proc. Natl. Acad. Sci. U.S.A.* **71**:4462.
9. SPERLING, W., P. CARL, CH. N. RAFFERTY, and N.A. DENCHER. 1977. Photochemistry and dark equilibrium of retinal isomers and bacteriorhodopsin isomers. *Biophys. Struct. Mech.,* **3**:79.

STUDIES ON PROTEINS AND tRNA WITH TRANSIENT ELECTRIC BIREFRINGENCE

MICHAEL R. THOMPSON, RAY C. WILLIAMS, AND CHARLES H. O'NEAL,
*Department of Biophysics, Medical College of Virginia,
Richmond, Virginia 23298 U.S.A.*

Transient electric birefringence was used in this study for the determination of size and shape of several native and sodium dodecyl sulfate (SDS)-denatured proteins, yeast and rat liver bulk tRNA, and purified yeast tRNA^{Phe}. An instrument constructed by R. C. Williams (1) with a resolution time of 8 ns for propylene carbonate allowed the observation of birefringence decay phenomena previously masked by instrumental noise. The light source was a 5 mW HeNe laser with a wavelength of 623.8 nm. The Kerr cell was designed to fit a 1 × 1 cm spectrophotometer cuvette and consisted of platinum sheet electrodes spaced 1.5 mm apart in a Teflon support. The high-voltage source was a 20 kV, 1.5 mA power supply. The high-voltage pulser used a switched charged line of 500 ft of RG-8A/U cable short-circuited by a triggered spark gap to effect a very fast decay time. A thin film high-frequency resistor capable of withstanding high current and voltage was mounted in a logarithmically tapered cavity designed for correct impedance match for the charged line. The photo detector was a RCA 8644 photomultiplier tube (RCA Solid State, Somerville, N.J.) with a Zener diode divider. A Tektronix type 454 oscilloscope was used to observe the voltage or optical pulse (Tektronix, Inc., Beaverton, Ore.). With this apparatus, two relaxation times were detected for a number of interesting proteins. The semimajor and minor axes of prolate ellipsoids may be determined from this data by using the numerical inversion procedure described by Wright (2, 3). This technique has also been shown to allow the rapid estimation of molecular weight of proteins in SDS solution.

Bovine serum albumin (BSA) obtained from a commercial supplier was shown to be homogenous by Sephadex chromatography, ultracentrifugal analysis, and gel electrophoresis. The two relaxation times observed were 28 and 78 ns for a 0.75% solution. The relaxation times extrapolated to zero concentration were 26.9 and 75.5 ns. These data, applied to the inversion procedure of Wright, yield the following dimensions for the equivalent prolate ellipsoid model for BSA: $2 a_3 = 140.9 \pm 4.9$ Å and $2 a_1$ 41.6 ± 3.6 Å. These values are in good agreement with data obtained by other procedures. Other proteins examined were human transferrin, ovalbumin, chymotrypsinogen A, lactoglobulin, lysozyme, cytochrome c, and rhodopsin. SDS-denatured proteins gave a slow relaxation time linearly related to log molecular weight (4).

Total tRNA and purified species of tRNA gave similar decay curves. tRNA in buffer, in high Mg^{++}, and in solutions of EDTA exhibited a decrease in relaxation time with increasing tRNA concentration, possibly indicating a more compact structure at higher concentrations. A 0.5% solution of purified yeast $tRNA^{Phe}$ gave relaxation times of 18 and 44 ns, corresponding to major and minor axes of 54.1 and 18.7 Å at 20°C. The magnitude and sign of the birefringence signal and the relaxation time for tRNA was found to vary considerably with temperature. Between 20°C and 35°C a dramatic drop in birefringence occurred with an increase in size of the molecule. This change in tRNA structure was also observed at similar temperatures with acridine conjugates of tRNA by Millar and Steiner (5) and was not altered by higher concentrations of Mg^{++}. Examination of the hyperchromicity curve shows only a few hydrogen bonds had been broken during this temperature change. At 38–40°C the birefringence disappears. Between 40 and 60°C another birefringent species of opposite sign appears, most probably due to the denatured tRNA. The birefringence of tRNA at room temperature is negative and can be obscured by the positive birefringence of water at low tRNA concentrations.

A 0.5% solution of total yeast deacylated tRNA in 0.001 M EDTA had a slow relaxation time of 84 ns at 20°C. The addition of Mg^{++} to a concentration of 0.5 mM decreased the relaxation time to 71 ns. After incubation of this solution at 60°C for 3 min, the relaxation time at room temperature fell to 51 ns. These changes are consistent for a molecule going to a more compact structure. Total yeast tRNA in 0.001 M Mg^{++} gave calculated semi-major and minor axes of 55 and 17 Å at 20°C, compared to 53 and 21 Å for purified acylated $tRNA^{Phe}$.

This work was supported in part by National Science Foundation grant GB 14046.

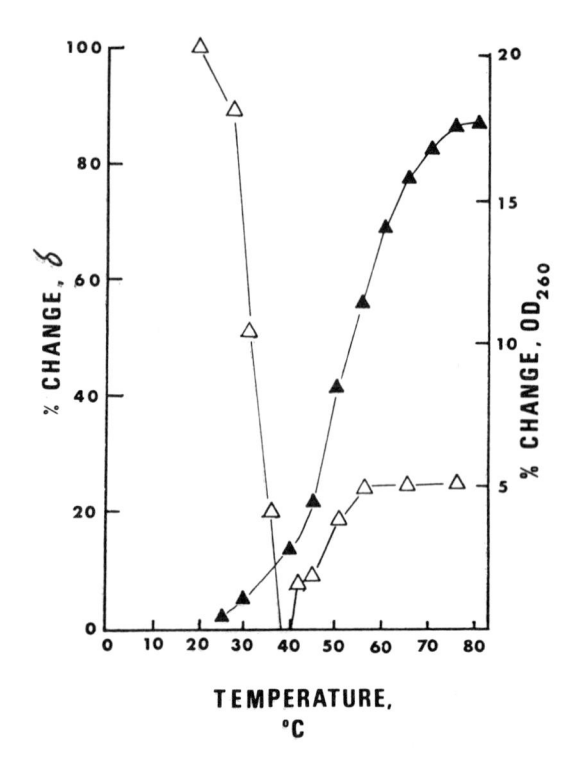

FIGURE 1   Birefringence △ and optical density ▲ as a function of temperature for yeast tRNA[Phe].

REFERENCES

1. WILLIAMS, R. C., W. T. HAM, and A. K. WRIGHT. 1976. Ultra high speed electro-optical system for transient birefringence studies of macromolecules in solution. *Anal. Biochem.* **73**:52–74.
2. WRIGHT, A. K., R. C. DUNCAN, and K. A. BECKMAN. 1973. Numerical inversion of the Perrin equation for rotational diffusion constants for ellipsoids of revolution by iterative techniques. *Biophys. J.* **13**:795–803.
3. WRIGHT, A. K., and M. R. THOMPSON. 1975. Hydrodynamic structure of bovine serum albumin determined by transient electric birefringence. *Biophys. J.* **15**:137–141.
4. WRIGHT, A. K., M. R. THOMPSON, and R. L. MILLER. 1975. A study of protein-SDS complexes by transient electric birefringence. *Biochemistry.* **14**:3223–3228.
5. MILLAR, D. B., and R. F. STEINER. 1966. The effect of environment on the structure and helix-coil transition of sRNA. *Biochemistry.* **5**:2289–2301.

METAL ION INTERACTIONS WITH FLUORESCENT DERIVATIVES OF NUCLEOTIDES

J. M. VANDERKOOI, C. J. WEISS, AND G. WOODROW III,
University of Pennsylvania, Philadelphia, Pennsylvania 19174 U. S. A.

The 1, N^6-ethenoadenosine phosphate derivatives have been shown to be useful probes of nucleotide binding sites in a variety of enzymes (1). Perturbation of the ethenoadenosine derivative with light results in formation of an excited-state compound that can undergo a variety of reactions within the time scale of its fluorescent lifetime, including deprotonation, rotational diffusion, and energy transfer to paramagnetic and colored metal cations near the chromophore. The rates of these reactions are a function of the excited-state lifetime and environmental factors such as the distance between the interacting molecules and the diffusional properties of the bound species. A comparison of the rates of the excited state decay when bound to the enzymatic site to those in solution reveals information of the binding site with respect to the spacial organization (distances between the metal and nucleotide) and the temporal organization (how fast the bound molecules diffuse).

In solution, high concentrations (>50 mM) of Ni(II), Co(II), and Mn(II) cations quench the fluorescence of ethenoadenosine derivatives by collisional quenching. At low concentrations (10^{-6}–10^{-3} M) quenching occurs when the metal binds to the phosphates. The affinity constant depends upon the metal ion and the pH. The increase in the decay rate of the excited-state molecule by the bound metal is primarily a function of the spectral overlap in the metal absorption and the ethenoadenosine fluorescence emission. The effect on intersystem crossing rates by the unpaired electrons of the metal appears to be less significant.

The pK of ethenoadenosine in the excited state is lower than in the ground state. Excitation of the acid form results in emission from the base form, indicating that deprotonation occurs rapidly. Measuring fluorescence decay rates of the ethenoadenosine derivative as a function of pH allows one to calculate the rate of protonation and deprotonation of the excited-state species and then to compare with the quenching rates of the metal cations. The collisional rates for quenching are of the order Ni $\simeq$ Co > H$^+$ > Mn. It is interesting to note that the pH dependence of ethenoadenosine, etheno AMP, etheno ADP, and etheno ATP differ, indicating that the conformation of the molecule allows interaction between the adenosine ring and the phosphate.

REFERENCES

1. SECRIST, J. A., J. R. BARRIO, N. J. LEONARD, and G. WEBER. 1972. *Biochemistry.* 11:3499–3506.

RUPTURE DIAPHRAGMLESS APPARATUS FOR PRESSURE-JUMP RELAXATION MEASUREMENT

TATSUYA YASUNAGA AND NOBUHIDE TATSUMOTO, *Department of Chemistry, Faculty of Science, Hiroshima University, Hiroshima, Japan 730.*

The kinetics of micelle formation and dissociation in solutions of anionic surfactants, such as sodium dodecyl and tetradecyl sulfates (SDS and STS), have been studied experimentally by various techniques. Recently a new theory of micelle formation mechanism was proposed together with experimental results obtained in the course of chemical relaxation studies of micellar solutions of ionic surfactants by Aniansson

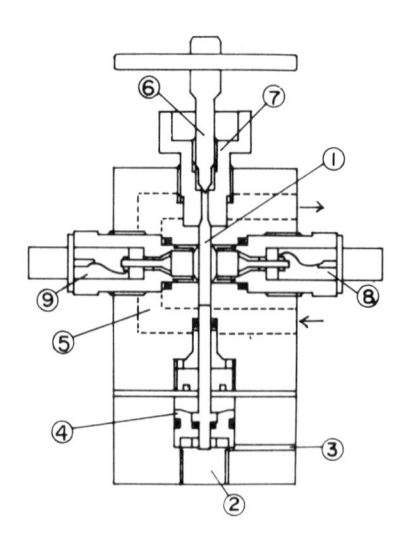

FIGURE 1 1, experimental chamber; 2, gas pressure inlet; 3, gas pressure outlet; 4, piston; 5, heat exchanger; 6, bubble remover; 7, socket; 8, 9, conductivity cells.

et al. (1). In their study the validity of the theory was maintained from the agreement between theoretical and experimental values. It seems to be a successful interpretation of their data. A large discrepancy, however, was found between their experimental data and ours, obtained from the surfactant synthesized extremely purely; i.e. their reciprocal relaxation time (τ_2^{-1}) is dependent upon the concentration of surfactant, whereas ours is independent of the concentration. This fact demanded the reexamination of their theory. Here, the value of τ_2^{-1} is vague near the concentration C^* for SDS and STS because the relaxation strength decreases. The purpose of this study is to develop a new pressure-jump apparatus capable of measuring the relaxation time for small relaxation strengths and to reconfirm the micellar relaxation mechanism by using the newly obtained data.

In the usual pressure-jump method, a single rectangular step-forcing function is realized by blowing out a rupture diaphragm. The decay time of pressure of this method is less than 100 µs, but the precise measurement of relaxation time is impossible in the case of a relaxation of a poor signal-to-noise (S/N). The repetitive application of pressure perturbation and signal averaging is used to enhance the S/N of relaxation curve. The apparatus was designed to produce 100–200 atm of water in an autoclave with a piston (surface area ratio 1:25) by 5–9 atm of air. A cross-sectional diagram of the autoclave is given in Fig. 1. To obtain a short rise time, an electric solenoid valve and gas reservoir were set close to a gas inlet 2. Air in the gas inlet 2 always went out an outlet, 3, but the pressure in the chamber 1 was kept constant as long as the valve was opened, since the air supply was much greater than the exhaust. The signal detected by an AC bridge was rectified by a ring demodulator[1] and averaged with a

---

[1]Tatsumoto, N., K. Takehara, and T. Yasunaga. In preparation.

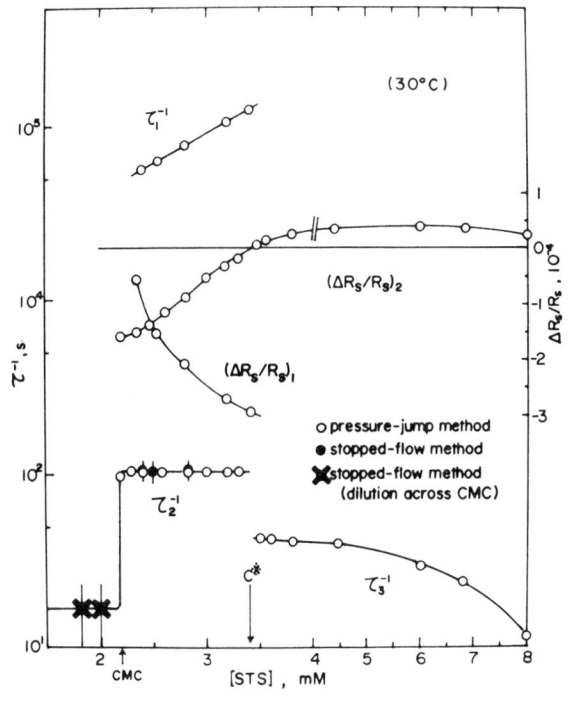

FIGURE 2

signal memory averager. The rise and decay times of pressure in the autoclave were determined to be within 5 and 40 ms, respectively, with use of $MgSO_4$ and NaCl solutions. Although this rise time is longer than that of the usual method, the new technique has the outstanding merit that the repetitive measurement can be performed in a short time. In this experiment, STS was studied. Since τ_2 is about 10 ms at 30°C, the pressure pulse-width was set at 100 ms. The repeating total time is about 25 s for signal averaging of 128 times. The relaxation strength $\Delta R_s/R_s$ and its characteristics were also measured with a usual pressure-jump system.

The relaxation effect characterized by τ_2 vanished around the concentration of C^*. With further increase of the concentration, another new relaxation effect characterized by τ_3 appeared. This phenomenon was also observed in the relaxation strength; i.e., the sign of $\Delta R_s/R_s$ changed from negative to positive with increase of the STS concentration at C^*, as shown in Fig. 2. τ_2^{-1} is clearly independent of and τ_3^{-1} decreases with the concentration. The electroconductometric stopped-flow measurement yielded the same result for τ_2 as that in the pressure-jump method above the critical micelle concentration (CMC) and new relaxation effect, designated by (X) in Fig. 2, was also observed in diluting across the CMC.[2] Taking account of these new

---

[2]Tatsumoto, N., T. Tasunaga, and T. Nagamura. In preparation.

experimental results, the mechanism of the micelle formation of surfactants will be discussed.

REFERENCE

1. ANIANSSON, E. A. G., S. N. WALL, M. ALMGREN, H. HOFFMANN, I. KIELMANN, W. ULBRICHT, R. ZANA, J. LANG, and C. TONDRE. 1976. *J. Phys. Chem.* **80**:905.

Large Perturbations

L. M. DORFMAN *and* P. M. RENTZEPIS, *Chairmen*

KEYNOTE ADDRESS

PROBING ULTRAFAST BIOLOGICAL PROCESSES BY PICOSECOND SPECTROSCOPY

P. M. RENTZEPIS, *Bell Laboratories, Murray Hill, New Jersey 07974 U. S. A.*

ABSTRACT A brief discussion of the initial events leading to the visual transduction process will be presented to illustrate the capabilities of picosecond spectroscopy.

INTRODUCTION

In view of the fact that picosecond spectroscopy is a new field, at least relative to nuclear magnetic resonance (NMR) and mixing, a brief discussion of the experimental techniques seems necessary despite the wide exposure it has received since its genesis 12 years ago.

The goal of this optical system is to produce a high-quality picosecond pulse, well defined in bandwidth, energy, and time duration. This pulse will be utilized to generate the continuum that will probe the changes in absorption taking place after excitation in picosecond time intervals. The changes will then be detected, analyzed, and displayed by the diagnostics system. This apparatus is used to probe the events leading to visual perception after the absorption of a photon by the chromophore rhodopsin.

THE OPTICAL SYSTEM

Fig. 1 displays these components and the mode of utilization of each component. One of the important aspects of the system is in the generation of high-quality picosecond pulses, as developed by Huppert and myself (1). The major advantages are in the narrow bandwidth, <3 cm^{-1}, coupled to a short duration, 4–6 ps, and the ease in construction. The oscillator cavity consists of a $7\frac{1}{2} \times \frac{1}{2}$ inch Nd^{3+} silicate glass rod, mode-locked by a 62% T, Kodak 9860 dye (Eastman Kodak Co., Rochester, N.Y.) in a 1-cm optical path length cell placed at the Brewster angle close to the rear mirror. The output of this laser consists of ~100 pulses per train, detected by an ITT F-4000S-1 photodiode (ITT, Electro-Optical Products Div., Roanoke, Va.) and displayed by a 519 Tektronix oscilloscope (Tektronix, Inc., Beaverton, Ore.).

From this train, a single pulse was extracted by means of a Pockels cell. The extracted single pulse was then amplified by the Nd^{3+}-yttrium aluminum garnet (YAG)

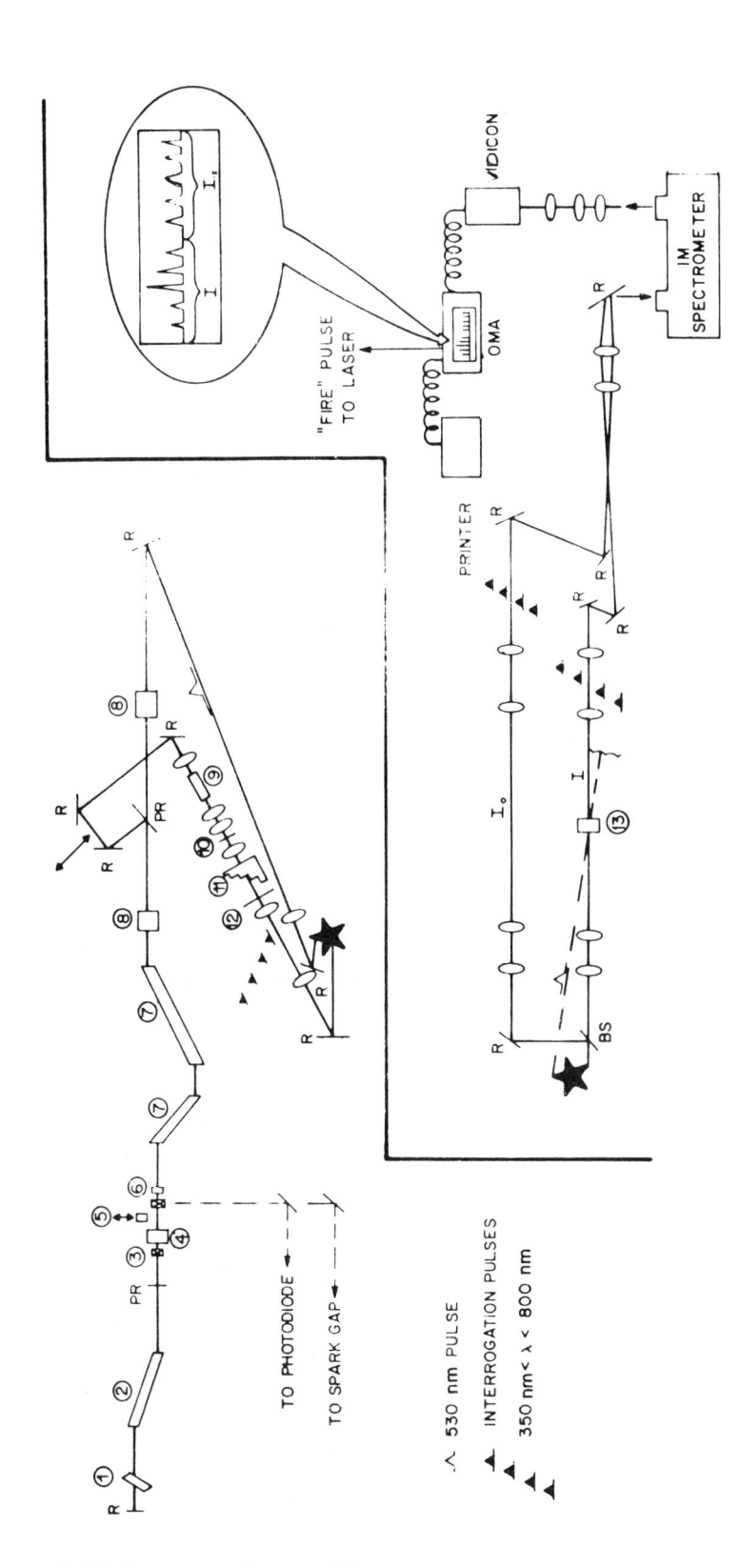

FIGURE 1 Double-beam picosecond spectrometer utilizing a silicon vidicon detector. Components: 1, mode-locking dye cell; 2, laser oscillator; 3, calcite polarizer; 4, Pockels cell; 5, translatable 90° polarization rotator for 1,060-nm radiation; 6, fixed-position 90° polarization rotator; 7, YAG amplifier rod; 8, second harmonic (530-nm) generating crystal (KDP); 9, 20 cm octanol cell for generating the interrogation wavelengths; 10, ground glass diffuser; 11, index matched glass echelon for producing picosecond optical delays between the stacked interrogation pulses; 12, vertical polarizer; 13, sample cell; R, reflector, PR, partial reflector, BS, beam splitter; OMA, optical multichannel analyzer.

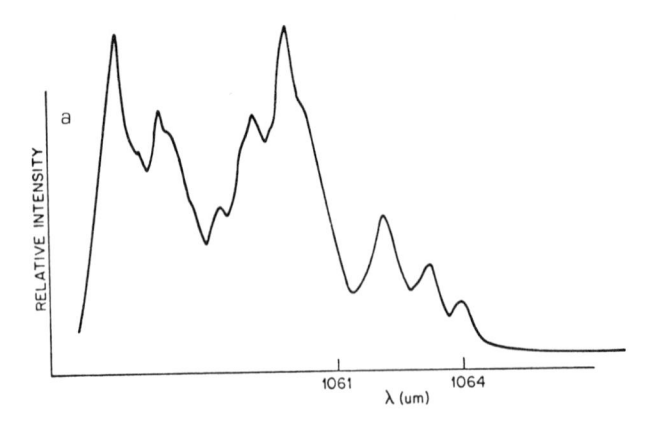

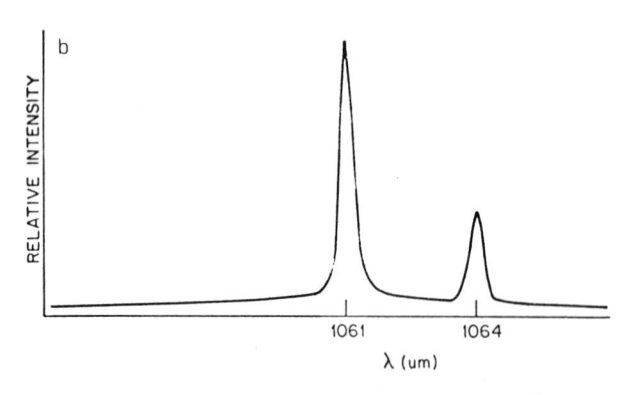

FIGURE 2 (a) Spectra of single-picosecond pulse emitted by the Nd^{3+}-silicate glass oscillator. The spectral structure varies within the pulses of a single train and from one shot to another. The average spectral width of Nd^{3+}-ED-2 silicate glass oscillator pulse is ~100 cm^{-1}. (b) Spectra of a Nd^{3+}-glass picosecond pulse amplified by Nd^{3+}-YAG. The right-hand peak corresponds to the well-known Nd^{3+}-YAG laser line at 1,064 nm. The center peak is ~30 cm^{-1} shifted from the 1,064 nm, i.e., ~1,061 nm. The spectral width at half-maximum of each of these two lines is ~3 cm^{-1}.

amplifier element (Fig. 1) consisting of a 6⅛ × ⅜-inch rod pumped by a 4-inch flash lamp. Additional amplification was provided by a 12 × ¾-inch Nd^{3+}/glass rod (Fig. 1).

The advantage of this system, devised by Huppert and myself (1), is the generation of a 3 cm^{-1}, 6 ps pulse. The spectrum of this pulse is shown in Fig. 2. Notice that the spectral width of the Nd^{3+} glass is larger than 100 cm^{-1} (Fig. 2a) and exhibits a nonsymmetric structure. Each "peak" of the spectrum varies in its relative intensity, band width, and position, from one laser "shot" to another, and even from one pulse to another within a single pulse train.

The addition of the Nd^{3+} YAG rod as the amplifier, results in a frequency-selective amplification (Fig. 2b), which corresponds to the emission spectrum of Nd^{3+}/YAG, ~3 cm^{-1} and the pulse width of the Nd^{3+}/glass of 6 ps and ~800 mJ in energy. It is obvious that amplification takes place at two wavelengths, 1,064 and

1,061 nm. The full width at half-maximum (FWHM) of each of these bands is <3 cm^{-1}, corresponding to a broadening of 1–2 cm^{-1} above the inherent Nd^{3+}/YAG pulse. In contrast, the Nd^{3+}/YAG-amplified pulse of the Nd^{3+}-glass laser has experienced a band width narrowing by at least a factor of 30, compared to the original Nd^{3+}-glass pulse, while retaining the short time width.

The 6 ps, <3 cm^{-1} pulse passes through a KD*P crystal generating an approximately 10 mJ, 530-nm pulse or a 640-nm stimulated Stokes-Raman pulse in 1 cm liquid. These pulses are used for excitation of the sample while a broad-band continuum born in a 10-cm cell of H$_2$O or alcohol(s) as a result of self-phase modulation of the laser pulse is used for interrogating the spectroscopic changes in the sample as a function of time, after passing through the echelon. The paths of these pulses in the experiment are shown in Fig. 1.

The echelon transforms the continuum into seven pulses with an interpulse separation of either 20 or 7 ps. This echelon-induced train is then split into two sets of seven segments each by a pellicle beam splitter (Fig. 1); these sets form the reference I$_0$ and interrogating (I) beams of the double-beam picosecond spectrometer (2). In a particular experiment, a segment of the I beam entered the sample simultaneously with the excitation, whereas the rest of the I segments entered subsequently, while the I$_0$ train traversed through the air or a cell containing the solvent only.

For all experiments, the echelon segments were sharply imaged onto distinct spatial regions of a N$_2$-cooled silicon vidicon. The horizontal vidicon axis (x axis) provided

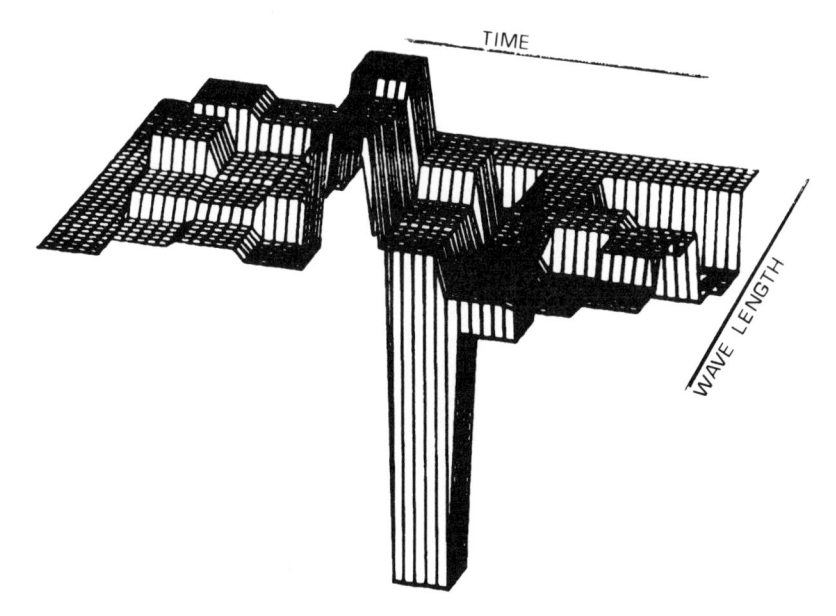

FIGURE 3 Three-dimensional display (τ, λ, ΔOD) of the optical density changes after excitation with a picosecond pulse at 530 nm. The wavelength range displayed is between 570 and 550 nm, the time in 10-ps segments covers 100 ps, and the optical density ranges from −0.6 to 0.3 OD.

the wavelength coordinate, while the vertical axis (along the spectrometer entrance slit, y axis) formed the time element.

The data are displayed either in a plot of τ (in picoseconds) vs. I, or three-dimensionally as shown in Fig. 3, in the form of time in picoseconds (x axis) versus wavelength in nanometers (y axis) versus intensity (z axis). ΔA was calculated for each time (echelon) segment by evaluating $\Delta A + \log(I^w/I^n)$, in which I^w and I^n refer to the intensity of the interrogating pulse in the presence and in the absence of the excitation pulse, respectively. The effect of excitation intensity and reliability of the data becomes more evident by considering that in most experiments discussed, the ratio of ΔA with and without excitation is ~ 10 – i.e., $\Delta A_{(with)} = 0.3$ and $\Delta A_{(without)} = 0.03$. This system, with obvious variations, was utilized in the study of many ultrafast processes, including the primary processes in vision.

PRIMARY PROCESSES IN RHODOPSIN

Prior stationary studies at low temperature have shown that the intermediate, prelumirhodopsin (prelumi), is formed first, decaying into lumirhodopsin, followed by the intermediates meta I and meta II (3–5). The first intermediate, prelumirhodopsin, was identified at low temperatures; however, its physiological temperature kinetics were unknown because of its rapid formation and decay. Later-forming transients were found to be *trans* isomers of the original *cis*-rhodopsin, and therefore a number of workers (6) assumed that the initial action of light caused isomerization of the retinal and thus the prelumirhodopsin intermediate was a *trans* isomer of rhodopsin.

The earliest species observed in the sequence, prelumirhodopsin, was unique among the intermediates, in that it was the only one with an absorption maximum shifted to the red (λ_{max} 543 nm) of normal dark-adapted rhodopsin (λ_{max} 500 nm). Thus, by using a probe pulse at a Stokes-Raman shifted wavelength (561 nm) within the absorption band for prelumirhodopsin, it was possible to monitor for its appearance and disappearance.

With this approach, an absorbing intermediate was observed (7) and interpreted to be prelumirhodopsin. It appeared within 6 ps after excitation and decayed with a time constant of about 50 ns. This result provided strong evidence for the production of prelumirhodopsin at physiological temperatures and also permitted conclusions concerning the types of structural changes that might be occurring in the molecule that would be consistent with this time scale.

To verify that this absorption band (7) monitored at 560 nm was prelumirhodopsin, the entire difference spectra were recorded of the prelumirhodopsin spectrum at room temperature and were measured in the time range 6–300 ps (8). This difference spectra (Fig. 4) shows (*a*) the bleached rhodopsin band between 430–510 nm with a peak ΔA at 485 nm, (*b*) the formation of a new absorption band at 530–680 nm with a maximum at 580 nm, (*c*) an isobestic point at 525 nm, (*d*) the rhodopsin bleached within 6 ps and the prelumirhodopsin band also formed within 6 ps. To elucidate the mechanism and resolve the rate of formation of prelumirhodopsin, we repeated these studies at lower temperatures. Such experi-

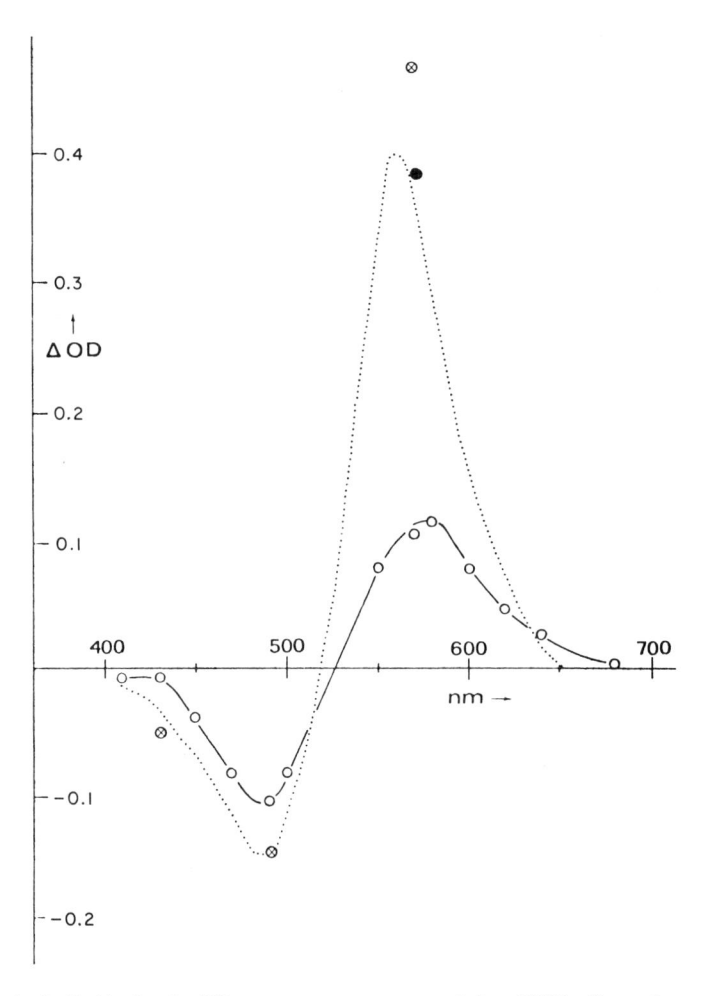

FIGURE 4 Prelimirhodopsin difference spectrum. ○, recorded at 298°K, 60 ps after excitation with a 5-mJ pulse at 530 nm (4); ●, 77°K, and ⊗ 4°K (this work); ...; difference spectrum generated by photostationary studies of low-temperature glasses for 77° and 7°K. Photostationary studies at 77° and 7°K give identical difference spectra. The data is normalized to concentrations used for this kinetic study. Rhodopsin was solubilized in 0.3 M Ammonyx (cetalkonium chloride, Onyx Chemical Co., Jersey City, N.J.)-0.01 M Hepes at pH 7.0; A_{500} = 0.73 in a 2 mm path.

ments between 300° and 4°K reveal that at as low as 77°K the rise time of the band at 570 nm is still less than 10 ps and only below 30°K could a resolvable lifetime be observed (9). Even at 4°K the rise time was only 36 ps as shown in Fig. 5.

To elucidate these data, which were not expected if one assumed that the isomerization mechanism were responsible for the formation of prelimirhodopsin at low temperatures, the rhodopsin was immersed in D_2O to substitute all exchangeable protons with deuterium. It is well known that the hydrogens of the retinal chromophore do not exchange, except for the proton of the protonated Schiff base, which easily exchanges

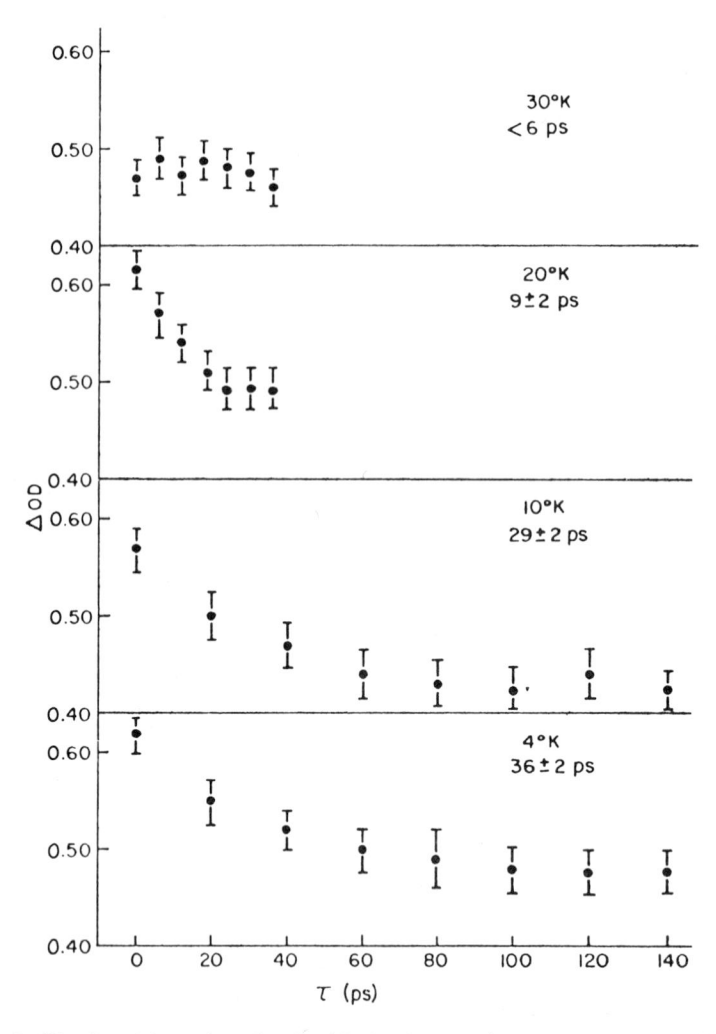

FIGURE 5 Kinetics of formation of prelumirhodopsin at various temperatures monitored at 570 nm. Excitation of rhodopsin was with a 5-mJ, 530-nm, 6-ps pulse. The glass for low-temperature study was formed by mixing one part rhodopsin, solubilized in 0.3 M Ammonyx–0.01 M Hepes at pH 7.0, A_{500} = 10.0, with two parts distilled ethylene glycol. The lifetime for formation, given in the upper right of each panel, is the reciprocal of the rate constant obtained by a least-square fit of $I_n (A_t - A_\infty)$ versus time, t, in picoseconds; A is absorbance at 570 nm.

for deuterium in D_2O (10). The rationale for the deuterated experiment was that if isomerization of the retinal is responsible for the observed kinetics, then deuterium substitution will not cause a major change in the rates, since at the point of isomerization, the retinal protons have not been replaced by deuterium.

However, the deuterated rhodopsin showed rates for the formation of prelumirhodopsin much slower than its protonated homologue. For example, at 570 nm D-rhodopsin exhibits a formation lifetime for prelumi of 17 ps at 4°K and, as the temperature decreases. this rate achieves values ranging from 51 ps at 30°K to 256 ps at

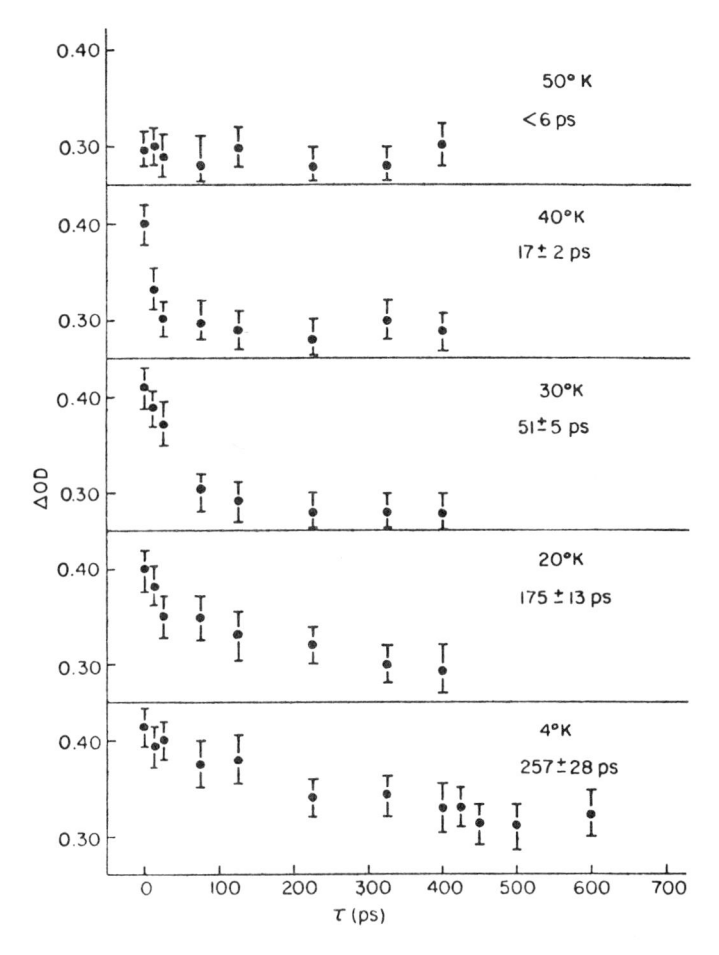

FIGURE 6 The kinetics of formation of deuterium-exchanged prelumirhodopsin at various wavelengths. The excitation of deuterium-exchanged rhodopsin was with a 5-mJ, 530-nm, 6-ps pulse. The glass for low temperature study was formed by mixing one part D-rhodopsin in deuterium-exchanged 0.1 M Ammonyx/0.01 M Hepes at pH 7.0, A_{500} = 7.0, with two parts deuterium-exchanged ethylene glycol. The lifetime of formation was calculated as for Fig. 5.

4°K (Fig. 6). This behavior is in clear contrast to the H-rhodopsin, prelumi formation with lifetimes of 9 ps at 30°K to 36 ps at 4°K (Fig. 5).

A transient species also appeared that decays fast into a long-lived species absorbing at 570 nm. We believe that this is an excited singlet state of rhodopsin prepared by the 530-nm excitation pulse.

PROTON TRANSLOCATION

The conversion of rhodopsin into prelumi via the excited state is a very efficient process, since no significant rhodopsin ground state repopulation is observed during the first 100 ps after excitation. Information concerning the mechanism responsible for the

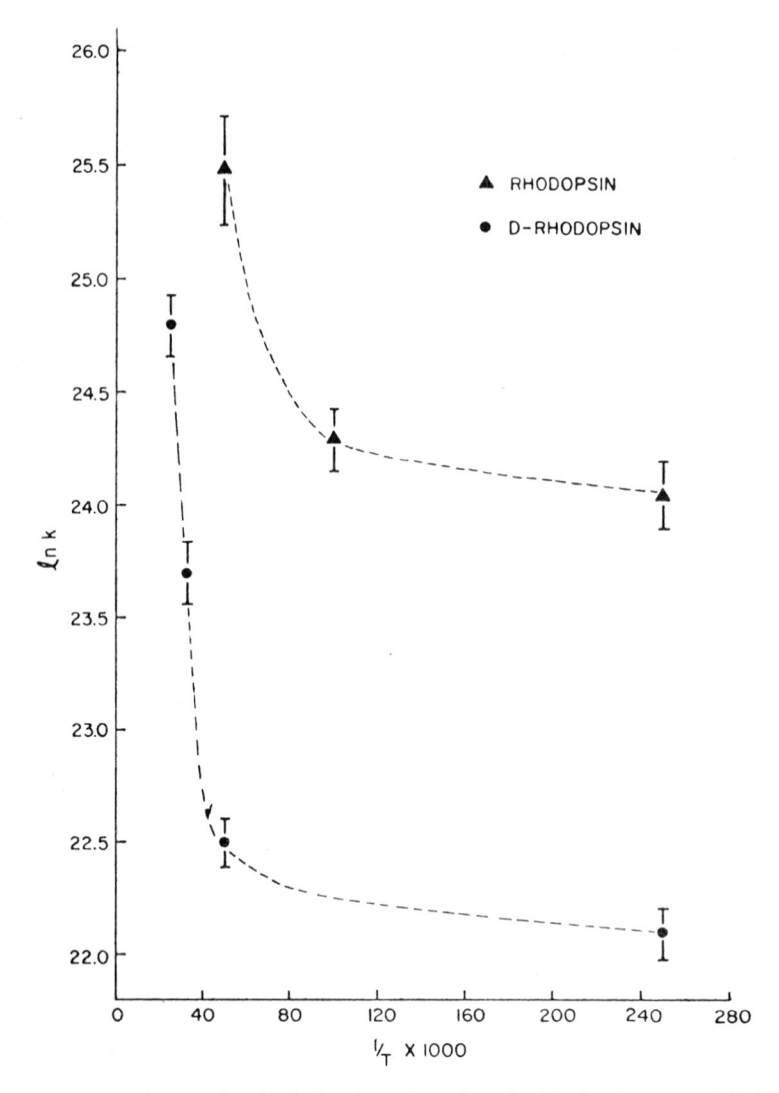

FIGURE 7 An Arrhenius plot, ln k for formation of prelumirhodopsin versus $1/T$ $(L) \times 10^3$, of the kinetic data in Figs. 5 and 6. The value of ln $k = 25.84$ corresponds to a lifetime of 6 ps.

formation of prelumi can be obtained by a closer examination of the temperature dependence of the formation rate constant K_{PL} of prelumi and its isotope dependence. Fig. 7 shows the temperature dependence of K_{PL} plotted in the Arrhenius form (ln k vs. $1/T$). Two striking features of this plot are immediately apparent: (a) the curve results from ln K_{PL} vs $1/T$ is not a straight line, as expected for an Arrhenius process. (b) At the limit of $T = 0$, K_{PL} is finite and temperature independent.

The effect of deuteration on K_{PL} is to make the rate slower in D- vs. H-rhodopsin with a ratio $K_{PL}^H/K_{PL}^D \approx 7$. Since only the proton of the protonated Schiff base is exchangeable in the retinal sequence, in the absence of drastic changes of the protein structure by D_2O, the observed isotope dependence suggests a tunneling process (11). Proton tunneling has indeed been proposed to take place in some excited state proton transfer reactions. The data on temperature and isotope dependence strongly suggest a proton transfer reaction as being mainly responsible for the formation of prelumi at low temperatures. That the absorption spectrum of prelumi is red-shifted with respect to the spectrum of rhodopsin itself suggests that prelumi may be a more tightly protonated Schiff base than rhodopsin. Model studies (12) indicate that translocating the proton towards the Schiff base nitrogen could account for the spectral red shift.

The mechanism of formation of prelumi at higher temperatures has been described as a *cis-trans* isomerization. The extremely fast formation of prelumi at 4°K, though, argues against such an isomerization process, at least at low temperatures. Several studies have shown (13) that as the temperature is decreased, the quantum yield of photoisomerization of retinal analogues and retinal Schiff bases decreases, and at 77°K photoisomerization ceases completely, as does the isomerization of stilbene (14). However, we must make it clear that the mechanism of the formation of prelumi at higher temperatures cannot be ascertained by the available results. Regarding the mechanism for the proton translocation process, the existing theories are not sufficiently refined to give an exact answer. However, assuming that the expression provided (11),

$$k_t = \nu_0 \exp \frac{-\pi^2 a_0 k_t}{b} (2ME_a)^{1/2}, \qquad (1)$$

is applicable to rhodopsin, we can calculate an activation energy or barrier height, E_a, and a width of the barrier or translocation distance a_0. Using the experimental rate constant of 2.8×10^{11} s^{-1} at 4°K, and the normal value ν_0, and the mass of the proton M, we calculate an E_a value of 1.4 kcal if we give ν_0 the N—H band frequency of 1,500 cm^{-1}. If we use instead the more reasonable value of 10^{13} for the frequency of the translocated proton, then we have a barrier of 740 kcal or a height ~100 cal above room temperature. This indicates that at room temperature the proton translocation could be an activated process proceeding without a barrier. This would explain the ultrafast rate observed for the formation of prelumi at room temperature. Similarly the distance, a_0, through which the proton tunnels is calculated to be ~0.5 Å.

MODELS FOR PROTON TUNNELING

The kinetic data reported originally by Peters et al. (9) are consistent with at least two proton translocation mechanisms: (*a*) A concerted double hydrogen transfer leading to a retro-retinal structure, a slightly different mechanism than the one proposed previously (15). (*b*) A single proton translocation to the Schiff base nitrogen generating a carbonium ion, as proposed by Mathies and Stryer (16). The proton translocation is thought to be facilitated by an active role of an amino acid, with histidine being a

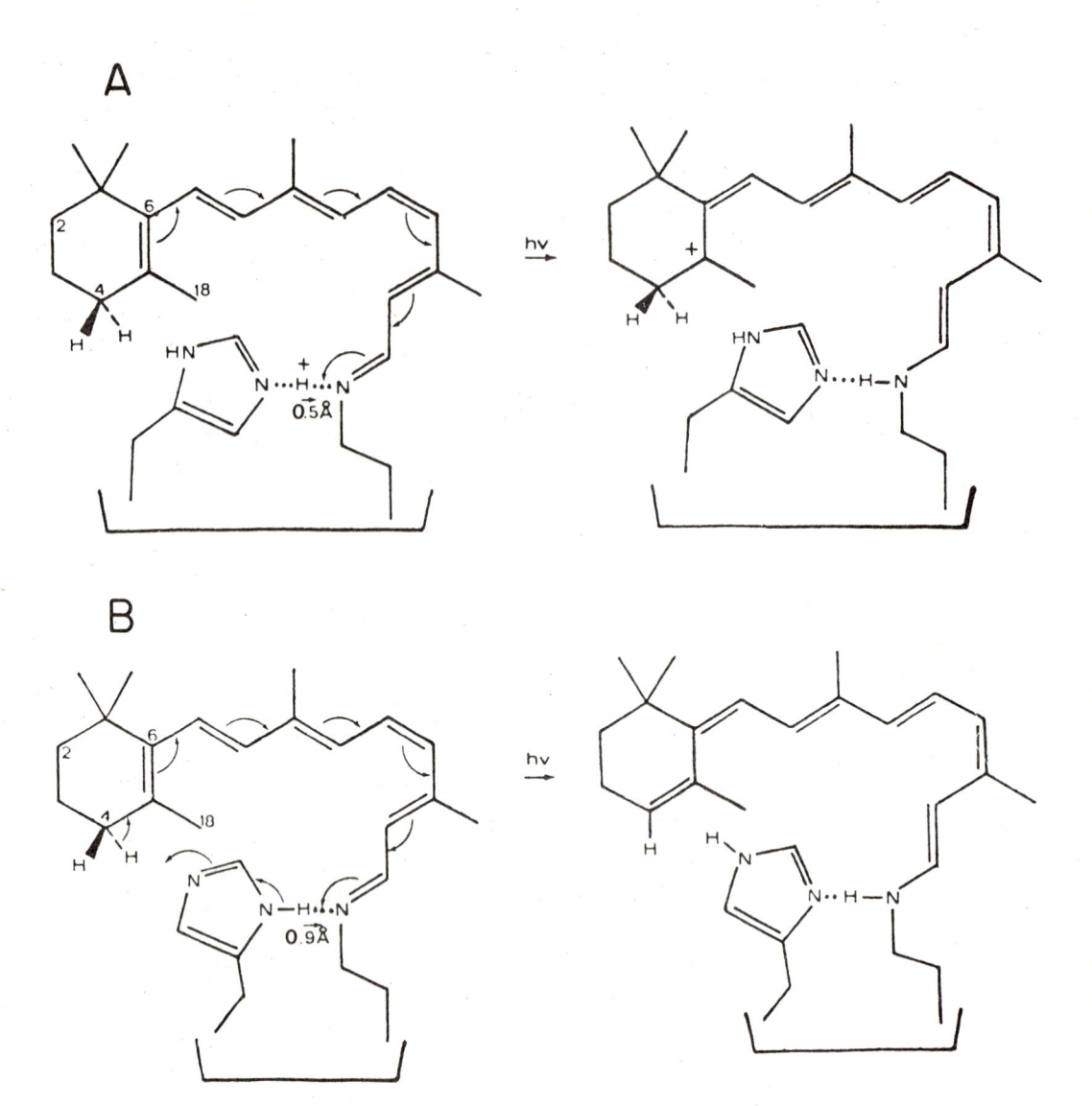

FIGURE 8 Models for proton translocation to form prelumirhodopsin: A, single proton translocation with carbonium ion formation; B, concerted double proton translocation with retro-retinal formation. The tunneling distances, a_0, through the barriers for the formation of prelumirhodopsin are 0.5 Å for single proton model A and 0.9 Å for the concerted double proton translocation.

strong candidate (Fig. 8). At the present time, we can postulate several other candidates and mechanisms involving the proton; however, until more definitive data is available, we shall limit our proposals to the above two models, presented schematically in Fig. 8. If the histidine proton is translocated, this would occur between the hydrogen-bonded nitrogens of the histidine and Schiff base, and the proton would translocate along the N—H stretch coordinate, moving near the Schiff base nitrogen, which has been said to have a "negative" character in the excited state and provide the force for the proton movement. If the proton is originally shared by the two nitrogens $N_h \cdots H \cdots N_S$, the translocation distance will be ~0.5 Å, while if the proton is held strongly by the histidine nitrogen $N_h—H \cdots N_S$, then the distance for translocation would be ~0.9 Å. Under these conditions there would be no need for an immediate hydrogen abstraction for carbonium ion stabilization. In the two-photon model, stabilization can be achieved by the removal of a hydrogen from an adjacent carbon. Protein conformational changes would be affected by the proton translocation; however, their magnitude and rate, especially at very low temperatures, would not be expected to be very high. I can only conclude from the available data that proton tunneling plays a dominant role in the formation of prelumirhodopsin, at least at low temperatures. The exchanges of D for H strongly support, if not confirm, our tunneling proposal and shows that proton tunneling indeed takes place. Which proton? The present data do not provide an unequivocal answer.

REFERENCES

1. HUPPERT, D., and P. M. RENTZEPIS. 1978. *Appl. Phys. Lett.* **32**:241.
2. NETZEL, T. L., and P. M. RENTZEPIS. 1974. *Chem. Phys. Lett.* **29**:337.
3. YOSHIZAWA, T., and Y. KITO. 1958. *Nature (Lond.).* **182**:1604.
4. YOSHIZAWA, T., and G. WALD. 1963. *Nature (Lond.).* **197**:1279.
5. YOSHIZAWA, T. 1972. *In* Handbook of Sensory Physiology. H. Dortnell, editor. Springer-Verlag KG. Berlin. Vol. VII/I, 146–179.
6. ROSENFELD, T., B. HONIG, M. OTTOLENGHI, and T. G. EBREY. 1977. *Pure Appl. Chem.* **49**:341.
7. BUSCH, G. E., M. L. APPLEBURY, A. A. LAMOLA, and P. M. RENTZEPIS. 1972. *Proc. Natl. Acad. Sci. U.S.A.* **69**:2802.
8. SUNDSTROM, V., P. M. RENTZEPIS, K. S. PETERS, and M. L. APPLEBURY. 1977. *Nature (Lond.).* **267**:645.
9. PETERS, K., M. L. APPLEBURY, and P. M. RENTZEPIS. 1977. *Proc. Natl. Acad. Sci. U.S.A.* **74**:3119.
10. OSEROFF, A., and R. CALLENDER. 1974. *Biochemistry.* **13**:4243.
11. LOWDIN, P. O. 1965. *Adv. Quantum Chem.* **2**:213.
12. WADDELL, W., and R. S. BECKER. 1971. *J. Am. Chem. Soc.* **93**:3788.
13. WADDELL, W., A. M. SCHAEFFER, and R. S. BECKER. 1973. *J. Am. Chem. Soc.* **95**:8233.
14. SALTIEL, J., J. D'AGOSTINO, E. D. MEGARITY, L. METTS, K. R. NEARBERGER, M. WRIGHTON, and O. I. ZAFIRIOU. 1973. *Org. Photochem.* **3**:1.
15. VON DER MEER, K., J. J. C. MULDER, and J. LUGSTENBURG. 1976. *Photochem. Photobiol.* **24**:363.
16. MATHIES, R., and L. STRYER. 1976. *Proc. Natl. Acad. Sci. U.S.A.* **73**:2169.

DISCUSSION

WELLER: With your diagram of the proton tunneling from excited rhodopsin to prelumi, do you indicate that an adiabatic proton transfer process occurs from excited singlet state to singlet excited prelumi, or do you consider it to be an adiabatic process leading from singlet excited rhodopsin directly to the ground state of prelumi?

RENTZEPIS: We believe that tunneling takes place in the excited state. Prelumi decays thereafter to lumi with a 50 ns lifetime.

FARRAGI: You've measured the pH effect?

RENTZEPIS: Yes, We've measured the relaxation as function of pH from low pH to a pH of 10, where rhodopsin is believed to denature. In this pH region there is no noteworthy effect on the lifetime.

FARAGGI: If imidazole is participating in the charge transfer mechanism, wouldn't you expect a change in the proton transfer reaction around the pK_a value of the imidazole?

RENTZEPIS: Yes, this is what one finds. However, the process that you call pH is a later process than the one we are considering. In the excited state the nitrogen of the Schiff base is believed to be negative, which provides the force for pulling of the proton. Afterwards, there can be other effects associated with the proteins.

AMEEN: Could you expect any cooperative phenomena to be involved other than the proton tunneling you have described here?

RENTZEPIS: What kind, or between what?

AMEEN: At low temperature, (e.g. 77°K), aggregation may exist or some molecules come much closer to interact with each other. The viscosity variation may cause rigidity of environment and energy exchange processes different from those at room temperatures.

RENTZEPIS: Anything is possible, I guess; however, we have no evidence for aggregation.

RUBIN: Why don't you discuss the process of electron tunneling, as well as proton tunneling? because when you exchange the proton for deuterium you can change the condition for electron tunneling as well.

RENTZEPIS: The reason we don't discuss electron tunneling in this particular case is that we have evidence of proton translocation. Of course you could have asked me another question, "Why not discuss nuclear tunneling?" After all, this would be the first process that could take place. Again may I repeat that we discuss proton tunneling because the strong evidence provided by the deuteration is for proton tunneling. Electron tunneling in this case probably also occurs, as it does in chlorophyll. Because of time limitations, I restricted myself to proton tunneling.

FAST ELECTRON TRANSFER PROCESSES IN CYTOCHROME C AND RELATED METALLOPROTEINS

MICHAEL G. SIMIC AND IRWIN A. TAUB, *Food Engineering Laboratory, U.S. Army Natick Research and Development Command, Natick, Massachusetts 01760 U.S.A.*

ABSTRACT Various free radicals formed on pulse radiolysis of aqueous solutions have been used to investigate the mechanisms of reduction of cytochrome(III) c by inter- and intramolecular electron transfer. The rapid formation of free radicals ($t < 1$ μs) and their high reactivity with cytochrome ($k \approx 10^8 - 5 \times 10^{10} M^{-1}s^{-1}$) make such studies feasible. Reduction of cytochrome by free radicals is monitored by optical methods. Fast optical changes in the 1–500-μs region correspond to reduction of the iron center; whereas the slower changes in the 10–500-ms region are attributed to postreduction conformational changes. It has been concluded that the reduction path is mediated through the crevice and that no reduction intermediates are being formed.

INTRODUCTION

Kinetic studies of electron transfer processes in heme proteins (1–3) have been one of the major approaches in determining the mechanistic behavior and subtle structural features of these large electron carriers found in living systems. These kinetic measurements made *in situ* and in isolated systems are continually improving our understanding of the electron transport systems in general and the mechanisms of energy conversion in particular.

The search for ever faster kinetic techniques brought about the development in the early 60s (4) of pulse radiolysis, which is one of the most convenient techniques in the study of fast electron transfer processes and in some applications has a time resolution approaching picoseconds (5).

The kinetic studies presented here of some free radical reactions with cytochrome c, cyt(III)-c, in aqueous solutions represent an attempt to ascertain the exact paths an electron takes toward its focal point, Fe(III). It has already been demonstrated (6) that the reactions of some free radicals with cyt(III)-c can lead to quantitative reduction, yielding an unperturbed cyt(II)-c. It has also been stated (6) that reduction of heme proteins by free radicals under controlled radiolytic conditions may have certain advantages over classic reducing agents, which may involve as intermediates either H atoms (e.g., H_2/Pt) or other free radicals derived from two electron reductants (e.g., dithionite, ascorbate).

METHODS

Primary water free radicals were generated by irradiating aqueous solutions with either Co-60 γ-rays, referred to as steady-state radiolysis, at dose rates 1.8 krad/min or by high energy (2 MeV) electron beams of short duration (50 ns), referred to as pulse radiolysis, at dose rates 1-2 krad/pulse. In the pulse experiments, a Febetron 705 was used as a generator of pulsed electron beams in conjunction with a kinetic spectrophotometric detection system with time resolution of 0.5 μs (7). Standard pulse radiolysis procedures were used to obtain absorption spectra of the transients and the products and to determine the corresponding rate constants.

High purity Millipore-filtered water (Millipore Corp., Bedford, Mass.) and chemicals were used in preparing solutions. Sigma cyt(III)-c, type VI (Sigma Chemical Co., St. Louis, Mo.), from horse heart prepared in the absence of trichloroacetic acid (a strong electron and free radical scavenger) was used without further purification. All solutions were 1 mM in phosphate buffer, and the pH was adjusted with KOH. Purified carboxymethylated cytochrome c (kindly supplied by Dr. R. E. Eakin), with both methionine moieties converted into $>\overset{+}{S}-CH_2CO_2^-$ was used (8). Whale myoglobin(III) was used without purification. Physiological integrity of both cyt(II)-c and myoglobin(II) was fully demonstrated.

RESULTS

Generation of Free Radicals

Free radicals were generated by reaction of added solutes with the primary water free radicals, which in turn were formed initially either on steady-state or pulse radiolysis of aqueous solutions (4):

$$H_2O \leadsto e_{aq}^-(2.8), OH(2.8), H(0.6), H_2(0.4), H_2O_2(0.8), \tag{1}$$

where G values (number of species formed/100 eV of energy absorbed) in parentheses, strictly speaking, refer to dilute solutions and may vary to a certain extent depending on the nature and concentration of the dissolved material.

Considering the probable connection between the protein free radicals presumably formed in the cytochrome and the redox processes, we generated model peptide free radicals for investigation, the choice being made on the basis of high reactivities of the cytochrome components for e_{aq}^- or OH. Related model radicals were also employed. All the free radicals used and the compounds from which they were generated are shown in Table I. The concentrations of the solutes were chosen on the basis of particular reaction rate constants (9, 10).

The reactions of OH radicals were always studied in the presence of N_2O because the hydrated electron, a strong reductant ($E^\circ = -2.8$ V), readily converts to OH radical, a strong oxidant ($E^\circ = +2$ V):

$$e_{aq}^- + N_2O \rightarrow OH + OH^- + N_2,$$
$$k_2 = 8.0 \times 10^9 \text{ M}^{-1}\text{s}^{-1}. \tag{2}$$

Electron adducts of imidazole and the peptide bond are easy to generate because of their high reactivity with the hydrated electron (9), e.g.,

TABLE I

REDUCTION RATE CONSTANTS FOR CYT(III)-c BY FREE RADICALS IN AQUEOUS SOLUTIONS AT 20°C

Solute	{S} mM	pH	Radical	k, M^{-1}s^{-1}
t-Butanol	10	7.0	e^-_{aq}	5.5×10^{10} (ref. 15)
t-Butanol*	10	7.0	(CH$_3$)$_2$ĊH$_2$COH	No reaction
Formate*	10	7.0	·CO$_2^-$	1.3×10^9‡
Gly anhydride*	10	6.8	CH$_2$CONHĊHCONH	$<10^7$
Gly anhydride	10	6.8	—ĊOHNH—	8.0×10^8
Pentaerythritol*	100	6.8	(CH$_2$OH)$_3$CĊHOH	$<10^6$
	100	9.8	(CH$_2$OH)$_3$CĊHOH	1.6×10^8
Benzoate*	10	6.8	·C$_6$H$_5$(OH)CO$_2^-$	‖
Benzoate	10	6.8	C$_6$H$_5$COOH$^{\overline{\cdot}}$	1.8×10^9
Imidozole*	3	7.0	·Im—OH	7.5×10^8
Imidazole§	3	6.3	Im$^{\overline{\cdot}}$(H$^+$)	$<10^7$
His-His*	1	7.1	·His—OH	2.1×10^8
His-His§	10	6-7	His$^{\overline{\cdot}}$(H$^+$)	$<10^7$
Met*	10	6.8	Met—OH¶	1.0×10^9
Phe*	10	6.8	Phe—OH	‖
Acetophenone	10	7.0	C$_6$H$_5$ĊOHCH$_3$	8.0×10^8
p-Nitroacetophenone	2	7.0	$^-$O$_2$Ṅ—C$_6$H$_4$COCH$_3$	1.2×10^8

*In the presence of 1 atm N$_2$O to convert e^-_{aq} into OH radicals.
‡Ref. 16 indicates a decrease in rate with increasing formate concentration.
§In the presence of 100 mM pentaerythritol to eliminate OH radicals—the resulting pentaerythritol radical does not show any reaction under pulse conditions and pH < 7.
‖ Poor reductant even under steady-state conditions, k probably <1 M^{-1}s^{-1}.
¶Or free radicals derived from it.

$$e^-_{aq} + \underline{CH_2CONHCH_2CONH}$$
$$\rightarrow \underline{CH_2\dot{C}(O^-)NHCH_2CONH} \underset{}{\overset{H^+}{\rightleftharpoons}} \underline{CH_2\dot{C}(OH)NHCH_2CONH}$$
$$k = 1.7 \times 10^9 \text{ M}^{-1}\text{s}^{-1}. \quad (3)$$

The corresponding electron adducts of Phe, Tyr, and Trp were not amenable to investigation because of their much lower k values ($\sim 10^8$ M^{-1}s^{-1}) relative to cyt(III)-c. The electron adduct of benzoate (II) was used as a substitute for Phe.

When electron adducts were generated, OH radicals were removed with solutes giving rise to nonreducing radicals, e.g., t-butanol radicals or to pentaerythritol radicals, which are unreactive under pulsed conditions in the time frame for observation.

One of the strongest reducing radicals ($E° = -2$ V), excluding e^-_{aq} and H atoms, is ·CO$_2^-$ (12). It is usually produced via

$$OH + HCO_2^- \rightarrow \cdot CO_2^- + H_2O$$

$$k = 2.0 \times 10^9 \, M^{-1}s^{-1} \, (10). \tag{4}$$

No reaction of $\cdot CO_2^-$ was observed with any of the protein components. The consequences of this observation for the mechanisms of reduction of the cytochrome are discussed below.

The α-peptide radicals are also reducing agents, though not as good as electron adducts of the peptide bond or $\cdot CO_2^-$. They can be conveniently produced via

$$OH + \underline{CH_2CONHCH_2CONH} \rightarrow \underline{\cdot CHCONHCH_2CONH} + H_2O$$

$$k = 1.2 \times 10^9 M^{-1}s^{-1}. \tag{5}$$

Besides the abstraction reaction, i.e., reactions 4 and 5, OH readily adds (10) to the benzene ring (Phe, Tyr), to heterocyclic systems (His, Trp) and to sulfur (Met, Cys). The resulting free radicals are shown in Table I.

In addition to the mechanism for generating a free radical, one has to consider its acid-base properties (13), because the electron transfer processes depend on the state of protonation of the donor as well as of the acceptor.

Reaction Rate Constants

The k values for reduction were derived primarily from the growth of the α-band at 550 nm. When possible, the decay of absorbance of the reducing radical was also followed. Unfortunately, the absorption changes produced upon reduction of the cytochrome in most cases masked the absorption bands of the free radicals. The absorption changes were followed over at least two half-lives. The resulting rate constants are shown in Table I.

The k values have been found to be pH-dependent at pH > 7. In Fig. 1 the pH dependence for $\emptyset NO_2^-$ and $(CH_3)_2\dot{C}OH$ radicals is shown. In solutions of nitrobenzene, the OH radical gives HO—$\emptyset NO_2$, which is unreactive with cyt(III)-c, i.e., $k < 10^7 M^{-1}s^{-1}$, allowing the $\emptyset NO_2^-$ reaction to be followed. The solutions of isopropanol were saturated with N_2O, hence the $(CH_3)_2\dot{C}OH$ radicals were generated exclusively.

The pentaerythritol radical $(CH_2OH)_3C\dot{C}HOH$ is formed by abstraction of hydrogen by OH (14). This radical quantitatively reduces the cytochrome under steady-state conditions (6) but fails to do so under pulsed conditions; at pH < 7 and because of high concentrations, these radicals preferentially recombine before having a chance to react with the cytochrome. An increase of pH above 7 makes the cytochrome more reactive, and $k = 1.6 \times 10^8 \, M^{-1}s^{-1}$ at pH 9.8 was measured (14). The increase in the extent of cytochrome reduction with increasing pH is shown in Fig. 2. Above pH 10 the extent of reduction begins to fall off, presumably as a result of some other changes in the cytochrome.

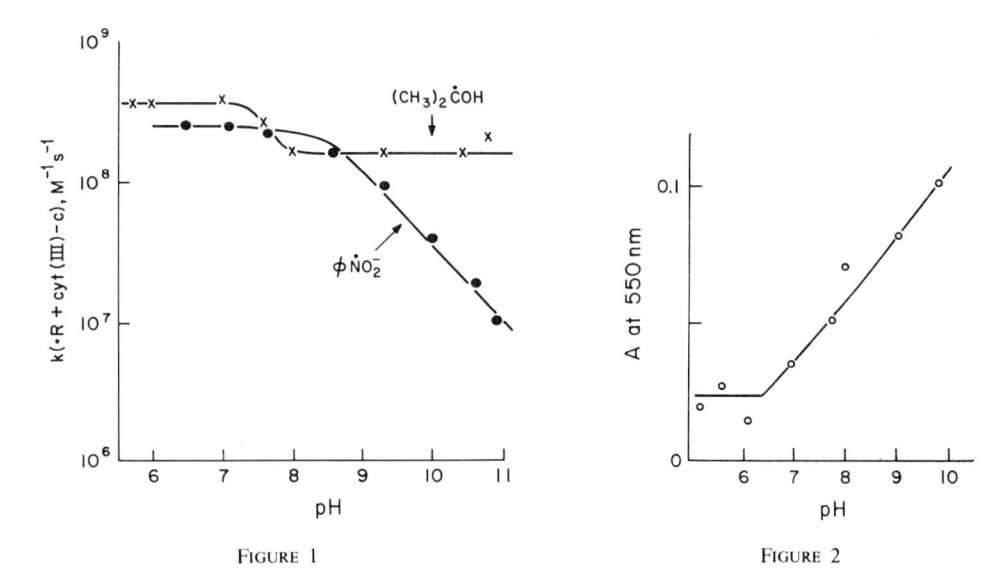

FIGURE 1 The pH dependence of reduction rate constants for cyt(III)-c by ϕNO_2^- and $(CH_3)_2\dot{C}OH$ radicals. Conditions: —O—; 2×10^{-5} M cyt(III)-c, 5×10^{-3} M nitrobenzene, 10^{-3} M phosphate buffer; —X—, 2×10^{-5} M cyt(III)-c, 5×10^{-2} M isopropanol, 10^{-3} phosphate buffer, 1 atm N$_2$O. Dose/pulse = 0.5 krad.

FIGURE 2 The pH dependence of the change in absorption at 550 nm corresponding to the reduction of cyt(III)-c by the pentaerythritol radical, $(CH_2OH)_3C\dot{C}HOH$. Conditions: 2×10^{-5} M cyt(III)-c, 2×10^{-2} M pentaerythritol, 10^{-3} phosphate buffer, 1 atm N$_2$O. Dose/pulse = 1.6 krad.

Reduction by the Hydrated Electron

The hydrated electron is one of the strongest reducing agents, and much attention has been paid to its reactions (4). The rate constant of e_{aq}^- reaction with cyt(III)-c is extremely high, $k = 5.5 \times 10^{10}$ M^{-1}s^{-1} (15), and it has been suggested that the full development of cyt(II)-c absorption is not concomitant with electron decay (15). The delayed development has a $t_{1/2} = 7$ μs, attributed to intramolecular electron transfer from the initial reaction site to the Fe center (15). These observations were made in the presence of 0.5 M t-butanol, but such a first-order change is not observed either when formate (16) or when 0.1 M pentaerythritol in this study were used as OH scavengers.

A 100% reduction yield by e_{aq}^- under steady-state conditions (6) and under pulsed conditions can be observed. However, the buildup of cyt(II)-c makes the yield vs. dose plot nonlinear because of the rather strong reactivity of cyt(II)-c toward e_{aq}^-, $k = 2.0 \times 10^{10}$ M^{-1}s^{-1}.

Reaction of the OH Radical

The OH radical has been observed to reduce cyt(III)-c (6). The following multistep sequence is suggested. First, OH abstracts H or adds to aromatic and sulfur-containing residues.

$$\text{OH} + \text{cyt(III)-}c \rightarrow \cdot\text{cyt(III)-}c \quad \text{and} \quad \text{HO-}\dot{\text{c}}\text{yt(III)-}c. \tag{6}$$

The $k_6 = 2.7 \times 10^{10}\,\text{M}^{-1}\text{s}^{-1}$ value was derived from competition studies.

Some of the resulting free radicals on this protein of the cytochrome appear to reduce Fe(III) intramolecularly,

$$\left.\begin{array}{r}\cdot\text{cyt(III)-}c \\ \text{HO-}\dot{\text{c}}\text{yt(III)-}c\end{array}\right\} \rightarrow \text{cyt}'(\text{II})\text{-}c. \tag{7}$$

This reduction was followed at 550 nm, and two distinct first-order rate constants were derived, $k_7' = 2 \times 10^5\,\text{s}^{-1}$ and $k_7'' = 3 \times 10^4\,\text{s}^{-1}$. Although the rates were relatively independent of the batch of the cytochrome, the extent of reduction ranged from 25 to 55%. Because of this apparent irreproducibility, consideration of the OH-induced reduction will be postponed. From Table I it is clear that some OH-induced protein radicals could very efficiently reduce the Fe(III) center.

Conformational Changes

Reduction of the cytochrome by most free radicals, as indicated by the growth of the α-band, takes place in the 20–500-μs time regime in these experiments. This change is then followed by a slower absorption change with $t_{1/2} = 0.1$ ms, which was attributed to conformational changes (16).

In this study the extent of the slow change was found to be variable (0–50% of the total ΔA) and dependent on the batch of cytochrome. It cannot be correlated to any free radical reaction, and it is most likely affected by the conformation of the cytochrome. No meaningful correlation can be drawn at present.

DISCUSSION

Most reducing free radicals ($E' < 0$ V) reduce cyt(III)-c with rate constants about 10^8–$2 \times 10^9\,\text{M}^{-1}\text{s}^{-1}$. The exceptions (14), such as $\cdot\text{O}_2^-$ ($k = 1.6 \times 10^6\,\text{M}^{-1}\text{s}^{-1}$ at pH <7) and flavine mononucleotide semiquinone radical ($k = 1.5 \times 10^7\,\text{M}^{-1}\text{s}^{-1}$ at pH 7), fall into a category of free radicals with redox potentials close to that of the cytochrome ($E' = +0.25$ V).

In addition to energetics, steric factors play a substantial role, as in the case of the pentaerythritol radical, $(\text{CH}_2\text{OH})_3\text{C}\dot{\text{C}}\text{HOH}$, which has a neopentane structure. The relatively low rate constant for reduction by this radical at pH <7 ($k < 10^6\,\text{M}^{-1}\text{s}^{-1}$) cannot be rationalized on purely energetic grounds because that radical reacts with hemin(III)-c ($E' \sim -0.2$ V) fast, $k = 3.0 \times 10^8\,\text{M}^{-1}\text{s}^{-1}$ (17). Though reduction rate at pH <7 is low, it increases at pH >7 ($k = 1.6 \times 10^8\,\text{M}^{-1}\text{s}^{-1}$ at pH 10), suggesting that the reduction path is through the crevice, which widens at higher pHs. The absence of initial reduction at external sites on the protein is apparent. This absence is consistent with the fact that the pentaerythritol radical does not react with either the peptide bond (e.g. glycine anhydride) or aromatic amino acids (Phe, His, etc.). The $\cdot\text{CO}_2^-$ radical, one of the strongest reducing radicals (12), also appears to be unreactive toward the peptide bond and the aromatic residue, indicating that operation of the so-called "Winfield mechanism" (18) would be highly improbable.

Most of the free radicals reduce the Fe center directly, e.g.,

$$\varnothing NO_2^- + cyt(III)\text{-}c \rightarrow \varnothing NO_2 + cyt(II)\text{-}c. \tag{8}$$

For protonated free radicals the electron transfer is associated with a release of a proton, e.g.,

$$(CH_3)_2\dot{C}OH + cyt(III)\text{-}c \rightarrow (CH_3)_2CO + H^+ + cyt(II)\text{-}c. \tag{9}$$

Despite the observations with the pentaerythritol radical, conformational changes induced by the increased pH (19–21) usually decrease the reduction rate constants as is found for classic reductants. These changes in k (see Fig. 1) fall into a similar pH range as the reported pK_a = 8.9 to 9.3 (22) for cytochrome conformers. The change that occurs at about pH 7.5 for the $(CH_3)_2\dot{C}OH$ radical may be the result of denaturing effect of the isopropanol from which the radical is derived. A drop in the k value for this radical was also observed at pH < 7 for solutions 2 M in isopropanol (14). The effect on k is therefore a composite one and is dependent on ΔE°, steric factors, and the influence of the medium.

TABLE II

COMPARISON OF REACTION RATE CONSTANTS FOR SOME FREE RADICALS WITH HEMIN C, WHALE MYOGLOBIN, HORSE CYTOCHROME C AND CARBOXYMETHYLATED CYTOCHROME C IN AQUEOUS SOLUTIONS AT pH = 7 AND 20°C

Free radical	k, M^{-1}s^{-1}*			
	Heme(III)-c‡	Mb(III)	cyt(III)-c	(cm)₂cyt(III)-c§
e_{aq}^-	2.1 × 10¹⁰	2.5 × 10¹⁰	5.5 × 10¹⁰ ‖	4.4 × 10¹⁰
·CO₂⁻ ¶	1.3 × 10⁹	2.0 × 10⁹	1.3 × 10⁹	1.4 × 10⁸
HOĊCO₂⁻ ‡‡	1.0 × 10⁸	3.5 × 10⁷	1.7 × 10⁸	2.8 × 10⁷
HOĊHCO₂⁻				
φCOOH⁻	2.1 × 10⁹	1.8 × 10⁹	1.8 × 10⁹	1.4 × 10⁹
CH₃N⟨·⟩⟨⟩NCH₃⁺ **	1.6 × 10⁹	2.3 × 10⁸	3.6 × 10⁸	5.2 × 10⁷
CH₃N⟨·⟩⟨⟩CONH₂ **	2.1 × 10⁹	8.7 × 10⁸	1.4 × 10⁹	1.1 × 10⁹

*All k values ±20%. Mb, myoglobin.
‡From ref. 17.
§M. G. Simic. Submitted for publication.
‖ From ref. 15.
¶10⁻² M formate.
**10⁻¹ M formate.
‡‡10⁻² tartrate.

An effect on free radical reduction kinetics can also be observed in chemically modified cytochromes. In carboxymethylated cyt(III)-c, Met 80 is modified and detached from Fe. The rate constants for reduction by $\cdot CO_2^-$, tartrate, and viologen radicals are consequently reduced by a factor of 6. Replacement of Met 80 sulfur ligand by an alternative protein ligand at pH > 7 (23) and the role of Met 80 in redox processes have already been recognized (24). On the other hand, k values for aromatic donors with low redox potentials such as electron adducts of benzoate, acetophenone, and 1-methyl nicotinamide are $\sim 10^9 M^{-1} s^{-1}$ for the heme compounds. The high k values for the reduction of cytochrome by aromatic electron donors suggest a mechanism involving $\pi - \pi$ interactions of the donors, the aromatic residues on the cytochrome and the porphyrin ring. Identical reduction rate constants of $k = 1.8 \times 10^9 M^{-1} s^{-1}$ for the reaction of the benzoate electron adduct with both the cytochrome and myoglobin, in spite of their totally different configurations, suggest that other interactions can also occur. A $\pi - \pi$ interaction between the donor and the porphyrin ring may play an important role as indicated by the equivalence of k values for the benzoate electron adduct reduction of cytochrome, carboxymethylated cytochrome, and myoglobin (see Table II).

Direct and stoichiometric reduction of the Fe in the cytochrome by e_{aq}^- that does not involve mediation by a longer-lived intermediate (although an intermediate with $t_{1/2} < 0.15 \mu s$ could be involved and not detected; 16) is surprising in view of a rather high reactivity of e_{aq}^- with many of the protein components. This can perhaps be rationalized, considering the effect of charge distribution (high positive charge around the cleft would attract e_{aq}^-) and the reluctance of e_{aq}^- to enter hydrophobic regions of the protein.

Received for publication 1 December 1977.

REFERENCES

1. BENNET, L. E. Metalloprotein redox reactions. 1973. *In* Current Research Topics in Bioinorganic Chemistry. S. J. Lippard, editor. John Wiley & Sons, Inc., New York.
2. SUTIN, N. 1977. Electron transfer reactions of cytochrome c. *Adv. Chem. Ser.* **162**:156.
3. HOLWERDA, R. A., S. WHEELAND, and H. B. GRAY. 1976. Electron transfer reactions of Copper Proteins. *Annu. Rev. Biophys. Bioeng.* **5**:363.
4. HART, E. J., and M. ANBAR. 1970. The hydrated electron. John Wiley & Sons, Inc., New York.
5. BRONSKILL, M. J., R. K. WOLF, and J. W. HUNT. 1970. Picosecond pulse radiolysis studies. I. The solvated electron in aqueous and alcohol solutions. *J. Chem. Phys.* **53**:4201.
6. SIMIC, M. G., and I. A. TAUB. 1977. Mechanisms of inter- and intra-molecular electron transfer in cytochromes. *Discuss. Faraday Soc.* **63**:270.
7. SIMIC, M., P. NETA, and E. HAYON. 1969. Pulse radiolysis on aliphatic acids in aqueous solutions. II. Hydroxy and polycarboxylic acids. *J. Phys. Chem.* **73**:4214.
8. MORGAN, L. O., R. T. EAKIN, P. J. VERGAMINI and N. A. MATWIYOFF. 1976. Carbon-13 nuclear magnetic resonance of heme carbonyls, cytochrome c and carboxymethyl derivatives of cytochrome c. *Biochemistry.* **15**:2203.
9. ANBAR, M., M. BAMBENEK, and A. B. ROSS. Selected specific rates of reactions of transients from water in aqueous solution. 1. Hydrated electron. *Natl. Stand. Ref. Data Ser., Natl. Bureau of Standards.* **43**.

10. DORFMAN, L. M., and G. E. ADAMS. 1973. Reactivity of the hydroxyl radical in aqueous solutions. *Natl. Stand. Ref. Data Ser., Natl. Bur. Stand.* **46**.
11. SIMIC, M. G., and M. Z. HOFFMAN. 1972. Acid-base properties of radicals produced on pulse radiolysis of aqueous solutions of benzoic acid. *J. Phys. Chem.* **76**:1398.
12. LILIE, J, G. BECK, and A. HENGLEIN. 1971. Pulsradiolyse und polarographie: halbstufen potentiale fur die oxydation und reduktion von kurzlebigen organischen radikalen an der Hg-elektrode. *Ber. Bunsenges. Phys. Chem.* **75**:458.
13. HAYON, E., and M. SIMIC. 1974. Acid-base properties of free radicals in solutions. *Acc. Chem. Res.* **7**:114.
14. SIMIC, M.G., I. A. TAUB, J. TOCCI, and P. A. HURWITZ. 1975. Free radical reduction of ferricytochrome c. *Biochem. Biophys. Res. Commun.* **62**:161.
15. LICHTIN, N. N., A. SHAFFERMAN, and G. STEIN. 1973. Reaction of hydrated electrons with ferricytochrome c. *Science (Wash. D. C.).* **179**:680.
16. LAND, E. J., and A. J. SWALLOW. 1971. One-electron reactions in biochemical systems as studied by pulse radiolysis. V. Cytochrome c. *Arch. Biochem. Biophys.* **145**:365.
17. GOFF, H., and M.G. SIMIC. Free radical reduction of hemin c. *Biochem. Biophys. Acta.* **392**:201.
18. DICKERSON, R. E. 1972. X-ray studies of protein mechanisms. *Annu. Rev. Biochem.* **41**:815.
19. STELLWAGEN, E., and R. CASS. 1974. Alkaline isomerization of ferricytochrome c from *Euglena gracilis. Biochem. Biophys. Res. Commun.* **60**:371.
20. GREENWOOD, C, and G. PALMER. 1965. Evidence for the existence of two functionally distinct forms of cytochrome c monomer at alkaline pH. *J. Biol. Chem.* **240**:3660.
21. CZERLINSKI, G. H., and K. DAR. 1971. On the electron transfer-coupled proton release of cytochrome c. *Biochem. Biophys. Acta.* **234**:57.
22. GREENWOOD, C., and M. T. WILSON. 1971. Studies on ferricytochrome c. 1. Effect of pH, ionic strength and protein denaturants on the spectra of ferricytochrome c. *Eur. J. Biochem.* **22**:5.
23. BLUMBERG, W. E., and J. PEISACH. 1971. Unified theory of low-spin forms of all ferric heme proteins as studied by EPR. *In* Probes of Structure and Function of Macromolecules and Membranes. B. Chance, T. Yonatani and A. S. Mildvan, editors. Academic Press, Inc., New York. 215.
24. SALEMME, F. R. 1977. Structure and function of cytochrome c. *Annu. Rev. Biochem.* **46**:299.

DISCUSSION

CZERLINSKI: The increase of the reduction of ferricytochrome c by the pentaerythritol radical (Fig. 2) is certainly interesting. You indicate in the text that this reduction decreases at pH 11. This effect is probably due to the second protonic dissociation described in the cited reference of Czerlinski and Dar (21). Your Fig. 1 points to similar deviations at pH 11. Do you have any new information on this point?

SIMIC: Yes. Referring to Fig. 2, the pentaerythritol radical has a neopentane structure and it has difficulties in reducing cytochrome(III)-c to cytochrome(II)-c. The pentaerythritol radical has absolutely no difficulty in reducing hemin-c, which has a redox potential of -0.2 V, much lower than that of cytochrome-c, which is $E^{01} = +0.250$ V. Hence there are no energetic considerations. Only the size and the structure of the radical is preventing it from reducing cytochrome(II)-c. Now, you are quite right about the high pH effect. I have not shown points beyond pH 10; I just didn't put them on. The k-values went considerably above pH 10, so something has been happening to cytochrome-c and also to the pentaerythritol radical in this pH region. Namely, it is deprotonating and is in the anionic form. The anionic form is an even better reductant than the protonated form, so the state should not stop it from reducing cytochrome-c. On the other hand, the increase in reduction above pH 7 (Fig. 2) is too far from the pK value of the radical, which is about 10.4, to be accounted for. Therefore the pH changes involving cytochrome-c here are something you have talked about in your paper

(21). You did also talk about the pK value around 7, I think, in another paper (Czerlinski, 1973). Right?

CZERLINSKI: The electron transfer-linked pK_H is near 9. One may wonder how this pK_H extends its effects all the way to pH = 7. I have another remark regarding the action of the pentaerythritol radical. Could the electron in the pentaerythritol radical be much more localized than in the others? The chemical structure seems to point in this direction. The electron of the pentaerythritol radical may need to get into close contact with iron(III) and cannot directly interact with the porphyrin system:

SIMIC: Yes, I fully believe in what you said. The thesis of the paper is that the delocalization of the electron, promoted by the aromaticity of the donor, substantially aids the electron transfer.

MAUZERALL: A simple explanation for the slow reaction of the pentaerythritol radical is possible in terms of electron tunneling. It is not so much a question of electron delocalization as it is of distance. In this radical the odd electron on the carbon may be held sufficiently further from the cytochrome by the —CH_2OH groups to slow the electron transfer below that of a radical without such groups. The decrease in rate would be fit by a reasonable tunneling parameter of 0.5 Å.

SIMIC: That's right. The cleft is hydrophobic in nature and these hydroxy derivatives are of course extremely hydrophilic and they might have problems in penetrating to a distance where it could tunnel. But we have not observed tunneling.

AUSTIN: Is it possible to measure the reaction rate of something like oxygen with the hemes in your experiments and get attachment rates?

SIMIC: Yes, you can. In the reaction rate of cytochrome(II)-c with oxygen, cytochrome(II)-c is extremely stable and does not react with oxygen. However, carboxymethylated cytochrome, which means binding —$CH_2CO_2^-$ to the sulfur of Met-80, results in the release of Met-80 from the iron, so that the iron is left with only five ligands instead of six. The redox potential is considerably reduced and you can measure the rate of oxygen with carboxymethylated cytochrome(II)-c. On the other hand, you can also use part of cytochrome-c, namely hemin(III)-c. The Fe(III) form can be rapidly reduced (1–10 μs) with an electron donor. You can then study the reactions of the reduced form with oxygen. This method is widely used in pulse radiolysis. For instance, we find that hemin(II)-c reacts with O_2 with $k = 4 \times 10^8 M^{-1} s^{-1}$. This is quite a fast rate constant, but not quite diffusion controlled.

KLAPPER: I have an answer to Dr. Austin's question on ligand binding to heme proteins previously reduced by the hydrated electron. Two papers have appeared recently (Ilan et al., 1978; Ho et al., 1978) on the reactions of CO and O_2 to singly reduced hemoglobin. The results from the two laboratories are comparable, though not identical. The rate constants for the binding of O_2 or CO, in the presence and absence of inositol hexaphosphate, onto hemoglobin with one iron reduced and the remaining three oxidized, are similar to those obtained by other techniques for the binding of the first O_2 or CO onto the fully reduced protein.

REFERENCES

CZERLINSKI, G. 1973. *Biochem. Biophys. Acta.* **275**:480.
HO, K., M. H. KLAPPER, and l . m . DORFMAN. 1978. *J. Biol. Chem.* **253**:238–241.
ILAN, Y. A., A. SAMUNI, M. CHEVION, and G. CZAPSKI. 1978. *J. Biol. Chem.* **253**:82–86.

EXPLORING FAST ELECTRON TRANSFER PROCESSES BY MAGNETIC FIELDS

KLAUS SCHULTEN AND ALBERT WELLER, *Max-Planck-Institut für Biophysikalische Chemie, Abteilung Spektroskopie, D 3400 Göttingen, Germany*

ABSTRACT Photoinduced electron transfer generates radical pairs which recombine within 10^{-9}–10^{-8} s by electron back-transfer to either singlet or triplet products. The product distribution determined by the spin motion of the unpaired electrons in the radical pairs is affected by external magnetic fields. The analysis of the magnetic field effect furnishes new information about electron transfer processes. Light-induced electron transfer in polar solvents and in the bacterial photosynthetic reaction center are discussed as examples.

INTRODUCTION

Unpaired electron spins in radicals experience various interactions (e.g. spin-orbit, exchange, hyperfine, Zeeman) and as a result carry out coherent and stochastic motions. In aromatic radicals without heavy atom substituents the electron spin motion over short periods of about 100 ns is coherent, because of the intramolecular hyperfine interaction between the unpaired electron spins and their surrounding nuclear spins. This motion entails the precession of the unpaired electron spins around an axis given by a combination of the nuclear magnetic moments with a frequency of 10^7–10^9 s^{-1}. An additional precession can be induced through the Zeeman interaction by an external magnetic field.

When radical pairs are generated in a pure quantum state, e.g. by a photon-induced electron transfer in a singlet spin state, this motion can be observed through the radical recombination products, the spin multiplicity of which is determined by the relative orientation of the electron spins at the instant of recombination (e.g. electron back-transfer). External magnetic fields alter the electron spin motion and, thereby, also the yields of (singlet versus triplet) recombination products (1). An analysis of the hyperfine coupling-induced recombination yields and their magnetic field modulation furnishes valuable information about the existence of a short-lived radical pair and its microscopic diffusion in a solvent-mediated force field, and about the reaction propensities to form singlet and triplet products (1–3).

The transformation of light into chemical energy in the photosynthetic apparatus of bacteria and plants originates also from a photoinduced electron transfer reaction. In the case of the bacterium *Rhodopseudomonas spheroides* under conditions that enforce electron back-transfer, the reaction processes are also affected by external magnetic fields (4–6). These magnetic field effects originate from the hyperfine coupling in the

FIGURE 1

FIGURE 2

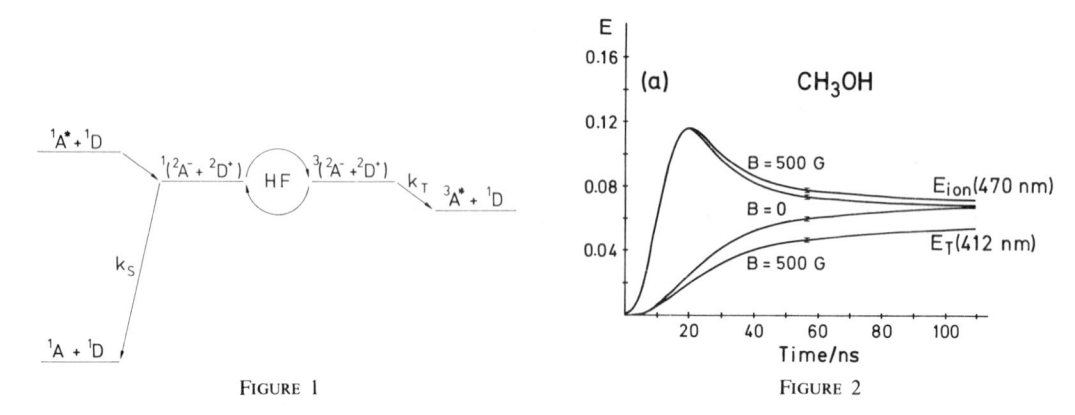

FIGURE 1 Reaction scheme for the primary electron transfer processes of electron donor-acceptor systems in polar solvents. Singlet, doublet, and triplet states are indicated by the left-hand side superscripts 1, 2, and 3, respectively.

FIGURE 2 Time evolution of the radical ion (E_{ion} = 10.8 × 10³ C_{ion}[mol/liter]) and triplet (E_T = 8.7 × 10³ C_T[mol/liter]) extinction of the system pyrene/N,N-dimethylaniline in methanol with and without an external magnetic field of 500 G. Error bars indicate standard deviations obtained from 8–10 measurements.

radicals but, more interestingly, depend also on other spin-dependent interactions characteristic of the primary electron transfer in photosynthesis.

ELECTRON TRANSFER PROCESSES IN SOLUTION

The electron transfer system employed in the study described in this chapter is presented in Fig. 1 by its relevant energy levels. In polar solvents (e.g. methanol, acetonitrile) electron acceptor molecules A (pyrene) are excited by a nanosecond laser flash from the singlet ground state ^{1}A to the singlet excited state ^{1}A*. Upon excitation the electron affinity of the acceptor molecules is increased so that when they collide with a donor molecule ^{1}D (N,N-dimethylaniline) an electron is transferred, resulting in the formation of a radical ion pair (^{2}A$^-$ + ^{2}D$^+$). In this pair the electron can be transferred back. For this process there are, however, two possibilities in which either the singlet ground state (^{1}A + ^{1}D) or the triplet excited state of the acceptor (^{3}A* + ^{1}D) is produced. The state obtained depends on the relative alignment of the two unpaired electron spins at the moment of the electron back-transfer: singlet alignment leads to the singlet ground state and triplet alignment to the triplet state.

Actually the recombination (i.e. electron back-transfer) takes place over two different time periods. A slow, so-called homogeneous, recombination occurs after separation of the initially formed radical ion pairs. Because of the small concentration of the initial pairs, the radicals have to diffuse for about 10 µs before they encounter other free radical ions with subsequent electron transfer. In these random encounters of radicals the electron back-transfer can result in singlet as well as triplet products. Since the relative orientation of the electron spins of ^{2}A$^-$ and ^{2}D$^+$ is random, one has a 25% probability for singlet and a 75% probability for triplet alignments. Hence predominantly triplet products are to be expected in the homogeneous recombination phase of the radical pairs.

A fast so-called geminate recombination results from the direct back-transfer of the electron within the initially formed radical pairs. These pairs, however, are generated from singlet precursors and it is, therefore, to be expected that they are produced in a singlet spin state, i.e. $^1(^2A^- + {}^2D^+)$. Hence, only singlet products $^1A + {}^1D$ should result from the electron back transfer, except if during the lifetime of the pairs the hyperfine interaction succeeds in bringing the electron spins to a triplet alignment, i.e. $^3(^2A^- + {}^2D^+)$.

The question of whether triplet products are formed in recombinations of radical ion pairs can be settled by observing the concentration of the ions $^2A^- + {}^2D^+$ and the triplet products $^3\overset{*}{A}$ by means of time-resolved absorption spectroscopy. Fig. 2 shows for the acceptor/donor system pyrene/N,N-dimethylaniline in the solvent methanol the observed radical ion and triplet concentrations obtained from the absorption of $^2A^-$ (at 470 nm) and $^3\overset{*}{A}$ (at 412 nm). The radical ion signal shows a very rapid rise after the laser flash and a decay in two time domains, a fast process lasting for several nanoseconds and a slow one extending over much longer times. The different decay modes of the radical ions entail the geminate and the homogeneous recombination (1). It is clearly seen in Fig. 2 that triplet products are formed at short as well as longer times. The early triplet products originate from the geminate phase of the radical ion pair recombination.[1] This implies that the hyperfine induced spin motion leading from singlet to triplet alignments takes place within the time the radicals spend in each other's neighborhood before they separate. Fig. 2 also demonstrates that the radical ion as well as the triplet product concentrations are affected by an external magnetic field. In a field of 500 G the triplet product concentration is reduced and the radical ion concentration slightly increased. This effect originates from the influence of the magnetic field on the electron spin motion and is demonstrated schematically in Fig. 3. For radical pairs with predominating hyperfine interaction (zero-field-splitting in the triplet radical pair and exchange interaction [J] negligible) the singlet and three triplet spin states are virtually degenerate at zero field and the hyperfine-induced transitions between these states occur at optimum rate. However, an applied external magnetic field lifts the degeneracy of the two triplet levels $T_{\pm 1}$ and, thereby, reduces the transitions from the initially occupied S_0 state to the $T_{\pm 1}$ states, until at fields which exceed appreciably the sum of the hyperfine coupling constants in the radicals (i.e. ca. 100 G for the pyrene/N,N-dimethylaniline system) these transitions are totally abolished.

The magnetic field reduction of the triplet yield is seen in Fig. 2 to build up during the geminate phase and to remain constant during the later homogeneous phase, i.e. the magnetic field modulation filters out the most interesting fast geminate process of the radical recombination. To abstract from the observations presented in Fig. 2 information on the detailed dynamics of geminate electron transfer processes, a theoreti-

---

[1] The fast decay of the ion signal at short times does also reflect the disappearance of an exciplex $^1(A^- D^+)$ by intersystem crossing to give $^3\overset{*}{A} + {}^1D$.

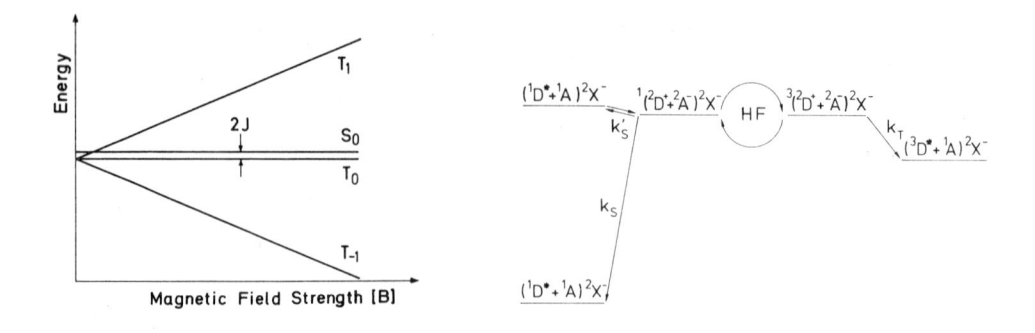

FIGURE 3 Magnetic field dependence of the radical ion pair singlet and triplet state energies (J = exchange interaction energy). The observed magnetic field effects demonstrate that the hyperfine interaction coupling, inducing coherent $S \rightleftarrows T$ transitions in the radical pairs, is predominant.

FIGURE 4 Reaction scheme for the primary electron transfer processes in bacteriochlorophyll reaction centers with reduced primary acceptor X. Singlet, doublet, and triplet states are indicated by the left-hand side superscripts 1, 2, and 3, respectively.

cal analysis is necessary (2, 3). Two results of such an analysis should be mentioned here. Firstly, one can show that any intrapair exchange interaction between the $^2A^-$ and $^2D^+$ radical ions exceeding the weak intramolecular hyperfine interaction abolishes singlet $\rightleftarrows$ triplet transitions. The observed magnetic field effect, hence, provides unequivocal evidence for the formation of free radical ions after the photo-induced electron transfer which engage in a Brownian motion in their respective Coulomb fields before they interact. Secondly, theory predicts that isotopic replacement changes the hyperfine interaction and shifts the magnetic field modulation (3). This could be confirmed experimentally by using the perdeuterated pyrene/N,N-dimethylaniline system, where the magnetic field modulation occurs at considerably lower fields than with the protonated system (7).

ELECTRON TRANSFER PROCESSES IN BACTERIAL PHOTOSYNTHESIS

The transformation of light energy in the photosynthetic apparatus of bacteria is based on a photoinduced electron transfer reaction in a membrane-bound complex of pigments and proteins, the reaction center. The reaction system is presented in Fig. 4. The primary electron transfer follows in this case excitation of the donor ^1D, probably a bacteriochlorophyll dimer. The electron acceptor ^1A is generally assumed to be bacteriopheophytin. Under normal conditions the electron is being transferred within 100–250 ps from $^2A^-$ to a second acceptor X, probably an iron-ubiquinone complex (8). If X is reduced chemically ($^1X \rightarrow {}^2X^-$) or removed, the electron transfer is blocked and the lifetime of the initial radical pair ($^2D^+ + {}^2A^-$) increases to about 10 ns, this time reflecting the electron back transfer to $^2D^+$. The spin multiplicity of the donor molecule in the reaction center, i.e. ($^1\overset{*}{D} + {}^1A$), ($^3\overset{*}{D} + {}^1A$), ($^1D + {}^1A$) depends in this case again on the relative alignment of the two electron spins at the instance of the electron jump, the system behaves very similarly to donor-acceptor pairs in solution. In fact, the yield of triplet products ($^3\overset{*}{D}$ + ^{1}A) generated in the reaction centers of *Rhodopseudomonas spheroides* with a reduced X is also lowered by external magnetic fields (4, 5). It was found, however, that the relative magnetic field effect depends sensitively on the preparation of the reaction center samples (5). This dependence originates from intermolecular interactions characteristic of the electron transfer processes in the reaction center.

The magnetic field effect on the geminate recombination process of radical pairs in solution reflects solely the intramolecular hyperfine coupling. Other interactions which influence the electron spin motion, the exchange interaction between the unpaired electrons, and the existence of spin-selective electron transfer channels come into play only during the short collision times of the diffusing radical pair and are therefore not influential. The photosynthetic reaction center is, however, a "solid-state" system, the ^{2}D$^+$ and ^{2}A$^-$ moieties are in permanent contact, and the above interactions influence the electron spin motion continuously and, therefore, contribute to the magnetic field dependence of the yield of triplet products $^3\overset{*}{D}$. For a theoretical demonstration, one can consider the case where the reversible electron transfer ($^1\overset{*}{D}$ + ^{1}A) $\rightleftarrows$ (^{2}D$^+$ + ^{2}A$^-$), as well as exchange interactions, can be neglected and where for the rates k_S and k_T of irreversible electron transfer holds: $k_S = k_T = k$. Fig. 5 displays the predicted magnetic field dependence of the relative triplet yield $\phi_T(B)/\phi_T(B = 0)$ for various values of k. One finds that with increasing k the magnetic field modulation shifts slightly to higher fields. The reason for this behavior is that fast electron transfer (large k) samples the short time domain of the spin motion when larger fields are needed to affect the spin motion. However, this effect cannot account solely for the high field effects observed by Hoff et al. (5).

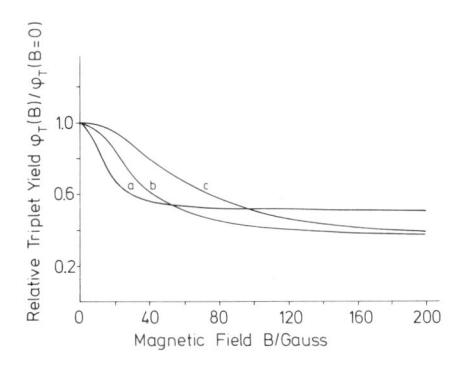

FIGURE 5 Magnetic field dependence of the relative triplet yield $\phi_T(B)/\phi_T(B = 0)$ from the (^{2}D$^+$ + ^{2}A$^-$) radical pairs calculated with the sums of the hyperfine coupling constants Σa_{D^+} ~24 G (for the bacteriochlorophyll radical) and Σa_{A^-} ~32 G (for the bacteriopheophytine radical) and with $k_S = k_T = k$ for $a, k = 0.1 \times 10^9$ s^{-1}; $b, k = 0.5 \times 10^9$ s^{-1}; $c, k = 1.0 \times 10^9$ s^{-1} (exchange interactions $J_{D,A} = J_{A,X} = 0$).

The irreversible electron transfer processes $^3(^2D^+ + {}^2A^-) \to {}^3(^3D + {}^1A)$ and $^1(^2D^+ + {}^2A^-) \to {}^1(^1D + {}^1A)$ bring about lifetime broadenings $\hbar k_S$ and $\hbar k_T$ of the energy levels of the singlet and triplet radical pair states, respectively. Higher magnetic fields are necessary in order for the Zeeman splitting to overcome the relative energy band width $\hbar(k_S - k_T)$, separating sufficiently the S_0 and $T_{\pm 1}$ states and inducing a reduction of triplet products. The reversible electron transfer $(^1\overset{*}{D} + {}^1A) \rightleftarrows (^2D^+ + {}^2A^-)$ has a similar effect as high $k_S = k_T = k$ values in that it restricts the spin motion to the short time regime, thereby shifting the magnetic field modulation up-field. Further effects on the electron spin motion are exerted by the exchange interaction between $^2D^+$ and $^2A^-$ and between $^2A^-$ and $^2X^-$. The first interaction tends to abolish the magnetic field modulation. The fact that a magnetic field modulation of the triplet yield is observed implies that this exchange interaction is smaller than the weak intramolecular hyperfine coupling, a finding with important ramifications with respect to the electron transfer mechanism. The exchange interaction between $^2A^-$ and $^2X^-$ has the net effect of donating a random spin to the $(^2D^+ + {}^2A^-)$ pair and, thereby, giving rise to a magnetic field independent formation of triplet pairs. From a comparison between observed yields of triplet states, their magnetic field modulation, and results of model calculations one can estimate electron transfer rate constants and the magnitude of the exchange interactions in the bacterial photosynthetic reaction centers (6).

Received for publication 23 December 1977.

REFERENCES

1. SCHULTEN, K., H. STAERK, A. WELLER, H.-J. WERNER, and B. NICKEL. 1976. Magnetic field dependence of the geminate recombination of radical ion pairs in polar solvents. *Z. Phys. Chem. (Frankfurt am Main).* **101**:371.
2. SCHULTEN, Z., and K. SCHULTEN. 1977. The generation diffusion, spin motion, and recombination of radical pairs in solution in the nanosecond time domain. *J. Chem. Phys.* **66**:4616.
3. WERNER, H.-J., Z. SCHULTEN, and K. SCHULTEN. 1977. Theory of the magnetic field modulated geminate recombination of radical ion pairs in polar solvents: application to the pyrene-N,N-dimethylaniline system. *J. Chem. Phys.* **67**:646.
4. BLANKENSHIP, R. E., T. J. SCHAAFSMA, and W. W. PARSON. 1977. Magnetic field effects on radical pair intermediates in bacterial photosynthesis, *Biochim. Biophys. Acta.* **461**:297.
5. HOFF, A. J., H. RADEMAKER, R. VAN GRONDELLE, and L. N. M. DUYSENS. 1977. On the magnetic field dependence of the yield of the triplet state in reaction centers of photosynthetic bacteria. *Biochim. Biophys. Acta.* **460**:547.
6. WERNER, H.-J., K. SCHULTEN, and A. WELLER. 1978. Electron transfer and spin exchange contribution to the magnetic field dependence of the primary photochemical reaction of bacterial photosynthesis. *Biochim. Biophys. Acta.* In press.
7. WERNER, H.-J., H. STAERK, and A. WELLER. 1978. Solvent, isotope and magnetic field effects in the geminate recombination of radical ion pairs. *J. Chem. Phys.*
8. PARSON, W. W., and R. J. COGDELL. 1975. The primary photochemical reaction of bacterial photosynthesis. *Biochim. Biophys. Acta.* **416**:105.
9. COGDELL, R. J., T. G. MONGER, and W. W. PARSON. 1975. Carotinoid triplet states in reaction centers from *Rhodopseudomonas sphaeroides* and *Rhodospirillum rubrum. Biochim. Biophys. Acta.* **408**:189.

NOTE ADDED IN PROOF

Photoinduced electron transfer processes produce in a picosecond to nanosecond time range high energy intermediates: exciplexes, radical ion pairs, and separated free radical ions. The fast processes involved are fundamental for oxidation-reduction processes as they occur in chemical, electrochemical, and biological (for example photosynthesis) systems. Since these reactions involve the motion of an electron, they should be extremely fast ($\sim 10^{-15}$ s) except that they are coupled to slower degrees of freedom, which are rate-limiting and, therefore, the focal point of most studies. Degrees of freedom coupled to electron transfer are those of the microenvironment, of internal vibrations, and of the relative distance of the reactants (Fig. 1).

Electron transfer reaction between electron donors, (e.g. dimethylamiline) in the ground state, D, and electron acceptors (e.g. pyrene) in the excited singlet state, $^1A^*$, produce the following intermediates (sequentially):

$$\underset{\text{singlet exciplexes}}{^1(A^-D^+)} \quad \rightarrow \quad \underset{\substack{\text{radical ion pairs} \\ \text{(in the overall singlet state with unrelated spins)}}}{^1(^2A^- + {}^2D^+)} \quad \rightarrow \quad \underset{\text{free radical ions}}{{}^2A^- + {}^2D^+}.$$

From these triplet products are formed via three different pathways: (*a*) intersystem crossing in the exciplex; (*b*) germinate (or intrapair) recombination of the solvated radical ion pairs; (*c*) homogeneous (or interpair) recombination of the free radical ions.

Only the germinate triplet production (pathway *b*), due to the hyperfine coupling between the

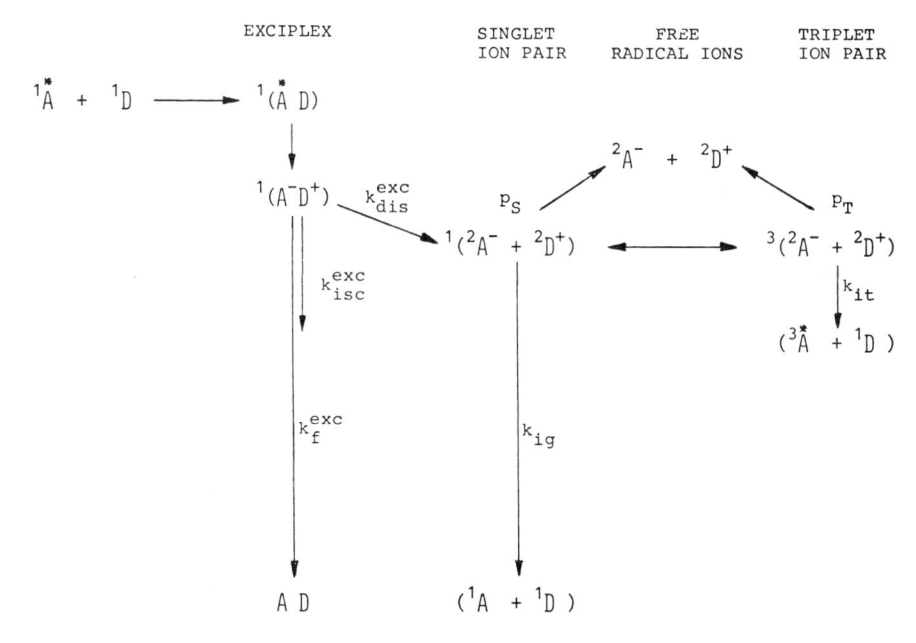

FIGURE 1 Reaction scheme for electron transfer involving exciplex formation followed by radical ion pair formation; only the electron spin motion in the radical ion pairs is influenced by external magnetic fields which modulate the hyperfine coupling-induced (singlet versus triplet) recombination yields.

FIGURE 2   FIGURE 3

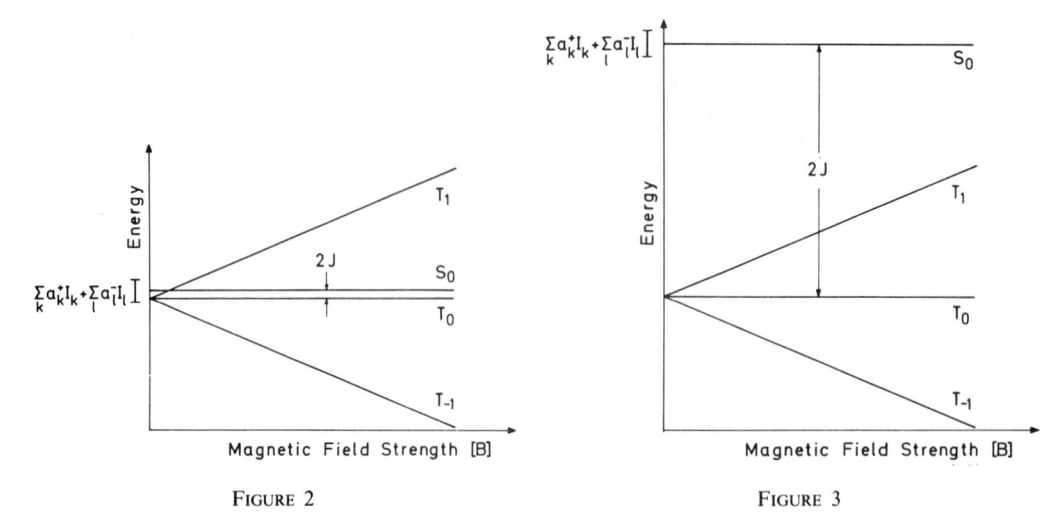

FIGURE 2 Magnetic field dependance of the radical ion pair singlet and triplet state energies (J = exchange interaction energy).

FIGURE 3 Magnetic field dependence of the radical ion pair singlet and triplet state energies under the condition that the exchange interaction energy, J, is much greater than the hyperfine coupling energy, E_{hfc} (indicated by the energy bar $\mathbf{I}$).

nuclear spins and the unpaired electron spin in each radical, is reduced by weak external magnetic fields between 0 and 100 G (Fig. 2).

This reduction of the germinate triplet production is based on the fact that at zero magnetic field strength the three triplet states T_{+1}, T_0, T_{-1} are degenerate and can all be populated within the hyperfine coupling energy range (indicated by the bar $\mathbf{I}$) from the initially populated singlet state S_0, which can lead to a maximum of 75% triplet population, whereas at higher magnetic field strengths only T_0 can be populated from S_0, which leads at the most to only 50% triplet population.

It is important to point out that triplet production from the exciplex and triplet production through homogeneous recombination of the free radical ions are not affected by weak magnetic fields. Therefore, from the magnetic field-modulated signal of the recombination products inferences can be drawn about the mechanistic aspects of the radical ion recombination process.

An important aspect is the magnitude of the exchange interaction between the radicals in the radical ion pair (Fig. 3).

The hyperfine interaction energy (indicated by the bar $\mathbf{I}$) may be typically of the order of 50 G, which corresponds to a rate of 1.4×10^8 s^{-1}. Now, if J is 100 times greater (i.e. 5,000 G), population of T_{+1}, T_0, T_{-1} from S_0 by the hyperfine mechanism is not possible anymore. The repopulation now requires that an energy of the order of $2J$ (10,000 G) be exchanged between the spin-system and the heat bath environment. The required spin-lattice relaxation time, T_1, to do that is of the order of 1 μs, so that the triplet population rate now is only $T_1^{-1} \approx 10^6$ s^{-1}. Thus no triplets are formed from the germinate radical pair because its lifetime is too short by at least one or two orders of magnitude.

It should be pointed out that the value assumed here for J, namely 5,000 G or 0.6×10^{-4} eV, is still very small. A very rough guess suggests that it corresponds to a radical pair center to center distance of about 5Å (Fig. 4).

Bacterial photosynthesis in three different preparations investigated by Hoff and co-workers (5) shows a magnetic field effect, in that the relative triplet yield decreases with increasing

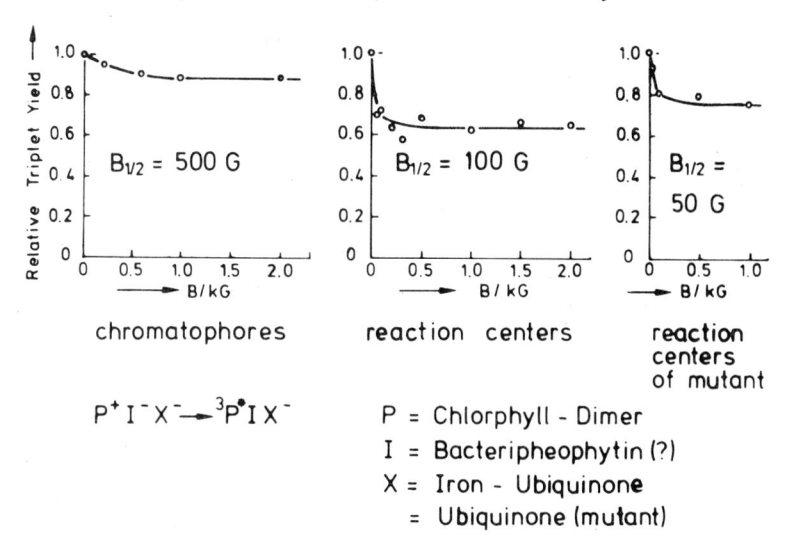

FIGURE 4 Magnetic field dependence of the yield of bacteriochlorophyll triplets generated in reaction centers of *Rhodopseudomonas spheroides* with reduced acceptor X (from Hoff et al., ref. 5).

magnetic field by 15–60%, depending on the type of the reaction center and its preparation. This shows that the electron exchange interaction $J_{P^+I^-}$ (between oxidized bacteriochlorophyl [dimer] and reduced bacteriopheophytin) must be smaller than 100 G or 10^{-6} eV. This small value is very remarkable in view of the "solid state" character in the membrane environment of the reaction centers, in which the electron cannot be transferred over large distances within the short times available except by tunneling.

In the exchange interaction $J_{I^-X^-}$ between reduced bacteriopheophytin (I^-) and the reduced iron-ubiquinone complex (X^-), the aligned electron spin on I^- (aligned with respect to the spin on P^+) is replaced by a random spin (from X^-) and thus introduces more triplet character to the P^+I^- systems thereby reducing the relative magnetic field effect. Since the chromatophore preparations still have an intact iron-ubiquinone complex, one may assume that with the chromatophores $J_{I^-X^-}$ is greater than with the other reaction center preparations so that the magnetic field effect is weaker, but that $J_{I^-X^-}$ is still not great enough to wipe out the magnetic field effect completely.

DISCUSSION

SWENBERG: I would like to know whether the overall photosynthetic yield is affected by an external magnetic field.

WELLER: Under normal photosynthetic conditions the previously formed radical ion pair in its overall singlet state decays in about 120 ps, according to Kaufmann, by donating the electron to a secondary acceptor, ubiquinone. There does not seem to be any substantial leakage through which the primarily formed radical ion pair would go over into the bacteriochlorophyll triplet state. In other words, the lifetime of the primary formed radical ion pair, 120 ps, is much too short for a magnetic field effect to be operative under normal photosynthetic conditions.

SWENBERG: I agree with your answer. There is no magnetic field effect on the photosynthetic yield. We may say that the triplet pair state is not operating in series to the product of the reaction. If there are ever any magnetic field effects in biological systems, whether they be birds migrating or photosynthesis, it is necessary for the triplet states to be direct precursors to the reaction.

This leads to my second question: Do you know of any biological systems in which triplets are precursers to products that will eventually form and have biological effects?

WELLER: Our explanations for the lack of the magnetic field effect on the overall yield are essentially similar. With respect to the question of triplet intermediates I am a bit at a loss. Perhaps in the magnetic sensory system of birds such intermediates may occur.

SWENBERG: I have one more question. It is clear that the exchange interaction has to be greater than 10^{-6} eV, yet if it is there is no magnetic field effect. I find 10^{-6} hard to reconcile with the fact that at short distances the tunneling has elements with comparable exchange values. From the theoretical point of view, there must be some sort of renormalization of the Heisenberg exchange term going on here, because, as Dr. Rentzepis just told us, the distances between the dimer and the acceptor are quite small. Do you have any comments on that?

WELLER: No.

GEACINTOV: From your Fig. 5, it appears that the relative yield of triplets in high magnetic field to that in zero magnetic field depends on the lifetime of the intermediate radical ion pair state. Are there any other parameters on which this relative yield depends, and if so what else can be learned about the bacteriochlorophyll reaction centers from the magnitude of the magnetic field?

WELLER: The relative yield of triplets in high magnetic field to that in zero magnetic fields depends indeed on the lifetime of the intermediate radical ion pair state. It also depends on the relative rates by which the intermediate ringlet radical ion pair state decays to the ground state and the triplet radical ion pair state decays to the local triplet state: the faster the former process, the smaller the relative triplet yield.

Similar arguments apply to the exchange interaction between bacteriochlorophyll radical cation and pheophytin radical anion, which, if large enough, can completely prevent any triplets from being formed. This shows that the electron exchange interaction (between oxidized bacteriochlorophyll dimer and reduced bacteriopheophytin) must be smaller than 10^{-6} eV and that the primary electron transfer must occur over comparatively large distances, probably by electron tunneling.

RODGERS: On the formation of radical ion-pair state in the pyrene-dimethyl aniline homogeneous system, your work and that of others have shown quite clearly that this requires a highly polar matrix to effect separation. In low polar media one gets exciplex formation mainly. In the bacterial situation, are you telling us now that charge-separation occurs in membrane lipids that are low-polar media? Can you rationalize this difference?

WELLER: This is a very important question and part of the answer is given in our paper. The situation is indeed different between the homogeneous part of the reaction solution and the electron transfer reactions, which probably occurred in the bacterial chlorophyll system in a lipid membrane. The main difference is that the lifetimes of the intermediate radical ion pair state singlet and triplex are softened by the rate by which the species disappear and this has to be taken into account. This argument differs slightly from that applied to the system in a cooler solvent.

MAUZERALL: With reference to Fig. 5, to what do you physically assign the small lag in the plot of triplet yield versus the magnetic field strength? And do your data actually show this lag?

WELLER: Fig. 5 shows calculated curves. They indicated what kind of calculations can be used to understand the behavior of the systems; the details are not important.

JAKOBSSON: Since the question of birds migrating in magnetic fields has been introduced, may I insert for the record and in the interest of ecumenicity that electrophysiologists do not know how to deal with it; we eliminate it from our formulas. In our experiments, we impose a kind of geometric symmetry and, using Maxwell's equation, we eliminate magnetic field effects. Yet of course magnetic fields are biologically important, not only in migration as cited, but also as associated with neural activity. So I would just like to reinforce the idea that this is a biological question of major importance, undealt with so far on the cellular level.

AUSTIN: What is the rate at which radical pairs diffuse away from one another? Have you tried solvents of differing viscosities?

WELLER: Well, it depends on course on viscosity, but it is a very complicated process that cannot be described by a single exponential, because the radicals in the solvent will separate to some extent and then re-encounter again.

DORFMAN: Is there anything analogous to the Onsager escape radius that comes out the data or in a calculation?

WELLER: This is indeed the first step: the escape radius describes the first separation of the radical ion pair. What Onsager has not taken into account is the re-encountering of the radical ions. One can probably forget the re-encountering for neutral species or for systems where only one species is charged.

MAUZERALL: We have analyzed in detail the probelm of the escape of ions from the initial pair at separation R_0 with reaction at a smaller R_m. Our equation for the yield of escaped or uncorrelated ions reduces to Onsager's, exp. $(-R_c/R_0)$ where R_c is the Coulomb radius $(e^2/\epsilon kT)$, when R_m is negligible. It reduces to the simple equation $1 - (R_m/R_0)$ when R_c is zero, i.e., an infinite dielectric constant. The yield for neutral species therefore need not be unity. It is typically 0.5 when the electron transfer distance, R_0, is about twice the collapse distance, R_m.

APPLICATION OF PULSE RADIOLYSIS TO THE STUDY OF PROTEINS

CHYMOTRYPSIN AND TRYPSIN

MOSHE FARAGGI, MICHAEL H. KLAPPER, AND LEON M. DORFMAN,
*Department of Chemistry, The Ohio State University,
Columbus, Ohio 43210 U.S.A.*

ABSTRACT The one-electron reduction of chymotrypsin, trypsin, and their zymogens have been studied by pulse radiolysis. The optical spectra of the transient products from the two active enzymes display a pH-dependent band at 360 nm, associated with the histidine-electron adduct. The yield of the histidyl radical as a function of pH is consistent with a $pK_a < 4.5$, which suggests that the radical is located at the enzyme active site. The histidines of the proenzymes chymotrypsinogen and trypsinogen are unreactive towards the hydrated electron. We conclude that formation of the histidine-electron adduct at the serine protease active site is sensitive to the physical alterations which accompany protease activation.

INTRODUCTION

The relationship between an enzyme's catalytic functions and its three-dimensional structure continues to be a preeminent problem in enzymology. The structural data obtained in recent years by X-ray crystallography have contributed greatly to the exposition of this relationship. However, these data have so far yielded primarily static pictures of enzymes, and ambiguities remain in understanding the dynamic basis of protein catalysis. In this report we shall show that the technique of pulse radiolysis can be utilized to obtain information about the dynamic structural properties of proteins— in particular, the serine proteases.

The serine proteases are among the most extensively investigated of all proteins. (For a most recent review see Kraut, 1.) The reaction pathway in the catalytic hydrolysis of peptides and esters is understood in general terms. The crystal structures of a number of enzymes in this class have been elucidated, and recent low-temperature work has allowed the isolation of reactive intermediates (2). The active sites of these enzymes are all thought to contain an aspartic acid-histidine-serine triad called the "charge relay system" (3). The dynamic consequences of this structural feature, found in chymotrypsin (3, 4), trypsin (5, 6), and elastase (7), are not adequately understood. Using ^{13}C nuclear magnetic resonance, (NMR) Hunkapiller and co-workers (8) have assigned the low pK_a of ~3.8 to the histidine side chain of this grouping in α-lytic protease, leaving the higher pK_a of 6.8, known from enzyme kinetic studies, to the aspartic acid side chain. These assignments can be rationalized in terms of the charge relay system. In support of this conclusion, Markley and Porubcan (9) have determined a pK_a of ~4.5 for the active site histidine of porcine trypsin. In contrast, Robillard and Shulman (10), on the basis of their proton nuclear magnetic resonance studies, have been led to the conclusion that the active site histidine of bovine chymotrypsin has a pK_a of 7.5.

Many of the serine proteases are found as inactive precursors which can be activated by limited proteolysis (11). The structural changes accompanying this activation have been investigated for chymotrypsin (12) and trypsin (13, 14). The results of these studies have been recently reviewed by Stroud et al. (15). While structural differences have been observed between the inactive proenzymes and their activated forms, changes at the active site appear to be minor, and not identical in the two systems. Thus, the correlation between structural changes and enzyme activation is not properly understood.

Pulse radiolysis (16, 17) permits production of hydrated electrons within less than a microsecond; combined with a fast response detection system it offers an effective tool for studying rapid electron transfer reactions, and the properties of transient radicals together with their parent molecules. The atomic groupings in proteins most reactive with the hydrated electron have been established with amino acids and small peptides; these are cystine disulfide bridge, the imidazolium side chain of histidine, the peptide carbonyl group, and the aromatic side chain of tyrosine and tryptophan.

The reactions of linear disulfides with the hydrated electron have rate constants in the range of 2×10^9–4×10^{10} $M^{-1} s^{-1}$ (18, 19). The radical anion $RSSR^-$ and its conjugate acid $RSSRH$ have absorption bands centered near 410 nm ($\epsilon_{max} \sim 1 \times 10^4$) and 385 nm ($\epsilon_{max} \simeq 7 \times 10^3$), respectively. The disulfide radical anion decays by the reaction $RSSR^- \rightarrow RS \cdot + RS^-$, with a first-order rate constant between 2×10^5 and 2×10^6 s^{-1} (19). The reaction of the hydrated electron with the protonated imidazolium of histidine is characterized by a rate constant of $\sim 4 \times 10^9$ $M^{-1} s^{-1}$. The neutral imidazole is much less reactive, $k < 1.4 \times 10^6$ $M^{-1} s^{-1}$ (20, 21). The histidine-electron adduct absorbs with a maximum ($\epsilon_{max} \sim 2 \times 10^3$ $M^{-1} cm^{-1}$) near 360 nm. Because the protonated imidazole reacts more rapidly, the absorbance yield at 360 nm is pH-dependent, and the pK_a of the histidine side chain can be determined even when other reactive sites are available on the molecule (21). The reactions of the peptide carbonyl group and of aromatic amino acid side chains with the hydrated electron are characterized by rate constants on the order of 5×10^8 $M^{-1} s^{-1}$ (22, 23).

Several groups have investigated the reaction of the hydrated electron with proteins (24–31). With disulfide-containing proteins a transient species is produced having an absorption band centered near 410 nm, indicating the radical $RSSR^-$. Bisby and coworkers (30), working with pancreatic ribonuclease, reported an additional band centered near 360 nm. It was observed at low pH, but was absent under alkaline conditions. They suggested that this absorbing species was the protonated radical RSSRH. We have concluded, however, that this 360-nm band is due to the histidine-electron adduct (31). The absorbance yield at 360 nm was found to be pH-dependent and in ribonuclease characterized by a pK_a of ~5.9, close to the acid dissociation constants of the two active-site histidines (32–34). The decay rate of the 360-nm band is approximately three orders of magnitude smaller than that expected of the radical disulfide, and that observed with the 410-nm band in the same protein. Finally, the transient spectra of α-casein, which contains histidine but no cystine, show the 360-nm band, while the 410-nm band characteristic of RSSR$^-$ is absent.

The pH-dependent formation of the histidine-electron adduct in ribonuclease A and α-casein suggests that pulse radiolysis is an additional tool for determining histidine pK$_a$'s in proteins. However, not all protein imidazole groups appear to be reactive toward the hydrated electron (31). Thus, there is no evidence for the reaction of the histidine in ribonuclease, which titrates with a pK$_a$ near 7. No 360-nm band was observed with lysozyme, known to have one histidine on the surface of that molecule (35). Nor have we observed the 360-nm band with lactalbumin, a protein containing three histidine residues (M. Faraggi, M. H. Klapper, and L. M. Dorfman, unpublished results). It appears, therefore, that the three-dimensional structure of the polypeptide chain is a controlling factor in the reactivity of the histidine residue.

In this report we shall extend our observations to the reactions of the hydrated electron with the serine proteases α-chymotrypsin, trypsin, and with their proenzymes chymotrypsinogen, trypsinogen.

METHODS

Bovine pancreatic three times crystallized salt-free α-chymotrypsin and trypsin, and five times crystallized chymotrypsinogen and trypsinogen were purchased from Worthington Biochemical Corp. (Freehold, N.J.) and used with no further purification. The pulse radiolysis instrumentation has been described elsewhere (36). Pulses of 500-ns duration (hydrated electron concentration ~5 μM) were used for the determination of transient spectra, while kinetic measurements of electron decay utilized 100-ns pulses. Irradiated solutions contained 1 mM phosphate buffer, 0.4 m t-butanol as the scavenger for the OH radical, and 0.05 – 0.8 mM protein. Oxygen was removed by gently bubbling argon through the solution for approximately 30 min. The protein concentrations were sufficiently high to insure that ≥90% of the hydrated electrons formed in solution reacted with the protein. To establish these combination yields, the decay rate of the electron was monitored at 550 nm (ϵ_{550} = 10^4 M^{-1} cm^{-1}), and comparisons were made between solutions with and without protein. The fraction of electrons reacting with the protein was calculated as the ratio $(k'_p - k'_b)/k'_p$, where k'_p and k'_b are the pseudo first-order decay constants obtained from sample and blank experiments, respectively. The second-order rate constants for the reaction of hydrated electron and protein were determined from the slope of the linear relationship between the pseudo first-order decay constant k'_p and protein concentration.

RESULTS

The reaction of the hydrated electron with the four proteins trypsin, α-chymotrypsin, trypsinogen, and chymotrypsinogen is diffusion-controlled, as indicated by the magnitudes of the second-order rate constants, 1.1–4.5 × 10^{10} M^{-1} s^{-1}. Others (24, 25, 37) have reported similar results for trypsin and chymotrypsin.

With all four proteins, at high and low pH the transient spectra obtained within 1 μs of the pulse showed a large band centered near 410 nm, indicating the formation of

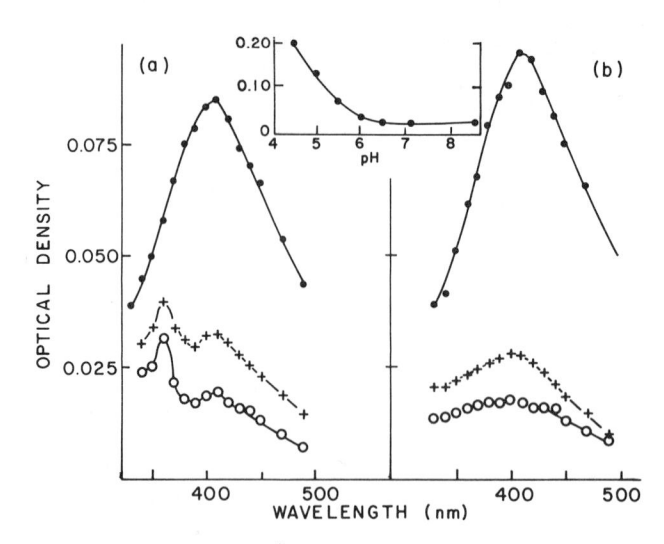

FIGURE 1 Transient spectra of 4×10^{-4} M α-chymotrypsin in argon-saturated solutions containing 0.4 M t-butanol and 10^{-3} M phosphate buffer. Initial concentration of the hydrated electron 6.1×10^{-6} M. Temperature $25 \pm 1°C$. (●—●), $t = 0$ μs after pulse; (+—+), $t = 200$ μs after pulse; (o—o), $t = 1,800$ μs after pulse. (a) pH 4.5. (b) pH 7.1. (insert) effect of pH on the 360 nm absorbance at 200 μs after subtraction of residual $RSSR^-$ absorbance.

$RSSR^-$ (Figs. 1–3). No distinct 360-nm band was observed, although a shoulder at this wavelength was seen in the low pH trypsin spectrum (Fig. 2). In the microsecond time range ($t \leq 10$ μs) the 410 nm band decreased in intensity with no observed shifts at any solution pH, suggesting the absence of the neutral radical RSSRH.

At longer times ($t \geq 200$ μs) the trypsin and chymotrypsin transient spectra displayed a distinct band centered near 360 nm, known from previous work to be due to the histidine-electron adduct (see Introduction). By assuming that both 410 and 360-nm bands are Gaussian with respect to wave number, the magnitude of the latter band could be estimated. The insets of Figs. 1 and 2 show the effect of pH on the 360-nm absorbance; the yield goes up as the pH is lowered. Unfortunately, the reaction could not be studied below pH 4.2 due to the competition of hydronium ion for the electron: $e_{aq}^- + H_3O^+ \rightarrow H + H_2O$. One way to minimize the importance of this competing reaction is to increase the protein concentration. We were, however, limited by the decreased intensity of the analyzing light at higher protein levels.

As discussed in the Introduction, the pH dependence of the 360-nm absorbance is due to the greater reactivity of the imidazolium ion relative to the uncharged imidazole. Thus, the pK_a of histidine can be determined by pulse radiolysis. Because we were not able to measure absorbance yields below pH 4.2, an unambiguous pK_a could not be assigned for trypsin or chymotrypsin. An estimate was based on the following assumptions: the extinction coefficient of the histidine-electron adduct in proteins is identical with that of small peptides (2,000 $M^{-1} cm^{-1}$; 20, 21); the maximum yield of the histidine adduct in trypsin and chymotrypsin is 50%. (The fraction of protein-attached electrons in the histidine-electron adduct was found in earlier experiments to be 85%

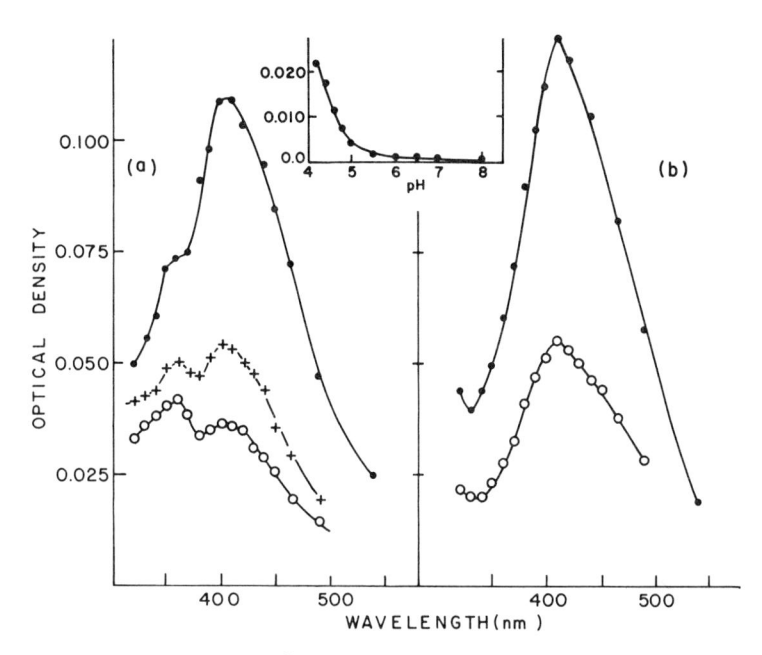

FIGURE 2 Transient spectra of 5×10^{-5}M trypsin: conditions were identical to those of Fig. 1 except the pH's of the two experiments were 4.4 and 8.4.

for pancreatic ribonuclease A at pH 4.3 (31). The fractions obtained with α-chymotrypsin at pH 4.5, and trypsin at pH 4.4 were 20% and 32%, respectively.) The pK_a's thus estimated for both proteins are ~4.3. Although the potential for error in this estimate is large, the data clearly indicate pK_a's below 4.5. Chymotrypsin and trypsin have 1 and 2 histidines, respectively, with titration midpoints near 7. We found no 360 nm absorption change associated with this pK_a.

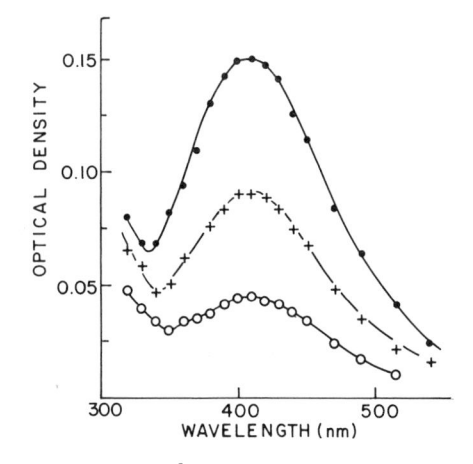

FIGURE 3 Transient spectra of 5×10^{-5}M chymotrypsinogen: conditions were identical to those of Fig. 2.

In contrast to the results obtained with the two active proteases, no 360-nm band was observed with either chymotrypsinogen (Fig. 3) or trypsinogen down to the lowest pH tested, 4.2. Since it is known that the histidines of both proenzymes titrate with midpoints near pH 7 (38), we conclude that our inability to see the histidine-electron adduct in the zymogens is not due to an acid pK_a shift of the histidines we observed in the active enzymes.

DISCUSSION

We have shown previously that the diffusion-controlled reaction of the hydrated electron with proteins results in the formation of disulfide and/or histidine adducts (31). Aromatic amino acid side chains may also react with the hydrated electron to produce species absorbing in the ultraviolet below 350 nm (21, 28). Experimental limitations precluded our observation of these reactions, if they occurred to the small extent expected from the 10–50-fold lower reactivities. Because the yield of the histidine adduct depends on solution pH, the pK_a of the reactive histidine group(s) in small peptides and proteins may be determined by pulse radiolysis (28, 31). The pK_a's for the histidines that react in bovine α-chymotrypsin and trypsin are below pH 4.5, comparable to the active site histidine pK_a's assigned by Hunkapiller et al. (8) in α-lytic protease, and by Markley and Porubcan (9) in porcine trypsin. These latter assignments were made on the basis of nuclear magnetic resonance experiments. Thus, the pulse radiolysis results represent an independent confirmation of the low apparent pK_a for the active-site histidine in serine proteases.

As mentioned in the Introduction, not all protein histidines are reactive toward the hydrated electron, and this is true for trypsin and chymotrypsin as well. Those histidines not at the active site, and which have pK_a's closer to 7, are not seen in the pulse radiolysis experiment. Chymotrypsinogen and trypsinogen contain no reactive histidines. Solvent inaccessibility of the imidazole side chain does not appear to be a reasonable explanation of the observed variability in histidine reactivity. The single histidine of lysozyme, solvent-accessible in the crystal structure (35), does not produce a histidine-electron adduct (31). In contrast to this observation, the yield of histidine plus disulfide electron adducts is greater in chymotrypsin than would be expected on the basis of intrinsic reactivities, amounts, and solvent accessibilities of those protein groups which are most reactive. If intrinsic reactivities are similar in proteins and small molecule models, and the hydrated electron reacts directly only with those groups accessible to the solvent, then the expected partitioning of the hydrated electron amongst the various groups may be computed (Table I). The calculated results imply that most of the electrons will react with peptide carbonyls, and that ~20% will combine directly with disulfides and imidazoles. However, we have observed approximately 60% bound to these two groupings.

It is known from studies with oligopeptides that electrons attached initially to the carbonyl oxygen can migrate to other loci (39, 40), and that the intramolecular hydrogen bond between peptide units may serve as a path for electron migration (41). We,

TABLE I

YIELDS OF ELECTRON ATTACHMENT TO DIFFERENT SITES IN α-CHYMOTRYPSIN

Protein component	M_i No. of the component in contact with the solvent*	k_i Second-order rate constant for the reaction of e_{aq}^- with the component	Fraction of e_{aq}^- located on the component. Calculated‡ Acid	Calculated‡ Alkaline	Observed pH = 4.5	Observed pH = 8.4
Carbonyls	104	5×10^8	71%	78%		
Disulfides	1	1.3×10^{10}	18%§	20%§	40%	60%
Histidines	1	4×10^9 ‖	6%	0	20%	0
Aromatic amino acids	5	3×10^8	2%	2%		

*Obtained from Birktoft and Blow (44).
‡Calculated by $f = k_i M_i / \Sigma k_i \cdot M_i$; k_i is the intrinsic rate constant of the ith component with M_i units accessible to the hydrated electron.
§Assuming one disulfide bridge.
‖ Value for the protonated imidazole.

therefore, propose that the hydrated electron attaches initially to the protein surface, then rapidly migrates into the protein, and that it settles finally into a potential sink provided by either imidazole or disulfide, with the yields of each determined competitively. Grossweiner (42) has also proposed internal electron migration in proteins to interpret photochemical data. Thus, two explanations may be forwarded for the variable reactivity of histidines in proteins. A particular histidine may be unreactive either because there is no "electron path" leading to it from the protein surface, or because the potential sink provided by the imidazole side chain is not low enough for effective competition with the disulfide bond, or with other potential sites.

The reactivities of the imidazoles in chymotrypsin and trypsin and their contrasting unreactivity in the corresponding proenzymes are particularly noteworthy. These results suggest that formation of the histidine electron adduct is sensitive to the physical changes at the serine protease active site, which accompany conversion of the proenzyme to its catalytically active form. Parenthetically, it is suggestive that the only other reactive histidine(s) we have encountered so far in disulfide-containing proteins appear to be at the active site of ribonuclease. We cannot now explain the observed reactivity alterations upon zymogen activation. However, Birktoft and co-workers (43) have proposed that upon activation of chymotrypsinogen the hydroxyl group of the active site serine 195, hydrogen-bonded to histidine 57, is rotated very close to the disulfide bond of cysteines 42 and 48. The resultant disulfide-serine-histidine triad, not present in the zymogen, may serve as an electron path in chymotrypsin.

Irrespective of the detailed mechanisms that will be required to explain the results we have obtained to date, the apparent reactivities of histidine residues in globular proteins clearly reflect some topological properties of the folded polypeptide chain. Thus, pulse radiolysis may serve as an additional tool for dynamic protein structure studies.

Received for publication 23 November 1977.

REFERENCES

1. KRAUT, J. 1977. Serine proteases: structure and mechanism of catalysis. *Annu. Rev. Biochem.* **46**:331-358.
2. FINK, A. L. 1977. Cryoenzymology: the study of enzyme mechanisms at subzero temperatures. *Acc. Chem. Res.* **10**:233-239.
3. SIGLER, P. B., D. M. BLOW, B. W. MATHEWS, and R. HENDERSON. 1968. Structure of crystalline α-chymotrypsin. II. A preliminary report including a hypothesis for the activation mechanism. *J. Mol. Biol.* **35**:143-164.
4. BLOW, D. M. 1969. The study of α-chymotrypsin by x-ray diffraction. *Biochem. J.* **112**:261-268.
5. STROUD, R. M., L. M. KAY, and R. E. DICKERSON. 1974. The structure of bovine trypsin: electron density maps of the inhibited enzyme at 5 Å and at 2.7 Å resolution. *J. Mol. Biol.* **83**:185-208.
6. BODE, W., and P. SCHWAGER. 1975. The refined crystal structure of bovine β-trypsin at 1.8 Å resolution. II. Crystallographic refinement, calcium binding site, benzamidine binding site and active site at pH 7.0. *J. Mol. Biol.* **98**:693-717.
7. SHOTTEN, D. M., and H. C. WATSON. 1970. Three dimensional structure of tosyl elastase. *Nature (Lond.).* **225**:811-816.
8. HUNKAPILLER, M. W., S. H. SMALLCOMBE, D. R. WHITAKER, and J. H. RICHARDS. 1973. Carbon nuclear magnetic resonance studies of the histidine residue in α-lytic protease. Implications for the catalytic mechanism of serine proteases. *Biochemistry.* **12**:4732-4742.
9. MARKLEY, J. L., and M. A. PORUBCAN. 1976. The charge relay system of serine proteases: proton magnetic resonance titration studies of the four histidines of porcine trypsin. *J. Mol. Biol.* **102**:487-509.
10. ROBILLARD, G., and R. G. SHULMAN. 1974. High resolution nuclear magnetic resonance studies of the active site of chymotrypsin. I. The hydrogen bonded proton of the "charge relay" system. *J. Mol. Biol.* **86**:519-540.
11. NEURATH, H. 1975. Limited proteolysis and zymogen activation. *In* Proteases and Biological Control. E. REICH, D. B. RIFKIN, and E. SHAW, editors. Cold Spring Harbor Laboratory Press, New York. 51-64.
12. BLOW, D. M. 1976. Structure and mechanism of chymotrypsin. *Acc. Chem. Res.* **9**:145-152.
13. KOSSIAKOFF, A. A., J. L. CHAMBERS, L. M. KAY, and R. M. STROUD. 1977. Structure of bovine trypsinogen at 1.9 Å resolution. *Biochemistry.* **16**:654-664.
14. FEHLHAMMER, H., W. BODE, and R. HUBER. 1977. Crystal structure of bovine trypsinogen at 1.8 Å resolution. II. Crystallographic refinement, refined crystal structure and comparison with bovine trypsin. *J. Mol. Biol.* **111**:415-438.
15. STROUD, R. M., A. A. KOSSIAKOFF, and J. L. CHAMBERS. 1977. Mechanisms of zymogen activation. *Annu. Rev. Biophys. Bioeng.* **6**:177-194.
16. MATHESON, M. S., and L. M. DORFMAN. 1969. Pulse Radiolysis. The M.I.T. Press, Cambridge, Mass.
17. DORFMAN, L. M. 1974. Pulse Radiolysis in Investigation of Rates and Mechanisms of Reactions. G. G. Hammes, editor. Part II. John Wiley & Sons, Inc., New York. 463-519.
18. BRAAMS, R. 1966. Rate constants of hydrated electron reactions with amino acids. *Radiat. Res.* **27**:319-329.
19. HOFFMAN, M. Z., and E. HAYON. 1972. One-electron reduction of the disulfide linkage in aqueous solution. Formation, protonation and decay kinetics of the RSSR$^-$ radical. *J. Am. Chem. Soc.* **94**:7950-7957.
20. RAO, P. S., M. SIMIC, and E. HAYON. 1975. Pulse radiolysis study of imidazole and histidine in water. *J. Phys. Chem.* **79**:1260-1263.
21. FARRAGI, M., and A. BETTLEHEIM. 1978. The reaction of the hydrated electron with amino acids, peptides, and proteins in aqueous solutions. Part III. Histidyl peptides. *Radiat. Res.* In press.
22. TAL, Y., and M. FARAGGI. 1975. The reaction of the hydrated electron with amino acids, peptides, and proteins in aqueous solution. I. Factors affecting the rate constants. *Radiat. Res.* **62**:337-346.
23. ANBAR, M., M. BAMBENEK, and A. B. ROSS. 1973. Selected Specific Rates of Reactions of Transients from Water in Aqueous Solutions. 1. Hydrated Electron. National Standard Reference System NSRDS-NBS 43. National Bureau of Standards, Washington, D.C.
24. BRAAMS, R., and M. EBERT. 1967. Reactions of proteins with hydrated electrons: the effect of conformation on the reaction rate constant. *Int. J. Radiat. Biol. Relat. Stud. Phys. Chem. Med.* **13**:195-197.

25. MASUDA, I., J. OVADIA, and L. I. GROSSWEINER. 1971. The pulse radiolysis and inactivation of trypsin. *Int. J. Radiat. Biol. Relat. Stud. Phys. Chem. Med.* **20**:447–459.
26. LICHTIN, N. N., J. OGDAN, and G. STEIN. 1972. Fast consecutive radical processes with the ribonuclease molecule in aqueous solution. II. Reaction with OH radicals and hydrated electrons. *Biochim. Biophys. Acta.* **276**:124–142.
27. ADAMS, G. E., J. L. REDPATH, R. H. BISBY, and R. B. CUNDALL. 1972. The use of free radical probes in the study of mechanism of enzyme inactivation. *Isr. J. Chem.* **10**:1079–1093.
28. FARAGGI, M., and I. PECHT. 1973. The electron pathway to Cu(II) in ceruloplasmin. *J. Biol. Chem.* **248**:3146–3149.
29. ADAMS, G. E., K. F. BAVERSTOCK, R. B. CUNDALL, and J. L. REDPATH. 1973. Radiation effects on α-chymotrypsin in aqueous solution: pulse radiolysis and inactivation studies. *Radiat. Res.* **54**:375–387.
30. BISBY, R. H., R. B. CUNDALL, J. L. REDPATH, and G. E. ADAMS. 1976. One electron reduction reactions with enzymes in solution. A pulse radiolysis study. *J. Chem. Soc. Farad.* II. **73**:51–63.
31. FARAGGI, M., M. H. KLAPPER, and L. M. DORFMAN. 1978. Fast reaction kinetics of one-electron transfer in proteins: the histidyl radical; mode of electron migration. *J. Phys. Chem.* **82**:508.
32. PATEL, D. J., C. WOODWARD, L. L. CANUEL, and F. A. BOVEY. 1975. Correlation of exchangeable (NH) and nonexchangeable (C_2H) histidine resonances in proton NMR spectrum of ribonuclease A in aqueous solution. *Biopolymers.* **14**:975–986.
33. MARKLEY, J. M. 1975. Correlation proton magnetic resonance studies at 250 MHz of bovine pancreatic ribonuclease I. Reinvestigation of the histidine peak assignments. *Biochemistry.* **14**:3546–3554.
34. WESTMORELAND, D. G., C. R. MATHEWS, M. B. HAYES, and J. S. COHEN. 1975. Nuclear magnetic resonance titration curves of histidine ring protons. The effect of temperature on ribonuclease A. *J. Biol. Chem.* **250**:7456–7460.
35. BLAKE, C. C. F., G. A. MAIR, A. C. T. NORTH, D. C. PHILLIPS, and V. R. SARMA. 1967. On the conformation of the hen egg white lysozyme molecule. *Proc. R. Soc. (Lond.) B. Biol. Sci.* **167**:365–377.
36. FELIX, W. D., B. L. GALL, and L. M. DORFMAN. 1967. Pulse radiolysis studies IX. Reactions of the ozonide ion in aqueous solution. *J. Phys. Chem.* **71**:384–392.
37. BRAAMS, R. 1967. Rate constants of hydrated electron reactions with peptides and proteins. *Radiat. Res.* **31**:8–26.
38. PORUBCAN, M. A., I. B. IBANEZ, and J. L. MARKLEY. 1977. The charge relay system of serine proteases: protonation behavior of histidine-57 in chymotrypsinogen and trypsinogen. *Fed. Proc.* **36**:763.
39. SIMIC, M., and E. HAYON. 1971. Reductive deamination of oligopeptides by solvated electrons in aqueous solution. *Radiat. Res.* **48**:244–255.
40. FARAGGI, M., J. L. REDPATH, and Y. TAL. 1975. Pulse radiolysis studies of electron transfer reaction in molecules of biological interest. I. The reduction of a disulfide bridge by peptide radicals. *Radiat. Res.* **64**:452–466.
41. FARAGGI, M., and Y. TAL. 1975. The reaction of the hydrated electron with amino acids, peptides, and proteins in aqueous solution. II. Formation of radicals and electron transfer reactions. *Radiat. Res.* **62**:347–356.
42. GROSSWEINER, L. I. 1976. Photochemical inactivation of enzymes. *Curr. Top. Radiat. Res. Q.* **11**:141–199.
43. BIRKTOFT, J. J., J. KRAUT, and S. T. FREER. 1976. A detailed structural comparison between the charge relay system in chymotrypsinogen and in α-chymotrypsin. *Biochemistry.* **15**:4481–4485.
44. BIRKTOFT, J. J., and D. M. BLOW. 1972. Structure of crystalline α-chymotrypsin. V. The atomic structure of tosyl-α-chymotrypsin at 2 Å resolution. *J. Mol. Biol.* **68**:187–240.

DISCUSSION

PECHT: In view of the recently determined three-dimensional structure of the copper-containing blue proteins—plastocyanin and azurin, where two of the metal ligands are histidines—would you also expect the imidazole residue to be an effective electron-transferring group in an outer sphere type of a reaction? Furthermore, if this is the case, would you expect to find intermediates with the electron on the imidazole?

FARAGGI: The question of how the electron probes the histidine is quite difficult. In a given protein our model assumes that the hydrated electron reacts firstly with the carbonyls on the surface of the protein. Secondly the electron migrates through the hydrogen bonds to the "electron sinks," such as disulfides or histidines. Our findings indicate that the hydrated electron probes those histidines in the active site of the protein. Why other histidines are not probed is a matter of speculation.

PECHT: In view of the fact that we know now from reduction potential and the spectrum of blue copper proteins that the only ligand of the corpuscle accessible to solvent is in the middle zone, would you comment on the possibility of transferring the electrons to the copper via the electron and middle zone?

FARAGGI: The model compound is the glycyl histidine-Cu(II) complex, which we studied with pH 5 and pH 8 solutions. In acid solutions, where the ligand reacts rapidly with the hydrated electron, a transient was observed, suggesting the formation of the imidazole electron adduct. This formation rate is bimolecular and concommitant with the bimolecular decay of the hydrated electron. The second stage of the reaction (delayed by two orders of magnitude) corresponds to the decay of the transient and the concommitant bleaching process of the Cu(II) absorption. These reactions have been shown to be monomolecular.

In alkaline solutions, where the glycyl histidine ligand is unreactive, the reaction of the hydrated electron seems to be related to the Cu(II) ion and no intermediate species could be detected.

The results of Cu(II) protein in stellacyanin at pH 5 are similar to those for copper complex in acid solution. Cu(I) shows a rapid formation of the imidazole electron adduct, followed by a decay concommitant to the bleaching process of the protein due to the absorption of the Cu(II) ion.

SIMIC: There is something very interesting in the reactions of hydrated electrons with proteins. Comparing the reaction of hydrated electron cytochrome-c and your proteins, you can see delays. Cytochrome-c has all those components except the disulfide, and yet the transmission of the electron to the ion center is absolutely synchronous with the appearance of the electron. Would you tell me why there is no electron loss in the cytochrome systems, even though 100% go to ions?

FARAGGI: First, I am surprised that you got 100% reduction yield with the hydrated electron. Yields of the order of 50% or less are normally found for the same system. With proteins containing disulfides and histidines the formation of the transients (imidazole electron adduct and RSSR$^-$) is within the 0.1-ms range without the observed existence of carbonyl electron adduct. The only case where we saw intermediate radical formation is the Cu(II) blue proteins. In these studies imidazole electron adduct and RSSR$^-$ are formed, then donate electrons to the Cu(II) ion. The fact that in cytochrome-c an intermediate radical is not observed could be explained by the direct reaction of hydrated electrons with the heme. One then should not expect 100% yield but rather a smaller one.

DORFMAN: I might just add that, with regard to anion formation by electron addition to chymotrypsin, what we find surprising is not the inefficiency of the process of electron access to the histidine and the disulfide bridge, but indeed how efficient it is. It turns out to be roughly three times more efficient than one might estimate from the random diffusion, rate constants, and accessibility.

FEE: You don't want to suggest then that mild reducing agents actually produce a side reaction?

FARAGGI: Not at all. A very mild reducing agent will have a direct reaction with the Cu(II) ion. I think it would be impossible to assume that with a mild reducing agent an energy adduct on the histidine could be formed.

FEE: Are you suggesting that goes on through the histidine or are you suggesting that one of the ligands comes off when we reduce the natural energy when it enters the formation stage?

FARAGGI: No. I assume that with a mild reducing agent you will have a direct reaction with the copper, not through the ligand as suggested for the cytochrome-c in the Winfield mechanism.

FEE: I see that as a contradiction in terms, that's all. How do you react directly with metal if you can't approach it with a reductant?

FARAGGI: Is it a fact that you cannot approach the metal?

FEE: From the crystal structure that I've seen, it would seem difficult for even a small reductant such as O_2^- to get directly into the metal.

BALLOU: I may have missed something, but in the slides you showed for the reaction with the blue proteins, it looked to me that the formation of the histidine radical lagged in the conversion of the copper protein to the reduced form. If that is the case, perhaps it is either coincidence that the rates of the radical decay and the copper reduction are the same or possibly that the histidine radical decays as a result of the reduction.

BERGER: I would just like to point out some work done by Charles Martin and Mario Marini, which is in press. They have redetermined the pK's and heats of ionization of chymotrypsin to a new accuracy and have reassigned some peak values. The metal ion is somewhat capped by the ligands, histidine being the one more or less on the outside. So direct contact with the metal seems to be excluded.

BRILL: Several kinds of experimental evidence from blue proteins in solution indicate that the metal ion is not accessible. Proton relaxation rates are not enhanced. Substitution of heavy water for ordinary water has very little effect upon electron nuclear double resonance and electron paramagnetic resonance spectra. Also, fluorescence measurements indicate that the copper is not accessible from the aqueous phase.

FARAGGI: It may be. All I can say at this moment is that pulse radiolysis can produce milder reducing agents. If you take CO_2^-, for example, it reacts directly with the metal ion and you do not see any histidine radical as an intermediate.

BRILL: You don't have to have an intermediate; there is the possibility of tunneling.

CHAIRMAN: On the subject of tunneling, maybe that is an appropriate time to ask for further questions.

PECHT: Without invoking the tunneling process we have recently been compelling systematically the electron transfer between several small blue proteins and cytochromes. Model theory analysis of the electron transfer reaction clearly shows that in all cases you are dealing with an outer sphere type of reaction. This, I think, is the simplest way to resolve your observations of the formations of this adduct with the so to speak thermal energies of electron transfers. Cytochrome-c, as well as hemoglobin, show that they can go from acidic to the alkaline and back again without any kind of hysteresis. I am saying this because using my potentiometric thermal titration apparatus you can titrate the whole protein in 120 s.

FAST REACTIONS IN CARBON MONOXIDE BINDING TO HEME PROTEINS

N. ALBERDING, R. H. AUSTIN, S. S. CHAN, L. EISENSTEIN, H. FRAUENFELDER,
D. GOOD, K. KAUFMANN, M. MARDEN, T. M. NORDLUND, L. REINISCH,
A. H. REYNOLDS, L. B. SORENSEN, G. C. WAGNER, AND K. T. YUE,
Departments of Physics and Chemistry, University of Illinois at Urbana-Champaign, Urbana, Illinois 61801 U.S.A.

ABSTRACT Using fast flash photolysis, we have measured the binding of CO to carboxymethylated cytochrome *c* and to heme *c* octapeptide as a function of temperature (5°–350°K) over an extended time range (100 ns–1 ks). Experiments used a microsecond dye laser (λ = 540 nm), and a mode-locked frequency-doubled Nd-glass laser (λ = 530 nm). At low temperatures (5°–120°K) the rebinding exhibits two components. The slower component (I) is nonexponential in time and has an optical spectrum corresponding to rebinding from an S = 2, CO-free deoxy state. The fast component (I*) is exponential in time with a lifetime shorter than 10 μs and an optical spectrum different from the slow component. In myoglobin and the separated α and β chains of hemoglobin, only process I is visible. The optical absorption spectrum of I* and its time dependence suggest that it may correspond to recombination from an excited state in which the iron has not yet moved out of the heme plane. The temperature dependences of both processes have been measured. Both occur via quantum mechanical tunneling at the lowest temperatures and via over-the-barrier motion at higher temperatures.

INTRODUCTION

The heme group, a planar organic molecule with a central iron atom, plays a crucial role as active center in biomolecules performing functions ranging from oxygen storage to redox reactions and catalysis. Since most heme proteins contain the same prosthetic group, iron protoporphyrin IX, its reactivity must be controlled by the structure of the protein. To explore the relation between structure and function, we studied the kinetics of ligand binding in heme proteins with flash photolysis over wide ranges in time and temperature. In previous papers we have shown that the overall features of the binding of CO to myoglobin (1), the isolated heme group (2), and the separated hemoglobin chains (3) possess striking similarities. Access to the active center is governed by sequential potential barriers. At sufficiently high temperatures, typically above 300°K, essentially all ligands initially bound at the heme iron overcome these

---

Dr. Austin and Dr. Chan's present address is: Max Planck Institut für Biophysikalische Chemie, D-3400 Göttingen-Nikolausberg, Germany. Dr. Nordlund's present address is: Biozentrum, CH-4056 Basel, Switzerland.

barriers after photodissociation and move into the solvent. All ligands in the solvent then compete for the vacant binding site, and the corresponding binding process, called IV, is exponential in time and proportional to the CO concentration in the solvent. At low temperatures, typically below 200°K, the photodissociated CO cannot leave the protein and binding occurs from the immediate vicinity of the heme iron, and is seen as an intramolecular process, called I. The time dependence of I is approximately a power law. We have explained the nonexponential binding by postulating that the activation enthalpy and entropy of the innermost barrier are not sharp, but distributed (1–2). At temperatures below about 30°K, process I does not satisfy an Arrhenius relation, but proceeds through quantum-mechanical tunneling (2–4).

In the present paper, we extend our studies to carboxymethylated cytochrome c and heme c octapeptide. The features first discovered in globins can again be seen, but additional ones are found, such as a new fast binding process at low temperatures. While all processes observed in the globins can be explained satisfactorily with a simple model, we have not yet been able to establish a simple and unambiguous picture that includes the new results. The present work therefore is written to present the new data and stimulate additional investigations.

EXPERIMENTAL

Cytochrome c, a heme protein of mol wt 12,400, is involved in electron transport and does not normally bind CO. The heme group is covalently linked to the protein and the iron is bound to the two axial ligands His[18] and Met[80] in addition to the four pyrrole nitrogens of the heme group. The protein can, however, be modified to have the sixth, axial, position available for CO binding. One such modification is the carboxymethylation of Met[80], the product of which will be referred to as cm cyt c (5). A more extensive modification is the enzymatic digest of cytochrome c by pepsin and trypsin to the heme c octapeptide (heme peptide) (6). In the latter, His[18] is retained as the fifth, axial ligand while the sixth position is free. Horse heart cytochrome c (type III) was purchased from Sigma Chemical Co., St. Louis, Mo., and used without further purification. Cm cyt c was prepared by the method of Schejter and Aviram (5), which results in carboxymethylation of both methionine residues (at positions 65 and 80) of cytochrome c. Heme peptide was prepared from the pepsin-trypsin digest of horse heart cyt c (6). Experiments were performed in ethylene-glycol-phosphate pH 7 buffer solutions (3:2 vol/vol) for cm cyt c and glycerol-phosphate pH 7 buffer solutions (3:1 vol/vol) for heme peptide. The solutions were saturated with 1 atm CO gas and the heme concentrations $\simeq 200$ μM. The final buffer concentrations were 50 mM for cm cyt c and 10 mM for heme peptide. The optical path length of the sample is 0.5 mm.

We investigated the binding of carbon monoxide (CO) to cm cyt c and to heme peptide with flash photolysis. Consider the heme with CO bound at the iron. The bond between Fe and CO was broken by a light pulse and the subsequent rebinding of CO to the heme followed by monitoring the optical absorbance, $A(t)$, at various wavelengths. The slower components were studied by initiating photodissociation with a 0.1J coumarin 6 dye laser with pulse width 1 μs and wavelength 540 nm. Rebinding was monitored from 2 μs to 1 ks with a transient analyzer with logarithmic time base (7), and below 10 μs also with a storage scope. The fast component was also explored with 20-ps, 530-nm pulses from a frequency-doubled mode-locked Nd-glass laser.

RESULTS

Fig. 1 shows the rebinding of CO to cm cyt c and heme peptide after photodissociation by 540 nm light. For comparison, myoglobin (Mb) is also included. The binding is described by plotting the change in absorbance, $\Delta A(t) = A(t) - A(t < 0)$, versus $\log t$. Here, $A(t < 0)$ is the absorbance of the sample before and $A(t)$ the absorbance

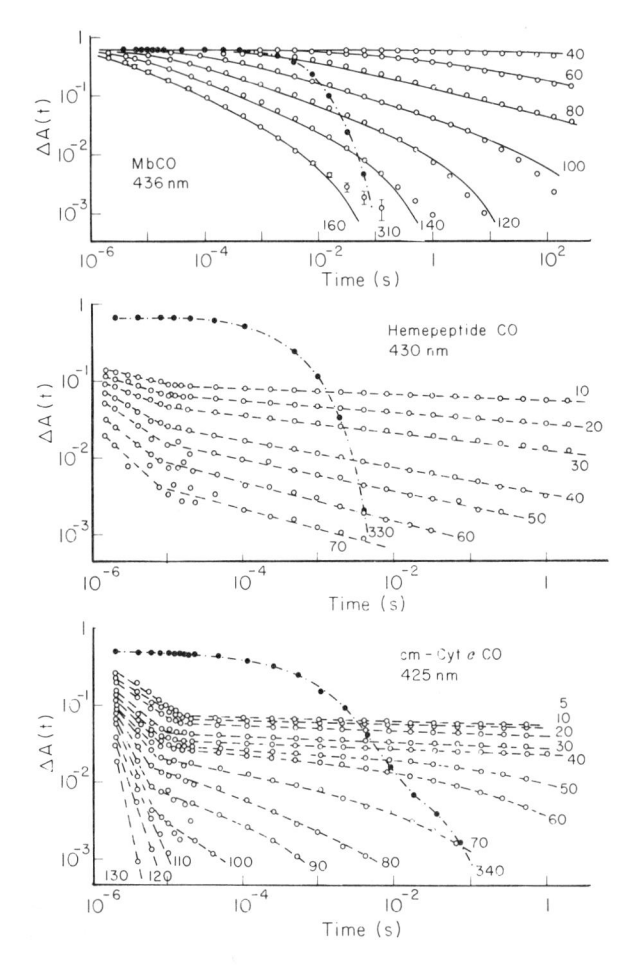

FIGURE 1 Rebinding of CO to myoglobin (Mb), heme peptide, and carboxymethylated cytochrome c (cm cyt c) after photodissociation by a microsecond laser flash at temperatures below 160°K. For Mb and heme peptide the solvent is glycerol-phosphate buffer, pH 7, (3:1, vol/vol). For carboxymethylated cytochrome c the solvent is ethylene glycol-phosphate buffer, pH 7, (3:2, vol/vol). The monitoring wavelengths are indicated in the figure. For heme peptide and carboxymethylated cytochrome c, processes I and I* are observed. In Mb only process I is seen. The dashed lines are drawn to guide the eye. The solid curves for Mb are a theoretical fit obtained with a temperature-independent enthalpy spectrum. The high-temperature limits of rebinding are given by the solid circles. For Mb the absorbance changes at high and low temperatures agree. Numbers at ends of lines represent degrees Kelvin.

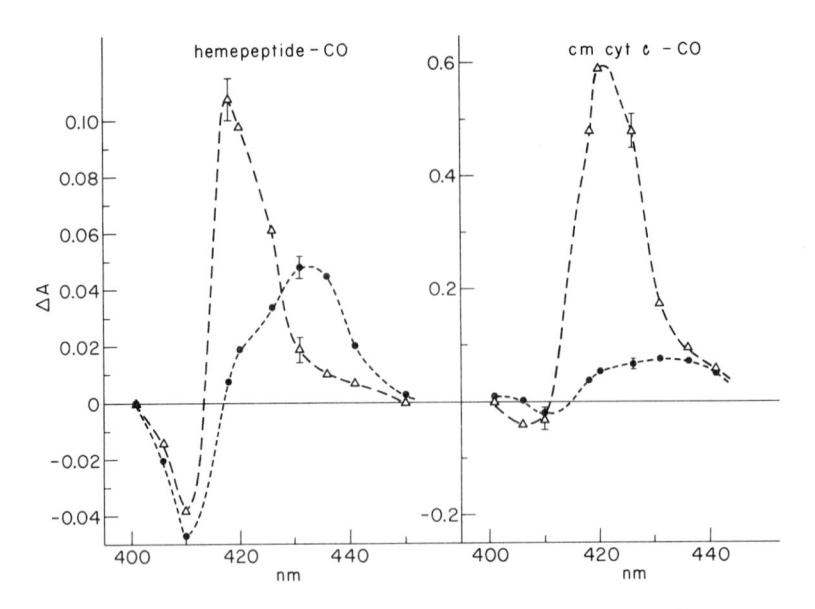

FIGURE 2 Soret spectral changes of processes I* (△) and I(●) at 40°K for heme peptide and carboxymethylated cytochrome c. If I follows a power-law behavior to microsecond times, the change in absorbance for I, ΔA_I, can be extrapolated and subtracted from the total absorbance to give ΔA_{I*} for I*. For heme peptide the absorbance changes are calculated at $t = 2$ μs, for cm cyt c at $t \approx 0.5$ μs. The optical interference filters for the monitoring light have a full width at half-maximum of 7 nm.

at time t after photodissociation. The data in Fig. 1 exhibit pronounced similarities and some characteristic differences among the three cases. At high temperatures, Mb and heme peptide show a single prominent process (IV), which is exponential in time and proportional to the CO concentration. Cm cyt c has two pH-dependent components corresponding to rebinding to low-spin and high-spin ferrous cm cyt c (8, 9). At low temperatures, all three show a process (I) that is nonexponential in time, but in addition cm cyt c and heme peptide display a fast process that we call I* (one-star). While for Mb the maximal absorption change is the same for processes I and IV, in cm cyt c and heme peptide I is much less intense than IV.

Fig. 2 gives the absorbance change for processes I and I* for cm cyt c and heme peptide at 40°K as a function of wavelength. The optical difference spectra for I and I* are clearly different.

The time dependence of I* is obtained by subtracting the extrapolated contribution of I from the total absorbance change. The resulting absorbance changes for cm cyt c at several temperatures are plotted versus time in Fig. 3. Within errors, process I* is exponential in time, $\Delta A_{I*}(t) = \Delta A_{I*}(0) \exp(-k^*t)$ and $\Delta A_{I*}(0)$ temperature-independent below about 130°K. In contrast to I*, process I approximately follows a power law in time, $\Delta A_I(t) \approx \Delta A_I(0)(1 + t/t_0)^{-n}$, where t_0 and n are temperature-dependent parameters. I* and I thus have very different time dependences.

FIGURE 3

FIGURE 4

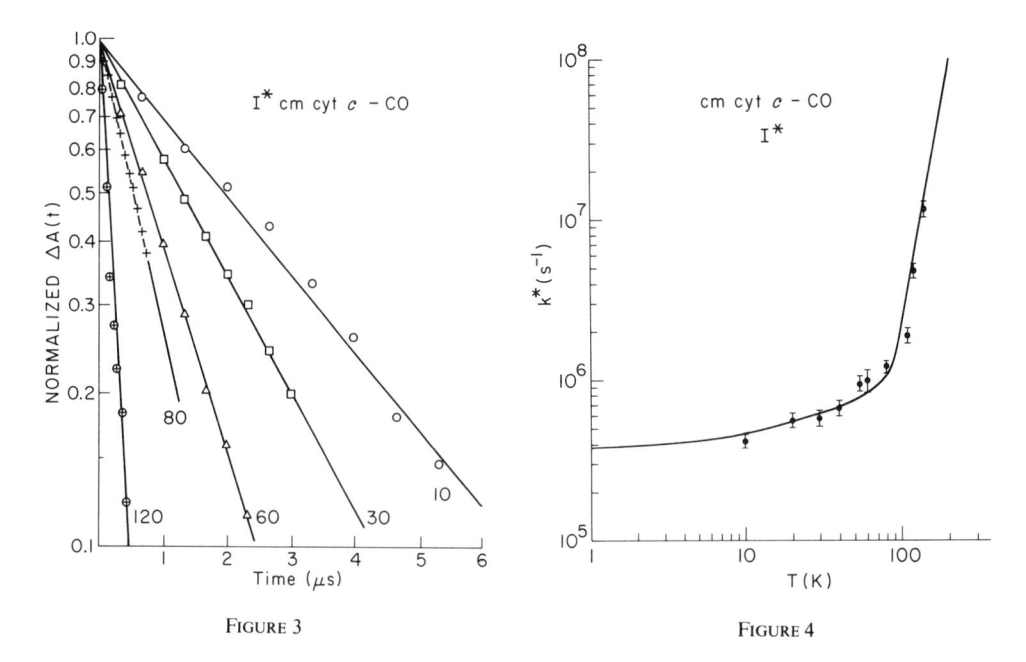

FIGURE 3 The absorbance change due to I* for cm cyt c plotted as a function of time. Photodissociation is induced by a 20 ps, 530 nm pulse from a frequency-doubled mode-locked Nd-glass laser. Monitoring wavelength is 425 nm. Because of pulse-to-pulse energy variation of the laser, the different temperatures have been normalized to the same change in absorbance at $t = 0$. Data taken with the microsecond dye laser at a constant energy indicate that the initial absorbance change due to I* is indeed independent of temperature below 130°K. Numbers at ends of lines represent degrees Kelvin.

FIGURE 4 Rate parameter k^* for rebinding of CO via process I* for cm cyt c as a function of temperature. The solid line is a fit to $k^* = A + BT^\alpha + Ce^{-E/k_BT}$, which yields $A = 3.8 \times 10^5$ s^{-1}, $B = 4.8 \times 10^3$ s^{-1} °K$^{-\alpha}$, $C = 7.8 \times 10^9$ s^{-1}, $E = 7.5$ kJ/mol, $\alpha = 1.1$. The data shown here have been obtained with the Nd-glass laser. Corresponding experiments with the dye laser (Fig. 1) give similar results and in particular show that k^* is temperature-independent below 10°K.

The temperature dependence of the rate parameter k^* for I* in cm cyt c is given in Fig. 4. The rate is approximately temperature-independent up to about 20°K, varies approximately linearly with temperature up to 100°K, and then appears to become exponential in $1/T$. Since process I is not exponential in time, it cannot be characterized fully by one rate parameter. The general features of the temperature dependence of I can be obtained by considering $k_{0.75} = 1/t_{0.75}$, where $t_{0.75}$ is the time at which $\Delta A(t)$ drops to $0.75 \Delta A(0)$. The parameter $k_{0.75}$ depends only weakly on temperature up to about 40°K and then becomes exponential. Up to at least 100°K, $k_{0.75}$ is much smaller than k^*. The rates and temperature dependences of processes I and I* thus are very different.

I and I* also differ in their dependence on the energy of the laser pulse. In Fig. 5 are plotted ΔA_I and ΔA_{I^*} for cm cyt c versus the energy of the laser flash. Both de-

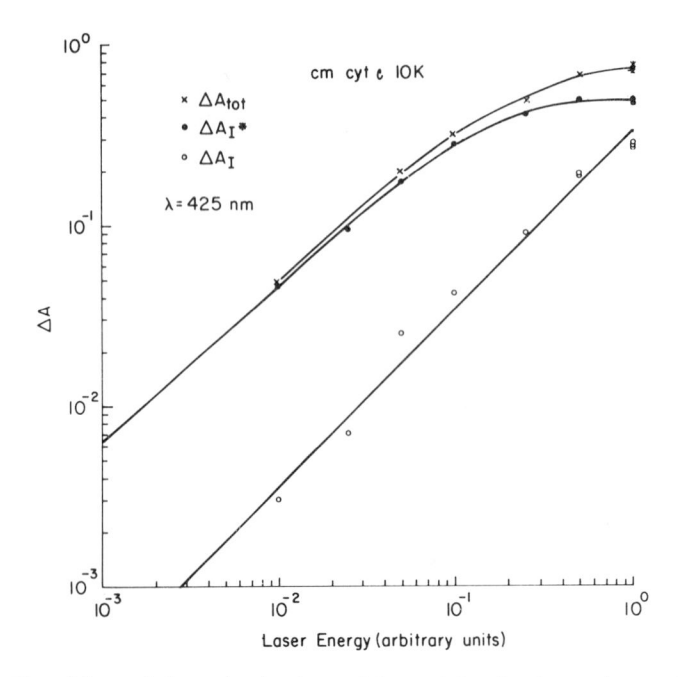

FIGURE 5 Plot of the total change in absorbance ΔA_{tot} and the absorbance changes ΔA_I and ΔA_{I^*} due to I and I* at a monitoring wavelength of 425 nm as a function of microsecond laser energy. The laser energy is varied by inserting neutral density filters in the laser beam. The experiment is performed at $T = 10°K$ with cm cyt c. I* saturates at a lower laser energy than I.

pend linearly on laser energy for low relative laser energies, but saturate at different values of laser energy. In the range of laser energies obtained in this experiment, the saturation of I* is observed, whereas I is not yet saturated.

The evidence presented so far indicates that two different processes, I and I*, occur in the low-temperature rebinding of CO to cm cyt c and heme peptide. Do both processes coexist and compete in the same system or do two classes of biomolecules exist, one with I and the other with I*? To decide between these alternatives, we have performed two experiments. In the first, we repeated photodissociation at time intervals such that all recombination via I* had taken place before the next flash. In the first alternative, such multiple flashing pumps CO into the state corresponding to process I; the intensity of I* should decrease, that of I increase with each successive flash. In the second alternative, no pumping occurs. Experiments on both heme peptide and cm cyt c show that the intensity of I* decreases and I increases with multiple flashes hence favoring the first alternative. To quantitate these results, a second experiment was performed on cm cyt c at 10°K. First the sample, monitored with a very weak light, was photodissociated with a flash from the dye laser. The very weak monitoring light causes no photodissociation and the absorbance of the sample before the flash is characteristic of the CO bound species. The absorbance changes of I and I* resulting from the flash are measured. Next the sample was allowed to come to equilibrium with a bright monitoring light and its absorbance recorded. Since the strong light photodissociates some of the cm cyt c-CO systems, the absorbance of the sample will move towards the value characteristic of the ligand-free protein. The sample was then exposed to a laser flash and the absorbance changes ΔA_I and ΔA_{I^*} were measured. This procedure was repeated for various monitoring light levels. ΔA_I and ΔA_{I^*} are plotted versus the value of the absorbance of the sample before the flash, A, in Fig. 6. Before the flash, the biomolecules are either in the CO-bound state or in the long-lived state giving rise to process I. The monitoring light is not intense enough to populate a state as short-lived as the one producing process I*. Hence if I and I* occur in separate biomolecules, we would expect ΔA_{I^*} to be independent of the monitoring light for the light levels used. Fig. 6 shows that ΔA_{I^*} decreases with increasing monitoring light intensity. If I and I* occur in the same biomolecule and are created from a common state by the laser flash we would expect the ratio $\Delta A_{I^*}/\Delta A_I$ to be independent of monitoring light. Fig. 6 shows that both ΔA_{I^*} and ΔA_I do decrease with increasing monitoring light levels, but not at the same rate. Thus processes I and I* must occur in the same biomolecule in at least a fraction of the sample.

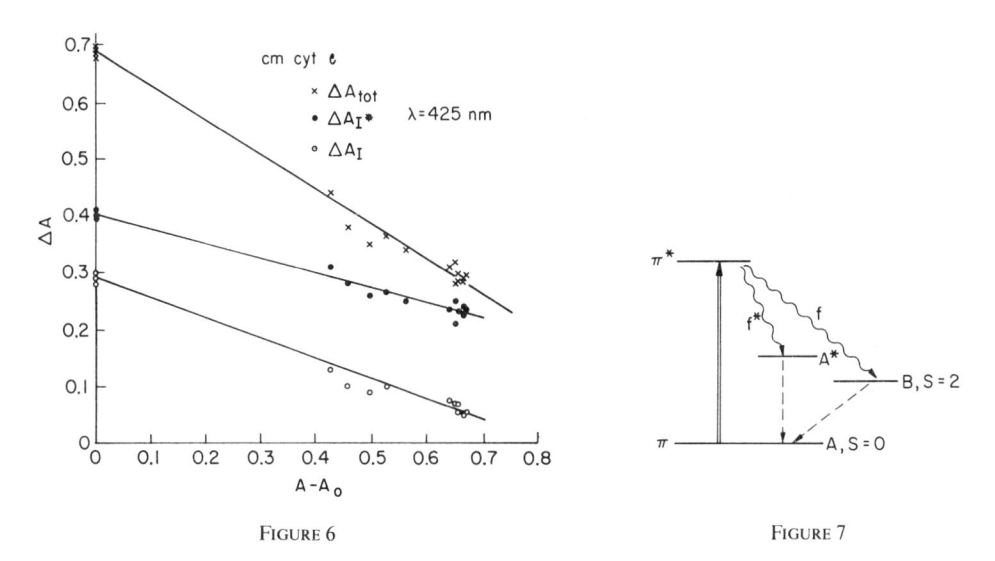

FIGURE 6

FIGURE 7

FIGURE 6 The total change in absorbance ΔA_{tot} and the absorbance change ΔA_I and ΔA_{I^*} due to I and I* resulting from a microsecond laser flash plotted versus $A - A_0$. Here, A is the absorbance of the sample before the flash and A_0 the absorbance before any photodissociation at all has occurred. The experiments are performed at $T = 10°K$ and a monitoring wavelength of 425 nm. Different values for A were obtained by equilibrating the sample with different intensities of light. Experimental details are given in the text. Note that $\Delta A_{I^*}/\Delta A_I$ is not independent of A. The solid lines drawn through the data indicate that ΔA_{I^*} and ΔA_I do not vanish for the same value of A.

FIGURE 7 States involved in the photodissociation of cm cyt c-CO and heme peptide-CO. The double lines indicate transitions involving radiation, the wavy lines radiationless transitions, and the dashed lines CO. The proposed mechanism is described in the text.

INTERPRETATION OF PROCESSES I AND I*

In previous papers (1–3), we have interpreted the kinetic data for the binding of CO to Mb, protoheme, and the separated hemoglobin chains in terms of a simple model. In particular, process I is assumed to represent direct rebinding after photodissociation and process IV the case where CO leaves the protein and moves into the solvent. Here, we will not be concerned with IV, but only with the intramolecular processes I and I*. Is it possible to construct a model in which both I and I* can be explained in a natural way? We have not found a model that explains all observed facts simply and convincingly, but we will present one that may be a good starting point for further improvements. We assume that two different types of molecules exist; in the first type, both I and I* coexist and compete, in the second, only process I takes place. We will restrict our attention to the first type because it is here where new features arise.

The observation of two processes with different optical spectra in the same protein implies the existence of three states. We denote the states, shown in Fig. 7, by A, A*, and B. The existence and properties of A and B are well established through X-ray, optical, and Mössbauer experiments. In A, CO is bound at the heme iron which is in a spin $S = 0$ ($d_{xy}^2 d_\pi^4$) state and lies in the heme (xy) plane. In B ("deoxy"), the ligand has moved away, the iron has changed to $S = 2(d_{xy}^2, d_\pi^2, d_{z^2}^1, d_{x^2-y^2}^1)$ and in most systems moved out of the heme plane by about 0.05 nm. In A and B, the electronic structure of the heme ring is in its ground state, denoted by π. The state A* is required to account for I*. The experimental results can now be interpreted as follows: Initially, the system is in the ground state A. The laser flash excites the electrons of the heme. At a laser wavelength of about 540 nm the excitation causes the promotion of an electron from the highest filled porphyrin π orbital to the lowest empty porphyrin π orbital ($\pi \rightarrow \pi^*$). Photoexcitation is followed by rapid radiationless transitions, leading to state B and to the transient state A* with probabilities f and f^*, respectively. Binding from A* produces process I*, from B process I.

The states and transitions in Fig. 7 can be interpreted in terms of the molecular orbital calculations of Zerner et al. (10). According to these authors, the transition $d_\pi \rightarrow d_{z^2}$ provides a plausible pathway for photodissociation. We thus tentatively identify state A* with the iron electronic configuration d_{xy}^2, d_π^3, $d_{z^2}^1$ which, by Hund's rule, has $S = 1$. The d_π in state A contributes to the bonding of CO to Fe, while the d_{z^2} in A* is antibonding. CO is thus less tightly bound in A* than in A and can dissociate. In A*, the orbital $d_{x^2-y^2}$ is empty; since it is occupation of this orbital that makes the out-of-plane position favorable for the iron, Fe will very likely still be in the heme plane in the $S = 1$ state. The CO can then rebind quickly via process I*. Some support for the assignment comes from the optical spectra, Fig. 2. For heme peptide the difference spectrum for I agrees with the difference spectrum calculated from the separately measured carbonmonoxy and high spin-deoxy states. For cm cyt c, the difference spectrum for I peaks at the same wavelengths as the separately measured carbonmonoxy and high spin-deoxy states, but is much too broad. The broadness of this spectrum, may, in part, be due to the existence of molecules with only a process I rebinding. The spectrum for I* in both heme peptide and cm cyt c resembles the difference spectrum between CO-bound and the low spin-deoxy conformations of the proteins.

Photodissociation, in the proposed model, involves a number of steps: first, excitation $\pi \to \pi^*$; second, radiationless transfer of energy from π^* to Fe; and third, dissociation of the CO. Details of the second and third steps are not yet fully understood. In the model of Zerner et al., the second step involves the transition $d_\pi \to d_{z^2}$, but recent calculations[1] and measurements (11) show that the $d_\pi \to d_{z^2}$ transition occurs at an energy at least an electron volt higher than the $\pi \to \pi^*$ transition at 540 nm. It is, however, still possible for the $\pi \to \pi^*$ excitation to couple to $d_\pi \to d_{z^2}$ via radiationless transfer of energy. It has also been suggested recently that the $\pi \to \pi^*$ transition itself directly leads to dissociation (12). In any case, in the scheme of Fig. 7, two different pathways are involved, one leading to A*, the other to B.

PROBLEMS

The results obtained so far with cm cyt c and heme peptide raise a number of questions that will require extension of the experiments to better samples, lower temperatures, shorter times, higher accuracy, and other systems, and will also call for other techniques. We briefly outline some of the open problems here.

How are I and I* related? In Fig. 7, we have proposed a simple scheme for the rebinding processes in cm cyt c and heme peptide. Is this scheme correct? Do transitions A* → B and B → A* exist? Another problem emerges from Fig. 1. While I is much slower than I* at low temperatures, it speeds up more quickly with increasing temperature. Will it overtake I* or will the two processes merge? Is the return B → A a one-step transition or is an intermediate state, which may be A*, involved?

Is the explanation invoking two types of biomolecules, one with I* and I, the other with I only, correct? Can the experimental data be explained with an alternate model, based on only one type, but with additional transitions among the various states?

At low temperatures, only a fraction of heme peptide or cm cyt c molecules move to the deoxy state B after photodissociation. At high temperatures, above about 340°K, all rebinding occurs via the solvent and in our model via state B. How and at which temperature does the switch from the predominant direct rebinding A* → A to the indirect sequence B → A occur? One possibility invokes the transition A* → B at higher temperatures; another that motion to B becomes dominant when the molecule can easily relax and change conformation.

In Mb and the separated α and β chains of hemoglobin, we have not observed process I* (1, 3). Why is I* absent, or very small, in these proteins? One explanation

---

[1]Eisenstein, L., D. R. Franceschetti, and K. L. Yip. Iterative extended Hückel studies of some pyridine-Fe(II)-porphin complexes. In preparation.

invokes the different anchorage of the proximal histidine that connects the iron to the protein backbone. In the globins, the histidine is connected to the F helix which may exert some tension on the iron and pull it out of the heme plane after photodissociation. In cm cyt c and heme peptide, the protein backbone is linked covalently to the heme and may thus hold the iron more tightly in the heme plane.

What are the properties of A*? In particular, what is its spin? Hopefully, fast susceptibility or pulsed Mössbauer experiments can provide more information.

It is interesting to speculate on why process I* has an exponential time dependence, whereas I is nonexponential. We attribute the nonexponential time dependence of process I to different conformational states of the protein, each with a different activation enthalpy for rebinding of CO via I. The different conformational states would then have the same value for the activation enthalpy for rebinding via process I*. According to our interpretation of the nature of processes I and I*, the iron $d_{x^2-y^2}$ orbital should then be more influenced by the protein state than the d_{z^2} orbital.

Fig. 1 shows that in cm cyt c the rebinding curve for I at 30°K is nearly parallel to the one at 5°K, but 5°K is shifted by about a factor of two towards larger values of ΔA. We have observed this behavior in various samples. If correct, it would indicate that the transition $\pi^* \rightarrow$ B becomes more favorable at very low temperatures.

Tunneling extends to remarkably high temperatures in process I*. The component proportional to T is measurable to about 120°K. It is possible that a component proportional to a higher power of T follows before the Arrhenius regime is reached. Why does tunneling extend to such temperatures?

We thank G. Careri, P. G. Debrunner, D. R. Franceschetti, I. C. Gunsalus, P. Hänggi, and T. Pederson for many stimulating discussions.

This work was supported in part by the U.S. Department of Health, Education and Welfare under grant GM 18051 and the National Science Foundation under grant PCM 74-01366.

Received for publication 12 December 1977.

REFERENCES

1. AUSTIN, R. H., K. W. BEESON, L. EISENSTEIN, H. FRAUENFELDER, and I. C. GUNSALUS. 1975. Dynamics of ligand binding to myoglobin. *Biochemistry.* **14:**5355–5373.
2. ALBERDING, N., R. H. AUSTIN, S. S. CHAN, L. EISENSTEIN, H. FRAUENFELDER, I. C. GUNSALUS, and T. M. NORDLUND. 1976. Dynamics of carbon monoxide binding to protoheme. *J. Chem. Phys.* **65:**4701–4711.
3. ALBERDING, N., S. S. CHAN, L. EISENSTEIN, H. FRAUENFELDER, D. GOOD, I. C. GUNSALUS, T. M. NORDLUND, M. F. PERUTZ, A. H. REYNOLDS, and L. B. SORENSEN. 1978. Binding of carbon monoxide to isolated hemoglobin chains. *Biochemistry.* In press.
4. ALBERDING, N., R. H. AUSTIN, K. W. BEESON, S. S. CHAN, L. EISENSTEIN, H. FRAUENFELDER, and T. M. NORDLUND. 1976. Tunneling in ligand binding to heme proteins. *Science (Wash. D. C.).* **192:**1002–1004.
5. SCHEJTER, A., and I. AVIRAM. 1970. The effects of alkylation of methionyl residues on the properties of horse cytochrome c. *J. Biol. Chem.* **245:**1552–1557.
6. HARBURY, H. A., and P. A. LOACH. 1960. Oxidation-linked proton functions in heme octa- and undecapeptides from mammalian cytochrome c. *J. Biol. Chem.* **235:**3640–3645.
7. AUSTIN, R. H., K. W. BEESON, S. S. CHAN, P. G. DEBRUNNER, R. DOWNING, L. EISENSTEIN, H. FRAU-

ENFELDER, and T. M. NORDLUND. 1976. Transient analyzer with logarithmic time base. *Rev. Sci. Instrum.* **47**:445–447.
8. BRUNORI, M., M. T. WILSON and E. ANTONINI. 1972. Properties of modified cytochromes. I. Equilibrium and kinetics of the pH-dependent transitions in carboxymethylated horse heart cytochrome c. *J. Biol. Chem.* **247**:6076–6081.
9. WILSON, M. T., M. BRUNORI, G. C. ROTILIO, and E. ANTONINI. 1973. Properties of modified cytochromes. II. Ligand binding to reduced carboxymethyl cytochrome c. *J. Biol. Chem.* **248**:8162–8169.
10. ZERNER, M., M. GOUTERMAN, and H. KOBAYASHI. 1966. Porphyrins. VIII. Extended Hückel calculations on iron complexes. *Theor. Chem. Acta (Berl.).* **6**:363–400.
11. CHURG, A. K., and M. W. MAKINEN. 1978. The electronic structure and coordination geometry of the oxyheme complex in myoglobin. *J. Chem. Phys.* In press.
12. HOFFMAN, B. M., and Q. H. GIBSON. 1978. On the photosensitivity of liganded hemeproteins and their metal-substituted analogues. *Proc. Natl. Acad. Sci. U.S.A.* In press.

DISCUSSION

BRILL: We would like to comment, first, upon the inner barrier energy distributions found to be a general property of heme proteins and heme derivatives (1, 2); second, on a possible relation between level broadening seen in frozen solutions and the activation energy of electron transfer reactions; and third, on the use of glass-forming solvent systems.

Analysis of electron paramagnetic resonance (EPR) orientation studies of ferric hemoglobin single crystals provides orientation disorder angles and hyperfine component line widths.[1] Agreement of the latter with the predictions of a dipolar broadening computation (based upon the three-dimensional structure of horse ferric hemoglobin) suggests that the various contributions to the line width in the crystals are known quantitatively within a small range of uncertainty.[2] Extension of these considerations to frozen solutions leads one to expect significant narrowing of the EPR lines, while the data show a broadening. On the assumption that the latter arises from energy level broadening associated with a distribution in structures, we have analyzed the frozen solution EPR spectra of many ferric heme-protein complexes in terms of Gaussian distributions in the spin-Hamiltonian parameters.[3] The internal consistency of the derived data suggests that this is the appropriate explanation of the phenomenon. The four parameters involved in fitting the spectra are the rhombic-to-axial symmetry ratio E/D and its root mean square (rms) deviation $\sigma_{E/D}$, the coefficient η for the admixture of quartet states (related to iron out-of-planarity) and σ_η. We view $\sigma_{E/D}$ and σ_η as reflecting conformational modes, with amplitudes given by the equipartition principle at the temperature at which they are frozen into metastable states (minima of secondary potentials which modulate the primary, approximately harmonic, potentials) (3). There is little leeway in specifying the four parameters; departures of 5% in E/D, 10% in $\sigma_{E/D}$, 4% in η, and 20% in σ_η significantly alter the quality of the fit. Likewise, the constant line width below 50°K in the frozen solution EPR spectra from low-spin ferric cytochrome-c has been attributed to a ±6% spread in rhombic potential (4). Distribution of spin-Hamiltonian parameters is not unique to heme proteins; this phenomenon has also been observed in iron-sulfur proteins (5) and copper blue proteins (6).

Information of this kind is related to protein reactivity. In the case of the cupric site in blue proteins, we have proposed that the EPR spectral distribution can be used to obtain the force

---

[1] Hampton, D. A., and A. S. Brill. 1978. Crystalline state disorder and hyperfine component line widths in ferric hemoglobin chains. To be published.

[2] Brill, A. S., and D. A. Hampton. 1978. Quantitative evaluation of contributions to electron paramagnetic resonance line widths in ferric hemoglobin single crystals. To be published.

[3] Brill, A. S., F. G. Fiamingo, and D. A. Hampton. 1978. Characterization of ferric energy levels in alcohol complexes of myoglobin and hemoglobin. To be published.

constant for the vibronic mode involved in electron transfer activation (3), and shown how this constant together with the measured enthalpy of activation provides an estimate of the angular displacement of the cuprous potential minimum from the cupric minimum (7). Electron transfer reactions of heme proteins can be considered similarly. For example, the displacement of the high-spin ferric ion from the plane of the porphyrin nitrogens is less than that of the ferrous ion, and the mean potentials for motion normal to the heme plane are involved in thermal activation of the redox reaction. Such considerations suggest broader application of the EPR spectral distribution analysis to heme enzyme systems. More generally, a new use of measurements on frozen protein solutions is suggested.

Along this line, attention should be drawn to the possibility that certain procedures often employed in investigation of frozen solutions may modify macromolecular systems. We refer to the use of glycerol, ethylene glycol, and other solvent modifiers to produce glasses. In early studies of the use of glycerol-aqueous buffers in optical measurements at liquid nitrogen temperature, concern was expressed that a reaction of reduced cytochrome-c with the glycerol might form a derivative discernible only at cryogenic temperatures (8). We have found that the binding of menthanol and ethanol to ferric hemoglobin increases by factors of 60 and 10, respectively, between room temperature and freezing (9). The EPR spectral distribution for ferric myoglobin in glycerol aqueous buffer mixtures is greater than in aqueous buffer alone. Glycols are known to denature proteins (10). The intensities of the Soret bands of cytochrome-c and myoglobin (in 0.1 M acetate, pH 5.7, 25°C) are reduced 50% by 75–80 vol % ethylene glycol. These observations suggest the need for careful investigation of the effects of such agents on protein structure, and point to the special influence of water on protein stability.

CHAIRMAN: Do you wish to make any comment?

EISENSTEIN: I'm pleased that there is another technique, as Dr. Brill has shown, that implies the existence of many conformational states of proteins, as our analysis at low temperature would indicate.

VANDERKOOI: What criteria do you have for the homogeneity of the carboxymethylated cytochrome-c preparation? For example, have you established the homogeneity of protein with regard to its redox properties and with regard to the polypeptide chain (one band on gel electrophoresis)? Let me point out that cytochrome-c dimerizes very easily, especially in the conditions under which the derivatization was done. Could your results be accounted for by a heterogeneity of cytochrome-c molecules?

EISENSTEIN: Yes indeed, that problem has worried us for sometime. Originally we prepared carboxymethylated cytochrome-c by the method of Schejter and Aviram, with no further purification. Recently we have obtained a sample of carboxymethylated cytochrome-c from Margoliash's group. It is homogeneous, in that they have determined that only methionine residue 80 is modified. There is no other modification of the protein. We have virtually identical results, in that the amount of I^* material is the same with our preparation as with theirs. Removing all the aggregates in our sample (less than 5%) did not change any of the kinetics.

LESTER: I am interested in what might happen between the time that the proton promotes new electron distributions in the ion and in the heme and the time that the carbon monoxide molecule actually leaves. First of all, what is the time scale of these kinds of processes? Is it nanoseconds, or microseconds? The reason I am interested, of course, is that we worry how long after our light flash does our perturbed molecule leave our acetyl choline receptor.

EISENSTEIN: Perhaps this is a question that Dr. Noe and Dr. Rentzepis could comment on, as they have recently done experiments on CO myoglobin with picosecond pulses of light. They have found that in the CO myoglobin there is a transient species. They have found that in less than 6 ps there is photo-dissociation, or at least there is a change in the optical spectrum. After that there is a species that exists for about 130 ps, and then goes to the deoxy state. There may be an intermediate. It will be interesting to see the spectrum of the intermediate produced in less than 6 ps; it decays to the deoxy spin state at room temperature.

NOE: Dr. Rentzepis and I did do the picosecond experiment, and have presented it here as a poster entitled "Picosecond Photodissociation and Subsequent Recombination Processes in HbCO, MbCO and MbO_2" (page 379). What we see is a transient species that relaxes after a photo dissociation of 130 ps. At a relaxed point, for example several hundred picoseconds, we have measured a difference spectrum and find that it roughly corresponds to the steady-state difference spectra between myoglobin and CO myoglobin. We are in the process of analyzing the maximum of optical density change to determine whether or not this reflects an electronic state different from the relaxed electronic state. The oxygen does not show this relaxation. One can speculate on various mechanisms for this.

LESTER: Could I then infer that you think the ligand leaves the protein less than 130 ps after the photon intercepts?

NOE: I believe that the CO probably dissociates in less than 6 ps. Our results show some interesting process going on and we have to do further experiments.

FERRONE: There also seems to be an inescapable homogeneity problem, in that carboxymethylated cytochrome-c titrates between several forms as one changes pH. The high-pH form, moreover, has a ligand which is not methionine—that is, which carboxymethylation doesn't block. Have you done these experiments at different pHs?

EISENSTEIN: We have done experiments as a function of pH. Most of the experiments reported here are done in ethylene glycol-phosphate buffer at pH 7. But we have done experiments at pH 6 and pH 8, and under these conditions at high temperatures, and have seen differences. For example, at high temperature, as the pH is lowered, you have more of a fast component for rebinding. This was shown in our Fig. 1, bottom. There are two components to the rebinding: a fast component and another component slower by a factor of 100. By changing the pH at high temperature you can vary the amount of fast and slow components. Looking at the low temperature kinetics, we do not see any difference in the amount of I^*.

FERRONE: Don't you also have to do the control experiment of determining how the titration curve goes as a function of pH and temperature, to be sure which forms remain as you go to lower temperatures?

EISENSTEIN: Yes, we have seen that the amount of this fast and slow component at high temperature is roughly temperature-independent. I can't tell you what happens when it freezes or when it forms a glass, but at least down to a temperature of 250°K, the ratio of the fast to the slow components does not seem to change very much with temperature.

FEE: May I make a brief comment that may extend the generality of the problem of conformational changes in proteins upon freezing? Professor Richard Sands and I have been looking at the high-potential iron protein from *Chromatium vinosum* and examining the EPR spectrum under a variety of different conditions. We find that some amazing changes occur. For example, in a very dilute solution the spectrum is broad and quite indistinct. If we put it into 0.2–2 M sodium chloride, a very well defined and reproducible structure appears. Now if we do a rapid (few millisecond) freeze-quench experiment, we find a very indistinct EPR spectrum. (The protein in the oxidized form, frozen by the normal procedure of slow immersion in liquid nitrogen, gives a distinct EPR spectrum.) The presumed conformational distribution depends not only on the nature of the solvent but also on the time in which it freezes. Perhaps Art Brill, whom I have to compliment for this beautiful insight, can also give us some insight why iron sulfur proteins as a rule display this behavior. If the conformation represents significant structural deviations around the active center region that you are concerned with, they must certainly play a part in the analysis.

BRILL: With respect to azurin, there is no effect upon the EPR spectra of the different rates of freezing in the cold gas over liquid nitrogen as compared with insertion directly into liquid nitrogen. With respect to ferric myoglobins and hemoglobins, there is no effect of the rate of freezing in the cold gas over liquid helium as compared with inserting into liquid helium directly. Catalase presents a very complicated freezing behavior; the conditions for obtaining reproducible EPR spectra are not yet known. With respect to the addition of methanol and ethanol to ferric myoglobins and hemoglobins, the EPR spectrum is independent of alcohol concentration once the alcohol complex is formed up until the concentration reaches a denaturing level. Undoubtedly there are metalloproteins where the EPR spectra strongly reflect the conditions of freezing. Consistent with the existence of metastable states is a narrowing of the structural spread with decreasing rate of freezing. In the systems we described, moderate changes in the freezing process do not significantly change the data.

KLAPPER: In terms of heterogeneity, this raises another question. I believe that these measurements are done in relative high gycerol-water ratios. I wonder about the possibility of local heterogeneity because of differences of solvent and solute concentrations after freezing. I understand that it is possible in fact to get patches of protein dissolved in glycerol, in no water.

EISENSTEIN: It's true that these experiments are done in high glycerol concentrations. Typical experiments are done in 75% glycerol so that we can obtain a good optical transmission through our sample. We have also done experiments in water. They are harder to do and we do see similar behavior at low temperatures. I think that it is possible that there are different environments around each protein molecule at low temperature. Another alternative, of course, is that the protein structure itself is different. Perhaps the experiments that we are planning to do at low and high temperature as a function of pressure will answer this. We think that applying pressure at high temperature changes the conformation, and it is a different conformation of the protein from what we are seeing at low temperature. However, I think that it is quite difficult to rule out the effect of the different solvent cages around each protein model.

BRILL: Sometimes there is even a problem at room temperature. If one uses ethylene glycol at significant concentration and does not wait long enough after putting it into the protein solution, the system may not be at equilibrium. For example, if an equal volume of ethylene glycol is added (with vigorous stirring) to a ferric myoglobin solution, a spectral (optical) change appears, which decays at a first-order rate of about 0.05 min back to the original spectral state. Local, partial denaturation has occurred which is reversible and goes away. (If more ethylene glycol is added, the optical change will not disappear completely.) One must be aware of possible effects of transient local variations in concentration whenever an agent is added to a protein solution.

PAVLOVIĆ: It was recently shown by William Smith and colleagues from Guy's Hospital in London, published in *Biochemical Journal*, that upon freezing, especially in phosphate buffers, a very drastic pH change (1–3 pH units) occurs. In another paper it is reported that the solution upon freezing passes through the glycerol transition and becomes solid around 200°K at this rate of cooling. Does this glass transition and transition into the completely solid state have any effect on the protein structure? In α chains or myoglobin it has been shown that changes occur around these temperatures.

EISENSTEIN: With regard to the first question about pH changes, in Douzou's laboratory in Paris, extensive work has been done on the changing effective pH in mixed solvents and as a function of temperature. If I remember correctly, at room temperature in 50% ethylene glycol-water solutions with 50% pH 7 phosphate buffer, the effective pH is 7.5. When you lower the temperature to 40°C, the effective pH goes to about 8. In a Tris buffer the change is much larger. To answer your second question, we have not seen any change in the kinetics at the glass temperature, and I would say that in a protein like myoglobin, where we can follow things right through as a function of temperature, we can do measurements over the entire temperature range. For some other proteins the processes are too fast for us to see. In other words, there is a certain temperature interval in some of the proteins over which we observe no signal with our current system. For these I can't tell you if anything dramatic happens at the glass temperature. But for proteins like myoglobin, I don't think anything dramatic happens.

BERGER: A number of years ago, we measured in 92% glycerol and 62 or 64% sucrose, 0.1 M, pH 7 buffer the reactions of oxygen and carbon monoxide with hemoglobin. If you correct for the nonlinearity of the diffusion coefficients and the reduction in solubility, the rate constants for oxygen and carbon monoxide do not change much. We did one other experiment that I would like to call to your attention. With J. J. Katz at Argonne, we took hemoglobin and dried it. We introduced it into anhydrous hydrogen chloride at liquid helium temperatures. Now obviously I didn't run a rapid reaction on that but I did oxygenate and deoxygenate it and, using a hand-held Martree spectroscope, found no significant changes in the optical spectra. Those of you who read science fiction will recall there are two standard systems, besides water, for creating biological worlds: anhydrous hydrogen chloride, because it is such a very good solvent and is polar, just like water, and liquid ammonia. You might wish to study it.

REFERENCES

1. AUSTIN, R. H., K. W. BEESON, L. EISENSTEIN, H. FRAUENFELDER, and I. C. GUNSALUS. 1975. Dynamics of ligand binding to myoglobin. *Biochemistry*. **14**:5355, 5373.
2. ALBERDING, N., R. H. AUSTIN, S. S. CHAN, L. EISENSTEIN, H. FRAUENFELDER, I. C. GUNSALUS, and T. M. NORLUND. 1976. Dynamics of carbon monoxide binding to protoheme. *J. Chem. Phys.* **65**: 4701–4711.
3. BRILL, A. S. 1978. Conformational distribution and vibronic coupling in the blue copper-containing protein azurin. *In* Tunneling in Biological Reactions. H. Frauenfelder et al., editors. Academic Press, Inc., New York. In press.
4. MAILER, C., and C. P. S. TAYLOR. 1971. The temperature dependence of the electron paramagnetic resonance spectrum of ferricytochrome c solutions between 4.2°K and 77°K. *Can. J. Biochem.* **49**: 695–699.
5. FRITZ, J., R. ANDERSON, J. FEE, G. PALMER, R. H. SANDS, J. C. M. TSIBRIS, I. C. GUNSALUS, W. H. ORME-JOHNSON, and H. BEINERT. 1971. The iron electron-nuclear double resonance (ENDOR) of two-iron ferredoxins from spinach, parsley, pig adrenal cortex and *Pseudomonas putida*. *Biochem. Biophys. Acta*. **253**:110–133.
6. BRILL, A. S. 1977. Transition metals in biochemistry. Springer-Verlag KG, Berlin, W. Germany. 76–77.

7. BRILL, A. S. 1978. Activation of electron transfer reactions of the blue proteins. *Biophys. J.* **22:** 139–142.
8. ESTABROOK, R. W. 1961. Spectrophotometric studies of cytochromes cooled in liquid nitrogen. *In* Haematin Enzymes. J. E. Falk et al., editors. Pergamon Press. London. Part 2, pp. 436–457.
9. BRILL, A. S., B. W. CASTLEMAN, and M. E. MCKNIGHT. 1976. Association of methanol and ethanol with heme proteins. *Biochemistry.* **15:**2309–2316.
10. HERSKOVITS, T. T., B. GADEGBEKU, and H. JAILLET. 1970. On the structural stability and solvent denaturation of proteins. I. Denaturation by the alcohols and glycols. *J. Biol. Chem.* **245:**2588–2598.

PHOTO-INITIATED ION FORMATION FROM OCTAETHYL-PORPHYRIN AND ITS ZINC CHELATE AS A MODEL FOR ELECTRON TRANSFER IN REACTION CENTERS

S. G. BALLARD AND D. MAUZERALL, *The Rockefeller University, New York 10021 U.S.A.*

ABSTRACT Ion formation from the reaction of triplet (T) and ground state (P) octaethyl-porphyrin (OEP) and zinc octaethyl porphyrin (ZnOEP) and the corresponding cross-reactions have been measured in dry acetonitrile. A uniquely sensitive and fast conductance apparatus and a pulsed dye laser allowed the measurements to be made at the necessarily very low concentrations of T. The homogeneous reaction of T (ZnOEP) and P (ZnOEP) occurs with rate constant $k_1 = 2.0 \times 10^8$ $M^{-1}s^{-1}$ and an ion yield of 67%. The similar homogeneous reaction of OEP has $k_2 = 1.3 \times 10^8$ $M^{-1}s^{-1}$ but an ion yield of only 3%. The cross-reaction of T (OEP) with P (ZnOEP) has $k_3 = 1.5 \times 10^8$ $M^{-1}s^{-1}$ and an ion yield of 27%, while the inverse cross-reaction of T (ZnOEP) with P (OEP) has $k_4 = 3 \times 10^8$ $M^{-1}s^{-1}$ and an ion yield of 20%. Thus, the rate constants are only slightly affected but the yields are sensitive to the porphyrin. The possible formation of the heterogeneous ions $ZnOEP^+ + OEP^-$, thermodynamically favored by 0.3 V over the homogeneous ions, has little influence on the observed yields. The data are explained by electron transfer and Coulomb field-electron spin-controlled escape of the initial ion-pair.

INTRODUCTION

Much progress has been made in our understanding of the primary photochemical reactions of bacterial photosynthesis (1). The availability of purified reaction centers has allowed the elucidation of the path of the electron transfer reactions and their description at the level of quantum mechanics. Thus, after excitation, the electron leaves the donor, a dimer of bacteriochlorophyll, in $<10^{-11}$ s and remains for 10^{-10} s on a bacteriopheophytin (2), before passing on to a ubiquinone molecule. The bacteriopheophytin anion acts as a bridge to the more stable quinone anion, and thus contributes a high-energy path through the barrier between the quinone acceptor and bacteriochlorophyll dimer donor. This barrier amounts to 30 ms at low temperatures (3). The thermodynamically highly favored reverse electron transfer reaction to the ground state is slowed sufficiently to allow useful work to be obtained by the biochemical electron transfer machinery. The usefulness of a primary charge transfer between the pigment molecules was pointed out by Kamen (4). The possibility of favoring this reaction by use of the metalloporphyrin-free base porphyrin to bridge the gap was also anticipated (5). Our work has shown that even in free solution well over 50% of the photon energy can be stored (for up to ~0.1 s) in the reactive free radicals after electron transfer reactions to the free base porphyrins or from the metalloporphyrins (6–7). These reactions have quantum yields near unity and so are very efficient. However, evidence for ion formation in the direct reaction of porphyrin and metalloporphyrin was lacking. Using an extremely fast and sensitive conduction apparatus (8), we have now obtained direct kinetic rate and yield measurements on precisely these reactions.

METHODS

A detailed description of the conduction apparatus (8) and its application to the study of the photochemistry of lumiflavin (9) have been published. The apparatus detects changes of 10^{-9} M of photogenerated univalent ions with a time constant of 0.3 μs and changes of 10^{-12} M at longer times. The anaerobic conductivity cell is operated by a voltage clamp in a positive pulse, negative pulse mode which assures long-term stability of the conductance. Gating and timing circuits allow the photogenerated conductance change to be measured after pulsing transients have relaxed. A high-speed digitizer and signal averager allow efficient data collection. The results were plotted by an x, y recorder and analyzed "by hand."

The light pulse is obtained either directly from a nitrogen laser emitting 7-ns full width at half maximum pulses, or via a dye laser tuned with an interference filter. The beam was spatially filtered and expanded to achieve homogeneous illumination (<5% variation) of the cell. Direct photoeffects at the platinum electrodes were always negligible. Absolute energies were determined at the position of the conductance cell by a YSI model 65 radiometer (Yellow Springs Instrument Co., Yellow Springs, Ohio). The conductance cell was absolutely calibrated with tetraethyl ammonium perchlorate.

The ion yield is defined in an absolute sense: ions formed per quantum absorbed. The former is obtained from the instrument sensitivity, X: $X = 1.8 \times 10^{-3} \cdot V \cdot s \cdot \lambda$, in millivolts per nanomolar concentration, where V is the polarizing voltage (typically 10 V), s is the cell constant for homogeneous illumination (2.14 cm), and λ the ion equivalent conductance. The latter value is assumed equal for P^+ and P^- and was calculated, with minor correction for oblateness, from the Stokes, Einstein, and Walden relations: 35 mho cm^2 mol^{-1} in acetonitrile. These relations are accurate for large ions immersed in small solvent molecules. The quanta absorbed were obtained from the absolute light intensity and the absorption of the solution at that wavelength. The absolute value of the ion yields thus defined may be in error by ± 10% because of the systematic errors. However, the relative error, important in the present context, is far less, and is determined by the signal-to-noise ratio.

Since oxygen quenches the triplet state at encounter-limited rates, and the reaction studied was 100 times slower, the requirement $O_2 < 0.01\ P_0$ was achieved by vigorous purging with O_2-free (<0.5 ppm) N_2. A typical concentration of O_2 was 10^{-9} M. Since the triplet-triplet reaction is also encounter-limited, it was necessary to use extremely weak excitation (saturation parameter, $1 - e^{-\sigma l}, < 10^{-3}$). Thus the ion concentration was 10^{-9}–10^{-11} M. This conductance measurement is the only feasible method of detecting such low concentrations of ions. It appears that most, if not all, data on porphyrin and chlorophyll photoreactions in the literature are highly contaminated with the voracious triplet-triplet reaction. The solvent was of spectroscopic quality and was doubly distilled, first from P_2O_5, then from highly activated molecular sieves directly into the reaction cell for several flushings before mixing in the small amount pigment. The background conductance typically corresponded to $< 10^{-6}$ M ions.

RESULTS

The determination of the rate constants and yields for the homogeneous reactions, i.e., T (OEP) + P (OEP) and T (ZnOEP) + P (ZnOEP), was relatively straightforward since the yields were independent of excitation wavelength. OEP is octaethylporphyrin, ZnOEP is its zinc chelate, T is the triplet state, and P the ground state of the pigment. Since P was in large excess ($\sim 5 \times 10^{-6}$ M) over T ($<10^{-9}$ M) the kinetics of formation of (uncorrelated) ions was accurately first-order. Although the ion recombination rate is encounter-limited, the very low concentration of ions made the recombination rate negligible in the ion formation time range (>0.1 s vs. 0.5 ms). The order of the reaction was determined by varying the pulse energy (I) and the pigment concentration (P). It was accurately first-order in each and thus the second-order rate constant could be obtained from the measured pseudo first-order ion formation constant. Details will be given in a complete report of the photoreactions of the pigment (P + T, T + T, and T*) in a wide variety of solvents.[1] Evidence for electron tunneling through some 7 Å of solvent and for the striking effect of electron spin states (ion yield of P + T > T + T) was obtained (10). The results of these measurements are that the free base porphyrin reacts at about one-half the rate of the zinc porphyrin, but with a far smaller yield (ϕ) of ions:

$$\text{T (ZnOEP)} + \text{P (ZnOEP)} \xrightarrow[\phi_1 = 0.67]{k_1 = 2.0 \times 10^8} \text{P}^+ \text{(ZnOEP)} + \text{P}^- \text{(ZnOEP)}, \quad (1)$$

$$\text{T (OEP)} + \text{P (OEP)} \xrightarrow[\phi_2 = 0.03]{k_2 = 1.3 \times 10^8} \text{P}^+ \text{(OEP)} + \text{P}^- \text{(OEP)}. \quad (2)$$

The use of a rhodamine B dye laser, emitting at 618 nm, allowed the selective excitation of the free base porphyrin in a mixture of OEP and ZnOEP:

$$\text{T (OEP)} + \text{P (ZnOEP)} \xrightarrow[\phi_3]{k_3} \text{P}^+ \text{(ZnOEP)} + \text{P}^- \text{(OEP)}. \quad (3)$$

The product ions are written in their thermodynamically favored form, but we have no independent proof of this assumption. In any case, kinetic and yield analysis are not dependent on this assumption. The rate constant and yield are calculated from the observed values of the pseudo first-order formation rate constant, k, and the observed ion concentration C:

$$k \text{ (obs)} = k_2 \text{P (OEP)} + k_3 \text{P (ZnOEP)}, \quad (3a)$$

$$C \text{ (obs, mixture)}/C \text{ (obs, OEP)} = (k_2 \phi_2 + k_3 \phi_3)/(k_2 + k_3)\phi_2. \quad (3b)$$

The data and results are listed in Table I. The rate constant and yield for reaction 3 are intermediate to that of reactions 1 and 2.

---

[1] Ballard, S. G., and D. Mauzerall. 1978. Manuscript in preparation.

TABLE I
DATA OF REACTION: T (OEP + P (ZnOEP) → P$^+$ (ZnOEP) + P$^-$ (OEP)

Pigment	k (obs)	C (obs)	k_3	ϕ_3
	$s^{-1} \times 10^4$	nM	$M^{-1}s^{-1} \times 10^{-8}$	
5 μM OEP	6.3	0.059		
5 μM OEP + 5 μM ZnOEP	13.9	0.32	1.5	0.27

Excitation wavelength, 618 nm.

The problem of determining the characteristic parameters for the second heterogeneous reaction:

$$T\,(ZnOEP) + P\,(OEP) \xrightarrow[\phi_4]{k_4} P^+\,(ZnOEP) + P^-\,(OEP), \qquad (4)$$

is more complicated, since there is no region of the ZnOEP absorption spectrum where OEP absorption is negligible. Consequently, all four reactions proceed simultaneously.

We write the general equation for the concentration of ions as a function of time following impulse excitation:

$$C = T_Z^0(A'/A)[1 - \exp(-At)] + T_O^0(B'/B)[1 - \exp(-Bt)], \qquad (5)$$

$A = k_1 P_Z + k_4 P_O$ $B = k_2 P_O + k_3 P_Z,$

$A' = \phi_1' k_1 P_Z + \phi_4' k_4 P_O$ $B' = \phi_2' k_2 P_O + \phi_3' k_3 P_Z,$

$T_Z^0 = F_Z I_{abs} \theta_Z$ $T_O^0 = F_O I_{abs} \theta_O,$

$F_Z = \epsilon_Z P_Z/a$ $F_O = \epsilon_O P_O/a,$

$I_{abs} = I_0[1 - \exp(-al)]$ $a = \epsilon_Z P_Z + \epsilon_O P_O.$

The derivation is straightforward from the definitions. The quantum yield of triplet

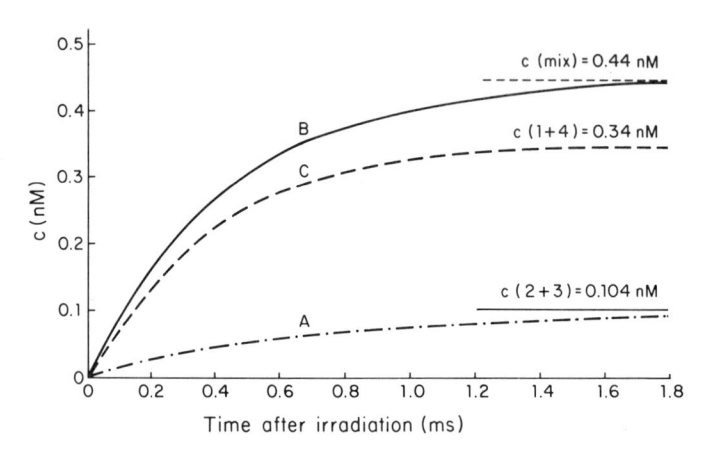

FIGURE 1 The increase in ion concentration (c) versus time for the mixed reactions of ZnOEP + OEP after pulse illumination (curve B). Curve A is the calculated component from reactions 2 and 3. Curve C is the difference of curve B and curve A, caused by reactions 1 and 4.

TABLE II
DATA OF REACTION: T(ZnOEP) + P(OEP → P⁺(ZnOEP) + P⁻(OEP)

	OEP		ZnOEP
P_i	5×10^{-6} M		5×10^{-6} M
$\epsilon_i(337)$	3.09×10^4		2.32×10^4
F_i	0.571		0.429
I_0		7.68 nM	
C_{tot}		0.444 nM	
C_i	0.104 nM		0.340 nM
	$B = 1.39 \times 10^3 \, s^{-1}$		$A = 2.48 \times 10^3 \, s^{-1}$
		$k_4 = 3.0 \times 10^8 \, M^{-1} s^{-1}$	
		$\phi_4 = 0.21$	

formation for ZnOEP is θ_Z and for OEP is θ_O, and the ion yield from the triplet for the specific reactions 1–4, is ϕ'_i. These factors are combined in the previous definition of ion yield, i.e., $\phi_1 = \theta_Z \phi'_1$, $\phi_2 = \theta_O \phi'_2$, $\phi_3 = \theta_O \phi'_3$ and $\phi_4 = \theta_Z \phi'_4$. T^0 refers to the initial triplet concentration, ϵ to the absorbancy index, l to the light path length, and I_0 is the einsteins of photons in a pulse passing through the cell volume.

Now the second term on the right of Eqs. 5 (ion formation from $T_O = C_O$) contains only known parameters. It may thus be computed (curve A, Fig. 1) and subtracted from the measured ion formation (curve B, Fig. 1). The resulting curve (C, Fig. 1) is just the first term on the right of Eq. 5 (ion formation from $T_Z = C_Z$) and thus is reduced to the previous case for the calculation of k_4 and $\theta_Z \phi'_4 = \phi_4$. The parameters of the calculation are given in Table II, and the results are summarized in Table III. The errors are best estimates of relative errors.

DISCUSSION

In principle the variation in the ion yield (Table III) could be caused by changes in the yield of triplet state (θ_O, θ_Z) or in the yield of ions (ϕ'_i). However, the yield of triplets for porphyrins is very high and quite constant. The triplet yield of etioporphyrin was 0.83 and of zinc etioporphyrin was 0.94 when measured by Gradyusko and Tsvirko (11, 12) under conditions similar to ours.

The redox properties of porphyrins have been well summarized by Fuhrhop (13).

TABLE III
SUMMARY OF RATE CONSTANTS AND ION YIELDS OF
T-P REACTIONS IN ACETONITRILE

Reaction		ϕ_i
	$M^{-1} s^{-1} \times 10^{-8}$	
1. T(ZnOEP) + P(ZnOEP)	2.0 ± 0.05	0.67 ± 0.01
2. T(OEP) + P(OEP)	1.3 0.1	0.03 0.002
3. T(OEP) + P(ZnOEP)	1.5 0.15	0.27 0.01
4. T(ZnOEP) + P(OEP)	3.0 0.2	0.21 0.02

The difference in the one-electron oxidation and reduction potentials of many porphyrins is constant, near 2.2 V, the energy of the lowest optical transition. This supports our contention that little "reorganization energy" is required for electron transfer reactions in these systems (7). Entropy effects prevent a simple comparison of absorbed photon energy and redox energy of the resulting photochemical reaction. We will discuss these effects elsewhere, and here turn to the relation between formation of porphyrin cations and anions and their redox potential. The oxidation potential of a free base porphyrin is about 0.2 V more positive (oxidizing) than that of a zinc porphyrin. The oxidation potential increases with decreasing electronegativity of the central substituent. Conversely, the reduction potential of a free base porphyrin is about 0.2 V less negative (reducing) than that of a zinc porphyrin. Thus the pair $ZnP^+ + P^-$ is favored over the pair $ZnP^- + P^+$ by 0.3 ± 0.1 V, and the cross-reactions to this pair (Eq. 3, 4) should be favored over the homogeneous reactions (Eq. 1, 2) by 0.15 V. We see from Table III that these thermodynamic expectations are poorly realized. Only the rate constant for the T (ZnOEP) cross-reaction is slightly larger than that of the corresponding homogeneous reaction, and the yields of both cross-reactions are intermediate to the high (ZnOEP) and low (OEP) yields of the homogeneous reactions. We believe the data are best explained on purely kinetic grounds.

Our detailed studies of the homogeneous reactions[1] show that both diffusion in the Coulomb field and the spin state of the electrons are crucial in determining the fate of the initial loose encounter complex. Similar considerations have been invoked by several workers (14–16). In the present case, the approach of triplet (T) and ground state (P) molecules will allow an electron to tunnel between the pair to form a triplet charge transfer species T $(P^+ P^-)$. Since this spin state cannot decay to the singlet ground state, the ion pair will tend to diffuse apart. It may regenerate T + P at the crossover point between ionic "charge transfer" and molecular "triplet" states. Thus the relatively slow rate constant for ion formation from T + P is caused by these unproductive collisions. A general mechanism for the decorrelation of the electron spins is through coupling with nuclear spins (14). Although the result is in general complex, for large molecules with many protons and nitrogens the relaxation from triplet to the equilibrium mixture of triplets and singlets can be approximated with a single time constant. Once a singlet ion pair, S $(P^+ P^-)$ is formed, the ions may diffuse together and collapse to the ground state at the reaction radius, or diffuse apart to be measured as uncorrelated ions. Using the Smoluchowski concept, the problem is that of diffusion of two interconverting species in the presence of a Coulomb potential and with boundary conditions different for each species. We have numerically integrated these equations for various conditions.[1] A quantitative calculation in the present case would require more information than we have available, e.g., ion yield as a function of dielectric constant. Our extensive data on the homogeneous reaction of ZnOEP can be fit with spin decorrelation times of about 20 ns and a formation radius of 20–24 Å(10).[1] This radius is larger than the measured reaction radius of $P^- + P^+$, which is about the sum of the molecular radii, 15 Å. The excess distance is attributed to electron tunneling in the excited state reaction. These calculations suggest that the 20-fold lower yield of OEP cannot be attributed only to a faster spin decorrelation time. The collision of OEP with its partner must be "sticky", i.e. the complex must have a sufficient lifetime for considerable spin decorrelation to occur, and the singlet state ion pair rapidly collapses to the ground state at this short (contact) distance.

Inspection of Table III yields some interesting information on the mechanism of the electron transfer reactions. The ion yields are far more sensitive to the particular reaction than are the rate constants. This accords with our interpretation of the rates as being determined by the spin decorrelation to the singlet charge transfer state. The twofold increase in rate constant when the triplet is ZnOEP instead of OEP, reactions 1 and 4, may be caused by increased spin orbit coupling in this excited state. The difference in rate constants and possibly in yields of reactions 3 and 4 show that the trival mechanism of triplet energy transfer to transform reaction 4 to 3 occurs to only a limited extent. This agrees with our hypothesis that electron transfer occurs over greater than nearest-neighbor distances.

These considerations suggest that the pheophytin in the reaction center of photosynthetic bacteria plays not so much a thermodynamic role as one of electron spin decorrelation. This is particularly important if, as is believed (1), the first electron transfer occurs from the singlet state, and thus is highly susceptible to loss to the ground state. Although this loss may be slowed by the energy level gap between the first excited and ground states of porphyrins (17), rapid spin decorrelation is critical. This will be favored by the increased distance and the increased contact with protons in the pheophytin reaction. Because of the anisotropy of the magnetic interactions, the orientation of the molecules in the rigid complex will be important. The presence of a high spin Fe^{+2} ion nearby at the quinone acceptor will also greatly facilitate spin uncoupling. It is known that if the quinone is reduced, the triplet state of bacteriochlorophyll (dimer) is rapidly formed from the charge transfer state involving bacteriopheophytin (1). Thus a consistent explanation of the detailed mechanism of the energy conversion step in photosynthesis is emerging, and is remarkably well based on quantum mechanics.

We thank Prof. J. H. Fuhrhop for gifts of the porphyrins.

This work was supported by National Science Foundation grant PCM74-11747 and by the Rockefeller University.

Received for publication 26 November 1977.

REFERENCES

1. PARSON, W. W., and R. J. COGDELL. 1975. The primary photochemical reaction of bacterial photosynthesis. *Biochim. Biophys. Acta.* 416:105–149.
2. FAJER, J., D. C. BRUNE, M. S. DAVIS, A. F. FORMAN, and L. D. SPAULDING. 1975. Primary charge separation in bacterial photosynthesis: oxidized chlorophylls and reduced pheophytin. *Proc. Natl. Acad. Sci. U.S.A.* 72:4956–4960.
3. MCELROY, J. D., D. MAUZERALL, and G. FEHER. 1974. Characterization of primary reactants in bacterial photosynthesis. II. Kinetic studies of the light-induced EPR signal (g = 2.0026) and the optical absorbance changes at cryogenic temperatures. *Biochim. Biophys. Acta.* 333:261–277.

4. KAMEN, M. 1963. Primary processes in photosynthesis. Academic Press, Inc., New York. 155 ff.
5. MAUZERALL, D. 1973. Photosynthesis and photochemistry of porphyrins. Symposium abstract, first meeting American Society for Photobiology. June 10–14. p. 161.
6. MAUZERALL, D. 1973. Why chlorophyll? *Ann. N.Y. Acad. Sci.* **206**:483–494.
7. MAUZERALL, D. 1976. Electron transfer reactions and photoexcited porphyrins. *Brookhaven Symp. Biol.* **28**:64–73.
8. BALLARD, S. G. 1976. Automatic conductimetric instrument for kinetic studies of photochemical ionization reactions in solution. *Rev. Sci. Instrum.* **47**:1157–1162.
9. BALLARD, S. G., D. MAUZERALL, and G. TOLLIN. 1976. Photochemical ion formation in lumiflavin solution. *J. Phys. Chem.* **80**:341–351.
10. BALLARD, S. G. and D. MAUZERALL. 1977. Kinetic evidence for electron tunneling in solution. Conference on Tunneling in Biological Systems, Philadelphia Nov. 3–5. In press.
11. GRADYUSKO, A. T., and M. P. TSVIRKO. 1971. Probabilities for intercombination transitions in porphyrin and metalloporphyrin molecules. *Opt. Spectrosc.* **27**:291–296.
12. SONG, P. S., and H. BABA. 1974. Exited states of biomolecules. II. *Photochem. Photobiol.* **20**:527–532.
13. FUHRHOP, J. H. 1975. Reversible reactions of porphyrins and metalloporphyrins and electrochemistry. *In* Porphyrins and Metalloporphyrins. K. M. Smith, editor. Elsevier's Scientific Publishing Company, Amsterdam. 593–614.
14. BROCKLEHURST, B. 1976. Spin correlection in the geminate recombination of radical ions in hydrocarbons. *J. Chem. Soc. Faraday Trans. II.* **72**:1869–1884.
15. GOUTERMAN, M., and D. HOLTEN. 1977. Electron transfer from photoexcited singlet and triplet bacteriopheophytin II. Theoretical. *Photochem. Photobiol.* **25**:85–92.
16. SCHULTEN, Z., and K. SCHULTEN, 1977. The generation, diffusion, spin motion, and recombination of radical pairs in solution in the nanosecond time domain. *J. Chem. Phys.* **66**:4616–4634.
17. MAUZERALL, D. 1977. Electron transfer photoreactions of porphyrins. *In* The Porphyrins. D. Dolphin, editor. Vol. V. Academic Press, Inc., New York. 29–52.

NOTE ADDED IN PROOF

Hoff, Rademaker, Grondelle, and Drysens (*Biochim. Biophys. Acta.* **460**:547. 1977) and Blankenship, Schaafsma, and Parson (*Biochim. Biophys. Acta.* **461**:297. 1977) have shown that magnetic fields of the order of 10^2 G decrease the yield of triplet bacteriochlorophyll from the bacteriochlorophyll cation, pheophytin anion state when further electron transfer to the quinone is blocked by previous reduction. Hoff et al. claim that a preparation low in iron shows a larger and more magnetically sensitive effect than reaction centers with iron or than chromatophores. All workers in this field have found a large variability of results among different preparations. We suggest that the replacement of iron by manganese, as observed by Feher, Isaacson, McElroy, Ackerson, and Okamura (*Biochim. Biophys. Acta.* **368**:135. 1974), may contribute to these heterogeneous results. As far as we know, the quantum yield of the formation of the initial ion-radical pair in the "iron-free" preparations has not yet been determined.

DISCUSSION

SWENBERG: Since you have created ion pairs, did you look for a magnetic field effect on the overall yield of the reaction?

MAUZERALL: Yes, but not on the yield of the free-base porphyrin, only of the metalloporphyrin (see ref. 1, below). We do see magnetic field effects as expected and we interpret them in a way similar to that discussed by Professor Weller.

SWENBERG: You have mentioned some extensive data on the homogeneous reaction of ZnOEP, which you say can be explained by extended correlation times of 20 ns. Your theoretical analysis does not say anything about the type of exchange interaction or the magnitude of the hyperfined terms.

MAUZERALL: Yes, we have explained the data in that way. The maximum yield of ions is fairly high and we have measured yields over a large range of solvent dielectric. We have used about 30 different solvents and we obtain yields from about 0.5 down to 10^{-6}, from which we obtain the radius of reaction. The single (S) plus triplet (T) reaction, energetically just about even with the ions, has a greater yield than the T plus T reaction, which has twice the amount of energy. This shows that these yields are not correlated with the thermodynamics of the process. By analysis of the higher-yield S plus T reaction we obtain the delay time for T to S spin dephasing caused by, we believe, the hyperfine interactions. At the moment we say that the numbers agree with the electron spin resonance (ESR) spectrum of the porphyrin radicals, which have been studied in great detail.

RODGERS: My question concerns the statement of ion recombination, said to be encounter limited. It seems to me that this should not necessarily be the case when you are using conductivity to monitor the system, but conductivity is sensitive only to uncorrelated ions. As soon as the ion motion becomes correlated and is within a critical distance, then it becomes a geminate pair and I don't think that the electrical system is sensitive to it. So the mutual ion diameter at which the signals disappear should be the Onsager escape distance in the solvent and not the ion contact distances. Can you comment on that?

MAUZERALL: In fact the Debye-Smoluchowski equation states that the capture radius changes from the reaction radius (R_m) at very high dielectric constant to the Coulomb radius (R_c) at low dielectric constant. I believe that by Onsager "escape distance" you mean $e^2/\epsilon kT$. If your point is that inside the Coulomb radius the ions always collapse, then this is not quite so. The situation is not the same as a black hole, because of the balance with thermal motions. The only way optical detection would give a different rate constant from that of conductivity (we exclude the trivial case of interionic interactions, which occur at much higher concentrations than $\leq 10^{-7}$ M of these experiments), is if the ions would just sit by each other for a long time; but then the reaction would no longer be encounter limited.

RODGERS: I don't see the reason for electrical and optical detection being equivalent.

MAUZERALL: There is no difference in these methods, since the homogeneous ion recombination is a random process in space and time and the encounter-limited second-order rate constant is all that is possible to measure. Essentially, the individual, stochastic events occur on a time scale shorter (≤ 1 μs even in low dielectric) than the average ion recombination time (typically ~1 ms). The only way to measure the individual, stochastic events directly is to have a phased, coherent process, just the opposite of the homogeneous random ion recombination process.

GEACINTOV: A singlet ion pair formed from a triplet charge transfer species is degenerate, with a highly excited vibrational electronic state of the system you are portraying. Because of the larger nuclear distortions in this situation the Franck-Condon factor is small. And I expect that the matrix elements connecting the singlet charge transfer ion pair state with the neutral vibrationally excited ground state are rather small. Therefore, the situation is different from the one described by Dr. Weller earlier, with degenerate singlet and triplet states, where the singlet excited states had a higher energy and therefore the Franck-Condon factors were quite large. Have you considered the role of Franck-Condon factors in your particular situation? And if so, how important are they?

MAUZERALL: We disagree with Dr. Geacintov's assumption that there are large nuclear distortions and therefore small Franck-Condon factors between the ion pair and the ground or excited state molecules, because the recombination of the ions is strictly encounter limiting, following the Debye-Smoluchowski equation, with a reaction radius about equal to the hydrodynamic radii. Further evidence for this view, also cited in the paper, is that the difference of the redox potentials of the cation and anion are equal, within error, to the optical excitation energy.

GEACINTOV: This is rather surprising, because we do expect the major exponents to be small, and I wonder if you find it so for the encounter-limited recombination for some other reason, perhaps the Casey factor or whatever.

MAUZERALL: We doubt that can be so because of the wide range of solvents that we have used, and because the rate constants depend on viscosity and electrostatics in the expected way.

LESTER: How does one go about proving that the reaction is strictly encountered limited?

MAUZERALL: That's a good question. First of all, it should be strictly inversely proportional to viscosity, within the errors of the measurement. That is the ultimate criterion of an encounter-limited reaction. In the case of ions there is in addition the Coulomb interreaction, for which one has a very good Debye factor.

LESTER: But does the Debye-Smoluchowski equation imply knowledge of the radii?

MAUZERALL: That is right. In fact, we showed that our data for the radius correspond closely to the sum of the radii of these large molecules.

LESTER: Thus, the viscosity is the important parameter in proving an encounter-limited reaction?

MAUZERALL: Yes, that is the critical parameter.

WELLER: I still wonder about the thermodynamic situation in your systems. You claim that the difference of oxidation potentials is of the order of 2.2 V. Now the energy of the triplet state from which the reaction starts out may be as small as 1.8 eV, which gives 0.4 eV too few in energy for a diffusion-controlled electron transfer reaction between the triplet state and the quencher, as you pointed out several years ago. We have studied a number of electron transfer reactions in the singlet excited state. The formula used contains the difference of the oxidation potential of the donor and the reduction potential of electron acceptor minus the energy of the excited state. This gives a very good measure of the energy used in the electron transfer reaction. Whenever this energy is negative by ~0.3 eV, the reaction is extremely fast, i.e., diffusion controlled. If the energy change is zero, the rate is already $\sim 4 \times 10^9 \, M^{-1} s^{-1}$. If the free energy change becomes positive, the rate constant drops drastically and becomes extremely slow. At 0.3 eV the value is only about $10^6 \, M^{-1} s^{-1}$. I would like to know what are the energies of your triplet states that initiate the electron transfer reaction.

MAUZERALL: The energies of the ZnOEP system are approximately as you quoted them and you have brought out an important point, which we also made in the paper. We do not believe that the thermodynamics of these reactions determine the rate constants, but thermodynamics enter as a limiting factor on the yields. The porphyrin triplet levels are such that the overall free energy would be zero or slightly positive. However, the absolute error associated with the triplet energy level and with the redox potentials is easily 0.2 V. Thus we claim one cannot use these numbers as decisive criteria. The point of our paper was that for a ΔE_0 of 0.15 V, the rate constant changes by <2, not 10 as expected from Professor Weller's view. The yield of ions is considerably lower for the free base prophyrins, which do have a lower triplet energy. We do not necessarily quarrel with Professor Weller's extensive and impressive work, but our case is very different. The second-order rate constant for the S-T reaction is about 10^8 M^{-1} s^{-1}, independent of solvent used. This is explained by our theory of electron transfer via tunneling.

WELLER: Tunneling never goes uphill.

MAUZERALL: I guess we do agree with that (barring anti-Stokes inelastic tunneling), and therefore there exists a limit with which we have no arguments.

WELLER: I have a comment on the triplet excitation energies of OEP and ZnOEP. The free energy change connected with photo-induced electron transfer reactions is given by $\Delta G = E^{ox} - E_A^{red} - \Delta^* E_{0,0}$ where E_D^{ox} and E_A^{red} are the oxidation and reduction potentials of the electron donor and acceptor, respectively, and $\Delta^* E_{0,0}$ is the zero-zero transition energy of the excited state. In your systems the difference in the one-electron oxidation and reduction potentials is 2.2 V (2.0 V for the mixed system), so that the zero-zero transition energy of the excited triplet state is decisive in determining whether ΔG is positive or negative.

It should be pointed out that our own measurements on electron transfer fluorescence quenching, carried out in 1969 investigating more than 150 electron donor-acceptor systems in acetonitrile, showed that entropy effects connected with excitation can be safely neglected and that for

$\Delta G < -0.3$ eV the rate constants are $k = 18 \times 10^9$ $M^{-1}s^{-1}$ (diffusion-controlled)
$\Delta G = \sigma$ $\qquad k = 4 \times 10^9$ $M^{-1}s^{-1}$
$\Delta G = 0.15$ eV $\qquad k \approx 1.5 \times 10^8$ $M^{-1}s^{-1}$
$\Delta G > 0.3$ eV $\qquad k < 1 \times 10^6$ $M^{-1}s^{-1}$

REFERENCE

1. BALLARD, S. G., and D. MAUZERALL. 1978. *Biophys. J.* **21**:107a. (Abstr.).

A PICOSECOND PULSE TRAIN STUDY OF EXCITON DYNAMICS IN PHOTOSYNTHETIC MEMBRANES

N. E. GEACINTOV AND C. E. SWENBERG, *Chemistry Department and Radiation and Solid State Laboratory, New York University, New York 10003, and*

A. J. CAMPILLO, R. C. HYER, S. L. SHAPIRO, AND K. R. WINN, *University of California, Los Alamos Scientific Laboratory, Los Alamos, New Mexico 87545 U.S.A.*

ABSTRACT The fluorescence decay time of spinach chloroplasts at 77°K was determined at 735 nm (corresponding to the photosystem I emission) using a train of 10-ps laser pulses spaced 10 ns apart. The fluorescence lifetime is constant at $\simeq 1.5$ ns for up to the fourth pulse, but then decreases with increasing pulse number within the pulse train. This quenching is attributed to triplet excited states, and it is concluded that triplet excitons exhibit a time lag of about 50 ns in diffusing from light harvesting antenna pigments to photosystem I pigments. The diffusion coefficient of triplet excitons is at least 300–400 times slower than the diffusion coefficient of singlet excitons in chloroplast membranes.

INTRODUCTION

The availability of picosecond laser pulses has prompted a number of research groups to study fast energy transfer processes and fluorescence kinetics in photosynthetic membranes (1–8). At low temperatures, e.g. at 77°K, the fluorescence of chloroplasts extracted from spinach leaves, or the fluorescence exhibited by whole algal cells such as *Chlorella pyrenoidosa*, is characterized by two prominent emission maxima at 735 and 685 nm. These maxima are due to emission from chlorophyll *a* molecules associated with pigment system I (PS I) and pigment system II (PS II), respectively (9). Recently, Butler and co-workers (10,11) have utilized standard fluorescence techniques to study the energy distribution and energy transfer pathways between the light harvesting antenna pigments and PS I and PS II. We have studied both the fluorescence yields at low temperatures (4,5,12,13) and fluorescence decay time (1,4; see also references 7 and 14), using single picosecond laser pulses as well as picosecond laser pulse trains.

Using picosecond pulse trains, Breton and Geacintov (5,13) noticed that the quenching of the fluorescence was more pronounced in PS I than in PS II. Furthermore, the ratio of the fluorescence yield at 735 nm relative to that at 685 nm decreased as the number of picosecond pulses in a train was increased from 4 to about 300. On the other hand, using a single picosecond laser pulse, the intensity of the fluorescence yield for both PS I and PS II decreases in a parallel manner with increasing pulse intensity (4,12). The increased quenching within PS I when trains of pulses are used was attributed to the preferential accumulation of long-lived quenchers within this pigment system. These quenchers were identified as triplet excitons and ions. Their concentration increases as a function of time within a pulse train because their lifetimes are long compared to the spacing of the picosecond pulses (5–10 ns), and the length of the train (typically <1 μs). Recent work, however, indicates that the predominant quenching effect is due to triplets.[1]

Other results (4,12,15) utilizing single picosecond pulse laser spectroscopy indicate that the PS I pigments derive their excitation energy mainly by exciton transfer from the light harvesting pigments rather than by the direct absorption of photons. The rise time of the PS I fluorescence is ≈ 140 ps (15), indicating that the singlet excitons diffuse within the light harvesting pigment matrix of the photosynthetic membranes before being captured by the PS I pigment system responsible for the 735-nm emission. Furthermore, the lifetime of the 685-nm fluorescence is strongly dependent on the intensity of the pulse, whereas the 735-nm emission is independent of intensity at least for incident intensities below 10^{15} photons cm^{-2} per pulse (4). These results indicate that singlet-singlet exciton annihilation takes place within the light harvesting system, that the actual optical cross section of the PS I pigment system responsible for the 735-nm fluorescence is relatively small at 530 nm, and that there is a time lag of $\simeq 140$ ps associated with the arrival of excitons from the light harvesting to PS I pigments. It should be noted that in this work we refer to the collection of pigments which give rise to the 735-nm fluorescence as the PS I pigments. However, Butler (16) has proposed that the 735-nm emission is due to a chlorophyll form C-705 which derives its energy by exciton transfer from PS I antenna pigments.

In this paper we have addressed ourselves to two questions: (*a*) Is there a time lag for the arrival of triplet exciton quenchers within PS I analogous to the 140 ps time lag for singlet excitons? In organic crystals the diffusion coefficients for triplets are frequently two orders of magnitude smaller than those of singlet excitons (17). In chloroplasts we expect this difference to be even larger because the triplet excitation is transferred between different chlorophyll-protein complexes whose average distances, and thus transfer times, are probably larger than those between chlorophyll molecules within a given chlorophyll-protein complex (18). Inasmuch as triplet energy transfer is due to an exchange interaction, whereas the singlet exciton transfer is a longer range process, the difference between singlet and triplet diffusion coefficients may be much larger in chloroplasts than in organic crystals. With the spacing between successive picosecond laser pulses being 10 ns with the Nd:YAG laser utilized in this work, such a slow buildup of long-lived quenchers within PS I might be detectable with a pulse train. (*b*) A decrease in the fluorescence yield may be due to a static or a

---

[1] J. Breton and N. E. Geacintov. Submitted for publication.

dynamic quenching process (20). In the latter case, which is characteristic of exciton annihilation described by kinetic rate equations (21), the lifetime of the fluorescence decreases with increasing intensity of the laser pulses. An example of this situation is the intensity-dependent fluorescence decay time at 685 nm (1,4). In static quenching, the fluorescence decay time is independent of intensity—there are no quenching pathways within the particular subsystem whose fluorescence is still observable, whereas the fluorescence from the remaining subsystems is completely quenched. The 735-nm fluorescence decay at 77°K, excited in the single pulse mode, is insensitive to the pulse intensity, indicating that the exciton-exciton annihilations which produce the quantum yield decrease take place within the light harvesting pigment system rather than in the PS I pigment system. A determination of PS I fluorescence lifetimes under conditions of quenching by long-lived quenchers (5,6,13) with a pulse train can thus provide further information on the mechanism of this quenching process.

We have utilized a train of 10 530-nm · s harmonic pulses derived from a Nd:YAG laser operating at 1,060 nm to determine the singlet exciton fluorescence (735 nm) decay times for pulse numbers 4, 6, 8, and 10 within this pulse train. The decay times decrease with increasing pulse number, whereas the decay profiles of the fluorescence are exponential, indicating that simple dynamic quenching decay kinetics are applicable for singlet-triplet exciton quenching within the PS I pigments. The results are interpreted quantitatively by integrating numerically the appropriate coupled kinetic equations, and it is shown that there is a time lag in the buildup of triplet excitons within PS I.

METHODS

The experiments were performed at the Los Alamos Laboratory using spinach chloroplast suspensions at 77°K. The fluorescence emission was viewed through a 735-nm (± 7 nm) interference filter. The fluorescence was excited with picosecond pulses from a frequency doubled and mode-locked Nd:YAG (Nd^{+3}:yttrium aluminum garnet) laser and a streak camera-optical multichannel analyzer combination. The train consisted of ten main pulses, each 20 ps in duration, each separated by 10 ns, and of variable intensity as described in the text. The triggering of the streak camera was appropriately delayed in order to view the fluorescence decay after the pulses selected for study. Additional experimental details may be found elsewhere (1).

RESULTS

The intensity profile of a representative pulse sequence is shown in graph form in Fig. 1 A and in numerical form in Table I. The fluorescence decay times for pulses 2, 4, 6, 8, and 10 are plotted in Fig. 1 B. Using single pulse excitation, repeated determinations of the fluorescence decay times at 735 nm show that the lifetimes are 1.5±0.4 ns. The lifetime values obtained for pulses 2 and 4 thus indicate that there is little or no dynamic quenching of the PS I fluorescence for these early pulses. Indeed, Breton and Geacintov (5) have shown that the preferential quenching within PS I attributable to long-lived quenchers within PS I is not very pronounced for sequences of

FIGURE 1

FIGURE 2

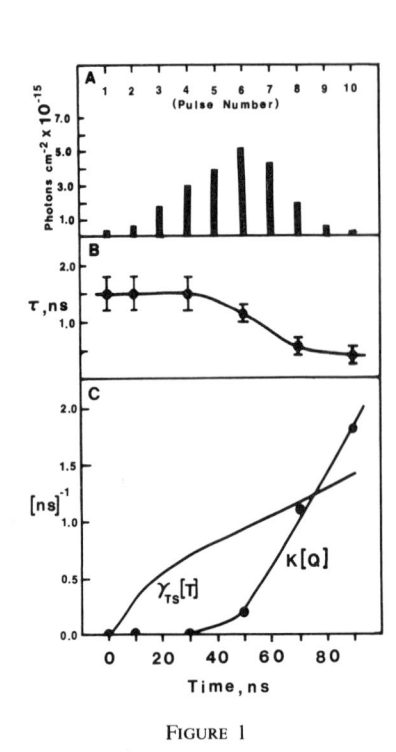

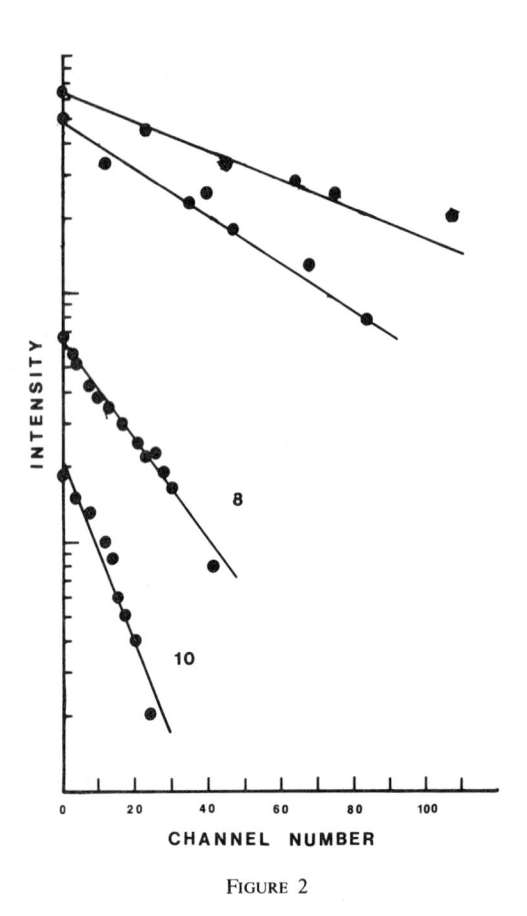

FIGURE 1 (A) Intensity profile of a typical pulse sequence. (B) Fluorescence (PS I) decay times measured at 77°K, 735-nm emission band. (C) $K[Q]$, quenching constant x quencher concentration within PS I calculated from experimental data using Eq. 1. $\gamma_{TS}[T]$—theoretical curve calculated from Eqs. 2–4. γ_{TS}— singlet-triplet exciton annihilation constant, $[T]$ is the triplet exciton density within the light harvesting pigments; the following parameters were used in this calculation: $\alpha = 1,400$ cm^{-1}, experimental I_i values from Table I, $\beta_S = 1.3 \times 10^9$ s^{-1}, $\beta_T = 0.6 \times 10^5$ s^{-1}, $\gamma_{SS} = 5 \times 10^{-9}$ cm^3 s^{-1} (from ref. 12) $\gamma_{TS} = 7 \times 10^{-9}$ cm^3 s^{-1} $k_{is} = 1.3 \times 10^8$ s^{-1} (Geacintov et al. Submitted for publication).

FIGURE 2 Examples of semilogarithmic plots of the fluorescence decay. Data taken from digital readout of the optical multichannel analyzer display was utilized to process the streak camera traces. Scale is 20.8 ps per channel. Typical fluorescence decay curves for pulses 4, 6, 8, and 10 are shown. The lifetimes calculated from these particular traces are 1.5, 0.96, 0.47, and 0.25 ns, respectively.

four pulses or less. The fluorescence decay times decrease with increasing pulse number for pulses 6, 8, and 10. Fluorescence decay data taken from the digital readout of the optical multichannel analyzer are plotted on a semilogarithmic scale in Fig. 2. It is evident that within experimental error, the fluorescence decay is exponential. Under these conditions, the fluorescence decay times τ_i for pulses $i = 6, 8$, and 10, can be related to the decay time τ_o in the absence of quenching (for pulses 2 and 4, or single

TABLE I

INTENSITY OF INDIVIDUAL PULSES, FLUORESCENCE DECAY TIMES, $K[Q]$ (EQ. 1) AND CALCULATED VALUES $\gamma_{TS}[T]$ USING EQS. 2–4, $\gamma_{TS} = 7 \times 10^{-9}$ cm^3 s^{-1} AND $k_{is} = 1.3 \times 10^8$ s^{-1}.

Pulse number	Intensity, photons	Fluorescence lifetime	$K[Q]$	$\gamma_{TS}[T]$
	cm^{-2}	ns	s$^{-1} \times 10^{-9}$	s$^{-1} \times 10^{-9}$
1	0.23×10^{15}	1.5 ± 0.3	0	0.13
2	0.54×10^{15}	1.5 ± 0.3	0	0.32
3	1.8×10^{15}	—	—	0.55
4	2.9×10^{15}	1.5 ± 0.3	0	0.71
5	3.9×10^{15}	—	—	0.83
6	5.0×10^{15}	1.2 ± 0.2	0.17	0.90
7	4.3×10^{15}	—	—	1.0
8	1.8×10^{15}	0.57 ± 0.2	1.1	1.2
9	0.53×10^{15}	—	—	1.3
10	0.24×10^{15}	0.40 ± 0.15	1.8	1.4

The values given are averages of five or six determinations with their standard deviations.

pulse values of 1.5 ns) by the well-known Stern-Volmer equation (20):

$$\tau_i^{-1} - \tau_o^{-1} = K[Q]_i, \tag{1}$$

where $[Q]_i$ is the concentration of quenchers for pulse i, and K is the dynamic quenching constant.

DISCUSSION

The absorption coefficient of 530 nm can be estimated from the data of Schwartz (22) and the known absorption spectrum of chloroplasts. We estimate that the absorption coefficient α at 530 nm is about 1,200–1,400 cm^{-1}. With such an absorption coefficient, about 10% of the total chlorophyll molecules, on the average, are excited at an intensity of 5×10^{15} photons cm^{-2}/ps pulse (12), e.g. pulse 6. At these intensities, there is a severe decrease in the lifetime of the PS II fluorescence (1,4); even for pulse 1 (2.3×10^{14} photons cm^{-2} per pulse), the PS II fluorescence is already quenched (2). The fluorescence lifetime for PS I, on the other hand, begins to decrease significantly only with pulse 6. Using Eq. 1, we can calculate the product of the quenching constant K and the quencher concentration $[Q]$ in PS I. The results are plotted in Fig. 1 C. It is evident that the quencher concentration $[Q]$ in PS I begins to build up rapidly after pulse 6, whereas pulses 3–5 do not appear to give rise to any discernable quenching, at least immediately after the occurrence of these pulses. It, therefore, seems that the appearance of the quenchers Q within PS I has a time lag of about 50 ns.

To show that this time lag is real, we will now calculate, using the standard set of kinetic equations (1, 2, 4) below, the concentration of the quenchers in the light-harvesting antenna pigment system as a function of time. We assume that the quenchers $[Q]$ are triplet excitons (T), which are formed from singlet excitons (S) by intersystem crossing with a rate constant k_{is} and eventually migrate to PS I. The singlet excitons can annihilate each other with a rate constant γ_{ss}, and the singlet excitons can also be annihilated by triplet excitons with a rate constant γ_{TS}. Rahman and Knox (23) estimate that $\gamma_{TS} = 6 \times 10^{-9}$ cm^{3}s^{-1} (23). The simple basic set of equations in this model then are

$$\frac{d[S]}{dt} = G(t) - \beta_S[S] - \gamma_{TS}[T][S] - \gamma_{ss}[S]^2 \tag{2}$$

and

$$\frac{d[T]}{dt} = k_{is}[S] - \beta_T[T]. \tag{3}$$

β_S(s^{-1}) denotes all unimolecular decay constants of singlets including photochemical pathways, where β_T denotes all unimolecular decay rates of triplets. The bimolecular triplet-triplet annihilation rates are neglected here. This approximation is justified on the time scales used because of the low probability of this process on the nanosecond time scale (this is shown for high triplet exciton densities in organic crystals in reference 24). $G(t)$ is the laser generating function

$$G(t) = \alpha \sum_{i=1}^{10} I_i \delta(t_i - t), \tag{4}$$

where α(cm^{-1}) is the absorption coefficient and I_i the intensity (photons cm^{-2}) of the pulse incident on the sample, and the delta function takes into account the discontinuous nature of the excitation.

We have recently shown that the sets of Eqs. 2 and 3 can quantitatively describe the time dependence of the fluorescence quantum yield and triplet concentrations in spinach chloroplasts using an approximately square wave continuous (0.50 μs) laser pulse excitation.[2] These quantities were experimentally measured with a 50-ns resolution within the excitation pulse using a gated optical multichannel analyzer arrangement, whereas the $[S]$ and $[T]$ concentrations were calculated using Eqs. 2 and 3 utilizing the Hartree approximation. Excellent agreement between theory and experiment is obtained if the $\gamma_{ss}[S]^2$ term in Eq. 2 is neglected (justifiable when a microsecond pulse excitation is used because the lifetime of singlet excitons is <1 ns), and if values of $\gamma_{TS} = (7 \pm 2) \times 10^{-9}$ cm^3 s^{-1} are taken. These values of γ_{TS} are incidentally in excellent agreement with those calculated by Rahman and Knox (23).

Using the same approach, but retaining the $\gamma_{ss}[S]^2$ in Eq. 2 because of the picosecond nature of the excitation used in this work, we now calculate the triplet concentration for each of the pulses. Because the unimolecular triplet decay time $\beta_T^{-1} > 10^{-6}$ s(19), the triplets produced by the pulses accumulate from pulse to pulse during the train of pulses. In these calculations, we assumed that during the first pulse,

---

[2]Geacintov, N. E., J. Breton, and C. E. Swenberg. Submitted for publication.

$[T] = 0$ and thus the $\gamma_{TS}[T][S]$ is zero for this initial point. Then we calculate $[S]$ and, using 10^{-10}-s intervals, we substitute this value into the equations and calculate $[T]$ for subsequent 10^{-10}-s intervals; the total triplet concentration produced by pulse 1 is then added to the triplets produced by pulse 2, etc. Thus, by using stepwise numerical integration, using the values of $[T]$ obtained in the previous 10^{-10}-s interval, and treating these values of $[T]$ as constants in the succeeding interval, both $[S]$ and $[T]$ can be calculated for each of the 10 pulses. The values of $\gamma_{TS}[T]$ thus calculated, using $\gamma_{TS} = 7 \times 10^{-9} \text{cm}^3\text{s}^{-1}$, are plotted in Fig. 1 C.

Numerically, the magnitudes of the experimental quantities $K[Q]$ and $\gamma_{TS}[T]$ agree fairly well; however, the vertical scale for the $\gamma_{TS}[T]$ curve can be shifted up or down, depending on the exact value of γ_{TS} chosen. Thus, for example, using $\gamma_{TS} = 10^{-8} \text{cm}^3\text{s}^{-1}$, $\gamma_{TS}[T]$ can be made equal to $K[Q]$ for the last pulse (number 10). However, more important, there is little change in the shape of the theoretical curve $\gamma_{TS}[T]$ when γ_{TS} is changed. Furthermore, the vertical scales of the two curves in Fig. 1 C should not be compared in any case because the calculation refers to the concentration of $[T]$ in the light harvesting pigments, whereas $[Q]$ is identified here with the triplet exciton concentration within PS I. Using single picosecond laser fluorescence quantum yield studies, we have shown that the concentration of singlet excitons in PS I is proportional to the singlet exciton concentration in the light harvesting antenna pigment system (12); this determination was made by integrating the fluorescence yield produced by single picosecond pulses of different intensities and showing that the ratio of the yields at 685 and 735 nm was independent of the pulse intensity (12). The results shown in Fig. 1 indicate that the concentration of triplets in PS I does not follow the calculated concentration of triplets in the light harvesting pigments. It should be emphasized that this conclusion does not depend on the nature of the model embodied by Eqs. 2 and 3. It is qualitatively evident from the simple fact that pulses 3, 4, and 5, which are not significantly less intense than the subsequent pulses, do not produce any concentration of $[Q]$ within the PS I pigment system. The calculation, whose results are plotted in Fig. 1 C and are shown in Table I, merely confirms this qualitative conclusion.

We propose that the time dependence of the accumulation of the quenchers within PS I is due to the time it takes for triplet excitons to diffuse to the PS I pigment system from the light harvesting pigments where most of the photons are absorbed. We estimate this time lag to be of the order of 50 ns. This time is about $50/0.14 = 360$ times longer than the diffusion time of singlet excitons to PS I (15). Assuming that the average diffusion lengths for triplet and single excitons from light harvesting to PS I pigments are about the same, we can estimate the relative diffusion coefficients of singlet and triplet excitons D_S and D_T in chloroplasts at 77°K using the expression $D_S t_S = D_T t_T$, where t_S and t_T are 0.14 and 50 ns, respectively. The ratio of D_T/D_S thus obtained is 1/360. This is a reasonable result, as discussed in the Introduction.

We finally make a comment on the exponentiality of the fluorescence decay which, as pointed out above, indicates the dynamic nature of this quenching. If PS I is a pigment system consisting of a small number of chlorophyll molecules (so that the radius of a unit is comparable to a singlet-triplet interaction distance $\simeq$40–50Å), and if a triplet exciton is present in a given PS I unit, one would expect any singlet exciton arriving at this unit to be quenched with nearly 100% efficiency. The quenching then would be static, because those PS I units (or pigments associated with the PS I fluorescence) without triplets would display the normal 1.5-ns fluorescence decay time, whereas those with triplets would exhibit little or no fluorescence. Because this is not the case, PS I must be large enough so that diffusion of excitons and dynamic quenching are both operative; as indicated in Eq. 1, the monomolecular decay rate is competitive in magnitude with the quenching rate. Thus, it is estimated that the radius of such a typical pigment unit is larger than about 40–50 Å and therefore has a cross-section of $\geq 75 \times 10^{-14}$ cm^2. Using a value of $\simeq 2.2 \times 10^{-14}$ cm^2 for the area of a chlorophyll molecule, we estimate that there are at least 30–40 chlorophyll molecules associated with a PS I unit (or pigment system associated with the 735-nm fluorescence). On the other hand, we have shown that the optical cross-section of PS I is small relative to the cross-section of the light-harvesting pigments (12). Inasmuch as one photosynthetic unit comprising both PS I and PS II consists of $\simeq$600 chlorophyll molecules (18), the value of at least 30–40 molecules in a PS I unit derived here is still consistent with the relatively small optical cross-section of the PS I pigments (4, 12).

CONCLUSIONS

Picosecond laser spectroscopy, using a combination of single pulse and pulse train excitation, can provide important information about triplet and singlet exciton dynamics and exciton distributions in photosynthetic membranes. The mutual annihilation of excitons in chloroplasts provides a dynamic probe for the topology of the membranes. Because singlet and triplet excitions have widely different lifetimes and migration velocities, they can be utilized as intrinsic dynamic probes on widely different time scales, by selectively chosing the appropriate mode of laser excitation.

The portion of this work performed at New York University (N. E. Geacintov and C. E. Swenberg) was supported by National Science Foundation Grant PCM 76-14359. Partial assistance from a United States Department of Energy Contract to the Radiation and Solid State Laboratory is also acknowledged. The portion of this work performed at Los Alamos Scientific Laboratory was supported by the United States Department of Energy.

Received for publication 7 December 1977.

REFERENCES

1. CAMPILLO, A. J., V. H. KOLLMAN, and S. L. SHAPIRO. 1976. Intensity dependence of the fluorescence lifetime *in vivo* chlorophyll excited by a picosecond light pulse. *Science (Wash. D.C.).* **193**:227–229.
2. CAMPILLO, A. J., S. L. SHAPIRO, V. H. KOLLMAN, K. R. WINN, and R. C. HYER. 1976. Picosecond exciton annihilation in photosynthetic systems. *Biophys. J.* **16**:93–97.
3. YU, W., P. O. HO, R. R. ALFANO, and M. SEIBERT. 1975. Fluorescent kinetics of chlorophyll in photosystems I and II enriched fractions of spinach. *Biochim. Biophys. Acta.* **387**:159–164.
4. GEACINTOV, N. E., J. BRETON, C. E. SWENBERG, A. J. CAMPILLO, R. C. HYER, and S. L. SHAPIRO. 1977. Picosecond and microsecond pulse laser studies of exciton quenching and exciton distribution in spinach chloroplasts at low temperatures. *Biochim. Biophys. Acta.* **461**:306–312.

5. BRETON, J., and N. E. GEACINTOV. 1977. Quenching of fluorescence of chlorophyll *in vivo* by long-lived excited states. *FEBS (Fed. Eur. Biochem. Soc.) Lett.* **69**:86–89.
6. PORTER, G., J. A. SYNOWIEC, and C. J. TREDWELL. 1977. Intensity effects on the fluorescence of *in vivo* chlorophyll. *Biochim. Biophys. Acta.* **459**:329–336.
7. SEARLE, G. F. W., J. BARBER, L. HARRIS, G. PORTER, and C. J. TREDWELL. 1977. Picosecond laser study of fluorescence lifetimes in spinach chloroplast photosystem I and photosystem II preparations. *Biochim. Biophys. Acta.* **459**:390–401.
8. PASCHENKO, V. Z., S. P. PROTASOV, A. B. RUBIN, K. N. TIMOFEEV, L. M. ZAMAZOVA, and L. B. RUBIN. 1975. Probing the kinetics of photosystem I and photosystem II fluorescence in pea chloroplasts on a picosecond pulse fluorometer. *Biochim. Biophys. Acta.* **408**:143–153.
9. GOVINDJEE, and L. YANG. 1966. Structure of the red fluorescence band in chloroplasts. *J. Gen. Physiol.* **49**:763–780.
10. BUTLER, W. L., and M. Kitajima. 1975. Energy transfer between photosystem II and photosystem I in chloroplasts. *Biochim. Biophys. Acta.* **396**:72–85.
11. SATOH, K., R. STRASSER, and W. L. BUTLER. 1976. A demonstration of energy transfer from photosystem II to photosystem I in chloroplasts. *Biochim. Biophys. Acta.* **440**:337–345.
12. GEACINTOV, N. E., J. BRETON, C. E. SWENBERG, and G. PAILLOTIN. 1977. A single pulse picosecond laser study of exciton dynamics in chloroplasts. *Photochem. Photobiol.* **26**:629–638.
13. GEACINTOV, N. E., and J. BRETON. 1977. Exciton annihilation in the two photosystems in chloroplasts at 100°K. *Biophys. J.* **17**:1–15.
14. HARRIS, L., G. PORTER, J. A. SYNOWIEC, C. J. TREDWELL, and J. BARBER. 1976. Fluorescence lifetimes of chlorella pyrenoidosa. *Biochem. Biophys. Acta.* **449**:329–339.
15. CAMPILLO, A. J., S. L. SHAPIRO, N. E. GEACINTOV, and C. E. SWENBERG. 1977. Single pulse picosecond determination of 735 nm fluorescence risetime in spinach chloroplasts. *FEBS (Fed. Eur. Biochem. Soc.) Lett.* **83**:316–320.
16. BUTLER, W. L. 1978. Energy distribution in the photochemical apparatus of photosynthesis. *Ann. Rev. Plant Physiol.* In press.
17. SWENBERG, C. E., and N. E. GEACINTOV. 1973. Exciton interactions in organic solids. *In* Organic Molecular Photophysics. J. B. Birks, editor. Wiley & Sons Ltd., Chichester, Sussex. 489–564.
18. SAUER, K. 1975. Primary events and the trapping of energy. *In* Bioenergetics of Photosynthesis. Govindjee, editor. Academic Press, Inc., New York. 115–181.
19. SHUVALOV, V. A. 1976. The study of the primary photo-processes in photosystem I of chloroplasts: recombination luminescence, chlorophyll triplet state and triplet-triplet annihilation. *Biochem. Biophys. Acta.* **430**:113–121.
20. BIRKS, J. B. 1970. Photophysics of Aromatic Molecules. Wiley & Sons Ltd., Chicester, Suxxes. Chap. 10.
21. SWENBERG, C. E., N. E. GEACINTOV, and M. POPE. 1976. Bimolecular quenching of excitons and fluorescence in the photosynthetic unit. *Biophys. J.* **16**:1447–1452.
22. SCHWARTZ, M. 1972. Quantum yield determination of photosynthetic reactions. *Adv. Enzym.* **24**:139–146.
23. RAHMAN, T. S., and R. S. KNOX. 1973. Theory of singlet-triplet exciton fusion. *Phys. Stat. Sol. (b).* **58**:715–720.
24. GEACINTOV, N. E., M. BINDER, C. E. SWENBERG, and M. POPE. 1975. Exciton dynamics in α-particle tracks in organic crystals: magnetic field study of the scintillation in tetracene. *Phys. Rev. B.* **12**:4113–4134.

DISCUSSION

CHAIRMAN: We begin with a question from an anonymous referee: The statement in your Results (page 349) that there is no difference between 685 and 735 nm fluorescence as a function of pulse intensity seems to contradict the earlier statement that 735 nm fluorescence is quenched more easily than the 685 nm fluorescence.

GEACINTOV: There is no difference in the fluorescence efficiencies at 685 and 735 nm when single picosecond laser pulses are used for excitation. When trains of picosecond pulses are used, or when microsecond excitation pulses are utilized, the quenching at 735 nm is more pronounced than at 685 nm due to triplet excitons, which appear to be more effective as quenchers in PS I than in the light-harvesting pigment system (refs. 4, 5, 12, and 13). Quenching of singlets by triplet excitons is not important when single picosecond pulses are used because of the short time scales involved.

BERGER: In Fig. 1 C you show two curves: one, $K[Q]$, is the quenching of PS I fluorescence determined experimentally; the other $\gamma_{TS}[T]$, is the theoretical curve calculated from Eqs. 2, 3, and 4. Can you explain the significance of these curves, and in particular, why their shapes are different?

GEACINTOV: First of all there is the experimental $K[Q]$ curve inferred by using dynamic quenching changes in the singlet's lifetime as given by the Stern-Volmer equation. The other curve is a one-compartmental model calculation of $\gamma_{TS}[T]$, using Eqs. 2 and 3 given in the text. We did not consider a two-compartmental model in which we have rates for singlet and triplet excitons migrating from the light-harvesting system (LH) to PS I. If singlet excitons had migrated from LH to PS I and then formed triplet excitons by intersystem crossing, we would not observe the large delay in quenching, since the rise time of the 735-nm emission is approximately 140 ps. The calculated curve is only semi-quantitative and demonstrates that the lag time is a real effect. The absolute scale on the $\gamma_{TS}[T]$ curve is unimportant. The density of triplet excitons is not known but from unpublished microsecond and pulse train studies we can infer a value of the singlet-triplet rate constant, γ_{TS}, of approximately 7×10^{-9} cm^3 s^{-1}.

KAUFMANN: Have you noticed on the triplet quenching PS I any effect of magnesium iron concentration?

SWENBERG: We have not done that experiment yet.

NASH: The data of decay lifetime vs. pulse number (Fig. 1 B) show that no quenching occurs before pulse 4, which indicates that the triplet formation occurs entirely in the light-harvesting pigments. However, the data of Fig. 1 B pulses 1, 2, and 4 have a large error and thus could be fit in several ways. A more pessimistic fit would indicate quenching as early as the second pulse. Two questions: (*a*) Does restricting triplet formation to the light-harvesting pigments seem reasonable, and do you have other convincing data indicating the delay of quenching to pulse 4? (*b*) Could the onset of quenching be pulse number-dependent rather than time-dependent, and have you varied pulse frequency to test this?

SWENBERG: Yes, you could take a pessimestic view. There are published experiments by Breton and Geacintov (ref. 5) where the wavelength dependence of the fluorescence after 1, 4, and 20 pulses has been measured. The data show that there is very little change during the first 4 pulses but very pronounced quenching at the end of 20 pulses. I think that strengthens our interpretation that there is a time lag.

GEACINTOV: For your second question, do you mean the spacing between the pulses?

NASH: Yes.

GEACINTOV: No. This spacing is fixed by the round-trip time of the pulses in the laser cavity.

NASH: It is reasonable to consider then that the pulse quenching is pulse number-dependent rather than time-dependent? For example, the number of quenchers could be incremented by a fixed amount in each pulse, and it would not necessarily be a time process.

GEACINTOV: Well, the data speak for themselves. There is a buildup of the quenchers with time, but there is also a lag in the buildup of quenchers in PS I. Thus there is a time dependence.

MAUZERALL: To amplify that suggestion given by Dr. Nash, we have models of these photosynthetic units where there are multiple traps, and in fact if there are three or four traps per unit and for some reasons you fill only one per pulse, then you would get in fact this delay of four and then fill up with the quenchers after the traps are filled. That's the model that amplifies.

SWENBERG: I will comment on that. I agree with Dr. Mauzerall that there are multiple-trap problems, even with regard to the singlet fluorescence quantum yield. I currently don't know how to include these effects in terms of a simple kinetic model, but I do agree with you on that particular point.

BADEA: I would like to ask three questions. (*a*) The fluorescence decay time of spinach chloroplasts has been determined to be 1.5 ± 0.4 ns. This ±24% interval of confidence is typical for the measurement of nanosecond relaxation times. What is it in picosecond technology? (*b*) Regarding the monoexponentiality of fluorescence decay, how accurately can it be ascertained at the present time with single picosecond pulses? How distinct should the two exponentials be to make possible their separation in the picosecond domain? (*c*) What is the minimum spacing of two pulses on the nanosecond time scale beyond which the fluorescence decay initiated by the first pulse is significantly perturbed by the second? Would the authors give a rough estimate for both mono- and biexponential decay in terms of the ratio: average lifetime/ spacing (or the inverse ratio)?

SWENBERG: With regard to the accuracy of ± 25%, I think that this is a problem of reproducibility. The errors in the paper are standard deviations from six independent measurements. It probably can't be done any better than that right now.

Concerning your second question: It is easy to add exponentials in fitting curves; however, considering the accuracy of the experiment, I call this an example of over-interpretation. Whenever you have any type of bimolecular processes that occur in a finite domain, you do not necessarily expect standard decays. Instead the decay curves are infinite-series sums of exponentials: the smaller the domain size, the more terms in the series.

With regard to the third question, do you want to place the second pulse at varying time delays after the first pulse and see when the effect of the first pulse occurs on the second pulse emission? Is that the question?

BADEA: No, what I was asking is how widely spaced do you have to have the pulses one after the other to measure, for example, nanosecond fluorescence with any degree of accuracy.

SWENBERG: Well, first of all there are instrumental problems. You are not automatically able to position the second pulse after the first one. Probably Dr. Rentzepis can talk about that. The other important point is that the quenching states produced by either autoionization after bimolecular singlet fusion or intersystem crossing into triplet manifolds need time to occur. Singlet intersystem crossing rates are the order of 10^{-8}/s; thus you need at least that much time before a sufficient number of triplet excitons accumulate. Furthermore, you don't get any of these nonlinear effects unless you have sufficient intensity. This is a crucial point and considerable disagreement exists in the literature because many investigators are not being careful about the number of pulses. Many of their experiments are in the intensity regime, where all sorts of nonlinear processes occur. I don't know if I have answered your question.

RENTZEPIS: The pulse separation is determined by twice the cavity length l, $2l/c$. Of course there are two ways to vary the pulse separation. One can enlarge or decrease the cavity length by moving the mirrors or one can place an optical modulator in it that transmits one every third pulse, etc. I think that Dr. Kaufmann is very anxious to make a comment.

KAUFMANN: I would just like to remind Dr. Rentzepis that a few years ago some elegant papers by Huppert and Rentzepis showed that you could easily use two pulses to do such experiments. They did things such as preparing solvated electrons in sodium methylamine solutions and showed a very nice way of varying the pulse rate. If I remember correctly, they varied the difference between pulses in the order of 300 ps or so but that technique obviously can be extended to 30 or 40 ns without too much trouble. This is especially apt for this kind of work, when you need pulses of very low intensity.

RENTZEPIS: I thank Professor Kaufmann for assisting my memory even when what he says is correct at least to a general extent. This method and one for nanosecond separation are described in Proc. Natl. Acad. Sci. U.S.A. 1972. **69**:2806. and *J. Chem. Phys.* 1976. **64**:191.

KAUFMANN: Another point on the technology is that in our laboratory in one system we have measured the singlet lifetime with the streak camera to be about 110 ps, and when we measured the singlet lifetime by absorption techniques, measuring ground state repopulation, it was on the order of 90 ± 20 ps. I think that there is reasonable agreement for this kind of time measurement.

RUBIN: What is the quantum yield conversion to triplet state?

SWENBERG: In our calculations we have taken the quantum yield for chlorophyll-*a* fluorescence and its lifetime in ethanol and have inferred an intersystem crossing rate of approximately 1.3×10^8/s. We assumed that this value is unchanged in vivo. No firm knowledge concerning this approximation is known. At least we don't know of any.

RUBIN: But is it a certain concentration of triplet state that contribute to fluorescence? I am talking about the quantum yield of the triplet state appearance in the light-harvesting system.

SWENBERG: Are you asking how many of the triplets formed in the light-harvesting unit actually make their way to PS I?

RUBIN: Yes.

SWENBERG: Right now I don't know the answer. However, since PS I emits at longer wavelengths then LH and PS II and has basically only chlorophyll-*a* molecules, I would assume that most triplets formed probably make their way there. That's why there is more quenching occurring in PS I.

RUBIN: Are you monitoring the triplet state concentration and its decay or chlorophyll ions?

SWENBERG: The triplet excitons are actually monitored by triplet-triplet absorption. Probably

Dr. Geacintov would wish to comment on this. Using microsecond pulse excitation and observing the actual fluorescence decay at various wavelengths, one can get some estimate of the number of triplet excitons present. This can then be compared directly with triplet-triplet absorption spectroscopy.

GEACINTOV: Breton and myself in France have done experiments of this type where we monitored both the ion information and the triplet formation during and after laser pulses of various lengths (not picosecond lasers). Within experimental error, which was rather large, we did not detect any chlorophyll ions. Our limit of detection was about 1 chlorophyll ion per 100 molecules. On the other hand we could easily detect and observe between 1 and 3 caroteroid triplets per 100 chlorophyll molecules.

RUBIN: Now I want to make some comments about the experiments in which we use the same high-intensity picosecond pulses as Geacintov. One problem arising is the nonlinear dependence of the lifetime of the chlorophyll in vivo on the intensity of light. This is known and everyone is careful now to use nonintense pulses. But there are other problems, such as the nonstability of the photosynthetic system under high intensity light. If you compare with the structure of the chloroplast before the laser pulse, you will see physical damage after a 10^{15} quanta pulse and then finally severe damage with the 10^{17}–10^{18} quanta pulse. If you look at the "softest" of the chloroplast membranes, you can see the structure well, with a lot of particles correlated with the photosynthetic activity of the chloroplast. You can see the surface of the chloroplast membrane and the drastic damage by the high intensity, 10^{18} quanta/cm^2 pulse. The damage threshold is $\sim 10^{15}$ quanta/cm^2. So you must be careful. Have you any evidence that you do not destroy your chlorophyll sample?

GEACINTOV: We have also been very successful in destroying chloroplasts at various times. This type of irreversible effect is easily observed. Your findings are very interesting. I would just like to say that we performed controls in all of our experiments; these controls consisted of monitoring the fluorescence yield under low excitation intensity conditions both before and after our high-intensity laser experiments. As far as the fluorescence yields are concerned, we did not observe any changes. Therefore if you do observe changes in the chloroplast structure in the low-intensity limits, these do not seem to have any influence on the fluorescence yield. But this needs to be investigated further.

SWENBERG: I do not expect this effect to be important over small domains but I do believe there is quite a bit of structural change on the stacking of the grana.

MAUZERALL: Yes, in our experiments on the algae we also noted that there is a great variability among the different plant material. Some are particularly resistant to our nanosecond pulses, but some blue-green algae are very rapidly damaged by such a pulse system and one must of course check the reversibility of these systems.

RUBIN: I can tell you that chloroplasts are not very resistant to laser radiation and the threshold of killing them is about 10^{15} quanta/cm^2.

MAUZERALL: That's just about what we had measured, too. That's why our data quit at that point.

MODULATION OF THE PRIMARY ELECTRON TRANSFER RATE IN PHOTOSYNTHETIC REACTION CENTERS BY REDUCTION OF A SECONDARY ACCEPTOR

M. J. Pellin, C. A. Wraight, and K. J. Kaufmann,
Departments of Chemistry, Botany, and Biophysics, University of Illinois, Urbana, Illinois 61801 U.S.A.

ABSTRACT Photosynthetic application of picosecond spectroscopic techniques to bacterial reaction centers has led to a much greater understanding of the chemical nature of the initial steps of photosynthesis. Within 10 ps after excitation, a charge transfer complex is formed between the primary donor, a "special pair" of bacteriochlorophyll molecules, and a transient acceptor involving bacteriopheophytin. This complex subsequently decays in about 120 ps by donating the electron to a metastable acceptor, a tightly bound quinone.

Recent experiments with conventional optical and ESR techniques have shown that when reaction centers are illuminated by a series of single turnover flashes in the presence of excess electron donors and acceptors, a stable, anionic ubisemiquinone is formed on odd flashes and destroyed on even flashes, suggesting that the acceptor region contains a second quinone that acts as a two-electron gate between the reaction center and subsequent electron transport events involving the quinone pool.

Utilizing standard picosecond techniques, we have examined the decay of the charge transfer complex in reaction centers in the presence of the stable semiquinone, formed by flash illumination with a dye laser 10 s before excitation by a picosecond pulse. In this state the decay rate for the charge transfer complex is considerably slower than when no electron is present in the quinone acceptor region. This indicates fairly strong coupling between constituents of the reaction center-quinone acceptor complex and may provide a probe into the relative positions of the various components.

INTRODUCTION

Knowledge of the primary electron transfer events occurring during bacterial photosynthesis has been greatly enhanced through the application of picosecond absorption spectroscopy (1–4). This technique has shown that an electron is rapidly transferred (< 10 ps) from a "special pair" bacteriochlorophyll molecule (P870) to an intermediate species, I, known to involve bacteriopheophytin. Transfer to I is accompanied by a rise in absorption at 630 nm and longer wavelengths, and by a bleaching in absorption peaked at 544 nm (1, 2, 5). This intermediate then transfers an electron to a tightly bound quinone-iron complex (Q_1) with a characteristic half time of $\simeq 100$ ps.

K. J. Kaufmann is an Alfred P. Sloan Fellow.

Recently, a second quinone molecule (Q_{II}) has been implicated in the transfer scheme, and a periodicity of two has been reported for a flash formation of ubisemiquinone (6, 7). Q_{II} was suggested to act as a two electron gate between Q_I and a quinone pool. Electron transfer from Q_I to Q_{II} has recently been confirmed by time-dependent optical absorption spectroscopy (8), and the following schematic combines the picosecond events with the quinone acceptor region kinetics:

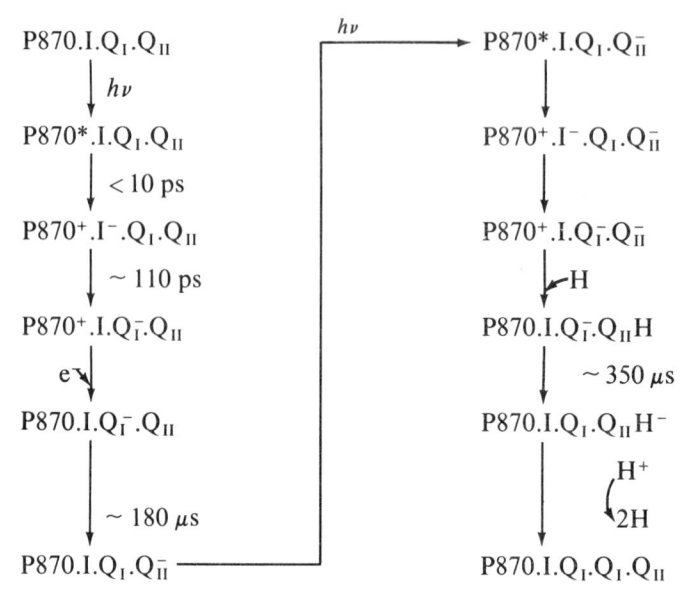

$$
\begin{array}{ccc}
P870.I.Q_I.Q_{II} & \xrightarrow{h\nu} & P870^*.I.Q_I.Q_{II}^- \\
\downarrow h\nu & & \downarrow \\
P870^*.I.Q_I.Q_{II} & & P870^+.I^-.Q_I.Q_{II}^- \\
\downarrow <10\,\text{ps} & & \downarrow \\
P870^+.I^-.Q_I.Q_{II} & & P870^+.I.Q_I^-.Q_{II}^- \\
\downarrow \sim 110\,\text{ps} & & \downarrow\, H \\
P870^+.I.Q_I^-.Q_{II} & & P870.I.Q_I^-.Q_{II}H \\
\downarrow e^- & & \downarrow \sim 350\,\mu\text{s} \\
P870.I.Q_I^-.Q_{II} & & P870.I.Q_I.Q_{II}H^- \\
\downarrow \sim 180\,\mu\text{s} & & \downarrow H^+,\, 2H \\
P870.I.Q_I.Q_{II}^- & \longrightarrow & P870.I.Q_I.Q_I.Q_{II}
\end{array}
$$

This report describes changes in the rate of electron transfer between I and Q_I due to the presence of ubisemiquinone (Q_{II}^-) as measured by picosecond absorption techniques. It also provides evidence for physical changes occurring in the reaction center in response to the two-electron gating process.

MATERIALS AND METHODS

The 640-nm absorption data were obtained with a standard picosecond absorption apparatus (1, 9). It consisted of an echelon to convert time to spatial information and a double beam to compensate for fluctuations in the laser intensity. Detection was accomplished with a vidicon camera, and the output was digitized and stored in a minicomputer for further data reduction.

For absorption measurements at 544 nm, the excitation light was Raman-shifted in methyl-cyclohexane from 528 to ~625 nm. This shift minimized scattering effects as well as reducing the background shift due to oxidized cytochrome c.

Reaction centers containing ubisemiquinone (Q_{II}^-) were prepared by excitation, with a saturating flash from a dye laser (300 ns, 10 mJ, λ_{max} = 590 nm), $\simeq$ 10 s before the picosecond measurement. This time was sufficient for transfer of an electron from Q_I to Q_{II} (~200 μs) but short compared to the lifetime of the semiquinone (Q_{II}^-), which is stable for many minutes (6).

Reaction centers were prepared from *Rhodopseudomonas sphaeroides,* R26, by detergent fractionation using lauryldimethylamine N-oxide (LDAO) as has been described previously (6, 7). Experiments utilized 200 μl of 80-μM reaction centers with 10 μl of 5 mM reduced cytochrome

c. The cytochrome c served to reduce P870$^+$ before the second flash. Diaminodurene (DAD), 10 μl of a 5 mM solution, was added every 60 min to insure that the cytochrome c remained reduced in the presence of LDAO and oxygen.

Samples were mounted in kinematic holders so that the 2-mm sample cells could be precisely replaced. This allowed alternation of the single-flash samples (i.e., those with no preflash and thus no ubisemiquinone) and the double-flash samples (ubisemiquinone present). Care was taken to insure that both single- and double-flash samples received equivalent exposure to intense light. 15 min of dark time was allowed (for each sample) between runs to obtain a consistent initial sample state.

RESULTS AND DISCUSSION

A rapid increase and subsequent decay of absorption at 640 nm has been observed by two groups (1, 2). This kinetic behavior has been attributed to the decay of the bacteriopheophytin anion as it donates an electron to the primary acceptor (Q_I). Fig. 1 shows this decay for reaction centers prepared with ubisemiquinone (double-flash) and without ubisemiquinone (single-flash). Similar results were obtained with three different reaction center preparations. Each curve is a compilation of at least 24 separate measurements. The average error bar for each point is ±0.018.

Application of a nonlinear curve fitting routine (11) to these data shows that the half time for electron transfer in the presence of ubisemiquinone is distinctly slower than

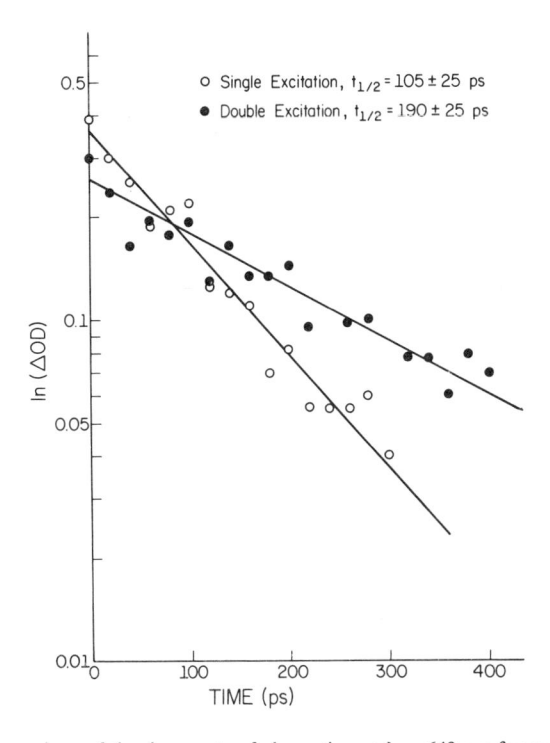

FIGURE 1 A comparison of the decay rate of absorption at λ = 640 nm for reaction centers in the presence of ubisemiquinone (●) and in the absence of ubisemiquinone (o). Each point has a standard deviation unit of 0.018.

FIGURE 2

FIGURE 3

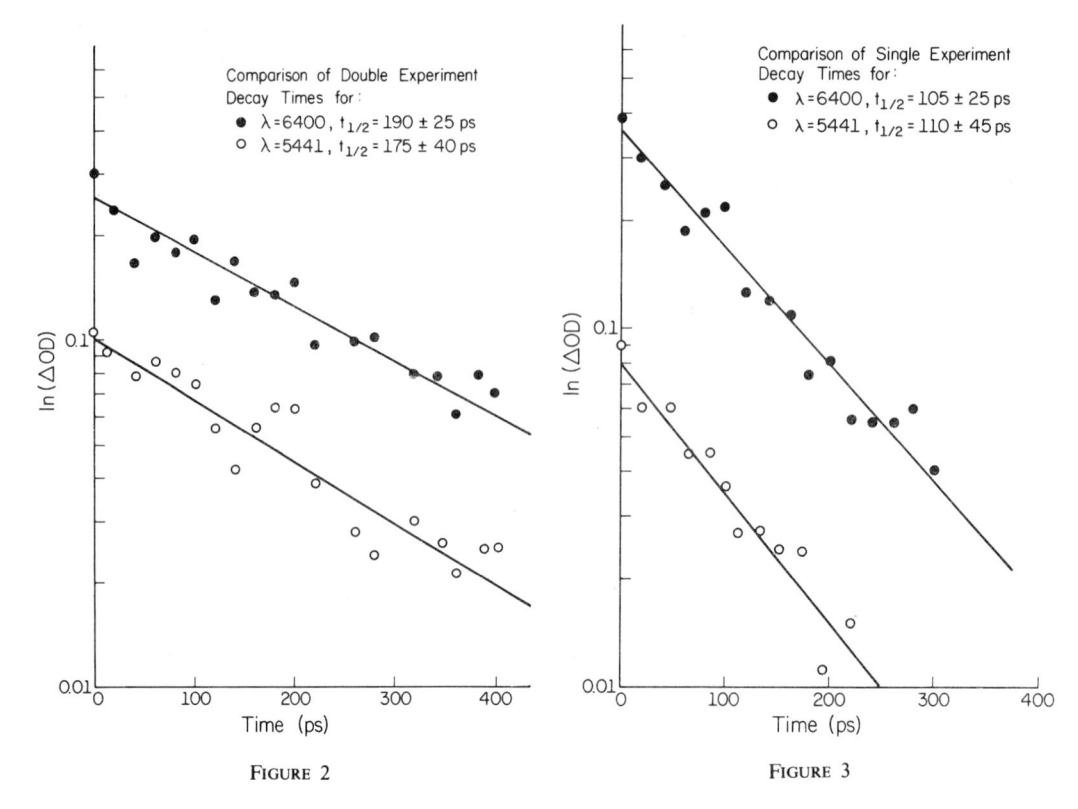

FIGURE 2 A comparison of the decay of absorption at $\lambda = 640$ nm (●) with the decay of bleaching at $\lambda = 544$ nm (o) for reaction centers in the presence of ubisemiquinone. Standard deviations ± 0.018 at 640 nm and ± 0.035 at 544 nm.

FIGURE 3 A comparison of the decay of absorption at $\lambda = 640$ nm (●) with the decay of bleaching at $\lambda = 544$ nm (o) for reaction centers in the absence of ubisemiquinone. See Fig. 2 for standard deviations.

with no ubisemiquinone present. Measurements at 700 ps (data not shown) are in close agreement with the decay measured both in the presence and in the absence of ubisemiquinone. For these samples only a slight negative (0.005) OD change was observed for single flash (no ubisemiquinone) experiments. For samples prepared with ubisemiquinone (double-flash), a residual positive OD change (0.05) was observed.

There is some disagreement in the literature on the precise value of the transfer time from I^- to Q_I (1, 2). We have consistently observed a value for the reaction half time of 105 ± 25 ps for single-flash samples, in agreement with results obtained at the Bell Telephone Laboratories (1). Our new results show that this transfer time is sensitive to local perturbations, and the discrepancies in this measurement may reflect the state or integrity of the preparation.

Bleaching at 544 nm and absorption at 640 nm are both taken to indicate I^- (1, 2). Figs. 2 and 3 compare kinetic measurements at 544 and 640 nm in the presence and absence of ubisemiquinone, respectively; the two wavelengths show similar effects. However, the optical density changes at 544 nm are somewhat smaller, and there is a large, variable contribution to the absorbance arising from the photooxidized cytochrome. The standard deviation is thus larger (± 0.035), the signal-to-noise ratio smaller, and the difference between the single- and double-pulse kinetics less dramatic.

The occurrence of certain photosynthetic electron transfer events at very low temperatures has led to the application of established theoretical descriptions of electron transfer to these processes (12–17). For example, electron transfer in photosynthetic systems has been extensively studied between cytochrome c and the bacteriochlorophyll dimer in the purple sulfur bacterium, *Chromatium vinosum* (12), and several mechanisms have been proposed to explain the temperature dependence of the observed electron transfer rate. Although the importance of Franck-Condon factors in controlling electron transfer processes has been known for some time, only recently has it been applied to biological systems (13–17). A vibronically assisted tunneling mechanism, which incorporates the Franck-Condon factors in a manner analogous to the Förster-Dexter theory for energy transfer, has been used by Hopfield to account for both the temperature dependence of the transfer reaction in *Chromatium* and the rates for the forward and backward electron transfer reactions in *R. sphaeroides* (14, 17). Jortner (16) has successfully applied nonadiabactic multiphonon theory to the temperature dependence of the electron transfer in *Chromatium*. Even though these approaches differ somewhat in their model for the electron transfer mechanism, they do contain the effect of Franck-Condon overlap on the transition rate. Differences in the two theories lead to disagreement at low temperatures, but at high temperatures both theories correspond to the Marcus theory for electron transfer (18). At high temperatures, the transfer rate is thus given by $K = \eta e^{-(\Delta U/RT)}$, where η is the probability for transition of the electron from the potential surface of the donor to the acceptor. For nonadiabatic reactions η must be small. ΔU is the height of the crossing point for the two curves above the equilibrium position of the donor and is therefore an energy of activation for the reaction. The rate of electron transfer can be modulated by a change in either η or ΔU. The transition probability, η, is highly sensitive to configurational changes in the reaction center but would remain constant if the energies of the donor and acceptor potential surfaces change relative to one another. The apparent activation energy, ΔU, can be modified either by configurational changes in the protein or by changes in the relative energies of the two potential surfaces.

If the modification of the electron transfer rate is the result solely of an increase in ΔU, one calculates a change in ΔU of 1.3×10^3 Jmol^{-1} (0.014 eV). A possible source of this increase is the electrostatic potential surface arising from the anionic semiquinone (Q_{II}^-). Assuming a colinear geometry for BPh, Q_I, and Q_{II}, a separation of 20 Å from the center of Q_{II} to the edge (near BPh) of Q_I, 3 Å as the edge-to-edge distance between BPh and Q_I, and a dielectric strength of three for the protein, a change in ΔU of 2.8×10^3 J·mol^{-1} (0.029 eV) can be calculated. In view of the arbitrary choice of geometry and distance parameters, this qualitative agreement indicates the feasibility of this interpretation.

A change in the barrier height need not arise solely from electrostatic interactions but could also result from a small increase in the distance between the donor and acceptor or modifications in the protein structure. Changes in protein configuration could also account for a decrease in the energy of I^-. Recent work shows that there is pH-dependent behavior within the acceptor-quinone complex giving rise to charge alterations within it,[1] which could also lead to modulation of the picosecond transfer rate. This pH dependence is currently under investigation.

CONCLUSION

Interaction of the secondary quinone acceptor with the reaction center unit is much stronger than might have been anticipated. Despite the fact that this ubiquinone is rather weakly bound to the reaction center, the reduction of Q_{II} to ubisemiquinone can induce a retardation in the rate of electron transfer between I and Q_I. This effect could be accounted for by the electrostatic influence of the anionic ubisemiquinone (Q_{II}^-) or by a configurational change in response to this species, causing a small extra separation between I and Q_I.

This work was supported by the National Science Foundation.

Received for publication 25 November 1977.

REFERENCES

1. KAUFMANN, K. J., P. L. DUTTON, J. L. NETZEL, J. S. LEIGH, and P. M. RENTZEPIS. 1975. Picosecond kinetics of events leading to reaction center bacteriochlorophyll oxidation. *Science (Wash. D.C.)*. **188**:1301.
2. ROCKLEY, M. G., M. W. WINDSOR, R. J. COGDELL, and W. W. PARSON. 1975. Picosecond determination of an intermediate in the photochemical reaction of bacterial photosynthesis. *Proc. Natl. Acad. Sci. U.S.A.* **72**:2551.
3. KAUFMANN, K. J., K. M. PETTY, P. L. DUTTON, and P. M. RENTZEPIS. 1976. Picosecond kinetics in reaction centers of *Rps. spheroides* and the effects of ubiquinone extraction and reconstitution. *Biochim. Biophys. Acta.* **70**:839.
4. DUTTON, P. L., K. J. KAUFMANN, B. CHANCE, and P. M. RENTZEPIS. 1975. Picosecond kinetics of the 1250 nm band of the *Rps. spheroides* reaction center: the nature of the primary photochemical intermediate state. *FEBS (Fed. Eur. Biochem. Soc.) Lett.* **60**:275.
5. PARSON, W. W., R. K. CLAYTON, and R. J. COGDELL. 1975. Excited states of photosynthetic reaction centers as low redox potentials. *Biochim. Biophys. Acta.* **387**:265.
6. WRAIGHT, C. A. 1977. Electron acceptors of photosynthetic bacterial reaction centers: direct observation of oscillatory behavior suggesting two closely equivalent ubiquinones. *Biochim. Biophys. Acta.* **459**:525.
7. VERMEGLIO, A. 1977. Secondary electron transfer in reaction centers of *Rhodopseudomonas spheroides*: out-of-phase periodicity of two for the formation of ubisemiquinone and fully reduced ubiquinone. *Biochim. Biophys. Acta.* **459**:516.
8. VERMEGLIO, A., and R. K. CLAYTON. 1977. Kinetics of electron transfer between the primary and the secondary electron acceptor in reaction centers of *Rhodopseudomonas spheroides*. *Biochim. Biophys. Acta.* **461**:159.
9. BUSCH, G. E., and P. M. RENTZEPIS. 1976. Picosecond chemistry. *Science (Wash. D.C.)*. **194**:276.
10. FAJER, J., D. C. BRUNE, M. S. DAVIS, A. FORMAN, and L. D. SPAULDING. 1975. Primary charge separa-

---

[1] Wraight, C. A. 1978. Manuscript in preparation.

tion in bacterial photosynthesis: oxidized chlorophylls and reduced pheophytin. *Proc. Natl. Acad. Sci. U.S.A.* **72**:4956.
11. BEVINGTON, P. R. 1969. *Data Reduction and Error Analysis for the Physical Sciences.* McGraw-Hill Book Company, New York. 180.
12. DEVAULT, D., and B. CHANCE. 1966. Studies of photosynthesis using a pulsed laser. *Biophys. J.* **6**:825.
13. LEVICH, V. G. 1966. Present state of the theory of oxidation-reduction in solution (bulk and electrode reactions). *Advan. Electrochem. Electrochem. Eng.* **4**:249.
14. HOPFIELD, J. J. 1974. Electron transfer between biological molecules by thermally activated tunneling. *Proc. Natl. Acad. Sci. U.S.A.* **71**:3640.
15. HOPFIELD, J. J. 1977. Photo-induced charge transfer: a critical test of the mechanism and range of biological electron transfer processes. *Biophys. J.* **18**:311.
16. JORTNER, J. 1975. Temperature-dependent activation energy for electron transfer between biological molecules. *J. Chem. Phys.* **64**:4860.
17. HOPFIELD, J. J. 1977. Fundamental aspects of electron transfer in biological membranes. *In* Electrical Phenomena at the Biological Membrane Level. E. Roux, editor. U.M. Elsevier, Amsterdam. 471.
18. MARCUS, R. A. 1960. Exchange reactions and electron transfer reactions including isotopic exchange. *Discuss. Faraday Soc.* **29**:21.

DISCUSSION

RENTZEPIS: To begin with an anonymous question: In your paper the use of nonlinear curve fitting obscures the basic question of how the base-line treatment affects the kinetic conclusion. What was the prepulse base line to the long-term base line used for Fig. 1?

KAUFMANN: The optical density changes in a sample as a result of excitation are measured with a dual-beam spectrometer. Each laser pulse is split into two beams, one passing through the air and one passing through the sample. The intensity of these two are recorded simultaneously, allowing a relative optical density (OD) to be derived. The optical density change (ΔOD) between excited and normal samples is measured by taking two successive laser shots. The first shot is taken in the absence of excitation and the second when the sample is excited. The ΔOD is, then, the ratio of the relative ODs found for these two shots.

The ratio of ODs used to determine the ΔOD of samples not prepulsed was developed using two shots, one without and one with excitation. The samples were dark-adapted for 15 min before each shot. The ΔOD for prepulsed samples was also calculated from the results of two shots, one without and one with excitation. However, after dark adaptation, the sample was illuminated with a dye laser pulse 10 s before measurements.

In this manner the ΔODs of samples in the presence and absence of a prepulse could be compared without interference from cytochrome and DAD changes arising from the prepulse. In spite of this care there was a difference in the residual ΔOD changes seen for the prepulsed and the nonprepulsed samples at 700 ps. In each case the OD measured at 700 ps was chosen as the base line for the calculation of lifetimes. One explanation for this relatively constant background is that it represents differences in the absorption spectrum of reaction centers containing $Q_I - Q_{II}$ and those containing $Q_I^- Q_{II}^-$.

The use of a nonlinear curve-fitting routine is standard in calculating exponential lifetimes. It favors points with the largest signal-to-noise ratio, providing a more accurate lifetime than a linear least squares analysis. In any case a linear least squares fit gives essentially the same lifetime as the nonlinear treatment, but the confidence limits are larger.

RENTZEPIS: Second question. What has the Arrhenius equation in your results section ($K = \eta e^{-(\Delta v/RT)}$) to do with the discussion preceding it?

KAUFMANN: Although of the form of an Arrhenius equation, this equation is the high-temperature limit for the rate of electron transfer derived by R. A. Marcus (ref. 18).

RENTZEPIS: Is it possible to make the electrostatic calculation of the last sentences in your Results section more comprehensible?

KAUFMANN: The electrostatic calculation was made to see if a simple change in the free energy of the transition state due to the electric field could account for the factor of two change in rate. It is not worth belaboring the electrostatic model, because the recent data of Rentzepis and co-workers now show that the transfer rate (in the absence of the electrostatic field) is independent of temperature down to 4°K. Therefore a much more sophisticated theory is needed to explain our results.

RENTZEPIS: What is the difference between Q_I and Q_{II}?

KAUFMANN: They are both believed to be quinone. In our experiments we purify the reaction centers and then add excess ubiquinone to be sure that we get association of the secondary quinone with the reaction centers. It is believed now that in vivo there indeed are two quinones associated with the reaction center. George Feher has demonstrated that both quinones are ubiquinones, at least in *Rhodopseudomonas sphaeroides*. It may be different in other species.

WELLER: I hate to give my name because this question may show my ignorance. Do you have one quinone per bacteriopheophytin, or a whole pool of quinones?

KAUFMANN: The excited bacteriochlorophyll dimer transfers an electron to bacteriopheophytin, which then transfers the electron to the primary quinone.

WELLER: But that is not your Q_{II}.

KAUFMANN: No, Q_I donates an electron to Q_{II}. Then there is a quinone pool, and it is believed that this two-electron gate couples to this quinone pool. The Q_{II} part of this work was pioneered by Colin Wraight at the University of Illinois, as well as Clayton and Vermiglio at Cornell. The belief is that there is a two-electron gate. Such oscillations are also seen in chromatophores, but only under certain redox and pH conditions. There is a cooperative phenomenon between reaction centers to donate their electrons to form sets of two.

MAUZERALL: Is it possible to see the two electrons reduce Q at any time—the diamagnetic species?

KAUFMANN: There are three ways one may observe those Qs. First is looking at the electron spin resonance (ESR) signal. The second way is to look around 430 nm for the semi-quinone anion signal or at 280 nm for the disappearance of the oxidized form. The third way is to look for electrochromic shifts in the bacteriopheophytins and the bacteriochlorophylls. And actually the degree of the shift is different for the different quinones—the shift is different for the electron on the Q_{II} than it is on the Q_I. Vermiglio has reported a fully reduced quinone species after even flashes, but there is no certainty that this is Q_{II} or a bulk quinone.

WELLER: I have a short comment. On the basis of our recent interpretation given for the magnetic field effect observed with bacterial photosynthetic reaction centers, I should like to point out that the term "charge transfer complex" for the primary reaction product is misleading because it implies the formation of a molecular sandwich-type complex whose exchange interaction ($\gamma > 10^{-3}$ eV) would be much too high for the hyperfine interaction to be operative.

The primary product formed should rather be called a "radical ion pair." The center-to-center distance of such a pair is expected to be at least 7 Å, because otherwise the exchange interaction never could be small enough.

KAUFMAN: I agree—that's perfectly fine nomenclature.

RENTZEPIS: Actually, what you said is correct; and most of the time when it is written in papers, you will see it written as a radical ion pair, i.e., $(BCh)_2^{-}$, BPh^{-}, written that way for expediency. With respect to the distance that you said, we have data in *Biophysical Journal* (1978. **23**:207.), which shows the BPh–Q distance to be 9–13 Å and the $(BCHl)_2$–BPh somewhat shorter.

KAUFMANN: We have some preliminary data on artificial reaction centers with a chlorophyll dimer pheophytin covalently linked to a dimer of chlorophyll. The distance may be as much as 15 Å, depending on the configuration, yet electron transfer appears to take place. We have not yet measured the rate of electron transfer. But it does appear to quench the fluorescence most efficiently.

RUBIN: I want to remind you that earlier I reported that if we oxidize a bacteriochlorophyll dimer, we increase the lifetime: we decrease the speed of the electron transfer from bacteriopheophytin to Q_I. The reduction of the Q_{II} also influences the speed of the electron transport from the bacteriopheophytin to the Q_I. The electric field within the reaction center has a strong influence on the probability of the speed of the electron transport between different components that organize the reaction center.

RENTZEPIS: Do you think of this transfer between the two as a tunneling process? Do you think it was electron tunneling?

KAUFMANN: I would have to agree, I think that electron tunneling is a general phenomenon. People forget what happens to the nuclei. They also may tunnel.

RENTZEPIS: I think that is a very nice answer, especially in view of our results.

EXTENDED ABSTRACTS

THE REACTION OF "BLUE" COPPER OXIDASES WITH O_2
A PULSE RADIOLYSIS STUDY

M. GOLDBERG AND I. PECHT, *Department of Chemical Immunology, the Weizmann Institute of Science, Rehovot, Israel.*

The mechanism of the reduction of O_2 to H_2O by oxidase enzymes is of particular chemical interest due to the ability of these enzymes to solve efficiently a number of major energetic and mechanistic problems. These include: the unfavorable first reduction step $O_2 + e^- \rightarrow O_2^-$, which is largely responsible for the "kinetic inertness" of O_2; the production of a potentially harmful intermediate species such as the OH radical; and the multiequivalent nature of the O_2 reduction taken as a whole. The latter process, which catalyzed by blue copper oxidases or cytochrome oxidase, takes place without the release of intermediate products, and is believed to involve multi-electron transfers. The first step in the reduction of the O_2 molecule is its binding to the enzyme. This raises the question as to the reduction state of the oxidase at which the interaction with O_2 takes place.

We have studied the blue copper oxidases tree laccase and human ceruloplasmin in this respect, using the large perturbation method of pulse radiolysis (Fig. 1). Blue copper oxidases contain four (laccase) or more (ceruloplasmin) Cu ions, bound in three different types of redox sites (type 1, "blue" copper absorbing around 600 nm, embedded deeply inside the protein; type 2, accessible to solvent and external ligands; type 3, pair of Cu ions, coupled in the native-oxidized state, accessible to external ligands). Controlled partial reduction of the oxidase in the presence of O_2 was achieved by pulse radiolysis of O_2-containing solutions, where within microseconds and in the presence of an OH scavenger (*t*-butanol) a part or all of the initially produced reducing species (e_{aq}^-, H) are converted into O_2^-. Both e_{aq}^- and O_2^- reduce the copper oxidases, as evidenced by the decrease of the type 1 Cu(II) absorption. The presence of O_2 then leads to the prompt reoxidation of this site.

Two distinct processes of reduction and reoxidation of type 1 Cu take place. This was most clearly demonstrated in experiments with laccase, where full reoxidation was also observed at low O_2 concentrations, conditions where most of the type 1 Cu(II)

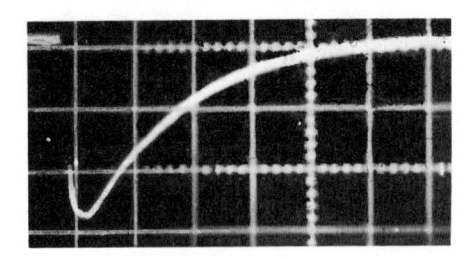

FIGURE 1

reduction is due to e_{aq}^-. However, when laccase or ceruloplasmin were reduced by e_{aq}^- or CO_2^- radicals under anaerobic conditions, no such reoxidation occurred. Nor was reoxidation observed with anaerobic solutions containing H_2O_2 (decay product of O_2^-), thus excluding this species as the oxidant.

The reduction of laccase type 1 Cu(II) by O_2^-. The absorption decrease was first-order under all conditions used. The observed rate constant, k_{red}, varied with the initial O_2^- concentration according to $k_{red} = 5 \times 10^1 + 3 \times 10^7 [O_2]_i$ s^{-1} (~25°C, pH ~6.6 – 6.8, $I < 0.001$ M). The small, but similar extent of reduction found for both O_2 and e_{aq}^- —≤3% of total type 1 Cu(II) reduced even at large excess of reductant ($I < 0.001$ M)—was explained in terms of most of the laccase being present in an "inactive state," which undergoes a transition to the "active" state at a slower rate than the self-decay of the reductants.

The reoxidation of laccase type 1 Ci(I) in the presence of O_2. (a) The absorption increase was first-order under all conditions used. The observed rate constant, k_{ox}, was independent of the O_2 or any other concentration; thus reoxidation is a true first-order process. (b) k_{ox} was 8.3 ± 0.4 s^{-1} (32 pulses) for a protein sample in all experiments where full reoxidation was obtained. In other protein samples where the relative extent of reoxidation varied from 60 to 100%, k_{ox} was 2.8 ± 1.0 s^{-1} (109 pulses). The degree of reoxidation did not depend on the O_2 concentration or on the amount of initially produced reducing equivalents. (c) In particular, full reoxidation was also obtained when $([e_{aq}^-] + [H])_i <$ [laccase], i.e. when almost no laccase molecules had accepted more than one electron. Furthermore k_{ox} was insensitive to the ratio $([e_{aq}^-] + [H])_i/$[laccase]. (d) F$^-$ ions (2.5 mM), known to bind to type 2 Cu(II) with a concomitant effect on the redox properties of the type 3 site, strongly inhibited the reoxidation, lowering k_{ox} by more than one order of magnitude. Also HCCO$^-$ exerted an inhibitory effect on the reoxidation process. Essentially similar results were obtained with ceruloplasmin.

Discussion and conclusions. Since the type 1 site is reoxidized only in the presence of O_2, it follows from (c) that laccase in the singly reduced state is able to interact with O_2. This agrees with other findings showing that a long-lived structural change is induced in the enzyme by the reaction of singly reduced laccase with O_2 (to be published). Since under turnover conditions the laccase molecule assumes the same state, as judged from spectral data, the catalytic pathway might involve binding of O_2 to singly reduced laccase.

It is widely accepted that the type 1 Cu is inaccessible to external ligands, and that there exist intramolecular redox equilibria between the different copper sites. Based on this and on (a), we propose that O_2 reacts with a solvent accessible redox site (type 2 Cu(I) or half-reduced type 3 site), in an electron exchange equilibrium with the type 1 site:[1]

$$\begin{Bmatrix} \text{type 1 Cu(I)} \\ X(\text{ox}) \end{Bmatrix} \underset{k_{-1}}{\overset{k_1}{\rightleftharpoons}} \begin{Bmatrix} \text{type 1 Cu(II)} \\ X(\text{red}) \end{Bmatrix} + O_2 \xrightarrow{k_2} \begin{Bmatrix} \text{type 1 Cu(II)} \\ X(\text{red}):O_2 \end{Bmatrix} \downarrow$$

$$K_1 = k_1/k_{-1} \ll 0.1$$

X = type 2 or type 3 site.

The rate-determining step is probably the intramolecular electron transfer from the type 1 site to the site of O_2 interaction ($k_{ox} = k_1 \ll k_2 [O_2]$). The fluoride effect is consistent with the type 2 site being involved directly or indirectly, e.g. via allosteric modulation of the type 3 electron acceptor properties.

---

[1] A complex and as yet unresolved transient pattern was observed around 330 nm (type 3 band).

VOLTAGE-INDUCED CHANGES IN THE CONDUCTIVITY OF ERYTHROCYTE MEMBRANES

KAZUHIKO KINOSITA, JR., AND TIAN YOW TSONG, *Department of Physiological Chemistry, The Johns Hopkins University School of Medicine, Baltimore, Maryland 21205 U. S. A.*

Previous reports (1-3) have shown that exposure of an isotonic suspension of erythrocytes to an electric field of a few kilovolts per centimeter for a duration in microseconds dramatically increases the permeability of the cell membranes. As a result, the erythrocytes undergo hemolysis through colloid osmotic swelling. Since the enhanced permeability is limited to small ions or molecules, the effect has been attributed to the formation of aqueous pores in the membranes. These pores are formed when the transmembrane potential induced by the externally applied field exceeds 1 V. The effective radius of the pores is several Ångstroms and can be varied by the adjustment of field intensity, field duration, or the ionic strength of the medium. Here we report a study of the kinetics of the pore formation, where the increase in permeability was detected by conductivity measurements.

Erythrocytes suspended either in isotonic NaCl or in a 1:9 mixture of isotonic NaCl

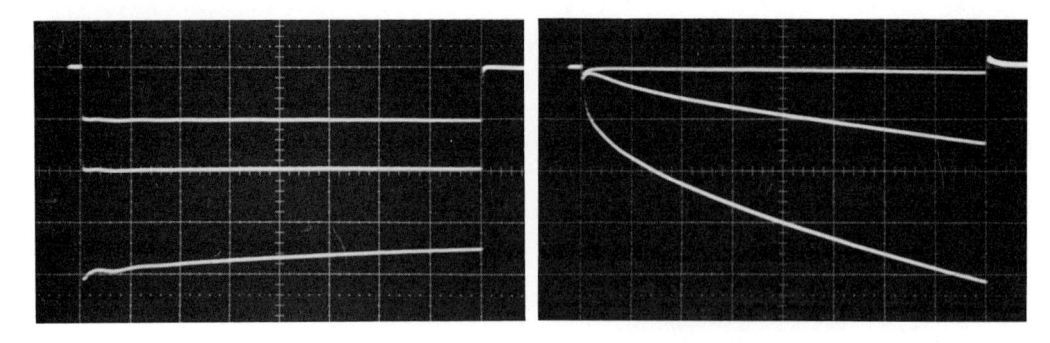

FIGURE 1 Increase in the conductivity of an erythrocyte suspension (20% vol/vol in isotonic NaCl) during voltage pulsation. Left traces, electric field (1 kV/cm per major div.); right traces, relative conductivity increment (5% per div.); time scale, 10 μs/div. Both traces are positive downward. Similar changes in conductivity, but with approximately three times greater amplitudes, were observed for the NaCl-sucrose medium. Part of the conductivity increase in NaCl is due to the Joule heating.

and sucrose solutions were exposed to a square-wave voltage-pulse, and change in the conductivity (current density divided by field intensity) of the total suspension was measured with a differential circuit. As shown in Fig. 1, time-dependent increases in the conductivity were observed when field intensity exceeded about 1.2 kV/cm. This value coincides with the intensity at which the most susceptible cells in the suspension are perforated with the pores under an 80-μs pulse (3). Thus, the observed increase in conductivity is due to the formation of pores through which current can penetrate the cell interior. At field intensities where most of the cell population undergoes the perforation ($\gtrsim$3 kV/cm), the time-course of the conductivity change consists of a rapid ($\approx$1 μs) and a slow ($\approx$100 μs) phase, as seen in the bottom traces of Fig. 1. The rapid phase corresponds to initial perforation, while the slow phase probably detects the expansion of the pores (3). Although an increase in conductivity might be expected if the cells reoriented or underwent a shape change, erythrocytes made spherical by an external agent or by hypotonic conditions gave essentially similar results. When a linearly increasing field was applied to the suspension, the conductivity started to increase at a field intensity of about 1.2 kV/cm, indicating that the cell membranes respond to the field intensity itself and not to its time derivatives.

Conductivity of the suspension after the removal of the applied field was estimated either by applying a second pulse or by a conventional measurement using a small-amplitude alternating voltage. In the NaCl-sucrose medium, the results are summarized as follows: (a) under an 80-μs pulse of 3.1 kV/cm, conductivity increased by 60%. (b) After the field was removed, the conductivity returned toward the original value until about 100 ms. (c) It then sharply rose again to an almost 100% increase at 1.5 min. (d) Thereafter it gradually decreased over a period of 1 h. Comparison with the previous results (1, 3) indicates that phase c is due to the leakage of intracellular ions into the low-ionic-strength external medium, and that the decrease in phase d is due to the swelling of the cells, which reduced the volume of extracellular space. Phase b is tentatively attributed to the shrinkage of widely opened pores.

Results in isotonic NaCl were qualitatively similar, except that the increase in phase c was very small, or almost absent.

This work was supported by a National Institutes of Health grant.

REFERENCES

1. KINOSITA, K., JR., and T. Y. TSONG. 1977. *Proc. Natl. Acad. Sci. U.S.A.* **74**:1923-1927.
2. KINOSITA, K., JR., and T. Y. TSONG. 1977. *Nature (Lond.).* **268**:438-441.
3. KINOSITA, K., JR., and T. Y. TSONG. 1977. *Biochim. Biophys. Acta.* **471**:227-242.

FAST BIOCHEMICAL REACTIONS IN THIN FILMS INDUCED BY NUCLEAR FISSION FRAGMENTS

R. D. MACFARLANE, *Department of Chemistry, Texas A&M University, College Station, Texas 77843 U. S. A.*

Fission track dynamics in thin films. The passage of nuclear fission fragments through thin solid films produces a fission track characterized by a high power density ($\sim 10^{13}$ W/cm^2), a diameter of ~ 100 Å, and a length of ~ 10 μm (1). Measurement of the kinetic energy and angular distribution of ion (2) emitted from the track give evidence for the formation of a superradiant state containing a high density of molecules simultaneously excited by the intense electromagnetic field associated with the fission fragment (3). The radiation emitted in two narrow cones at 0° and 180° to the fission fragment direction develops strong hypersonic pulses by stimulated Brillouin scattering (4). These nonlinear processes have two effects: molecules in the fission track are electronically and vibronically excited and can undergo fast chemical reactions in the excited state, and reaction products formed on the surface have a high probability of being emitted when the hypersonic pulse reaches the surface. If the reaction products are charged, they can be characterized by mass spectrometry.

Electron transfer reactions. This reaction has been studied with chlorophyll-*a* as a model compound. Molecular aggregates of chlorophyll produce a singlet exciton state characterized by an electron exchange current within the aggregate (5). We have verified this directly by observing the breakup of this state into (Chl-a)$^+$ and (Chl-a)$^-$ ions. We also have detected dimers, trimers, and tetramers of chlorophyll and have shown that the presence of Mg is essential for the stability of oligomer formation.

Proton transfer reactions. Molecules that form aggregates in the solid state, mediated by hydrogen bonding where the H$^+$ is weakly bound (acidic), give evidence of proton charge transfer in the singlet exciton states. Reaction products emitted from those states included (M + H)$^+$ and (M - H)$^-$ ions. Amino acids, peptides, and small oligonucleotides exhibit this reaction.

Cationization. Molecules forming aggregates that do not produce a charge delocalization do not form separated ion pairs in the excited state. The reactions that do occur involve charge exchange ($-OH + Na^+ \rightarrow -ONa + H^+$) and cation attachment ($M + Li^+ \rightarrow MLi^+$). Alpha cyclodextrin has been a good model molecule for these studies. These processes occur for large biomolecules that do not form charge transfer bonds, either because of absence of acidic hydrogens or random orientation of molecular aggregates. Maytansine, a tumor inhibitor that contains sugar moieties and a peptide chain, is an example of a molecule that undergoes this reaction.

Summary. Fission fragments produce tracks in thin films of biological molecules that form a superradiant state because of the high excitation density. Molecular excitation is similar to picosecond laser irradiation, in that nonlinear effects are observed. The added feature is that the acoustic pulse that follows the excitation gives a means of directly identifying products of fast chemical reactions that occur.

REFERENCES

1. MACFARLANE, R. D., and D. F. TORGERSON. 1976. Californium-252 plasma desorption mass spectroscopy. *Science (Wash. D.C.).* **191**:920–925.
2. FURSTENAU, N., W. KNIPPELBERG, F. R. KRUEGER, G. WEIZ, and K. WIEN. 1977. Experimental investigation about the mechanism of fission-fragment induced desorption. *Z. Naturforsch.* **32a**:711–719.
3. MACGILLIVRAY, J. C., and M. S. FELD. 1976. Theory of superradiance in an extended, optically thick medium. *Phys. Rev.* **14**:1169–1188.
4. CHIAO, R. Y., C. H. TOWNES, and F. P. STOICHEFF. 1964. Stimulated brillouin scattering and coherent generation of intense hypersonic waves. *Phys. Rev. Lett.* **12**:592–595.
5. SHIPMAN, L. L., J. R. NORRIS, and J. J. KATZ. 1976. Quantum mechanical formalism for computation of the electronic spectral properties of chlorophyll aggregates. *J. Phys. Chem.* **80**:877–881.

NONHOMOGENOUS CHEMICAL KINETICS IN PULSED PROTON RADIOLUMINESCENCE

JOHN HOWARD MILLER AND MARTIN L. WEST, *Battelle Northwest Laboratories, Richland, Washington 99352 U. S. A.*

The spatial distribution of absorbed energy influences the yield of chemically active species in a medium exposed to radiation. Diffusive motion of molecules rapidly destroys the initial distribution of primary species in most liquids; consequently, stroboscopic techniques must be used to observe the effects of nonhomogenous chemical kinetics directly. Pulsed radiolysis with electrons has been used extensively to prove the subnanosecond time region. However, the application of this method to other types of radiation, in particular radiations with high linear energy transfer (LET), is limited by the requirement of a large dose per pulse to achieve significant absorption by the chemical species under investigation (1). The use of fluorescence, rather than absorption, to detect the presence of chemical species circumvents this difficulty. By time-resolve emission spectroscopy, the evolution of a small population of excited states can be studied under varied radiation conditions with subnanosecond time resolution. Through fluorescence quenching, the time evolution of the concentration of nonradiative species can also be investigated.

Samples of 0.04 M benzene in cyclohexane (Nanograde, Mallinckrodt Inc., St. Louis, Mo.) were deaerated by helium purging and placed in an irradiation cell consisting of a cylindrical stainless steel chamber with 2.5 µm Havar foil (Hamilton Technology, Inc., Lancaster, Pa.) for proton beam entrance and a quartz window for light collection. An adequate flow rate was maintained to prevent buildup of stable radiolysis products in the irradiated volume. A scanning spectrometer was used to select a 4-nm band centered about a wavelength of 285 nm. The liquid was irradiated with subnanosecond pulses of protons from a 2 MV accelerator with a 3.33 MHz high voltage radio frequency oscillator for beam chopping. Fluorescence decay was measured by the single photon counting method with a signal derived from the chopper voltage providing zero time reference. Overall time resolution of the system is limited to about 1.5 ns full width at half maximum by response of the photomultiplier. The sample reservoir and irradiation cell were maintained to ±0.2°C over a temperature range of 15–50°C by means of a thermistor-controlled cooling-heating system.

Fig. 1 compares time-resolved emission from benzene in cyclohexane excited by pulsed proton irradiation at three energies with the fluorescence decay observed with ultraviolet (UV) irradiation. The nonexponential character of the fluorescence decay

FIGURE 1

FIGURE 2

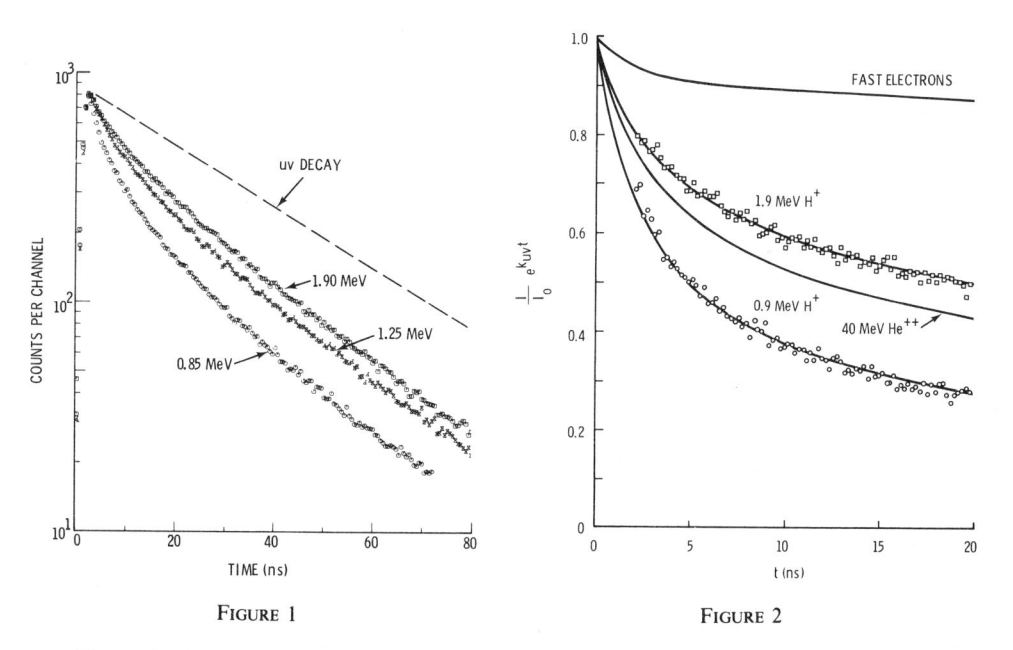

FIGURE 1 Fluorescence of benzene in cyclohexane irradiated with protons at several energies. Dashed curve represents exponential decay observed with UV irradiation.

FIGURE 2 Fluorescence predicted by the intratrack quenching model for several types of radiation. Ordinate is the ratio of fluorescence intensity with ionizing and UV irradiation.

Extended Abstracts

with proton irradiation results from a time-dependent concentration of quenching species not present under UV excitation (2). To develop a model for this effect, we assume that excited singlet states of benzene and the radiochemical quenchers are formed in spurs centered about the initial sites of ionization of the liquid. The distribution of chemical species within each spur is assumed to be Gaussian with a variance $\sigma^2(t) = r_0^2 + 2Dt$, where r_0 is the initial spur radius and D is the diffusion coefficient of quenchers and excited states. The model predicts a fluorescence decay of the form

$$-\frac{1}{N_*}\frac{dN_*}{dt} = k_{UV} + k_{spur}(1 + \langle g(\mathbf{r};t)\rangle), \qquad (1)$$

where k_{UV} is the quenching rate by solvent perturbations observed with UV irradiation, k_{spur} is the rate of radiochemical quenching in isolated spurs, and

$$g(\mathbf{r};t) = \sum_{\mathbf{r}_j \neq \mathbf{r}} \exp(-(\mathbf{r} - \mathbf{r}_j)^2/4\sigma^2(t)), \qquad (2)$$

is the probability that an excited state at position $\mathbf{r}$ will be quenched by radiochemical species from other sites of ionization. We call this the "spur overlap."

Due to the low beam fluence, the probability of chemical reaction between species formed in different proton tracks is small during the time interval over which the fluorescence decay is observed. Hence, we refer to the quenching by radiochemical species as an "intratrack" quenching. As a first approximation, we may neglect the radial dispersion of ionization in the proton track and use the result of Ganguly and Magee (3) to estimate the spur overlap. In this approximation, the predicted fluorescence decay is

$$-\frac{1}{N_*}\frac{dN_*}{dt} = k_{UV} + \frac{k_Q G_Q}{8\sigma^3(t)} + \frac{k_Q G_Q S}{4\sigma^2(t) W}, \qquad (3)$$

where k_Q is the second-order quenching rate constant, G_Q is the yield of radiochemical quenchers, S is the stopping power, and W is the energy absorbed per ion. Fig. 2 illustrates the amount of intratrack quenching predicted by Eq. 3 for various types of radiation. Parameters of the model were determined by best fit to the data for 1.9 MeV protons. For fast electrons, the last term in Eq. 3 is small and the fluorescence decay is nearly exponential beyond 5 ns. For proton irradiation from our 2 MV accelerator, the last term in Eq. 3 predominates. Fig. 2 shows that without further adjustment of parameters, the model correctly predicts the fluorescence decay with 0.9 MeV protons.

The fluorescence decay predicted by Eq. 3 for 40 MeV alpha particle irradiation is also shown in Fig. 2. For these energetic heavy ions, transport of energy away from the track core by high-energy secondary electrons should result in a more diffuse pattern of energy deposition than for proton irradiation of the same stopping power. Hence, Eq. 3 may overestimate the intratrack quenching in this case. Calculations are currently in progress to estimate the effect of the radial distribution of ionization on the fluorescence decay. Monte Carlo techniques are being used to determine the initial spatial distribution of spurs, and the spur overlap is calculated from Eq. 2. The results of these calculations will be compared with measurements of the fluorescence decay with proton and alpha irradiation at the same stopping power. These experiments should provide a crucial test of our intratrack quenching model.

This paper is based on work performed under U. S. Department of Energy (formerly Energy Research and Development Administration) Contract EY-76-C-06-1830.

REFERENCES

1. MATHESON, M. S., and L. M. DORFMAN. 1969. Pulsed Radiolysis. The M.I.T. Press, Cambridge, Mass.
2. MILLER, J. H., and M. L. WEST. 1977. Quenching of benzene fluorescence in pulsed proton irradiation: dependence on proton energy. *J. Chem. Phys.* **67**:2793.
3. GANGULY, A. K., and J. L. MAGEE. 1956. Theory of radiation chemistry. III. Radical reaction mechanisms in tracks of ionizing radiation. *J. Chem. Phys.* **25**:129.

PICOSECOND PHOTODISSOCIATION AND SUBSEQUENT RECOMBINATION PROCESSES IN CARBOXYHEMOGLOBIN, CARBOXYMYOGLOBIN, AND OXYMYOGLOBIN

L. J. NOE, W. G. EISERT, AND P. M. RENTZEPIS, *Bell Laboratories, Murray Hill, New Jersey 07974 U. S. A.*

The central problem in the study of hemoglobin is to understand the mechanism of cooperativity among the four subunits of the molecule. This cooperativity is evident in the sigmoidal nature of the oxygen saturation (equilibrium) curve and in the Bohr effect. Paramount in the understanding of cooperativity is the study of the trigger mechanism of ligand release and the tertiary and quaternary protein structural changes subsequent to this. Picosecond time-resolved spectroscopy is a relatively new experimental method capable of critically examining the dynamics of ultrafast molecular events in proteins—processes serving as precursors to other extensive protein structural changes. The application of this method with regard to photodissociation measurements on heme proteins permits one to obtain rate data on such processes in a heretofore inaccessible time region. This information, in turn, allows an analysis of the allosteric mechanism(s) of cooperativity from a totally different experimental perspective.

Dr. Noe is on leave of absence at the Department of Chemistry, University of Wyoming, Laramie, Wyoming 82070; Dr. Eisert's permanent address is: Institut für Strahlenbotanik, 3000 Hannover-Herrenhausen, Herrenhauser Str. 2, Germany.

Since it is of interest to detect the primary structural events cited above, we felt that a picosecond kinetic study of carboxymyoglobin (MyCO) and oxymyoglobin (MyO$_2$) would be desirable for several reasons. First, it would provide information on protein structural relaxations immediately after photodissociation (known to be facile [1], ≤10 ps) in a picosecond time window where such changes are likely to be seen. Second, the structural motions are limited to those of the tertiary type confined to one heme unit, thereby simplifying the interpretation of the effect. In other words, the interpretation of the relaxation results for tetrameric deoxyhemoglobin is not straightforward because of the necessity of differentiating between tertiary and quaternary protein motions and of accounting for the relaxation of individual subunits. Finally, picosecond photodissociation experiments on MyO$_2$ and MyCO should prove to be of great value since the electronic destabilization of the heme pocket caused by facile ligand detachment must be followed by the evolution of tertiary structural changes until a stable form of the deoxyhemoprotein results, the extent of and the type of destabilization reflecting the differences in the bonding of O$_2$ and CO to the heme.

Excitation of MyCO and MyO$_2$ by a single 530 nm, Nd^{+3}-glass laser, 8-ps pulse results in photodissociation monitored by following absorption changes at 440 nm in the Soret band as a function of time (2). Photodissociation occurs in less than 6 ps for both molecules. MyCO shows an additional decay immediately after photodissociation, having a rate $(6.7 \pm 2.7) \times 10^9$ s^{-1} ($\tau = 130 \pm 50$ ps) to a steady state that is the stable form of deoxymyoglobin. The presence of such relaxation in MyCO and the absence of relaxation in MyO$_2$ is a strong indication that the electronic destabilization after ligand detachment is much greater for MyCO than for MyO$_2$. Based on existing theory (3, 4), we attribute the relaxation in MyCO to tertiary structural changes in the heme pocket.

Because the photodissociation of HbCO provides information on the trigger mechanism for the cooperative binding of ligands to Hb (*vide supra*), we felt that a more complete, wide-range, kinetic investigation of the dissociation and recombination in HbCO would complement the myoglobin study. Excitation of carbon monoxide hemoglobin, HbCO, by a single 530-nm, 6-ps pulse results in photodissociation with a first-order constant of 0.89×10^{11} s^{-1} ($\tau = 11$ ps). The kinetics of photodissociation, monitored by following absorbance changes in the Soret band at 440 nm, are interpreted as corresponding to predissociation followed by a crossing into a dissociative state. Subsequent recombination of CO with the porphyrin system and protein structural transformations are monitored by use of a continuous He-Cd laser beam spacially coincident with the photolysis and Soret interrogation beams at the sample. We find that the latter events take place in three distinct time regions depending on excitation pulse energy and repetition rate. Excitation of HbCO with a single pulse (0.8–5 mJ) results in a relaxation to the ground state with an associative first-order constant of 5×10^3 s^{-1}. With a 100-pulse train (~7.5 mJ), a new decay grows with a rate constant of 63 s^{-1}. For a pulse train energy of 12 mJ or higher, a delay occurs at the onset of the second (slower) recombination. Our pulse train experiments, depending on the pulse energy, reflect higher degrees of complete photodissociation of the heme sites, followed by possible tertiary and quaternary structural changes on recombination.

An important conclusion is that structural inferences derived from our picosecond experiments on MyCO, and particularly those on HbCO with regard to conformational and tertiary content, may not necessarily be transferrable to the conjugate molecules, MyO_2 and HbO_2, respectively.

REFERENCES

1. NOE, L. J., W. G. EISERT, and P. M. RENTZEPIS. 1978. Picosecond photodissociation and subsequent recombination processes in carbon monoxide hemoglobin. *Proc. Natl. Acad. Sci. U.S.A.* In press.
2. EISERT, W. G., L. J. NOE, and P. M. RENTZEPIS. 1978. Tertiary structural changes in carboxymyoglobin and oxymoglobin studied by picosecond spectroscopy. *Biophys. J.* In press.
3. OLAFSON, B. D., and W. A. GODDARD, III. 1977. Molecular description of dioxygen bonding in hemoglobin. *Proc. Natl. Acad. Sci. U.S.A.* **74**:1315–1319.
4. GODDARD, W. A., III, and B. D. OLAFSON. 1975. Ozone model for bonding of O_2 to heme in oxyhemoglobin. *Proc. Natl. Acad. Sci. U.S.A.* **72**:2335–2339.

TIME-RESOLVED MAGNETIC SUSCEPTIBILITY

A NEW METHOD FOR FAST REACTIONS IN SOLUTION

J. S. PHILO, *Department of Physics, Stanford University, Stanford, California 94305 U.S.A.*

The high sensitivity and rapid response of superconducting magnetometers now make possible the measurement of changes in magnetic susceptibility during fast biochemical reactions in room-temperature samples. This new method has been demonstrated by measuring the recombination kinetics of hemoglobin and carbon monoxide after flash photolysis of HbCO at 20°C. The rate constants so determined are in excellent agreement with those obtained by photometric techniques.

A unique capability of this method is the determination of the magnetic susceptibilities of short-lived reaction intermediates. In partial photolysis experiments the magnetic moment of the intermediate species $Hb_4(CO)_3$ was found to be 4.9 ± 0.1 μB in phosphate buffer. This value compares to 5.3 μB per heme for deoxyhemoglobin under these conditions (1), and the difference indicates the change in quaternary structure when three ligands are bound. It is important to note that this method can be used to determine magnetic moments of reaction intermediates such as ferrous heme, which are paramagnetic but show no electron spin resonance signals.

Major improvements in the sensitivity and time resolution of this technique are expected. At present the time resolution is limited by eddy currents to 300 μs, and the

---

Dr. Philo's present address is: Superconducting Technology Division, United Scientific Corp., Mountain View, California 94043.

noise level is equivalent to a change in concentration of a spin half species of 7×10^{-7} $M(Hz)^{-1/2}$. Both higher sensitivity and 1-μs time resolution are possible with existing technology. Furthermore, in the near future magnetometers with nanosecond time resolution are expected. Such improvements should greatly extend the utility of this method for studies on biochemical systems.

REFERENCE

1. ALPERT, Y., and R. BANERJEE. 1975. *Biochim. Biophys. Acta.* **405**:144.

CARBOXYLATION KINETICS OF HEMOGLOBIN AND MYOGLOBIN

LINEAR TRANSIENT RESPONSE TO STEP PERTURBATION BY LASER PHOTOLYSIS

D. D. SCHURESKO AND W. W. WEBB, *School of Applied and Engineering Physics, Cornell University, Ithaca, New York 14853 U.S.A.*

The photochemical kinetics of the reactions of myoglobin and hemoglobin with carbon monoxide, in the time domain 10^{-4}–10^2 s, have been measured with high precision using a step-modulated continuous wave argon ion laser as the photolytic source. A steady state of the chemical system is fixed by the DC component of the amplitude-modulated laser beam; the oscillatory (square wave) component of the beam induces small perturbations of this steady state. The system's CO binding response is followed by monitoring optical absorbance at 435.6 nm. Digital transient recording on a quasi-logarithmic time scale enables single-sweep measurement of a decay with (typically) 10 decades of rate resolution; transient averaging provides the desired signal enhancement. Preparation of controlled protein-CO solutions was carried out in a closed mixing cell integrally connected with an optical measurement cell, designed to overcome sample heating and convection.

The linear kinetic response to small perturbations consists of a superposition of a set of eigenmodes, each an exponential relaxation. The CO-myoglobin response transients are single-mode (single exponential) in character; the rate constants vary linearly with DC laser intensity and free CO concentration, yielding at 20°C $k_F = 2.12 \times 10^5 M^{-1}s^{-1}$ for the combination kinetic constant, and $Q = 0.97 \pm 0.04$ for the photolytic quantum yield. The hemoglobin kinetic response transients are multi-component in character. They have been adequately fitted, using nonlinear least squares fitting methods, by a response function consisting of the sum of three exponentials.[1] The two

---

Dr. Schuresko's present address is: Oak Ridge National Laboratory, Oak Ridge, Tenn. 37830.

[1] Additional eigenmodes which contribute less than 5% of the total transient amplitude are not detectable by our analysis.

fast relaxations have rate constants which vary approximately linearly with the free CO concentration and the DC laser intensity; the slow relaxation rate constant is independent of both variables. The amplitude of the faster of the two CO-bonding modes predominates at both ligand saturation extremes, as has been previously observed for O_2-hemoglobin (1). Computer eigenmode calculations based on either simple sequential or allosteric schemes fail to reproduce all of this fitted data; hybrid models featuring weakly coupled parallel reaction pathways do reproduce the observed data features.

We wish to thank Professor Q. H. Gibson of Cornell for his suggestions and encouragement.

REFERENCE

1. ILGENFRITZ, G., and T. M. SCHUSTER. 1974. *J. Biol. Chem.* **239**:2959-73.

List of Participants

Indexes

LIST OF PARTICIPANTS

SAYED AMEEN
Formerly at Max-Planck-Institute for Biophysical Chemistry, Göttingen, W. Germany

ALAN AQUALINO
Physics Department, University of Virginia, Charlottesville, Va. 22901

R. H. AUSTIN
Max-Planck-Institute for Biophysical Chemistry, Göttingen, W. Germany

MUGUREL G. BADEA
Department of Biology, The Johns Hopkins University, Baltimore, Md. 21218

BODHAN BALKO
Laboratory of Technical Development, National Heart, Lung, and Blood Institute, National Institutes of Health, Bethesda, Md. 20014

DAVID P. BALLOU
Department of Biological Chemistry, University of Michigan, Ann Arbor, Mich. 48109

B. GEORGE BARISAS
Department of Biochemistry, St. Louis University School of Medicine, St. Louis, Mo. 63104

MARY D. BARKELEY
Department of Biochemistry, University of Kentucky Medical Center, Lexington, Ky. 40506

FRANCISCO J. BARRANTES
Department of Molecular Biology, Max-Planck-Institute of Molecular Biology am Fassberg, Göttingen, W. Germany

JOSEPH F. BECKER
Laboratory of Chemical Biodynamics, University of California at Berkeley, Berkeley, Calif. 94720

ROBERT L. BERGER
Laboratory of Technical Development, National Institutes of Health, Bethesda, Md. 20014.

SIDNEY A. BERNHARD
Institute of Molecular Biology, University of Oregon, Eugene, Or. 97403

YOAV BLATT
Department of Chemical Immunology, Weizmann Institute of Science, Rehovot, Israel

MANJA BLAZER
American Instrument Company, Silver Spring, Md. 20910

CELIA BONAVENTURA
Department of Biochemistry, Duke University Marine Laboratory, Beaufort, N.C. 28516

JOSEPH BONAVENTURA
Department of Biochemistry, Duke University Marine Laboratory, Beaufort, N.C. 28516

JOHN M. BREWER
Department of Biochemistry, University of Georgia, Athens, Georgia 30602

ARTHUR S. BRILL
Department of Physics, University of Virginia, Charlottesville, Va. 22901

RODNEY D. BROWN
IBM, T.J. Watson Center, Yorktown Heights, N.Y. 10598

PETER BUNGAY
Biomedical Engineering and Instrument Branch, National Institutes of Health, Bethesda, Md. 20014

JONATHAN B. CHAIRES
Biochemical and Biophysics, University of Connecticut, Storrs, Conn. 06268

SHIRLEY S. CHAN
Karl-Friedrich-Bonhoeffer Institute, Max-Planck-Institute for Biophysical Chemistry Göttingen-Nikolausberg, W. Germany

RAYMOND F. CHEN
National Heart, Lung and Blood Institute, National Institutes of Health, Bethesda, Md. 20014

COLIN F. CHIGNELL
Laboratory of Environmental Biophysics, National Institutes of Environmental Health Science, Research Triangle Park, N.C. 27709

P. BOON CHOCK
National Heart, Lung, and Blood Institute, National Institutes of Health, Bethesda, Md. 20014

PAUL W. CHUN
Department of Biochemistry, College of Medicine, University of Florida, Gainesville, Fl. 32610

WILLIAM CLARK
Spectra-Physics Inc., Mountain View, Calif. 94042

ROBERT CLEGG
Max-Planck-Institute for Biophysical Chemistry, Göttingen, W. Germany

ALAN H. COLEN
Laboratory of Molecular Biology, Veterans Administration Hospital, Kansas City, Mo. 64128

STANLEY CORRSIN
Department of Mechanics and Materials Science, Johns Hopkins University, Baltimore, Md. 21218

CHARLES L. COULTER
Division of Research Resources, National Institutes of Health, Bethesda, Md. 20014

EDWARD D. CRANDALL
Department of Physiology, University of Pennsylvania, Philadelphia, Pa. 19104

GEORGE H. CZERLINSKI
Department of Biochemistry, Northwestern University School of Medicine, Chicago, Ill. 60611

NANIBHUSHAN DATTAGUPTA
Department of Chemistry, Yale University, New Haven, Conn. 06520
NORMAN DAVIDS
Department of Engineering Science and Mechanics, Pennsylvania State University, University Park, Pa. 16802
DANIEL DERVARTANIAN
Department of Biochemistry, University of Georgia, Boyd Graduate Studies Research Center, Athens, Georgia 30602
ROBERT PAUL DETOMA
Department of Chemistry, University of Richmond, Richmond, Va. 23173
LEON M. DORFMAN
Department of Chemistry, The Ohio State University, Columbus, Ohio 43210

WILLIAM A. EATON
Laboratory of Chemical Physics, National Institute of Arthritis, Metabolism, and Digestive Diseases, National Institutes of Health, Bethesda, Md. 20014
BENJAMIN EHRENBERG
School of Applied and Engineering Physics, Cornell University, Ithaca, N.Y. 14853
LAURA EISENSTEIN
Physics Department, University of Illinois, Urbana, Ill. 61801

MOSHE FARAGGI
Chemistry Department, Nuclear Research Center, Negev, Beer Sheva, Israel
JAMES A. FEE
Institute of Science and Technology, The University of Michigan, Ann Arbor, Mich. 48105
FRANK A. FERRONE
Laboratory of Chemical Physics, National Institute of Arthritis, Metabolism, and Digestive Diseases, National Institutes of Health, Bethesda, Md. 20014
RICHARD W. FESSENDEN
Radiation Laboratory, University of Notre Dame, Notre Dame, Ind. 46556
DUANE P. FLAMIG
Department of Chemistry, University of Nebraska, Lincoln, Neb. 68588
DAVID C. FOYT
Center for Fast Kinetics Research, University of Texas, Austin, Tex. 78712
HANS FRAUENFELDER
Department of Physics, University of Illinois, Urbana, Ill. 61801
JEFFREY P. FROEHLICH
National Institutes of Health, National Institute on Aging, Baltimore City Hospitals, Baltimore, Md. 21224

NICHOLAS E. GEACINTOV
Chemistry Department, New York University, New York, N.Y. 10003

MARGARET W. GEIGER
Department of Molecular Biophysics and Biochemistry, Yale University, New Haven, Conn. 06520
BRUNO GIARDINA
Institute of Chemistry, Faculty of Medicine, University of Rome, Rome, Italy
M. C. GOODALL
Laboratory of Membrane Biology, University of Alabama, Birmingham, Ala. 35294
DIXIE J. GOSS
Department of Chemistry, University of Nebraska, Lincoln, Neb. 68588

HERBERT R. HALVORSON
Department of Biochemistry, Edsel B. Ford Institute, Detroit, Mich. 48202
EUGENE HAMORI
Department of Biochemistry, Tulane Medical School, New Orleans, La. 70112
RAYMOND E. HANSEN
Update Instrument Incorporated, Madison, Wis. 53719
DAVID M. HANSON
Department of Chemistry, State University of New York, Stony Brook, N.Y. 11794
LOUIS HELLEMANS
Department of Chemistry, University of Leuven, Heverlee, Belgium
GEORGE P. HESS
Cornell University, Ithaca, N.Y. 14853
TOSHIHIRO HIRAI
Department of Biochemistry, Tulane Medical School, New Orleans, La. 70112
H. J. HOFRICHTER
National Institutes of Health, Bethesda, Md. 20014
MICHAEL EDWARD HOGAN
Department of Chemistry, Yale University, New Haven, Conn. 06520
VIRGINIA H. HUXLEY
Department of Physiology, University of Virginia, Charlottesville, Va. 22901

M. K. JAIN
Department of Chemistry, University of Delaware, Newark, Del. 19711
ERIC JAKOBSSON
Department of Physiology and Biophysics, University of Illinois, Urbana, Ill. 61801
OLEG JARDETZKY
Magnetic Resonance Laboratory, Stanford University, Stanford, Calif. 94305

JOYCE J. KAUFMAN
Department of Chemistry, The Johns Hopkins University, Baltimore, Md. 21218
KENNETH J. KAUFMAN
Department of Biophysics, University of Illinois, Urbana, Ill. 61801

ROY KING
: Stanford Magnetic Resonance Laboratory, Stanford University, Stanford, Calif. 94305

DONALD S. KIRKPATRICK
: Department of Opthalmology, Baylor College of Medicine, Houston, Tex. 77025

LOUIS J. KIRSCHENBAUM
: Chemistry Department, University of Rhode Island, Kingston, R.I. 02881

MICHAEL H. KLAPPER
: Chemistry Department, Ohio State University, Columbus, Ohio 43210

VAUGHN J. KOESTER
: Physiology Department, University of Texas Health Sciences Center, Dallas, Tex. 75235

BRYAN E. KOHLER
: Department of Chemistry, Wesleyan University, Middletown, Conn. 06457

RUTH KOREN
: Biophysics Section-Botany, The Hebrew University, Jerusalem, Israel

MAURI E. KROUSE
: Department of Biology, California Institute of Technology, Pasadena, Calif. 91125

HOWARD KUTCHAI
: Physiology Department, University of Virginia Medical School, Charlottesville, Va. 22901

JOSEPH R. LAKOWICZ
: Department of Biochemistry, Freshwater Biological Institute, University of Minnesota, Navarre, Minn. 55392

DORON LANCET
: Harvard Biological Laboratory, Cambridge, Mass. 02138

GLEN LEHMAN
: Princeton Applied Research, Princeton, N.J. 08540

HENRY A. LESTER
: Biology Division, California Institute of Technology, Pasadena, Calif. 91125

AARON LEWIS
: Department of Applied Physics, Cornell University, Ithaca, N.Y. 14850

GERARD W. LIESEGANG
: Department of Chemistry, Harvard University, Cambridge, Mass. 02138

GEORGE W. LOWY
: American Instrument Co., Silver Spring, Md. 20910

JERRY LUCHINS
: Department of Biological Sciences, Columbia University, New York, N.Y. 10027

WILLIAM W. MANTULIN
: Baylor College of Medicine, Methodist Hospital, Houston, Tex. 77030

DAVID MAUZERALL
: The Rockefeller University, New York, N.Y. 10021

R. MENDELSON
: University of California Medical School, San Francisco, Calif. 94143

EUGENIE V. MIELCZAREK
: Laboratory of Technical Development, National Heart and Lung Institute, National Institutes of Health, Bethesda, Md. 20014

JOHN H. MILLER
: Battelle, Pacific North West Laboratory, Richland, Wash. 99352

HIROSHI MIZUKAMI
: Department of Biology, Wayne State University, Detroit, Mich. 48202

WILLIAM D. MOORHEAD
: Department of Physics, Youngstown State University, Youngstown, Ohio 44555

R. D. MACFARLANE
: Department of Chemistry, Texas A & M University, College Station, Tex. 77843

GREGORY MCCLUNE
: University of Michigan, Ann Arbor, Mich. 48109

JAMES A. MCCRAY
: Physics Department, Drexel University, Philadelphia, Pa. 19104

JAMES W. NASH
: Department of Physics, University of Maryland, College Park, Md. 20742

MENASCHE M. NASS
: Biology Division, California Institute of Technology, Pasadena, Calif. 91125

LEWIS J. NOE
: Department of Chemistry, University of Wyoming, Laramie, Wy. 82071

JON A. NOONAN
: Princeton Applied Research, Princeton, N.J. 08540

SEIJI OGAWA
: Bell Laboratories, Murray Hill, N.J. 07974

CHARLES H. O'NEAL
: Biophysics Department, Virginia Commonwealth University, Medical College of Virginia, Richmond, Va. 23298

EMILIA R. PANDOLFELLI
: Duke University Marine Laboratory, Beaufort, N.C. 28516

LAWRENCE J. PARKHURST
: Chemistry Department, University of Nebraska-Lincoln, Lincoln, Neb. 68588

V. ADRIAN PARSEGIAN
: Physical Sciences Laboratory, Division of Computer Research and Technology, National Institutes of Health, Bethesda, Md. 20014

ISRAEL PECHT
: Department Chemical Immunobiology, Weizmann Institute, Rehovot. Israel

ANDRÉ PERSOONS
: Department of Chemistry, Laboratory of Chemical and Biological Dynamics, Catholic

University, Leuven, B-3030 Heverlee-Leuven, Belgium

JOHN S. PHILO
Physics Department, Stanford University, Stanford, Calif. 94305

FRANKLYN G. PRENDERGAST
Department of Pharmacology, Mayo Medical School, Rochester, Minn. 55901

B. D. NAGESWARA RAO
Department of Biochemistry and Biophysics, University of Pennsylvania School of Medicine, Philadelphia, Pa. 19104

PETER M. RENTZEPIS
Bell Laboratories, Murray Hill, N.J. 07974

PETER RIESZ
National Cancer Institute, National Institutes of Health, Bethesda, Md. 20014

MICHAEL A. J. RODGERS
Center for Fast Kinetics Research, University of Texas, Austin, Tex. 78712

WILLIAM O. ROMINE, JR.
University of Alabama Medical School, Birmingham, Ala. 35213

LEONID RUBIN
Department of Physics, Lomonosov State University, Lenin Hills, Moscow 117234 USSR

WARREN RUDERMAN
Interactive Radiation Incorporated, Northvale, N.J. 07647

CHARLES A. SAWICKI
Biochemistry Department, Cornell University, Ithaca, N.Y. 14853

ALAN SCHECHTER
Laboratory of Chemical Physics, National Institute of Arthritis, Metabolism, and Digestive Diseases, National Institutes of Health, Bethesda, Md. 20014

WALTER SCHEIDER
Biophysics Research Division, Institute of Science and Technology, University of Michigan, Ann Arbor, Mich. 48109

Z. A. SCHELLY
Chemistry Department, University of Texas at Arlington, Arlington, Tex. 76019

D. D. SCHURESKO
Chemical Technology Division, Advanced Technology Section, Oak Ridge National Laboratory, Oak Ridge, Tenn. 37830

TODD M. SCHUSTER
Biological Sciences, University of Connecticut, Storrs, Conn. 06268

ARINDAM SEN
School of Life Sciences, Jawalarlal Nehru University, New Delhi 110057 India

JOSEPH D. SHORE
Biochemistry Department, Edsel B. Ford Institute, Detroit, Mich. 48202

MICHAEL G. SIMIC
United States Army NARADCOM, Food Engineering Laboratory, Natick, Mass. 01760

PAUL D. SMITH
Laboratory of Technical Development, National Institutes of Health, Bethesda, Md. 20014

L. B. SORENSEN
Department of Physics, University of Illinois, Urbana, Ill. 61801

JOHN R. SUTTER
Chemistry Department, Howard University, Washington, D.C. 20059

ROBERT C. SUTTON
Update Instrument Incorporated, Madison, Wis. 53719

CHARLES E. SWENBERG
Radiation and Solid State Laboratory, New York University, New York, N.Y. 10003 and Department of Physics, University of Puerto Rico, Rio Piedras, Puerto Rico

MEI-SHENG TAI
Section of Biochemistry and Biophysics, University of Connecticut, Storrs, Conn. 06268

YAIR TALMI
Princeton Applied Research, Princeton, N.J. 08540

DAVID G. TAYLOR
Chemistry Department, University of Virginia, Charlottesville, Va. 22901

RONALD PAUL TAYLOR
Department of Biochemistry, University of Virginia, Charlottesville, Va. 22901

JAMES TERNER
Department of Chemistry, University of California at Los Angeles, Los Angeles, Calif. 90024

H. L. TOORS
Department of Engineering, Carnegie-Mellon University, Pittsburgh, Pa. 15216

DENNIS A. TORCHIA
National Institute of Dental Research, Laboratory of Biochemistry, National Institutes of Health, Bethesda, Md. 20014

TEDDY G. TRAYLOR
Chemistry Department, University of California at San Diego, La Jolla, Calif. 92093

TIAN YOW TSONG
Physiological Chemistry, Johns Hopkins School of Medicine, Baltimore, Md. 21205

MEMPHIS TUFTS
Department of Physiology, Case Western Reserve University School of Medicine, Cleveland, Ohio 44106

JANE VANDERKOOI
Department of Biochemistry, University of Pennsylvania School of Medicine, Philadelphia, Pa. 19174

STANIMIR VUK-PAVLOVIĆ
Macromolecular Biophysical Laboratory, Institute of Immunology, 41000 Zagreb, Yugoslavia

HAREL WEINSTEIN
  Department of Pharmacology, Mount Sinai School of Medicine, New York, N.Y. 10029
JON WEBER
  Princeton Applied Research, Princeton, N.J. 08540
ALBERT WELLER
  Department of Spectroscopy, Max-Planck-Institute for Biophysical Chemistry, Göttingen, W. Germany
DABNEY K. WHITE
  Chemistry Department, University of California at San Diego, La Jolla, Calif. 92093

WILLIAM T. YAP
  Analytical Chemistry Division, National Bureau of Standards, Gaithersburg, Md. 20234
TATSUYA YASUNAGA
  Department of Chemistry, Hiroshima University, Hiroshima, Japan 730
WILLIAM YU
  Hamamatsu Company, Middlesex, N.J. 08846

GEORGE ZAVOICO
  Department of Physiology, University of Virginia, Charlottesville, Va. 22901

INDEX TO AUTHORS

ALBERDING, N., R. H. AUSTIN, S. S. CHAN, L. EISENSTEIN, H. FRAUENFELDER, D. GOOD, K. KAUFMANN, M. MARDEN, T. M. NORDLUND, L. REINISCH, A. H. REYNOLDS, L. B. SORENSEN, G. C. WAGNER, and K. T. YUE. Fast reactions in carbon monoxide binding to heme proteins   319

ANTONINI, E., M. BRUNORI, B. GIARDINA, P. A. BENEDETTI, G. BIANCHINI, and S. GRASSI. Single cell observations on gas reactions and shape changes in normal and sickling erythrocytes   187

AUSTIN, ROBERT H., and SHIRLEY SUI LING CHAN. The rate of entry of dioxygen and carbon monoxide into myoglobin   175

—, and W. VAZ. The motion of cytochrome b_5 on lipid vesicles measured via triplet absorbance anisotropy   49

—. See ALBERDING, AUSTIN, CHAN, EISENSTEIN, FRAUENFELDER, GOOD, KAUFMANN, MARDEN, NORDLUND, REINISCH, REYNOLDS, SORENSEN, WAGNER, and YUE.   319

BADEA, MUGUREL G., ROBERT P. DETOMA, and LUDWIG BRAND. Nanosecond relaxation processes in liposomes   197

BALKO, BOHDAN, and EUGENIE V. MIELCZAREK. The application of selective excitation double Mössbauer to time-dependent effects in biological materials   233

BALLARD, S. G., and D. MAUZERALL. Photo-initiated ion formation from octaethyl-porphyrin and its zinc chelate as a model for electron transfer in reaction centers.   335

BENEDETTI, P. A. See ANTONINI, BRUNORI, GIARDINA, BENEDETTI, BIANCHINI, and GRASSI   187

BERGER, ROBERT L. Some problems concerning mixers and detectors for stopped flow kinetic studies. Keynote address   2

BERNHARD, SIDNEY A. Spectral intermediates in the activation of glyceraldehyde-3-PO_4-dehydrogenase catalyzed reactions   49

BIANCHINI, G. See ANTONINI, BRUNORI, GIARDINA, BENEDETTI, BIANCHINI, and GRASSI   187

BLATT, Y. See VUK-PAVLOVIĆ, BLATT, GLAUDEMANS, LANCET, and PECHT   161

BONAVENTURA, CELIA. See PANDOLFELLI, BONAVENTURA, BONAVENTURA, and BRUNORI   257

BONAVENTURA, JOSEPH. See PANDOLFELLI, BONAVENTURA, BONAVENTURA, and BRUNORI   257

BRAND, LUDWIG. See BADEA, DETOMA, and BRAND   197

BREWER, JOHN M. Studies of the activation of yeast enolase by metals using a "transition state analogue"   53

BRUNORI, MAURIZIO. See ANTONINI, BRUNORI, GIARDINA, BENEDETTI, BIANCHINI, and GRASSI   187

—. See PANDOLFELLI, BONAVENTURA, BONAVENTURA, and BRUNORI   257

CAMPILLO, A. J. See GEACINTOV, SWENBERG, CAMPILLO, HYER, SHAPIRO, and WINN   347

CHAN, SHIRLEY SUI LING. See ALBERDING, AUSTIN, CHAN, EISENSTEIN, FRAUENFELDER, GOOD, KAUFMANN, MARDEN, NORDLUND, REINISCH, REYNOLDS, SORENSEN, WAGNER, and YUE   319

—. See AUSTIN and CHAN   175

CHRISTIAN, S. T. See ROMINE, WATKINS, CHRISTIAN, and GOODALL   76

CHUN, PAUL W., and MARK C. K. YANG. Scanning molecular sieve chromatography of interacting protein systems. Effect of kinetic parameters on the large zone boundary profiles for local equilibration between mobile and stationary phases   56

CLEGG, ROBERT, and MACK J. FULWYLER. A solution mixer with 10-μs resolution   57

COHN, MILDRED. See RAO and COHN   258

COLEN, ALAN H. Kinetic transients. A wedding of empiricism and theory   58

CONNOLLY, JOHN S. See FOYT and CONNOLLY   60

CONOVER, W. W. See KING, MAAS, GASSNER, NANDA, CONOVER, and JARDETZKY   103

CORRY, PETER M. See KIRKPATRICK, MCGINNESS, MOORHEAD, CORRY, and PROCTOR   243

CRANDALL, EDWARD D., A. L. OBAID, and R. E. FORSTER. Bicarbonate-chloride exchange in erythrocyte suspensions. Stopped-flow pH electrode measurements   35

CROTHERS, D. See DATTAGUPTA, HOGAN, and CROTHERS   238

CZERLINSKI, G. H. On magnetically induced temperature jumps   234

DA SILVA E WHEELER, MARIA F. See RODGERS, FOYT, and DA SILVA E WHEELER   75

DATTAGUPTA, N., M. HOGAN, and D. CROTHERS. Kinetic and transient electric dichroism studies of the irehdiamine-DNA complex   238

DEMAS, J. N. See TAYLOR, DEMAS, TAYLOR, and ZENKOWICH   77

DETOMA, ROBERT P. See BADEA, DETOMA, and BRAND   197

DORFMAN, LEON M. See FARAGGI, KLAPPER, and DORFMAN   307

DOWBEN, ROBERT M. See KOESTER and DOWBEN   245

EISENSTEIN, L. See ALBERDING, AUSTIN, CHAN, EISENSTEIN, FRAUENFELDER, GOOD, KAUFMANN, MARDEN, NORDLUND, REINISCH, REYNOLDS, SORENSEN, WAGNER, and YUE 319
EISERT, W. G. See NOE, EISERT, and RENTZEPIS 379
EL-SAYED, M. A. See TERNER and EL-SAYED 262

FARAGGI, MOSHE, MICHAEL H. KLAPPER, and LEON M. DORFMAN. Application of pulse radiolysis to the study of proteins. Chymotrypsin and trypsin 307
FEE, JAMES A. See MCCLUNE and FEE 65
FESSENDEN, RICHARD W., and NARESH C. VERMA. A time-resolved electron spin resonance study of the oxidation of ascorbic acid by hydroxyl radical 93
FLAMIG, DUANE P. See PARKHURST and FLAMIG 71
FORSTER, R. E. See CRANDALL, OBAID, and FORSTER 35
FOYT, DAVID C., and JOHN S. CONNOLLY. A method for determining the kinetic type of fast kinetic data 60
—. See RODGERS, FOYT, and DA SILVA E WHEELER 75
FRAUENFELDER, H. See ALBERDING, AUSTIN, CHAN, EISENSTEIN, FRAUENFELDER, GOOD, KAUFMANN, MARDEN, NORDLUND, REINISCH, REYNOLDS, SORENSEN, WAGNER, and YUE 319
FROEHLICH, JEFFREY P. The effect of pretreatment with calcium and magnesium ions on phosphoenzyme formation by sarcoplasmic reticulum ATPase 61
FULWYLER, MACK J. See CLEGG and FULWYLER 57

GASSNER, M. See KING, MAAS, GASSNER, NANDA, CONOVER, and JARDETZKY 103
GEACINTOV, N. E., C. E. SWENBERG, A. J. CAMPILLO, R. C. HYER, S. L. SHAPIRO, and K. R. WINN. A picosecond pulse train study of exciton dynamics in photosynthetic membranes 347
GIARDINA, B. See ANTONINI, BRUNORI, GIARDINA, BENEDETTI, BIANCHINI, and GRASSI 187
GIBSON, QUENTIN H. See SAWICKI and GIBSON 21
GLAUDEMANS, C. P. J. See VUK-PAVLOVIĆ, BLATT, GLAUDEMANS, LANCET, and PECHT 161
GOLDBERG, M., and I. PECHT. The reaction of "blue" copper oxidases with O_2. A pulse radiolysis study 371
GOOD, D. See ALBERDING, AUSTIN, CHAN, EISENSTEIN, FRAUENFELDER, GOOD, KAUFMANN, MARDEN, NORDLUND, REINISCH, REYNOLDS, SORENSEN, WAGNER and YUE 319
GOODALL, M. C. See ROMINE, WATKINS, CHRISTIAN, and GOODALL 76
GRASSI, S. See ANTONINI, BRUNORI, GIARDINA, BENEDETTI, BIANCHINI, and GRASSI 187

HALVORSON, HERBERT R. Optical detection of compressibility dispersion. Relaxation kinetics of glutamate dehydrogenase self-association 239

HAMORI, EUGENE, and TOSHIHIRO HIRAI. Recording of fast biochemical reactions using a logarithmic time sweep 63
HELLEMANS, L. See PERSOONS and HELLEMANS 119
HIRAI, TOSHIHIRO. See HAMORI and HIRAI 63
HOGAN, M. See DATTAGUPTA, HOGAN, and CROTHERS 238
HYER, R. C. See GEACINTOV, SWENBERG, CAMPILLO, HYER, SHAPIRO, and WINN 347

JAKOBSSON, ERIC. Creation of a nonequilibrium state in sodium channels by a step change in electric field 240
JARDETZKY, O. See KING, MAAS, GASSNER, NANDA, CONOVER, and JARDETZKY 103

KAUFMANN, K. J. See ALBERDING, AUSTIN, CHAN, EISENSTEIN, FRAUENFELDER, GOOD, KAUFMANN, MARDEN, NORDLUND, REINISCH, REYNOLDS, SORENSEN, WAGNER, and YUE 319
—. See PELLIN, WRAIGHT, and KAUFMANN 361
KING, R., R. MAAS, M. GASSNER, R. K. NANDA, W. W. CONOVER, and O. JARDETZKY. Magnetic relaxation analysis of dynamic processes in macromolecules in the pico- to microsecond range 103
KINOSITA, KAZUHIKO, JR., and TIAN YOW TSONG. Voltage-induced changes in the conductivity of erythrocyte membranes 373
KIRKPATRICK, DONALD S., JOHN E. MCGINNESS, WILLIAM D. MOORHEAD, PETER M. CORRY, and PETER H. PROCTOR. High-frequency dielectric spectroscopy of concentrated membrane suspensions 243
KLAPPER, MICHAEL H. See FARAGGI, KLAPPER, and DORFMAN 307
KOESTER, VAUGHN J., and ROBERT M. DOWBEN. Subnanosecond fluorescence lifetimes by time-correlated single photon counting using synchronously pumped dye laser excitation 245
KROUSE, MAURI E. See NASS, LESTER, and KROUSE 135
KUYEL, BIROL. See LINDLEY and KUYEL 254

LAKOWICZ, J. R., and F. G. PRENDERGAST. Detection of hindered rotations of 1,6-diphenyl-1,3,5-hexatriene in lipid bilayers by differential polarized phase fluorometry 213
LANCET, D., A. LICHT, and I. PECHT. Allostery in an immunoglobulin light-chain dimer. A chemical relaxation study 247
—. See VUK-PAVLOVIĆ, BLATT, GLAUDEMANS, LANCET, and PECHT 161
LESTER, HENRY A. See NASS, LESTER, and KROUSE 135
LEWIS, AARON. The structure of the retinylidene chromophore in bathorhodopsin 249
LICHT, A. See LANCET, LICHT, and PECHT 247
LINDLEY, BARRY D., and BIROL KUYEL. Contractile deactivation by rapid, microwave-induced temperature jumps 254

LUCHINS, JERRY. Far-ultraviolet stopped-flow circular dichroism 64

MAAS, R. See KING, MAAS, GASSNER, NANDA, CONOVER, and JARDETZKY 103
MACFARLANE, R. D. Fast biochemical reactions in thin films induced by nuclear fission fragments 375
MARDEN, M. See ALBERDING, AUSTIN, CHAN, EISENSTEIN, FRAUENFELDER, GOOD, KAUFMANN, MARDEN, NORDLUND, REINISCH, REYNOLDS, SORENSEN, WAGNER, and YUE 319
MAUZERALL, D. See BALLARD and MAUZERALL 335
MCCLUNE, GREGORY J., and JAMES A. FEE. A simple system for mixing miscible organic solvents with water in 10–20 ms for the study of superoxide chemistry by stopped-flow methods 65
MCGINNESS, JOHN E. See KIRKPATRICK, MCGINNESS, MOORHEAD, CORRY, and PROCTOR 243
MIELCZAREK, EUGENIE V. See BALKO and MIELCZAREK 233
MILGRAM, JEROME H. Fluid mechanics of rapid mixing 69
MILLER, JOHN HOWARD, and MARTIN L. WEST. Nonhomogenous chemical kinetics in pulsed proton radioluminescence. 376
MOORHEAD, WILLIAM D. Calculation of dielectric parameters from time domain spectroscopy data 256
—. See KIRKPATRICK, MCGINNESS, MOORHEAD, CORRY, and PROCTOR 243

NANDA, R. K. See KING, MAAS, GASSNER, NANDA, CONOVER, and JARDETZKY 103
NASS, MENASCHE M., HENRY A. LESTER, and MAURI E. KROUSE. Response of acetylcholine receptors to photoisomerizations of bound agonist molecules 135
NOE, L. J., W. G. EISERT, and P. M. RENTZEPIS. Picosecond photodissociation and subsequent recombination processes in carboxyhemoglobin, carboxymyoglobin, and oxymyoglobin 379
NORDLUND, T. M. See ALBERDING, AUSTIN, CHAN, EISENSTEIN, FRAUENFELDER, GOOD, KAUFMANN, MARDEN, NORDLUND, REINISCH, REYNOLDS, SORENSEN, WAGNER, and YUE 319
NOWICKI, T. See POPPER, NOWICKI, RUDERMAN, and RAGAZZO 73

OBAID, A. L. See CRANDALL, OBAID, and FORSTER 35
O'NEAL, CHARLES H. See THOMPSON, WILLIAMS, and O'NEAL 264

PANDOLFELLI, EMILIA R., CELIA BONAVENTURA, JOSEPH BONAVENTURA, and MAURIZIO BRUNORI. Light-jump perturbation of carbon monoxide binding by various heme proteins 257
PARKHURST, LAWRENCE J., and DUANE P. FLAMIG. Kinetics of association and dissociation phenomena in human hemoglobin studied in a laser light-scattering stopped-flow device 71
PARSEGIAN, V. ADRIAN. Editor's foreword v
PECHT, I. See GOLDBERG and PECHT 371
—. See LANCET, LICHT, and PECHT 247
—. See VUK-PAVLOVIĆ, BLATT, GLAUDEMANS, LANCET, and PECHT 161
PELLIN, M. J., C. A. WRAIGHT, and K. J. KAUFMANN. Modulation of the primary electron transfer rate in photosynthetic reaction centers by reduction of a secondary acceptor. 361
PERSOONS, A., and L. HELLEMANS. New electric field methods in chemical relaxation spectrometry 119
PHILO, J. S. Time-resolved magnetic susceptibility. A new method for fast reactions in solution 381
POPPER, R. A., T. NOWICKI, W. RUDERMAN, and J. RAGAZZO. Extending the wavelength range of fundamental laser sources 73
PRENDERGAST, F. G. See LAKOWICZ and PRENDERGAST 213
PROCTOR, PETER H. See KIRKPATRICK, MCGINNESS, MOORHEAD, CORRY, and PROCTOR 243

RAGAZZO, J. See POPPER, NOWICKI, RUDERMAN, and RAGAZZO 73
RAO, B. D. NAGESWARA, and MILDRED COHN. Measurement of interconversion rates of bound substrates of phosphoryl transfer enzymes by ^{31}P nuclear magnetic resonance 258
REINISCH, L. See ALBERDING, AUSTIN, CHAN, EISENSTEIN, FRAUENFELDER, GOOD, KAUFMANN, MARDEN, NORDLUND, REINISCH, REYNOLDS, SORENSEN, WAGNER, and YUE 319
RENTZEPIS, P. M. Probing ultrafast biological processes by picosecond spectroscopy. Keynote address 272
—. See NOE, EISERT, and RENTZEPIS 379
REYNOLDS, A. H. See ALBERDING, AUSTIN, CHAN, EISENSTEIN, FRAUENFELDER, GOOD, KAUFMANN, MARDEN, NORDLUND, REINISCH, REYNOLDS, SORENSEN, WAGNER, and YUE 319
RODGERS, MICHAEL A. J., DAVID C. FOYT, and MARIA F. DA SILVA E WHEELER. Kinetic studies of fast reactions at water-micelle interfaces 75
ROMINE, W. O., C. WATKINS, S. T. CHRISTIAN, and M. C. GOODALL. A macroscopic approach to fluctuation analysis. Solution of the phase problem of relaxation spectrometry 76
RUBIN, ANDREW B. See RUBIN and RUBIN 84
RUBIN, LEONID B., and ANDREW B. RUBIN. Picosecond fluorometry in primary events of photosynthesis. Keynote address 84
RUDERMAN, W. See POPPER, NOWICKI, RUDERMAN, and RAGAZZO 73

SAWICKI, CHARLES A., and QUENTIN H. GIBSON. The relation between carbon monoxide binding and the conformational change of hemoglobin 21
SCHEIDER, WALTER. Dissociation rate of serum albumin-fatty acid complex from stop-flow dielectric study of ligand exchange 260

SCHULTEN, KLAUS, and ALBERT WELLER. Exploring fast electron transfer processes by magnetic fields 295
SCHURESKO, D. D., and W. W. WEBB. Carboxylation kinetics of hemoglobin and myoglobin. Linear transient response to step perturbation by laser photolysis 382
SHAPIRO, S. L. See GEACINTOV, SWENBERG, CAMPILLO, HYER, SHAPIRO, and WINN 347
SIMIC, MICHAEL G., and IRWIN A. TAUB. Fast electron transfer processes in cytochrome c and related metalloproteins. 285
SORENSEN, L. B. See ALBERDING, AUSTIN, CHAN EISENSTEIN, FRAUENFELDER, GOOD, KAUFMANN, MARDEN, NORDLUND, REINISCH, REYNOLDS, SORENSEN, WAGNER, and YUE 319
SWENBERG, C. E. See GEACINTOV, SWENBERG, CAMPILLO, HYER, SHAPIRO, and WINN 347

TATSUMOTO, NOBUHIDE. See YASUNAGA and TATSUMOTO 267
TAUB, IRWIN A. See SIMIC and TAUB 285
TAYLOR, D. G., J. N. DEMAS, R. P. TAYLOR, and M. J. ZENKOWICH. An inexpensive microcomputer-based stopped-flow data acquisition system 77
TAYLOR, R. P. See TAYLOR, DEMAS, TAYLOR, and ZENKOWICH 77
TERNER, JAMES, and M. A. EL-SAYED. Time-resolved resonance Raman characterization of the intermediates of bacteriorhodopsin 262
THOMPSON, MICHAEL R., RAY C. WILLIAMS, and CHARLES H. O'NEAL. Studies on proteins and tRNA with transient electric birefringence 264
THUSIUS, DARWIN. On the integration of coupled first-order rate equations 79
TSONG, TIAN YOW. See KINOSITA and TSONG 373

VANDERKOOI, J. M., C. J. WEISS, and G. WOODROW III. Metal ion interactions with fluorescent derivatives of nucleotides 266
VAZ, W. See AUSTIN and VAZ 49
VERMA, NARESH C. See FESSENDEN and VERMA 93
VUK-PAVLOVIĆ, S., Y. BLATT, C. P. J. GLAUDEMANS, D. LANCET, and I. PECHT. Hapten-linked conformational equilibria in immunoglobulins XRPC-24 and J-539 observed by chemical relaxation 161

WAGNER, G. C. See ALBERDING, AUSTIN, CHAN, EISENSTEIN, FRAUENFELDER, GOOD, KAUFMANN, MARDEN, NORLUND, REINISCH, REYNOLDS, SORENSEN, WAGNER, and YUE 319
WATKINS, C. See ROMINE, WATKINS, CHRISTIAN, and GOODALL 76
WEBB, W. W. See SCHURESKO and WEBB 382
WEISS, C. J. See VANDERKOOI, WEISS, and WOODROW 266
WELLER, ALBERT. See SCHULTEN and WELLER 295
WEST, MARTIN L. See MILLER and WEST 376
WILLIAMS, RAY C. See THOMPSON, WILLIAMS, and O'NEAL 264
WINN, K. R. See GEACINTOV, SWENBERG, CAMPILLO, HYER, SHAPIRO, and WINN 347
WOODROW, G., III. See VANDERKOOI, WEISS, and WOODROW 266
WRAIGHT, C. A. See PELLIN, WRAIGHT, and KAUFMANN 361

YANG, MARK C. K. See CHUN and YANG 56
YASUNAGA, TATSUYA, and NOBUHIDE TATSUMOTO. Rupture diaphragmless apparatus for pressure-jump relaxation measurement 267
YUE, K. T. See ALBERDING, AUSTIN, CHAN, EISENSTEIN, FRAUENFELDER, GOOD, KAUFMANN, MARDEN, NORDLUND, REINISCH, REYNOLDS, SORENSEN, WAGNER, and YUE 319

ZENKOWICH, M. J. See TAYLOR, DEMAS, TAYLOR, and ZENKOWICH 77

INDEX TO SUBJECTS

Acetylcholine receptors
  response to photoisomerization of bound agonists
      135
Acyl-enzyme
  spectral intermediates in activation of   49
Agonists
  acetylcholine receptors, photoisomerization of
      135
Allostery in immunoglobulins
  chemical relaxation studies   161, 247
Aminoenolpyruvic acid-2-phosphate
  enolase activation by metals and   53
Anion exchange
  erythrocytes, rapid pH measurements   35
Anisotropy decay
  diffusion-limited rate from   175
  of emission   197
  fluorescence emission   213
  triplet state, cytochrome motion studies and   49
Ascorbic acid oxidation
  by OH radicals, electron spin resonance study   93
ATPase, sarcoplasmic reticulum
  catalysis of phosphoenzyme formation, role of Ca and Mg ions   61

Bacteria
  DNA source, (irehdiamine-DNA complex) studies   238
  electron transfer in photosynthesis   295
  photosynthetic, electron spin decorrelation and   335
  photosynthetic, electron transfer rate in reaction centers   361
  photosynthetic, picosecond fluorometry studies   84
Bacteriochlorophyll
  electron transfer in, magnetic field effects   295
  picosecond absorption studies of electron transfer   361
  picosecond laser studies   84
  porphyrins and   335
Bacteriopheophytin
  electron spin decorrelation and   335
  picosecond absorption studies of electron transfer   361
  picosecond laser studies   84
Bacteriorhodopsin
  resonance Raman spectra of intermediates   262
Ball mixer
  properties of   2
Bathorhodopsin
  retinylidene chromophore structure   249

Benzene
  radioluminescence in cyclohexane   376
  tetrabutylammonium picrate solution, electric field effects   119
Bicarbonate-chloride exchange
  erythrocytes, rapid pH measurements   35
Biochemical reactions
  kinetic transients and   58
  logarithmic recording of   63
Birefringence
  transient electric, protein and nucleic acid studies   264
Bis-Q
  photoisomerization of at acetylcholine receptors   135
Bovine serum albumin
  iodoacetamide-blocked, stopped-flow study   77
  transient electric birefringence   264

Calcium ions
  effect upon phosphoenzyme formation   61
Carbon monoxide
  myoglobin entry rate by   175
Carbon monoxide binding
  heme proteins, flash photolysis studies   319
  heme proteins, light-jump perturbation by   257
  hemoglobin, conformational change   21
  hemoglobin, flash photolysis of   381
  hemoglobin and myoglobin, laser photolysis of   382
  photodissociation and recombination processes   379
  single cell studies in erythrocytes   187
Carbon monoxide photochemistry
  single cell studies of normal and sickling erythrocytes   187
Charge transfer
  photosynthetic reaction centers and   361
  in thin films induced by nuclear fission fragments   375
Chemical relaxation
  allostery in an immunoglobin light-chain dimer studied by   247
  electric field methods of study   119
  glutamate dehydrogenase self-association   239
  hapten-linked conformational equilibria in immunoglobulins observed by   161
  pressure-jump apparatus, rupture diaphragmless   267
Chemical relaxation spectrometry
  electric field methods   119

Chlorophyll
  bacterio-, *see* Bacteriochlorophyll
  energy migration between light-harvesting antenna and reaction centers   84
  photosystem I fluorescence, picosecond laser studies   347
  thin film reactions induced by nuclear fission fragments   375
Chloroplasts
  fluorescence, picosecond laser studies   84, 347
Chromatography
  molecular sieve, interacting protein system studies   56
Chymotrypsin
  pulse radiolysis study, one-electron reduction   307
Compressibility dispersion
  optical detection of   239
Conformational changes
  acyl-enzyme activation and   49
  bacteriorhodopsin, resonance Raman spectra   262
  carboxyhemoglobin binding   21
  cytochrome c, postreductional   285
  electric field effects   119
  far-ultraviolet stopped-flow circular dichroism studies   64
  first-order rate equations and   79
  heme proteins, photodissociation   379
  immunoglobulins, hapten-induced   161, 247
  myoglobin, nanosecond fluctuations in   175
  in photosynthetic reaction centers   84
  in photosynthetic reaction centers, electron transfer rate and   361
  proton translocation and   272
  visual excitation and   249
Cytochrome
  tunneling mechanism during photosynthesis and   84
Cytochrome b_5
  motion on lipid vesicles measured via triplet absorbance anisotropy   49
Cytochrome c
  CO binding, flash photolysis studies   319
  electron transfer   285

Data processing
  microcomputer system and stopped-flow instrument   77
  stopped flow systems and   2
Dehydrogenase
  spectral intermediates of acyl-enzyme and   49
Depolarization
  sodium channel response to   240
Detectors
  problems of for stopped flow studies   2
Dichroism
  transient electric studies of irehdiamine-DNA complex   238
Dichroism, circular
  stopped-flow instrument, heme studies by   64

Dielectric absorption, high frequency
  membrane-aqueous interface studies   243
Dielectric parameters
  stopped-flow study of ligand exchange   260
  from time domain spectroscopy   256
Dielectric relaxation
  hydration dependent   243
Dimethyl sulfoxide
  stopped-flow method for mixing with H_2O solutions   65
Dimyrystroyllecithin
  nanosecond relaxation processes   197
Dioxygen
  myoglobin entry rate by   175
Diphenylhexatriene
  hindered rotations in lipid bilayers detected by differential polarized phase fluorometry   213
DNA
  -irehdiamine complex, dichroism and temperature jump studies   238

Electric absorption
  magnetically induced temperature jumps and   234
Electric field effects
  chemical relaxation spectrometry and   119
  membrane conductivity and   373
  primary events of photosynthesis and   84
  sodium channels, nonequilibrium state and   240
Electron spin
  ion formation from porphyrins and triplet states   335
  magnetic field effects upon hyperfine interactions   295
Electron spin resonance
  time-resolved, ascorbic acid oxidation by OH radicals   93
Electron transfer
  cytochrome c and related metalloproteins   285
  magnetic field effects   295
  photo-initiated ion formation from porphyrins as model   335
  photosynthetic, theories of   361
  photosynthetic, tunneling mechanisms and   84
  rate modulation in photosynthetic reaction centers   361
  in thin films induced by nuclear fission fragments   375
  water-micelle interfaces, kinetics of   75
***Electrophorus* electroplaque**
  kinetics of acetylcholine receptors in   135
Energy transfer
  in chloroplast membranes, picosecond laser studies   347
Enolase
  activation by metals, stopped-flow measurements   53
Enzymes, *see also* specific enzymes
  glutamate dehydrogenase self-association   239
  pulse radiolysis study, one-electron reduction   307

Enzyme catalysis
  first-order rate equations and   79
Enzymatic reactions
  blue copper oxidases with O_2 371
  dehydrogenase   2
  effect of metal ions   53, 266
  first-order rate equations and   79
  interconversion rates, nuclear magnetic resonance studies   258
  kinetic transients, empiricism and theory   58
  phosphoenzyme formation, effect of Ca and Mg ions   61
  spectral intermediates in activation of acyl-enzyme   49
Erythrocytes
  bicarbonate-chloride exchange, stopped-flow pH electrode measurements   35
  high-frequency dielectric absorption   243
  membrane conductivity, voltage-induced changes   373
  single cell studies   2
  single normal and sickling cells, gas reactions and shape changes   187
Exciton dynamics
  annihilation in photosynthetic membranes   347

Fatty acids
  dissociation rate of serum albumin complex   260
Ferrichrome A[5]
  selective excitation double Mössbauer studies   233
First-order rate equations
  integration of coupled   79
First-order reactions
  coupled, enzyme catalysis and   79
Fission fragments
  track dynamics in thin films   375
Flash photolysis
  CO binding to heme proteins studied by   319
  carboxyhemoglobin study   381
Flash photolysis
  hemoglobin reaction studies   2
Flow-flash
  carboxyhemoglobin binding studies   21
Fluctuation analysis
  macroscopic approach to   76
Fluorescence
  nucleotide derivatives, metal ion interactions   266
  photosystems I, II; picosecond laser studies   84, 347
  quenching rate of O_2 175
  radioluminescence of benzene in cyclohexane   376
  role of in solution mixer technique   57
  studied by microcomputer interfaced stopped-flow instrument   77
Fluorescence lifetimes
  by single-photon counting and synchronous laser excitation   245

Fluorescence probes
  liposomes, relaxation studies   197
  rotation in lipid bilayers   213
Fluorescence temperature jump
  hapten-induced conformational changes in immunoglobulins studied by   161
Fluorometry
  differential polarized phase   213
Fluorometry, picosecond
  photosynthesis, primary events   84
Free radicals, *see also* individual radicals
  reactions with cytochrome *c*   285

Glutamate dehydrogenase
  self-association, relaxation kinetics   239
Glyceraldehyde-3-PO₄-dehydrogenase
  spectral intermediates in activation   49

Haptens
  induced conformational changes in immunoglobulins   161, 247
Hartridge-Roughton mixer   2
Heme, *see also* Hemoglobin
  CO and O_2 entry rates   175
  CO binding, flow-flash studies   21
  conformational changes, stopped-flow circular dichroism instrument studies   64
  cytochrome motion studies   49
Heme proteins
  CO binding, flash photolysis studies   319
  CO binding, light-jump studies   257
  photodissociation and recombination in carboxyhemoglobin   379
Hemin *c*
  electron transfer   285
Hemoglobin, *see also* Heme
  carboxylation kinetics, laser photolysis studies   382
  carboxylation kinetics, magnetic susceptibility   381
  CO binding, flash photolysis studies   319
  CO binding, light-jump studies   257
  conformational change with CO binding   21
  dissociation studied by laser light-scattering stopped-flow device   71
  oxygen reaction with   2
  photodissociation and recombination in carboxyhemoglobin   379
  reactions, extended model of   2
Hemoglobin S
  single cell studies of sickling erythrocytes   187
Histidyl radicals
  pulse radiolysis study of serine proteases   307
Hydrated electrons
  cytochrome *c* reduction by   285
  reactions with serine proteases, pulse radiolysis study   307

Hydroxyl radicals
  ascorbic acid oxidation by, electron spin resonance studies 93
  cytochrome c reduction 285
Hyperfine interactions
  magnetic field effects 295
  studied by selective excitation double Mössbauer techniques 233

Ions
  cluster, fission track dynamics and 375
  copper oxidases, reaction with O_2 371
  Fe, cytochrome c reduction by free radicals and 285
  metal, interactions with fluorescent derivatives of nucleotides 266
  metal, yeast enolase activation by 53
  micelle binding, kinetics of 75
  radical pairs, magnetic field effect upon electron transfer 295
Ion channels
  and acetylcholine receptors 135
  sodium, nonequilibrium state 240
Ion formation
  photo-initiated in porphyrins 335
Immunoglobulins
  hapten-linked conformational equilibria observed by chemical relaxation 161
  light-chain dimer, allostery in 247
Irehdiamine
  -DNA complex, dichroism and temperature jump studies 238

Kinetics
  bicarbonate-chloride exchange in erythrocytes 35
  carboxyhemoglobin binding, flow-flash studies 21
  carboxylation of hemoglobin 381
  carboxylation of hemoglobin and myoglobin 382
  chemical, nonhomogeneous in pulsed-proton radioluminescence 376
  coupled first-order reactions 79
  electric field effects upon chemical relaxation 119
  of fast reactions at water-micelle interfaces 75
  of glutamate dehydrogenase self-association 239
  of hapten-induced conformational changes in immunoglobulins 161
  hemoglobin, association and dissociation phenomena 71
  interacting protein systems, molecular sieve chromatography studies 56
  irehdiamine-DNA complex studies 238
  ligand exchange, stopped-flow dielectric studies 260
  radioluminescence of benzene in cyclohexane 376
  stopped flow studies, problems of mixers and detectors for 2
Kinetic data
  logarithmic recording of biochemical reactions 63
  method to determine kinetic type 60
Kinetic transients
  empiricism and theory 58

Laser, *see also* Pulse radiolysis
  flash-flow studies of hemoglobin reactions 2, 21
  light-scattering stopped-flow device, hemoglobin studies by 71
  picosecond pulses, wavelength range extension and 73
  picosecond spectroscopy, chloroplast fluorescence studies 84, 347
Laser detector systems 2
Laser photolysis
  carboxyhemoglobin and carboxymyoglobin studies 379, 382
Laser sources
  wavelength range extension 73
Laser, synchronous excitation
  subnanosecond fluorescence lifetimes and 245
Ligand exchange
  serum albumin-fatty acid complex, dissociation rate 260
Light flash
  photoisomerization of bound acetylcholine receptor agonists 135
Light-jump
  carboxyhemoglobin studies 257
Lipid bilayers
  differential polarized phase fluorometry 213
  nanosecond relaxation processes 197
Lipid vesicles
  cytochrome b_5 motion measured via triplet absorbance anisotropy 49
Lipid vesicles, synthetic
  differential polarized phase fluorometry 213
  nanosecond relaxation in 197
Liposomes
  nanosecond relaxation processes 197

Macromolecules
  nuclear magnetic resonance relaxation analysis of dynamic processes 103
Magnesium ions
  effect upon phosphoenzyme formation 61
  yeast enolase activation by 53
Magnetic field
  electron transfer affected by 295
  generator of alternating 234
Magnetic relaxation
  analysis of dynamic processes in macromolecules 103
Magnetic susceptibility
  hemoglobin carboxylation and 381
Magnetometry
  flash photolysis of carboxyhemoglobin 381
Membranes
  cytochrome motion studies 49
  dielectric behavior, hydradion dependent 243
  erythrocytes, anion exchange 35
  erythrocytes, voltage-induced changes in conductivity 373
  high frequency dielectric spectroscopy of 243
  microviscosity, differential polarized phase fluorometry 213

permeability, electric field effects	119
photosynthetic, exciton annihilation in	347

Metal ions
interaction with fluorescent derivatives of nucleotides	266

Metalloproteins
electron transfer in cytochrome c	285
electron transfer in porphyrins	335

Metals
activation of yeast enolase by	53

Micelle formation
pressure-jump apparatus for study of	267

Micelle-ion binding
kinetics of	75

Microcomputer
acquisition of stopped-flow data	77

Microspectrophotometry
single cell studies in nomal and sickling erythrocytes	187

Microviscosity
differential polarized phase fluorometry measurements	213

Mixers
problems of for stopped flow studies	2

Mixing
fluid mechanics of rapid	69
organic solvents and H_2O, stopped-flow methods	65
solution mixer with 10 μs resolution	57

Molecules, thin film
fission track dynamics	375

Mössbauer techniques
selective excitation double, study of hyperfine interactions	233

Muscle cells
microwave temperature jump studies	254

Myeloma proteins
hapten-linked conformational changes	161

Myoglobin
carboxylation kinetics, laser photolysis studies	382
CO and O_2 entry rates	175
CO binding, flash photolysis studies	319
electron transfer	285
photodissociation and recombination in carboxymyoglobin and oxymyoglobin	379

Nanosecond relaxation
processes in liposomes	197

Nuclear fission fragments
reactions in thin films induced by	375

Nuclear magnetic resonance
enzyme reactions, interconversion rates studied by	258
problems of detectors for stopped flow studies	2
relaxation models, analysis of	103

Nucleic acids
transient electric birefringence	264

Nucleotides
fluorescent derivatives, metal ion interactions	266

Oxidases
blue copper, reaction with O_2	371

Oxidation
ascorbic acid by OH radicals, electron spin resonance study	93

Oxygen
hemoglobin reaction with	2
myoglobin entry rate by	175

Parshcall flume 2

pH
rapid measurements, stopped-flow apparatus	35

pH dependency
of ascorbic acid oxidation	93
carboxyhemoglobin binding and	21
cytochrome c reduction by free radicals	285
histidyl radical yield	307
light-jump studies of carboxyhemoglobin	257

pH electrodes
problems as detectors for stopped flow studies	2

Phosphatidylcholines
polarization, differential phase fluorometry studies	213

Phosphoenzyme formation
effect of Ca and Mg ions	61

Phosphoryl transfer enzymes
interconversion rates, nuclear magnetic resonance studies	258

Phosphorylation
role of Ca and Mg ions	61

Photochemistry, CO
single cell studies of normal and sickling erythrocytes	187

Photodissociation
CO binding to heme proteins after	319
intraerythrocytic hemoglobin	187
and recombination in carboxyhemoglobin, carboxymyoglobin, and oxymyoglobin	379

Photon, single counting
and synchronous laser excitation, fluorescence lifetimes and	245

Photoisomerization
of Bis-Q, response of acetylcholine receptors	135

Photoreceptors
excitation mechanism	249
rod outer segments, dielectric behavior	243

Photoreceptor membranes
high frequency dielectric absorption	243

Photosynthesis
bacterial, electron transfer rate in reaction centers	361
bacterial, magnetic field effects upon electron transfer	295
electron spin decorrelation and	335
primary events, picosecond fluorometry of	84

Photosystems I, II fluorescence
picosecond laser studies	84, 347

Picosecond fluorometry
photosynthesis, primary events	84

Picosecond pulse trains
photosynthetic membrane studies	347

Picosecond spectroscopy, *see* Spectroscopy, picosecond
Polarizaton
   diphenylhexatriene, differential phase fluorometry   213
Porphyrins
   octaethyl- and zinc octaethyl-, photo-initiated ion formation as electron transfer model   335
Pressure-jump
   apparatus, rupture diaphragmless   267
   relaxation kinetics and   239
Proteins, *see also* individual proteins
   interacting systems, molecular sieve chromatography studies   56
   ligand exchange, stopped-flow dielectric study   260
   metallo-, electron transfer   285, 335
   myeloma, hapten-linked conformational changes   161
   pulse radiolysis study, one-electron reduction   307
   refolding kinetics studied by microcomputer interfaced stopped flow instrument   77
   transient electric birefringence   264
Proton
   radioluminescence induction   376
   transfer reactions induced by fission fractions   375
Proton translocation
   prelumirhodopsin formation and   272
   visual excitation and   249
Proton tunneling
   models for   272
Protoporphyrin
   cytochrome motion studies   49
   triplet state quenching by O_2 175
Pulse radiolysis
   benzene in cyclohexane studies   376
   free radical reactions with cytochrome *c*   285
   OH radicals, electron spin resonance study of ascorbic acid oxidation   93
   proteins studied by   307
   reaction of oxidases with O_2 371
   water-micelle interfaces studied by   75

Quinones
   oxidation of photosynthetic reaction centers by   361

Radical pairs
   magnetic field effect upon electron transfer   295
Radicals
   histidyl, pulse radiolysis study of serine proteases   307
   hydroxyl, ascorbic acid oxidation by   93
   hydroxyl, cytochrome *c* reduction   285
Radioluminescence
   pulsed-proton induced   376
Reaction centers
   electron transfer model, photo-initiated ion formation from porphyrins   335

fluorometry studies   84
photosynthetic, electron transfer rate modulation   361
photosynthetic, magnetic field effect upon electron transfer   295
Relaxation, *see also* Chemical relaxation
   dielectric, hydration dependent   243
   magnetic, analysis of dynamic processes in macromolecules   103
   muscle cell, microwave temperature jump studies   254
   nanosecond processes in liposomes   197
   photodissociation effects upon heme proteins   379
Relaxation spectrometry
   fluctuation analysis   76
Resonance
   time-resolved electron spin resonance, ascorbic acid oxidation studies   93
Resonance, nuclear magnetic
   enzyme reactions, interconversion rates studied by   258
   relaxation models   103
Resonance Raman studies
   bacteriorhodopsin spectra   262
   bathorhodopsin, excitation mechanism   249
Retinal chromophore
   proton tunneling and   272
   resonance Raman studies   262
   structure in bathorhodopsin   249
Retinylidene chromophore
   structure in bathorhodopsin   249
Rhodopseudomonas sphaeroides
   picosecond fluorometry studies   84
   reaction centers, electron transfer and   361
Rhodopsin, *see also* Bathorhodopsin
   picosecond spectroscopy of primary processes in   272
tRNA
   transient electric birefringence   264
Rod outer segments
   dielectric behavior, hydration dependent   243

Sarcoplasmic reticulum
   ATPase catalysis of phosphoenzyme formation, effect of Ca and Mg ions   61
Secondary acceptors
   oxidation of photosynthetic reaction centers by   361
Serum albumin, *see also* Bovine serum albumin
   dissociation rate of fatty acid complex   260
Sickling
   microspectrophotometry of single *SS* erythrocytes   187
Single cells
   erythrocytes, stopped flow studies   2
   microspectrophotometry of normal and sickling erythrocytes   187
   striated muscle, microwave temperature jump studies   254

Sodium channels
   nonequilibrium state after step change in electric field   240
Solutions
   electron transfer, magnetic field effects   295
   mixing technique with 10 μs resolution   57
   stopped-flow mixer for dimethyl sulfoxide and H_2O 65
   time-resolved magnetic susceptibility   381
   viscous, mixing of   2
Solvents, organic
   stopped-flow method for mixing with H_2O solutions   65
Spectrometry
   chemical relaxation, electric field methods   119
   relaxation, fluctuation analysis   76
Spectroscopy
   high frequency dielectric, of concentrated membrane suspensions   243
   light-jump, carboxyhemoglobin studies   257
   single cell, normal and sickling erythrocytes   187
   stopped-flow, acyl-enzyme spectral intermediates studied by   49
   stopped-flow, enolase activation by metals   53
   time domain data, calculation of dielectric parameters   256
   time-resolved, micelle-ion binding studies   75
   time-resolved, photodissociation of carboxyhemoglobin, carboxymyoglobin, and oxymyoglobin   379
   time-resolved emission, liposome studies   197
   time-resolved electron spin resonance, ascorbic acid oxidation studied by   93
Spectroscopy, picosecond
   chloroplast fluorescence studies   84, 347
   electron transfer rate in photosynthetic reaction centers   361
   photodissociation of carboxyhemoglobin, carboxymyoglobin, and oxymyoglobin   379
   ultrafast biological processes probed by   272
Stopped-flow
   apparatus, hemoglobin dissociation studied by   71
   carboxyhemoglobin binding, flow-flash studies   21
   circular dichroism instrument to study conformational changes   64
   data, microcomputer-based acquisition   77
   dielectric study of ligand exchange   260
   kinetic studies, problem of mixers and detectors for   2
   method for mixing organic solvents with H_2O solutions   65
   pH electrode measurement; bicarbonate-chloride exchange in erythrocytes   35
Superoxides
   stopped-flow method for mixing with H_2O solutions   65
Techniques
   apparatus to study transient electric birefringence   264
   compressibility dispersion, optical detection of   239
   differential polarized phase fluorometry, diphenylhexatriene studies   213
   electric field methods in chemical relaxation spectrometry   119
   far-ultraviolet stopped-flow circular dichroism instrument   64
   kinetic data, type determination   60
   laser light-scattering stopped-flow device, hemoglobin studies   71
   laser wavelength range extension   73
   logarithmic recording of biochemical reactions   63
   magnetometric, flash photolysis of carboxyhemoglobin   381
   microcomputer-based stopped-flow data acquisition system   77
   mixers and detectors, problems of   2
   mixing apparatus   69
   pH, rapid measurements of   35
   picosecond fluorometry   84
   picosecond spectroscopy   272, 361
   pressure-jump apparatus, rupture diaphragmless   267
   proton-induced radioluminescence   376
   selective excitation double Mössbauer   233
   solution mixer with 10 μs resolution   57
   stopped-flow mixer for organic solvents and H_2O solutions   65
   stopped-flow, problems of mixers and detectors for   2
   stopped-flow rapid reaction apparatus   35
   time-resolved electron spin resonance   93
Temperature jumps
   hapten-induced conformational changes in immunoglobulins studied by   161
   magnetically induced, electric absorption and   234
   microwave-induced, contractile deactivation by   254
   relaxation kinetics of DNA-irehdiamine complex   238
Tetrabutylammonium picrate
   benzene solution, electric field effects   119
Thermistor probe   2
Time-domain dielectric spectroscopy
   data calculation   256
   of membrane-aqueous interfaces   243
Time-resolved emission spectra
   liposome studies   197
Triplet state
   chloroplast membranes, exciton dynamics   347
   cytochrome motion studies   49
   magnetic field effect upon yield   295
   photo-initiated ion formation from porphyrins   335
   protoporphyrin, quenching by O_2 175
Trypsin
   pulse radiolysis study, one-electron reduction   307

Tunneling
    CO binding to heme proteins   319
    electron transfer and   335
    mechanisms during photosynthesis   84
    models for proton   272

Ultraviolet
    far-, stopped-flow circular dichroism   64
    laser sources, wavelength range extension and   73

Vesicles. *see* Lipid vesicles

Voltage clamp
    acetylcholine receptor studies   135

Viscosity
    measurements of   2

Water
    -micelle interfaces, fast reactions at   75
    stopped-flow method for mixing with dimethyl sulfoxide   65

Water of hydration
    dielectric absorption studies   243

Wavelengths
    laser sources, range extension of   73

Yeast
    enolase activation by metals   53

Yeast tRNA
    transient electric birefringence   264

Biophysical Discussions

FIRST BOOK OF A SERIES

Fast Biochemical Reactions in Solutions, Membranes, and Cells

Proceedings of a Meeting sponsored by

the Biophysical Society

2–5 April 1978 at Airlie, Virginia

---

18 Research Papers with Discussion
43 Extended Abstracts

---

400 pages, *hard-covered book*, 7 x 10, 100 graphs and illustrations, fully indexed, $25.

Order from: The Rockefeller University Press, Order Service, 1230 York Avenue, New York 10021 U.S.A.